Statics, Formfinding and Dynamics
of Air-Supported Membrane Structures

Mechanics of Surface Structures*

Volume 5

1. Seide, P.: Small Elastic Deformations of Thin Shells. 1975. ISBN 90-286-0064-7
2. Panc, V.: Theories of Elastic Plates. 1975. ISBN 90-286-0104-x
3. Nowinski, T. L.: Theory of Thermoelasticity with Applications. 1978. ISBN 90-286-0457-x
4. Lukasiewicz, S.: Local Loads in Plates and Shells. 1979. ISBN 90-286-0047-7

* The first four titles in this series were published by Sijthoff and Noordhoff International Publishers, Alphen aan de Rijn, The Netherlands, Rockville, Maryland, U.S.A.

Statics, Formfinding and Dynamics of Air-Supported Membrane Structures

VLADIMÍR FIŘT

1983

MARTINUS NIJHOFF PUBLISHERS
THE HAGUE / BOSTON / LONDON

Distributors:

for the United States and Canada
Kluwer Boston, Inc.
190 Old Derby Street
Hingham, MA 02043
USA

for all other countries
Kluwer Academic Publishers Group
Distribution Center
P.O. Box 322
3300 AH Dordrecht
The Netherlands

for Czechoslovakia, Albania, Bulgaria, Cuba, China, German Democratic Republic, Hungary, Northern Korea, Mongolia, Poland, Rumania, U.S.S.R., Vietnam and Yugoslavia

ACADEMIA, Publishing House of the
Czechoslovak Academy of Sciences,
P.O.B. 896
112 29 Prague 1
Czechoslovakia

Library of Congress Cataloging in Publication Data

Fiřt, Vladimír.
Statics, formfinding and dynamics of air-supported membrane structures.

(Mechanics of surface structures; v. 5)
Bibliography: p.
Includes index.
1. Air-supported structures. I. Title.
II. Series.
TH1099.F568 693'.98—82-6390
AACR2

ISBN 90-247-2672-7 (this volume)
ISBN 90-247-2673-5 (series)

Scientific Editor: Doc. Ing. Ladislav Frýba, DrSc.
Reviewer: Ing. Josef Henrych, DrSc.
Translation: Slavoš Kadečka
Edited by S. Swierczkowski

Joint edition published by
MARTINUS NIJHOFF PUBLISHERS
P.O. Box 566, 2501 CN The Hague, The Netherlands
and ACADEMIA, Publishing House of the Czechoslovak Academy of Sciences,
P.O.B. 896, 112 29 Prague 1, Czechoslovakia

PRINTED IN CZECHOSLOVAKIA

Contents

III Formfinding of air-supported structures

IV Natural vibrations of air-supported structures

Preface

In the present period, which can be characterized by an ever increasing usage of various plastic materials, a new branch of the building industry is developing in all technologically advanced countries: the erection of air-supported membrane structures. An essential feature of these structures is the very low weight of their airtight covering fabrics and speedy and easy erection and dismantling.

The successful solution of the problems connected with the life expectancy of these structures and their interior microclimate makes it possible to apply them on a wide scale as shelters for storage, recreation arenas, production units, exhibitions and for other enterprises. Air-supported structures are used with ever-growing frequency as forms for casting rigid shells of rapid-hardening concretes or plastics. Research has been carried out concerning the possibilities of using inflatable structures in space projects for the construction of certain types of satellites and re-entry vehicles (gliders), [35] to [38].

The present book describes the results of the author's theoretical and experimental research in the field of membrane structures. Its purpose is to further the development of air-supported structures by affording the designers and development workers useful theoretical data.

Most of the first chapter is devoted to a discussion of examples of the fabrics used for the jackets of air-supported structures and the self-supporting roofing of suspended structures in which internal overpressure or underpressure is used only to counteract unusual wind forces.

The second and most extensive chapter deals with the static structural design of various types of air-supported structures. Particular attention is given to finding the required magnitude of the internal overpressure under various loads.

The third chapter is concerned with the formfinding of the air-supported structures spanned over a circular, oval, rectangular or square contour as well as with the calculation of the lengths of the meridians of the surfaces of revolution and the shapes of entrance openings.

In the fourth chapter an analysis is made of the natural vibrations of air-supported structures of cylindrical form; this is done by means of the semi-deformation method which is illustrated by a numerical example. In Section 8 of this chapter we shall describe some results of the linear theory of inflatable plates.

The fifth chapter is devoted to the investigation of forced vibrations of air-supported structures by the semi-deformation method and by modal analysis.

To facilitate the solution of dynamic problems that may arise in practical applications, the book comprizes tables of the frequently used functions.

References to equations, sections and subsections of a given chapter will be made without quoting the number of this chapter if they are made in the same chapter.

It is my pleasant duty to thank most sincerely Professor William A. Nash of the University of Massachusetts, U.S.A., for acquainting me with the recent American results in the field of air-supported structures and for valuable critical comments which have greatly contributed to the improvement of this book. I should like to extend my thanks also to the scientific editor, Professor L. Frýba, and the reviewer of the book, Research Engineer Dr J. Henrych, for their helpful advice which I made use of in the final preparation of the manuscript, Mrs B. Stránská for the drawing of illustrations, and Mr S. Kadečka for translating the book.

My particular thanks are due to Sijthoff and Noordhoff International Publishers of the Netherlands (now incorporated into Martinus Nijhoff Publishers) who initiated the English edition of the book, and to the ACADEMIA Publishers of Czechoslovakia for many years of gratifying cooperation.

V. Fiřt

List of symbols used

a, b	constants, parameters defined by equation (2.3) in Chapter IV, lengths
c	width of membrane strip (Fig. III-2)
$c(\vartheta)$	coefficient of form in equation (6.75), Chapter II
e	base of natural logarithms
f	function
f_j	natural frequency
$f_x = \dfrac{\partial f}{\partial x}$,	
$f_y = \dfrac{\partial f}{\partial y}$,	
$f_{xx} = \dfrac{\partial^2 f}{\partial x^2}$,	
$f_{xy} = \dfrac{\partial^2 f}{\partial x\, \partial y}$,	
$f_{yy} = \dfrac{\partial^2 f}{\partial y^2}$	
$f(\alpha)$	function defined by (3.96), Chapter II
$\bar{f}(\alpha)$	function defined by (3.67), Chapter II
f_v	function defined by (3.72), Chapter II
g	dead weight of the membrane
g_{11}, g_{12}, g_{22}	coefficients of the first quadratic differential form of surface
h_{11}, h_{12}, h_{22}	coefficients of the second quadratic differential form of surface
$\mathrm{i} = \sqrt{-1}$	imaginary unit
$i, j,$	end-points of a circular arc
k	constant
l	natural number, length
l_j	half the length of a circle (Fig. III-1)

ln	natural logarithm
m	constant; parameter in equation (4.1), Chapter II
n	constant
p	wind pressure or suction affecting cylindrical surface
q	resultant static load affecting membrane
$q(\xi, \eta)$	continuous external load applied to the middle surface of an air-supported membrane structure
q_0	internal overpressure
$q_1(q_2)$	component of external continuous load $q(\xi, \eta)$ in the direction of the tangent to the curve ξ (curve η)
q_n	component of external continuous load in the direction of the normal to the surface
q_s	snow load at the apex of a surface of revolution or on a part of a cylindrical surface
q_v	the highest negative value of antisymmetric wind load component on the windward side of an air supported structure ($q_v<0$ means wind pressure)
q_0^*	symmetric wind load component ($q_0^*>0$ means wind suction)
r	radius of the parallel circle of a surface of revolution; coordinate
r_j	radius of a circle (Fig. III-2)
s	length of an arc on a surface
t	time
u (v)	displacement in the direction of the tangent to the principal curve $\xi=$const ($\eta=$const) on a surface
v	amplitude of tangential displacement
v_0	wind velocity
w	displacement in the direction of the normal to the surface (deflection); amplitude of radial displacement
$\bar{w}$, $\bar{v}$	radial and tangential displacement in time t
x, y, z	rectangular coordinates
A, B, C, D, E, F, G, H	functions defined by (3.7), Chapter IV
A_1, A_2	Lamé's parameters
C, D	integration constants
E_1, E_2, E_{12}	moduli of elasticity of an orthotropic industrial fabric in tension and shear, defined by (8.4), Chapter II

$E_{1,0}$, $E_{2,0}$, $E_{12,0}$	initial moduli of a viscoelastic orthotropic membrane (at the beginning of loading, as the time $t=0$)
$E_{1,\infty}$, $E_{2,\infty}$, $E_{12,\infty}$	long-term moduli of a viscoelastic orthotropic membrane (as the time $t\to\infty$)
F	function
H	mean curvature of a surface
K	Gaussian curvature
I, J,	integrals
O	origin of a rectangular system of coordinates; centre of a circular arc
P, R	points on a surface
R	radius of a circular arc
R_0	value of principal radii at the apex of a surface of revolution; radius of a cylindrical surface
R_1, R_2	principal radii of curvature
S	amplitude of axial force
$\bar{S}$	dynamic axial force
$T_1(T_2)$	unit normal membrane force in the direction of tangent to the meridian (parallel circle)
T_2	static axial (membrane) force in a circular arc
T_{12}	unit shear (tangential) membrane force
$T_{1,g}$, $T_{2,g}$	unit membrane forces due to the dead weight of the membrane
$T_{1,q}$, $T_{2,q}$	unit membrane forces due to the internal overpressure
$T_{1,s}$, $T_{2,s}$, $T_{12,s}$	membrane forces due to symmetric or one-sided snow load
V, Z	functions defined by (3.4), Chapter IV
$\boldsymbol{r}$	radius vector
!	factorial
α	angle between the normal to a surface of revolution and the axis of revolution; central angle of a circular arc
β	angle of rotation of the meridian about the axis of revolution; root of a characteristic equation
γ	parameter in equations (4.5), Chapter II; angle
γ_{12}	relative (unit) displacement (shear strain)
δ_1, δ_2	angles shown in Fig. II-4 and defined by (2.2), Chapter II
$\varepsilon_1(\varepsilon_2)$	relative (unit) elongation in the direction of membrane force T_1 (force T_2); meridian strain (hoop strain)

η	curvilinear coordinate on a surface
ϑ	position angle
$\varkappa_1$, $\varkappa_2$	principal curvatures of a surface
$\varkappa_j$	half the central angle of a circle (Fig. III-2)
$\varkappa_n$	curvature of a normal section of a surface
$\Delta\varkappa_1$, $\Delta\varkappa_2$	changes of principal curvatures
λ	root of the characteristic equation (2.20), Chapter II; dimensionless parameter defined by formula (1.11), Chapter IV
μ	mass per unit arc length
ν_1, ν_2	Poisson's ratios of an orthotropic membrane, defined by (8.4), Chapter II
$\nu_{1,0}$, $\nu_{2,0}$	Poisson's ratios of a viscoelastic orthotropic membrane at the beginning of loading
$\nu_{1,\infty}$, $\nu_{2,\infty}$	Poisson's ratios of a viscoelastic orthotropic membrane as the time $t \to \infty$
ξ	curvilinear coordinate on a surface
$\xi = x/l$	dimensionless coordinate
π	the number pi (the ratio of the circumference of a circle to its diameter)
ϱ	radius of curvature of the opening in the jacket (Fig. III-9)
ϱ_0	density of air
$\bar{\varrho}$	instantaneous radius of an arc
τ_j	angle between the axis of symmetry and the edge of a membrane strip (Fig. III-2)
φ	angle between the principal tangent and the tangent to the normal section; amplitude of the angle of rotation of the cross-section of an arc or a straight sector; angle between the membrane force T_1 and the warp
$\bar{\varphi}$	angle of rotation of a cross-section of an arc in time t
χ	distorsion of the middle surface
χ_1	angle according to Fig. II-17
ω	quantity defined by (1.23), Chapter II; circular frequency
ω_j	natural circular frequency; angle
ω_0	constant in (2.18), Chapter II
Δ	expression (3.5), Chapter IV; determinant
Λ	dimensionless parameter defined by (5.2), Chapter IV
Φ, Ψ	functions defined by (5.10), Chapter IV

Note:
The symbols which do not occur frequently and other meanings of the above symbols are explained either in the text or by illustrations.

Chapter I

DEVELOPMENT OF MEMBRANE STRUCTURES

Introduction

In technologically advanced countries the manufacture of impermeable fabrics such as polyamide, polyesther and glass fibre fabrics coated with PVC or polyurethane, has attained a very high standard. Consequently the application of these fabrics has become an integral feature of modern structures. This applies particularly to air-supported as well as suspended membrane structures in which the physical and mechanical properties of such airtight fabrics can be utilized to their utmost. The very low mass of the fabrics of the order of 0.5—1.0 kg/m^2, their considerable tensile strength (500—1500 N/cm) and adequate ductility make it possible to use them for covering or roofing large spans. The fabric structures are not sensitive to large deflections or elongations due to load or temperature changes.

At the present time mostly polyesther fabrics are used for construction purposes, as they are considerably more resitant to the effects of photodegragation than polyamide fabrics.

The air-supported and suspended membrane structures are discussed e.g. in [1], [28], [29], [47], [48], [50] and [51]. In this chapter we shall give only their brief description and concentrate chiefly on some development types of membrane structures. We shall mention the models of membrane structures created by the spiders and shall give an example of the application of a suspended membrane used for the damping of dynamic wind effects on tall structures of circular cross-section, such as radio masts and TV relay towers.

1

Some Types of Erected Membrane Structures

1.1 *Air-Supported Structures*

In air-supported structures of conventional type, the jacket is prestressed by tensile forces. These are due to an internal overpressure of 10—15 mm water column which must be increased twice or even several times for strong wind. Most frequently the air-supported structures are of spherical form over a circular plan or of cylindrical form over a rectangular plan (Fig. I-1a). The internal overpressure is maintained by a blower. Air-supported halls are heated in winter, most frequently by air heaters operating on liquid fuels, i.e. crude or heating oil. In summer the outer surface of the jacket is cooled with water on sunny days to keep down the temperature inside the building. For this purpose the upper part of the jacket is provided with hoses transporting water and distributing it uniformly over the surface.

Fig. I-1a. Air-supported recreation hall in Prague-Strahov produced by the Norvegian firm of Scanhover.

As a rule an air-supported structure is provided with a drive-in or entrance chamber with two sets of doors preventing undesirable excessive escape of air from its interior. The Swedish firm Barrakudaverken has replaced the drive-in chamber of air-supported structures subject to heavy traffic by the so-called air curtains produced by blowers in the drive-in opening. The fans are situated on the outside of the structure on both sides of the entrance opening along its whole height. The shafts of the blowers form an acute angle with the opening plane. The blowers are actuated automatically as the drive-in entrance opens by lifting of a motor-driver shutter (Figs. I-1b, I-1c).

Fig. I-1b. Drive-in openings with an air curtain of an air-supported structure produced by the Swedish firm Barracudaverken for a motorcar servicing depot.

Fig. I-1c. Jacket of an air-supported structure made of polyamide fabric, provided with translucent bands of polyesther fabric (Sweden).

Air-supported roofing over sport halls, swimming pools or ice arenas, where a great number of people gather, is provided on the interior surface with suspended troughs gathering condensed vapours. Intercepted water is drained outside the structure by means of draining tubes.

Jackets over large areas must be reinforced by a grid of steel ropes, since the membrane forces arising from the internal overpressure, wind load and other types of transverse loads in the jacket are increasing with diminishing curvature of the middle surface of the jacket (Fig. I-1d).

The thermally insulating properties of the jacket of an air-supported structure can be ensured by air cushions which form automatically due to the internal overpressure (Fig. I-1e), see Chapter III, Subsection 3.3.

A special type of air-supported structure is the structure of cushion type, suitable for small and medium spans (Fig. I-1f).

From unheated air-supported structures snow must be removed in winter. For this purpose a blade is used fastened to a rope which straddles the structure; snow is removed by alternate pulling of both rope ends. For the same purpose it is also possible to use the facility described by the

Fig. I-1d. View of the U.S.A. exhibition pavilon in Osaka of an elliptical plan of maximum dimensions 83 × 142 m roofed with an air-supported structure whose glass fabric is reinforced with steel ropes.

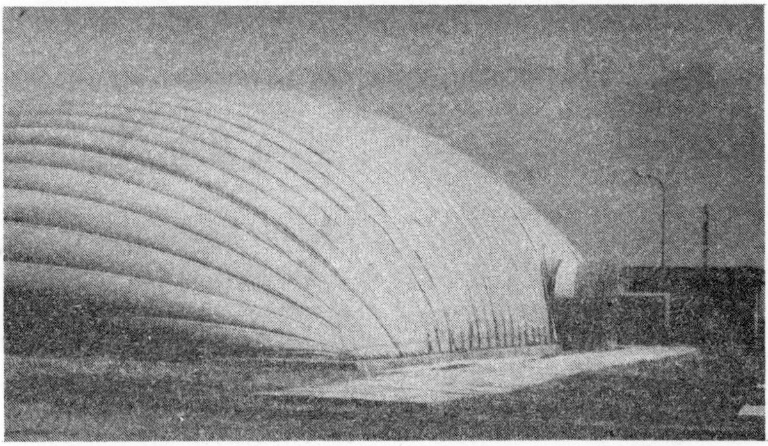

Fig. I-1e. Air-supported structure with thermally insulating jacket according to the author's Czechoslovak patent No. 155 310 [17] erected in 1978 in Prague (Czechoslovakia).

Fig. I-1f. Air-supported structure of cushion type according to the author's design, erected in 1978 in Hradec Králové (Czechoslovakia).

Czechoslovak patent No. 136 153, consisting of a harmonic vibration exciter with variable frequency, connected by means of a shaft with a plate fastened to the jacket [17]. Fig. I-2a shows the prototype of the facility which is most effective for cylindrical structures, since these structures vibrate easily in the resonance region. During the time of formation the effect of this instrument can be increased by the combination of dynamic effects with thermal effects, achieved by the preheating of air blown into the air-supported structure by the fan.

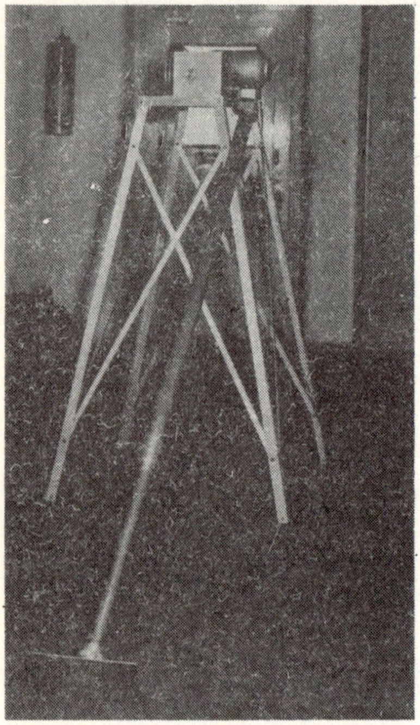

Fig. I-2a. Prototype of the device for the removal of snow from air-supported structures using the resonance effect (Czechoslovak patent No. 136 153 [17]).

In the case of some air-supported structure whose jacket has been made of translucent polyesther fabric coated with milkcoloured PVC no artificial lighting may be required at daytime. At night the interior of such structures is lit by spotlights and bulbs situated at the bottom, along the circumference of the structure, or on a lighting gallery suspended from the columns or from the jacket of the air-supported structure.

The jackets made of polyamide fabric are not translucent, as the fabric must be protected by a light barrier against the effect of light, particularly against the effect of ultraviolet rays. Therefore the upper part of these structures is provided with translucent bands of polyesther fabric or artificial lighting must be provided night and day (Fig. I-1c).

The bands of fabric are connected by seams or welds. PVC coated industrial fabrics are welded, as a rule, by means of a high frequency device or by a hot air welding apparatus. A weld some 4 cm wide has the same shear strength as the tensile strength of the fabric. The jackets of air-supported structures of medium and large size are made of one piece or of several pieces connected on the site.

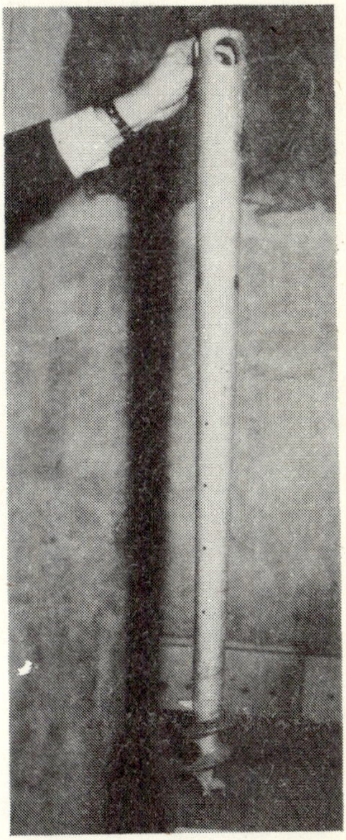

Fig. I-2b. Screw pile used for mobile anchorage of air-supported structures in Czechoslovakia.

The anchorage of the jacket is either mobile or stationary. For the purpose of mobile anchorage one uses most frequently screw piles (Fig. I-2b) or bags filled with water, sand or other material. Stationary anchorages are made of concrete ring beams laid along the circumference of the structure or of soil which covers a continuous band of the fabric around the whole circumference of the structure and which serves as a counterweight.

1.2 *Suspended Membrane Structures*

In the case of suspended structures the industrial fabric of which the membrane is made, must be usually prestressed by tensile forces to achieve sufficient rigidity and load bearing capacity.

For a suspended or air-supported membrane structure the are, in general, three equations of equilibrium determining the relations between external load and internal forces. They serve to determine, by integration, the internal membrane forces T_1, T_2 and T_{12} which must, under any load combination, satisfy the inequalities

$$T_1>0, \quad T_2>0, \quad T_1T_2>T_{12}^2 \tag{1.1}$$

following from the condition that the main membrane forces must be tensile forces (see Chapter II, Subsection 2.2).

In the direction of the normal to the surface whose principal directions are those of the tangents to the curvilinear coordinates, the equation of equilibrium has the form (Chapter II, Subsection 2.1)

$$\frac{T_1}{R_1}+\frac{T_2}{R_2}=q_n, \tag{1.2}$$

where q_n is the component of external load in the direction of the normal to the surface, R_1, R_2 are the principal radii of curvature of the surface, and T_1, T_2 are the membrane forces in the principal directions.

If $q_n=0$, then according to (1.2),

$$T_1=-\frac{R_1}{R_2}T_2 \tag{1.3}$$

which implies that for the prestressing of the membrane by a tensile force $T_2>0$ the force T_1 can also be a tensile force ($T_1>0$) only if $R_1/R_2<0$, i.e. when the surface has a negative Gaussian curvature.

In the case of suspended roofs consisting of ropes and rigid roofing the membrane forces due to prestressing attain as much as 60 % of their maximum value for the most unfavourable combination of prestressing and external load, i.e. the dead weight combined with snow and wind loads. In membrane structures it is possible to use lower prestressing, as the fabric will not fail under major deflections resulting from a lower rigidity of the structure. Decisive are only the conditions (1.1) and the critical wind velocity which produces the fluttering of the membrane.

Examples of erected suspended membrane structures with negative Gaussian curvature are shown in Figs. I-3a, b.

Fig. I-3a. Suspended membrane roof structure over an open-air café at Sopot (Poland).

Fig. I-3b. Roofing of a children's playground in Merseburg (German Democratic Republic) by an industrial fabric whose spatial form is obtained from a plane by elongation of the membrane by anchorage forces.

2

Air-Supported Structure with Automatic Control of Internal Overpressure

2.1 *Description of the Structure*

Fig. I-4 shows schematically in plan and Fig. I-5 in cross-section the air-supported structure with automatic control of internal overpressure according to the Czechoslovak patent No. 136 564 (see [17]). In the vertical parts of the membrane *2* the structure is provided with openings *1* at selected spacing and a suitable height (approximately 1—1.5 m) above ground level. All openings *1* are provided with flaps loosely fitting the openings *1* on the inside of the structure.

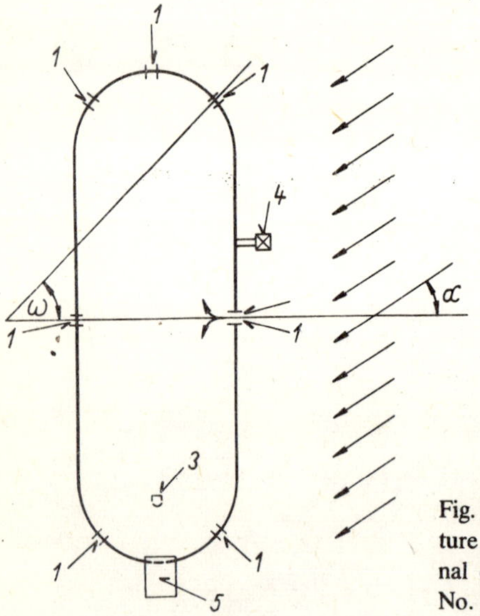

Fig. I-4. Plan of an air-supported structure with automatic control of the internal overpressure (Czechoslovak patent No. 136 564 [12], [17]).

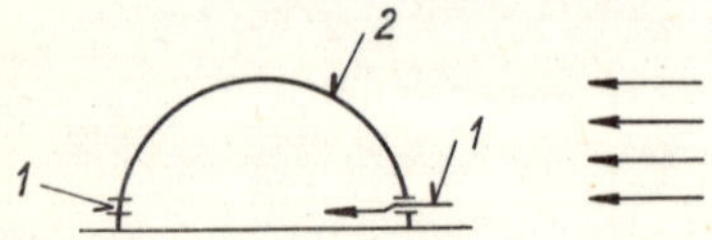

Fig. I-5. Cross-section of the jacket of an air-supported structure whose plan is shown in Fig. I-4.

Fig. I-6a shows the detail of the fastening of the flap *6* to the membrane *2* in a vertical section, Fig. I-6b shows the flap with a flange. The upper edge of the flap is fastened to the membrane *2* by means of the hinge *7*, socket *8* and flanges *9*, provided with sealing *10* on the face turned to the membrane. Also the whole edge of the socket *8* loosely matching the flap *6* is provided with the seal *11*. On the outside of the structure the openings may be provided with heaunches *12* forming an angle of approximately 22°30′ with the horizontal.

This design of the air-supported structure enables automatic control of the internal overpressure in the structure by the wind pressure, thus ensuring its rigidity and stability of form in stronger wind.

When there is no wind, all openings *1* are closed with flaps *6* pressed to the sockets *8* by a low internal overpressure maintained by the blower *4*. Under the effect of a stronger wind the flaps *6* on the windward side of the structure are lifted by wind pressure and the air flows into the interior of the structure, thus increasing the internal overpressure. The haunches *12* increase the air flow into the structure, serving simultaneously as gutters above the openings *1*. Elastic seals *10* and *11* and the entrance openings *5* with

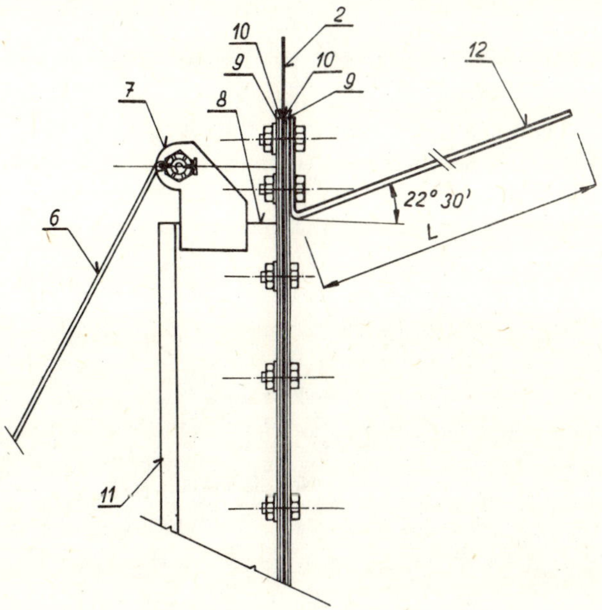

Fig. I-6a. Fastening of the flap to the membrane (vertical section).

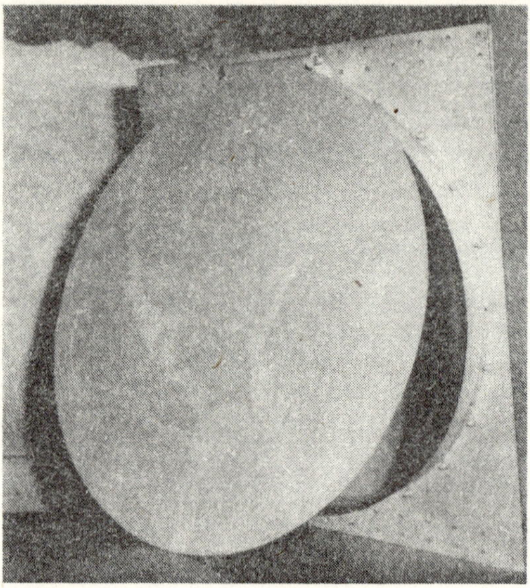

Fig. I-6b. View of the flanged flap.

double doors prevent undesirable air escape from the interior of the air-supported structure.

For circular openings of a diameter of up to approximately 80 cm the flaps *6* can be made, for example, from aluminium alloy sheet 1 mm thick. Instead of flaps *6* vaves can be used letting the air into the interior of the structure under strong wind and preventing the air from escaping due to the internal overpressure.

An air-supported membrane structure may be supplemented with a balanced flap *3* fitting the membrane opening on the outside of the structure. This is to ensure that when the wind velocity increases, the internal pressure does not exceed the permissible upper limit, related to the strength of the anchorage and the membrane.

Under normal weather conditions the stability of form of medium span (about 20 m) structures is ensured by an internal overpressure of about 100 N/m^2, i.e. about 10 mm of water column. For wind velocities of up to 130 km/h the internal overpressure must be increased to as many as 400 N/m^2 to ensure sufficient rigidity of the structure.

2.2 *Results of Model Tests in a Wind Tunnel*

The rigidity and the stability of form of an air-supported structure with automatic internal overpressure control was experimentally verified by the author on a model 90 × 30 cm in plan as on Fig. I-4, [12]. The PVC

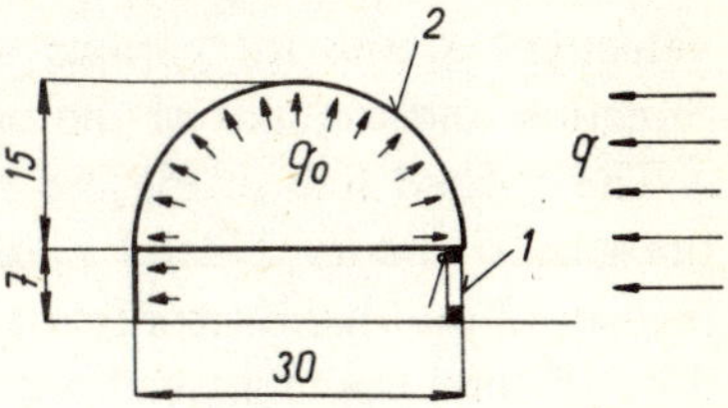

Fig. I-7. Cross-section of a model of an air-supported structure (dimensions in cm).

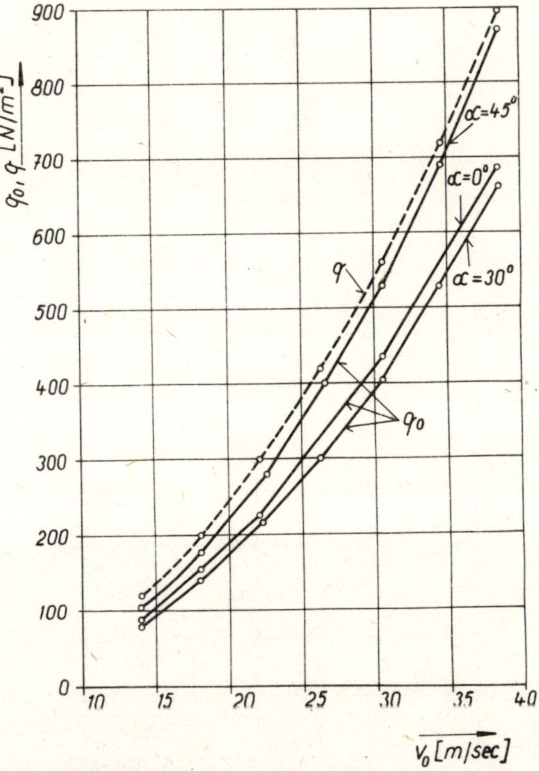

Fig. I-8. Air-pressure q on the plane perpendicular to the direction of air flow and the internal overpressure q_0 plotted against the wind velocity v_0.

membrane was fastened to a vertical wall 7 cm high (Fig. I-7) provided with circular openings of diameter 5 cm, equipped with flaps fitting the openings on the inside of the model. The angle between the axes of the openings in the longitudinal wall and those in the rounded part was $\omega = 45°$ (Fig. I-4).

In the course of the tests carried out in a wind tunnel of diameter 1.8 m the magnitude of the internal overpressure q_0 was measured for various wind velocities v_0 and for various values of the angle α between the air flow direction and the axis of the opening in the longitudinal wall (Fig. I-4).

The dashed line in Fig. I-8 shows the relation between the dynamic air pressure q and its velocity v_0, the solid lines show the relations between the measured internal overpressure q_0 and the velocity v_0 for the angles $\alpha = 0°$, 30°, 45° and openings without haunches *12*.

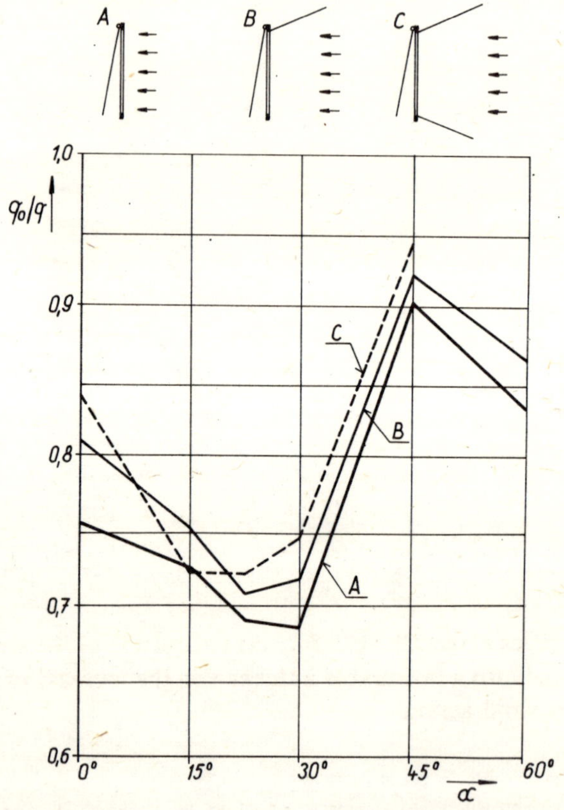

Fig. I-9. Experimentally ascertained dependence of the ratio q_0/q on the angle α.

Fig. I-9 shows the relations between the average measured values of the ratio q_0/q and the angle α. The line *A* holds for a model with openings without haunches, line *B* for a model with upper haunches *12* on the outside of the openings, and line *C* for a model with openings provided with upper and lower haunches on the outside of the openings.

In both latter cases all haunches formed an angle of 22°30′ with the horizontal and the ratio between the length of the haunch $L=3.3$ cm (Fig. I-6a) and the diameter of the opening $d=5$ cm was $L/d=0.66$. The width of the haunches measured in the direction of model wall was 7.5 cm.

Fig. I-10 shows the model in the wind tunnel.

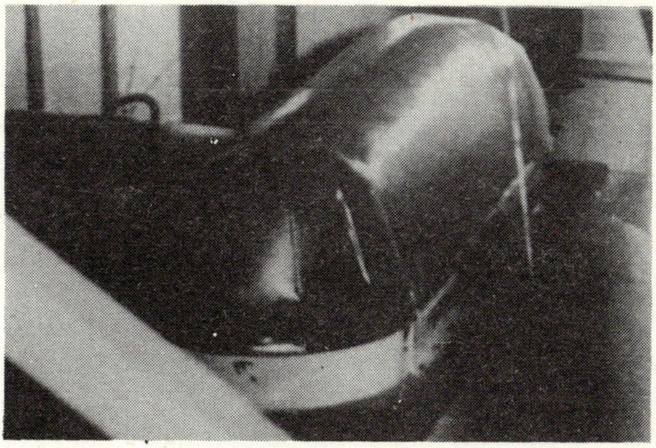

Fig. I-10. Model of the air-supported structure whose cross-section is shown in Fig. I-7 (Czechoslovak patent No. 136 564).

The wind tunnel model tests of the air-supported structure with automatic control of the internal overpressure have confirmed that the effect of wind pressure can be utilized, by means of a simple facility, to ensure sufficient rigidity and aerodynamic stability as well as the stability of form of the membrane under stronger wind attaining a velocity even up to 140 km/h. The upper haunches above the openings on the outside of the structure have been found to exercize a favourable influence on the increase of the internal overpressure by wind action.

3

Air-Supported Structure Consisting of Two or More Surfaces

In strong winds the jacket of air-supported structures of cupola form is stressed by considerable shearing forces in the vicinity of anchorage, which can break the fabric or impair its life. Apart from the strengthening of the jacket by additional bands of fabric it is also possible to use a suitable design of the membrane structure as shown on Figs. I-11 and I-12 [11]. Fig. I-11 shows the vertical section of the air-supported structure, Fig. I-12 the horizontal section on the level of pressure-control openings.

The jacket of the structure consists of the main membrane *1* inflated by the fans into a spherical form. The lower part of the main membrane *1* above the ground level is covered by the second membrane *2* which has the form of a band fastened along one edge *4* to the main membrane and along the other edge to the foundations. This second membrane generally acquires the form of a cylindrical surface.

Both the main membrane *1* and the second membrane *2* are provided along their circumference, between the connection of both membranes *4* and the ground level, with pressure control openings *3* covered with flaps or other one-way valves on the inside, letting the air under wind pressure only into the structure. The part of the main membrane *1* below the seam *4* may be conical or have the form of some other surface of revolution.

The wind blowing in the direction of the arrows *6* affects the windward side of the structure and its pressure opens the flaps of the pressure control openings *3*, allowing the air to flow into the space between both membranes. If the air pressure in this space attains a higher value than the internal overpressure inside the structure, the air begins to penetrate through the control openings *3* in the main membrane into the interior of the structure, thus increasing the internal overpessure. As a result of increased rigidity and aerodynamic stability the structure is capable to resist stronger wind.

The use of membrane *2* considerably reduces the stresses in membrane *1* due to shearing forces produced by a strong wind and the interior of the

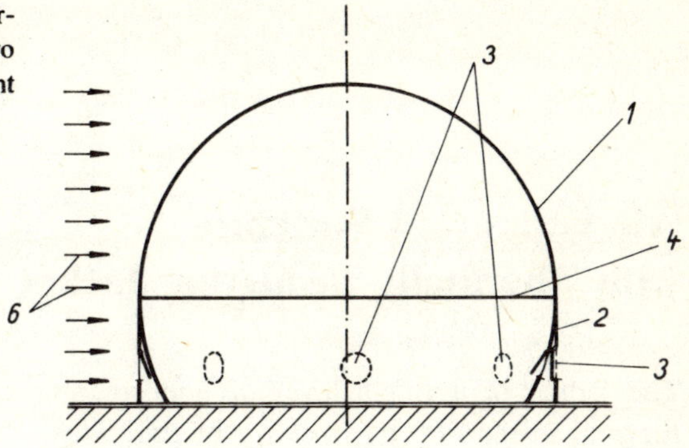

Fig. I-11. Vertical section of the air-supported structure consisting of two surfaces (Czechoslovak patent No. 160 953 [11]).

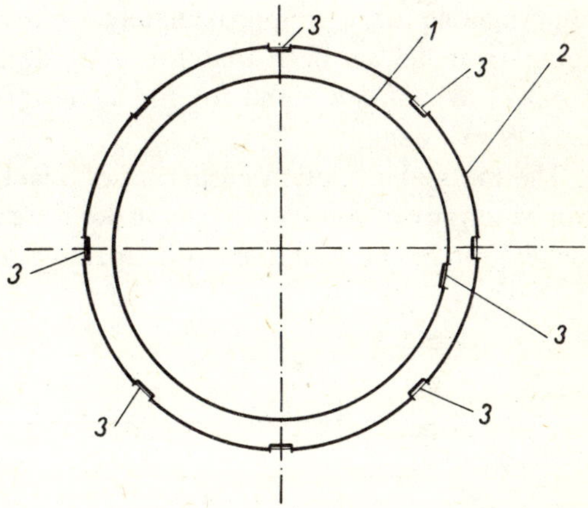

Fig. I-12. Plan of the air-supported structure according to Fig. I-11.

structure is filled with air under wind pressure more regularly. The air cushion between both membranes also plays a role of thermal insultation; the space can be also used, for example, for storage purposes or as entrance chamber, which makes it possible to close the entrance openings with single (and not double) doors only. Moreover, rain or snow which can be driven by wind through the pressure control openings will not penetrate the useful interior of the structure.

The described air-supported structure may be of any type. Its plan may be oval or of another shape.

4

Air-Supported Structure with Thermally Insulating Jacket

The jacket of a single layer of impermeable fabric is not satisfactory for thermal insulation, particularly in the cases when major requirements are imposed on the microclimate in the interior of the air-supported structure. The requirements on heat supply to ensure adequate temperature inside the structure in winter are considerable, as the coefficient of heat passage of the impermeable fabric is approximately $k = 6.19\ \mathrm{W\ m^{-2}\ deg^{-1}}$. For the sake of comparison let us note that for unplastered brick masonry 15 cm thick $k = 3.71\ \mathrm{W\ m^{-2}\ deg^{-1}}$ and for the same type of masonry 45 cm thick $k = 2.42\ \mathrm{W\ m^{-2}\ deg^{-1}}$.

The thermal insulation properties of a jacket of an air-supported structure can be improved considerably by making, according to the autor's design, the jacket out of air cushions with openings connecting the interiors of the

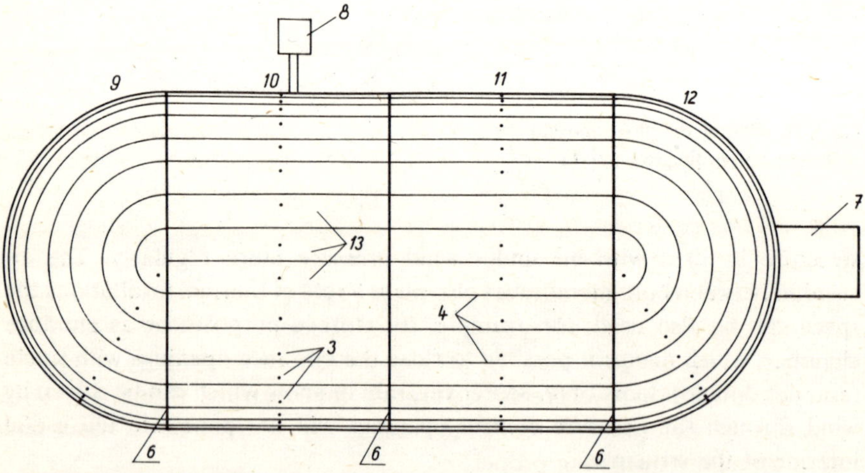

Fig. I-13. Plan of the air-supported structure with thermally insulating jacket (Czechoslovak patents Nos. 150 309 and 155 310 [17]).

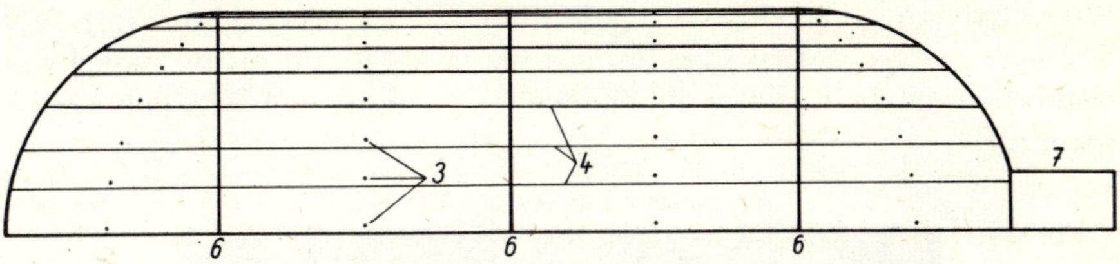

Fig. I-14. Longitudinal section of the thermally insulating jacket.

cushions with the interior of the whole structure with controlled atmosphere [11].

Fig. I-13 and Fig. I-14 show the plan and the elevation respectively of an air-supported structure with a thermally insulating jacket. Fig. I-15 showing the cross-section of this jacket. An alternative design of this jacket is described in Chapter III, Subsection 3.3.

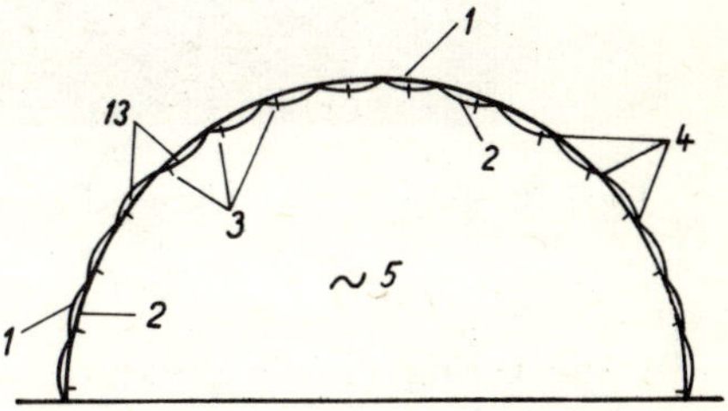

Fig. I-15. Cross-section of the thermally insulating jacket.

The jacket of the air-supported structure according to this example consists of four parts *9*, *10*, *11* and *12*, connected by an assembly seam *6*. Part *10* is connected with the blower *8*, in part *12* there is a drive-in chamber *7*.

Every part *9*, *10*, *11*, *12* consists of two membranes *1*, *2*, mutually connected by assembly seams *6* and by longitudinal joints (seams or welds) *4* to form a whole number of oblong air cushions *13* each of which is provided with an opening *3* in the inner membrane *2*.

When the air is blown into the interior *5* of the air-supported structure, it penetrates, due to overpressure, through the openings *3* into the voids between the membranes *1* and *2*, thus filling the air cushions *13* which have good thermally insulating properties.

For experimental verification a model of the above mentioned structure was made with a jacket 2.7 × 5 m in plan (Fig. I-16). Inside the structure three electric heating bodies of a power input of 0.5 kW each were installed. The heat passage was possible through the jacket mostly, as the model was clad below and on the sides with thermally insulating panels with polystyrene foam fill.

Fig. I-16. Model of an air-supported structure with a thermally insulating jacket [17].

By measuring the temperatures inside and outside the structure, the amount of heat supplied and the area of the jacket surface for a stationary thermal state it was ascertained that the jacket with air cushions has a heat passage coefficient $k = 4.16$ W m^{-2} deg^{-1}, and consequently, is equivalent to brick masonry 10 cm thick.

5

Suspended Structures and Roofs Stabilized by Wind Pressure

The system of flaps fitting openings on the inside of the structure is also suitable for the stabilization of suspended membrane structures.

Fig. I-17 shows a cross-section of a suspended membrane structure according to the Czechoslovak patent No. 136 566 [17], in which the membrane *2* with openings *1* provided with closing flaps is suspended from a rigid load bearing structure *3* and anchored into the foundations along its whole circumference. When the wind blows the flaps on the windward side of the structure lift and the air flows through the openings *1* into the interior of the structure, thus producing overpressure which stabilizes the structure by tensioning the membrane *2* [11].

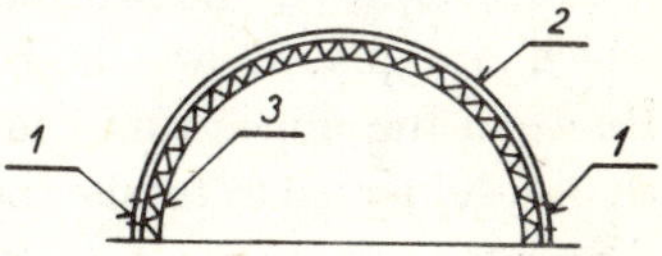

Fig. I-17. Cross-section of a suspended membrane structure (Czechoslovak patent No. 136 566 [17]).

Light-weight suspended membrane roofs must be provided with a special system of stabilization ropes to counter the static and dynamic effects of wind, particularly suction, pressure and uplift. Very often it is necessary to design roofs of double curvature (see Chapter I, Section 1). The installation of stabilization ropes, their thorough anchorage and the erection of the roofs of double curvature impose considerable requirements on construction and increase the overall costs of roofing.

These difficulties are eliminated by a light-weight suspended roof, according to the Czechoslovak patent No. 136 567, whose cross-section is illustrated together with the supporting structure in Fig. I-18 and whose plan is shown in Fig. I-19 [15], [17].

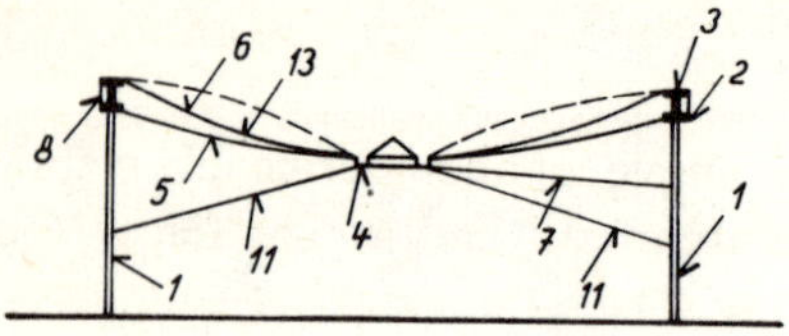

Fig. I-18. Cross-section of a suspended roof structure stabilized by wind pressure (Czechoslovak patent No. 136 567 [15], [17]).

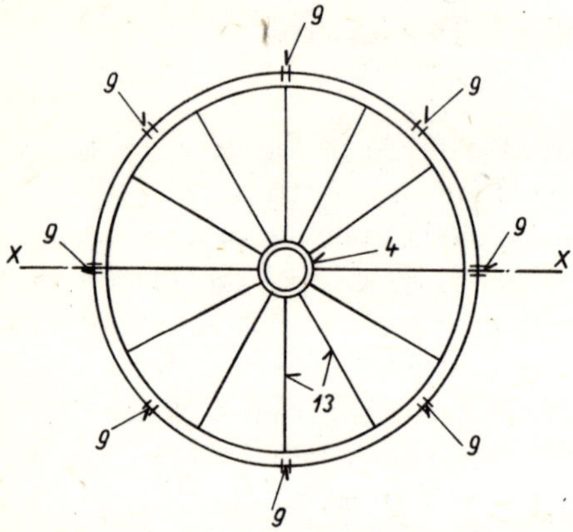

Fig. I-19. Plan of the suspended roof structure shown in Fig. I-18.

The columns *1* support the lower ring beam *2* and the upper ring beam *3* which are placed at suitable distance (about 1 m) above one another. Between the upper ring beam *3*, subjected to compression, and the inner rign *4*, subjected to tensile forces, radial ropes *13* are stretched together with the upper membrane *6* made, for example, of impermeable fabric or foil. Between the lower ring beam *2* and the ring *4* the lower membrane *5* is suspended. The upper membrane *6* may be connected with radial ropes *13* by means of sleeves. The inner ring *4* serves as a channel collecting rain water which is drained by the drain *7* to the edge of the structure. The ring *4* may be stabilized by anchorage ropes *11* which prevent the lifting of the centre of the roof during wind suction and uplift.

The ring beams *2* and *3* are connected to each other by an impermeable vertical wall *8* provided with openings *9* with closing flaps *10* loosely fitting the openings *9* on the inside of the wall *8* and hinged about their upper horizontal edge (Fig. I-20a). The openings are situated about the whole circumference of the structure.

Fig. I-20b shows the plan of an alternative design of the flap *10* hinged about its vertical edge and provided with a wire *12* which ensures the optimum opening of the flap *10* and its easy closing by draft of air in the opposite direction.

The vertical wall *8* can be made, for example, of an impermeable mebrane of the same material as that of the membranes *5* and *6*. The flaps *10* are made preferably of thin aluminium alloy sheet, the most suitable form of openings *9* being circular.

A detail of an alternative fastening of the flap *10* to the membrane wall *8* is shown in vertical section in Fig. I-20 c. The horizontal edge of the flap *10* is hinged to the wall *8* by means of a rigid strip *14* and angle *15* to which

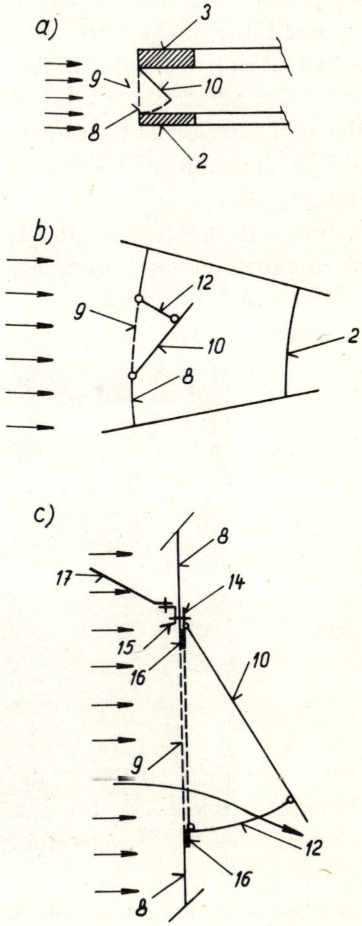

Fig. I-20. a) Flap hinged about a horizontal hinge. b) Flap hinged about a vertical hinge. c) Alternative way of fastening the flap to the membrane.

a metal sheet cover *17* is fastened on the outside face of the wall *8*. The other edge of the flap *10* which deviates most from its position under wind pressure, is connected with the wall *8* by means of two or more cables *12* of adequate length. On the inside every opening *9* is provided with elastic sealing *16*. The arrows in Fig. I-20 a, b, c indicate the direction of wind.

This type of light-weight suspended roof is automatically stabilized by wind pressure as follows:

Under the pressure of a stronger wind the flaps *10* lift on the windward side of the structure and the air flows through the openings *9* in the vertical wall *8* into the space between the lower membrane *5*, the upper membrane *6* and the vertical wall *8*. This produces an overpressure between the membranes *5* and *6* which lifts the upper membrane *6* into the position shown by a dashed line in Fig. I-18, thus producing an air cushion whose rigidity and weight, either alone, or in combination with the anchorage ropes *11*, stabilize the roof position in conditions of wind suction and uplift.

It is desirable that there be no obstacles to the wind on the level of openings *9*.

Just as the air cushion, the membranes *5* and *6* have good thermal as well as sound insulating properties.

The required magnitude of the internal overpressure between the mem-

Fig. I-21. Model of a suspended roof structure without the outside wall according to the Czechoslovak patent No. 136 567.

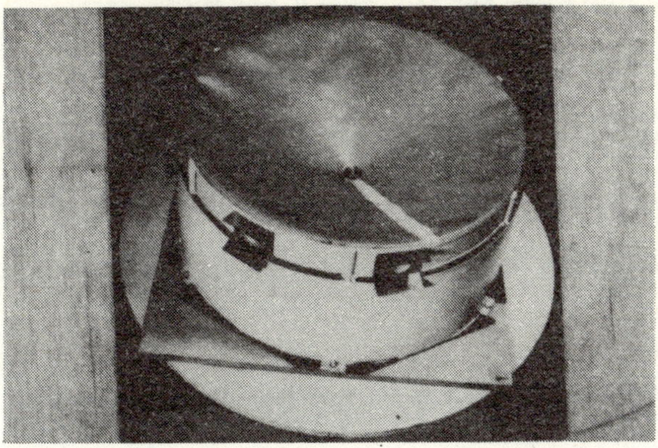

Fig. I-22. Model of the suspended roof according to Fig. I-21 provided with the outside wall and haunched openings with flaps.

branes *5* and *6* can be achieved by a suitable selection of the size of the openings *9* and their number. Also the cover *17* contributes to the achievement of a higher internal overpressure.

The described roof structure is particularly suitable for the buildings of circular, elliptical, oval or polygonal plan intended for most varied industrial, sport, exhibition and cultural purposes.

Fig. I-21 shows a model of the described roof structure without the vertical jacket, Fig. I-22 the same model with a vertical jacket and haunched openings.

6

Membrane Roof Structure Stabilized by Wind Suction

According to the author's design the membrane forming the covering of a suspended roof structure can be stabilized against the effect of a stronger wind also by wind suction [15]. Fig. I-23 shows the plan of one of the designs of a suspended roof structure using wind suction for stabilization purposes. Fig. I-24 shows a schematic vertical section of a structure with this roof, Fig. I-25 representing a detail of the shutter.

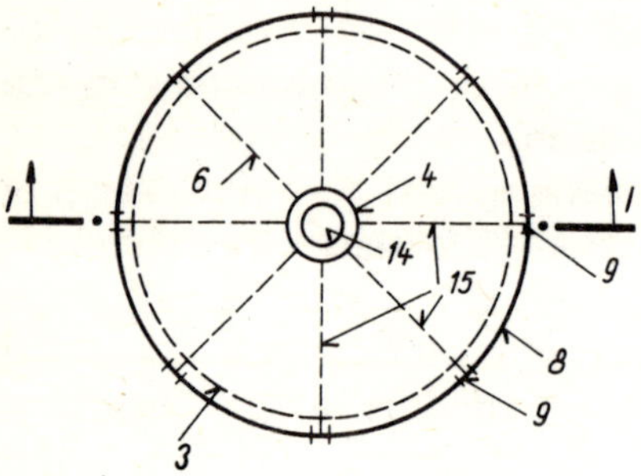

Fig. I-23. Plan of a suspended roof structure stabilized by wind suction (Czechoslovak patent No. 153 671 [15], [17]).

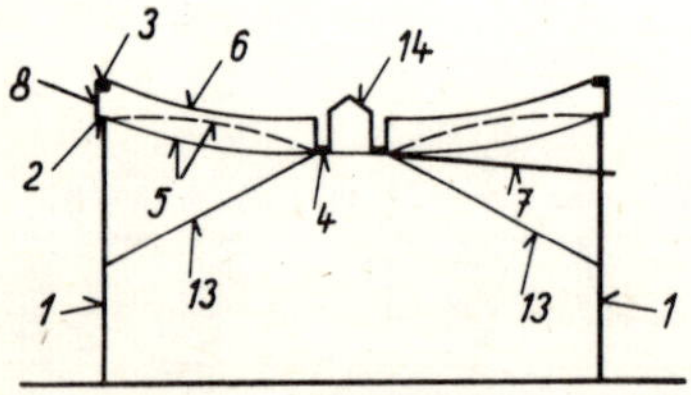

Fig. I-24. Cross-section of the roof according to Fig. I-23.

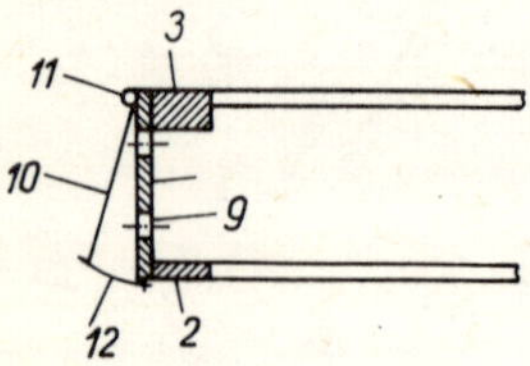

Fig. I-25. Detail of the flap with openings for the roof structure shown in Figs. I-23 and I-24.

The distance between the lower ring beam *2* and the upper ring beam *3*, supported on columns *1*, must be sufficiently large to ensure that during underpressure between the membranes the lower membrane *5* does not touch the upper membrane *6*. The drainage by means of a collector channel above the inner ring *4* and the drain *7* is the same as in the case of the structure shown in Fig. I-18.

Both ring beams *2* and *3* are interconnected along their whole circumference by a wall *8* provided with openings *9*. A group of these openings has a common flap *10* forming the actual shutter in this particular case. This flap *10* hinges about a horizontal hinge *11* and is provided in this particular case with a cable *12* limiting its deviation from the wall *8* with openings *9*. The flap *10* is made of thin aluminium alloy sheet sufficiently resistant to deformation, but also capable of closing the openings *9*.

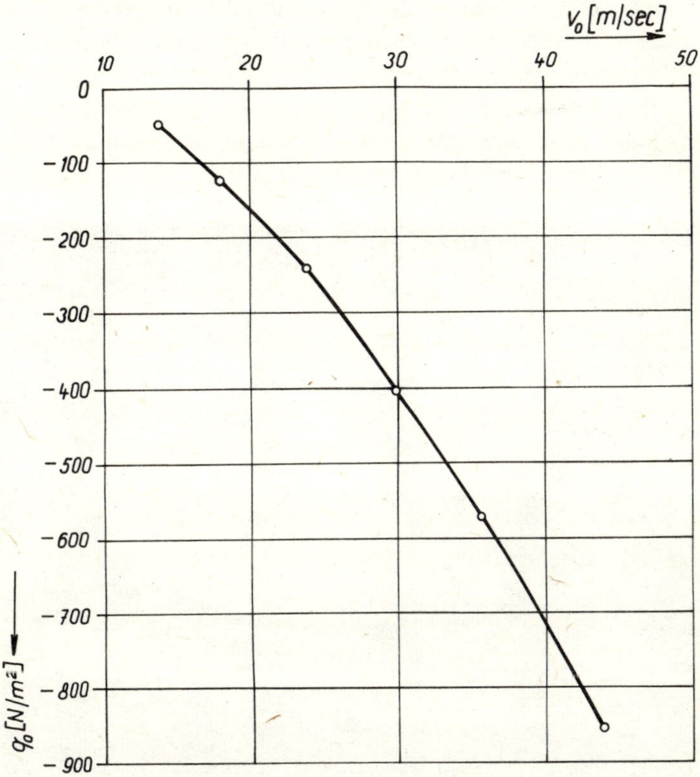

Fig. I-26. Experimentally ascertained dependence of the internal underpressure on wind velocity for the model of the roof structure shown in Figs. I-23 and I-24.

The inner ring *4* is fastened to the columns *1* by means of slanting ropes *13* and is provided with a central light *14*.

The upper membrane *6* is mounted on radial ropes *15* and possibly fastened to these by means of sleeves.

If the wind is stronger, underpressure arises on the major part of circumference of the wall *8* which lifts the flaps *10*, thus causing the suction of the air from the space between both membranes *5* and *6* through the openings *9*. Then the lower mambrane *5*, becoming convex, approaches the upper membrane *6*, which is schematically shown by a dashed line in Fig. I-24.

The required magnitude of the internal underpressure can be achieved by the selection of suitably reliable and effective shutters. It is sometimes advisable to provide the shutters with simple devices fostering the draught, for example, as those used on chimney tops. It is also possible to provide every opening *9* with an independent shutter of other known design.

Fig. I-26 shows the relation between the internal underpressure q_0 between the membranes and the wind velocity v_0 determined by the author experimentally in a wind tunnel on a model of the described type of suspended roof structure. The model had a circular plan of diameter 60 cm

Fig. I-27. Model of a suspended roof structure according to the Czechoslovak patent No. 153 671.

(Fig. I-27). The tests showed satisfactory rigidity of the upper membrane *6* achieved by the pretensioning due to the internal underpressure.

The most advantageous form of plan is circular, as the suction due to the air flow affects the major part of the circumference of the wall *8* which is greater than that affected by the pressure of the roof structure, and then the stabilization of the membrane roof structure is most effective.

Membrane Roof Stabilized by Wind Pressure and Suction

The suspended roof structure according to the author's design [11], [15] which is described in this part, is stabilized simultaneously by wind pressure and wind suction.

The wind tunnel tests have proved that the effect of both the internal overpressure and the internal underpressure is sufficient to ensure the stabilization of the membrane roofing. However, in a suspended roof with internal overpressure between both membranes unstabilized deflections of the upper membrane appear in the part near the leeward side of the structure, due to the fact that the external wind pressure on this part of the membrane is larger than the overpressure. In a suspended roof with internal underpressure between both membranes unstabilized deflections of the outer membrane appear in the part adjoining the windward side of the structure, because the wind suction affecting this part of the membrane exceeds the internal underpressure. Although the unstabilized deflections are of local character only and do not endanger the stability of the membrane roof as a whole, their unfavourable effect may manifest itself after a longer period by the fatigue of the material.

These shortcomings can be eliminated by the division of the inner space between both membranes into chambers by means of partitions. Each of these must be provided with at least one opening with a shutter allowing the air to flow inside and at least one opening with a shutter allowing the air to flow out of the chamber.

In this way the wind effect creates internal overpressure in the chambers adjoining the windward side of the roof structure and internal underpressure in the chambers adjoining the leeward side of the structure. Consequently, the leeward part of the upper membrane is affected by external wind pressure together with internal underpressure, while the windward part of the upper membrane is affected by external wind suction and internal overpressure. As the wind pressure and the internal underpressure act identically, and also the wind suction affects the outer membrane in the same way as the internal overpressure, no undesirable untabilized deflections of the upper membrane are produced by dynamic wind effect.

The design of the shutters and openings may be further simplified by providing one type of shutters with an opening allowing air flow in opposite direction.

Fig. I-28 gives a schematic upper view of the membrane roof with partly removed upper membrane, Fig. I-29 showing its vertical section.

Fig. I-30a shows a detail of the openings and shutters in the outside wall in one of the possible variants comprizing two openings in the wall provided with the respective shutters. Figs. I-30 b, c show two other possibilities of

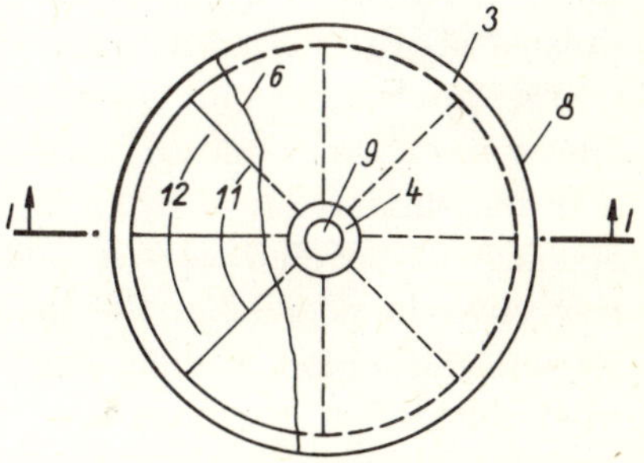

Fig. I-28. Plan of the suspended roof structure stabilized by wind pressure and suction (Czechoslovak patent No. 155 372 [15], [17]).

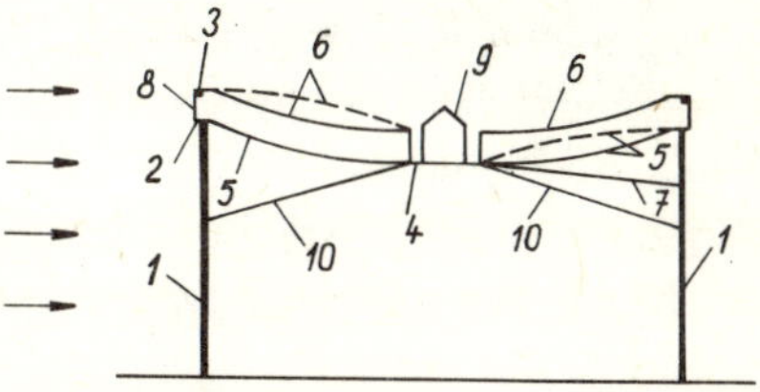

Fig. I-29. Cross-section of the roof structure according to Fig. I-28.

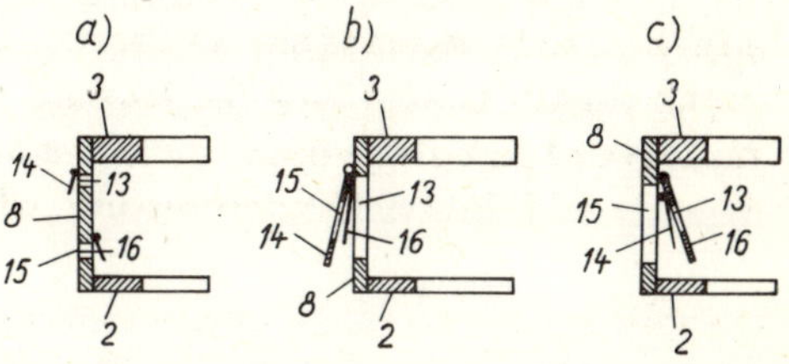

Fig. I-30. Openings with flaps for the roofs according to Figs. I-28 and I-29. a) Flaps fitting the openings in the vertical wall from opposite sides. b) Major flap closing the opening in the vertical wall from the outside, minor flap closing the opening in the major flap from the inside. c) Variant of the design according to b) with opposite directions of flap opening.

design of the respective openings and shutters in which one shutter allows the air flow in the opposite direction. For the sake of clarity all shutters are oversized to show distincly the sense of their opening and closing. The principal structural members, i.e. the columns *1*, the ring beams *2* and *3*, the inner ring *4* with the lantern *9*, the membranes *5* and *6* as well as the drain pipe *7* are situated in the same way as in the structure shown in Fig. I-18.

On Fig. I-30a the openings and the respective shutters in the wall *8* are placed above one another, but this is not necessary. On top there is the opening *13* with the shutter *14* allowing the air flow out of chamber *12*, formed between the membranes *5* and *6* by means of radial partitions *11*. This is illustrated in Fig. I-28, where there are eight such chambers and where the respective partitions *11* between the membranes *5* and *6* are situated above the slanting ropes *10*. Below the upper opening *13* there is another opening *15* with the shutter *16* allowing the air flow in the opposite direction, i.e. from outside into the chamber *12*.

In the variant shown in Fig. I-30b the wall *8* is provided with a single opening *13* covered with a shutter *14* allowing the air flow from the respective chamber *12*. In this shutter *14*, however, there is another opening *15* with the shutter *16*, allowing the air flow in the opposite direction, i.e. from outside into the respective chamber *12*.

The variant illustrated in Fig. I-30c is analogous that shown in Fig. I-30b but in this variant the opening *15* in the wall *8* is provided with a shutter *16* allowing the air flow from the outside into the respective chamber *12*, while the opening *13* in this shutter *16* is provided with a shutter *14* allowing the air flow from chamber *12*.

When the wind, blowing in the direction from left to right in Figs I-28 and I-29, affects the roof structure, all shutters *16*, allowing the air flow from outside into the respective chambers *12*, open on the windward side, while on the leeward side all shutters *14* open to allow the air flow from the respective chambers *12*. Due to the wind pressure and wind suction the air flows through the openings *15*, *13* so as to form the roof shape shown by a dashed line in Fig. I-29. On the windward side the upper membrane *6* deflects upwards and on the leeward side the lower membrane *5* deflects upwards.

The roof is provided with a system of radial ropes (not shown in the drawing) fastened to the upper ring beam *3* and the inner ring *4*.

8

Models of Suspended Membrane Structures Made by Spiders

In the design of the load-bearing systems of membrane structures it is possible, in the author's opinion, to use the models of cobwebs naturally created by some types of spiders (Fig. I-31). On the basis of the ascertained load-bearing capacity of the cobweb structure (Fig. I-32), the strength of

Fig. I-31. A cobweb representing a model of a suspended membrane structure [11].

Fig. I-32. A cobweb loaded by a piece of a white PVC foil.

cobweb samples (Fig. I-33) and the strength of the structural material it is possible to determine the permissible geometric magnification of the cobweb model [11].Using, for example, stereophotogrammetry it is possible to prepare the drawings of a suspended membrane structure which is merely a geometric magnification of the cobweb and its fastening. This makes it

Fig. I-33. Loading test of a cobweb sample (strip).

possible to abandon the contemporary conventional method of design of load-bearing systems of suspended roof structures, i.e. their static and dynamic analyses followed by experimental tests on models and their combinations. A test of a cobweb of a suitable form makes it possible to ascertain, how the load-bearing system of a suspended membrane structure is capable of transferring the decisive types of load, i.e. the vertical snow load, its own weight and the static and dynamic wind effects.

9

Device for the Damping of Dynamic Wind Effects

The Czechoslovak patent No 136 564 describes a device for the dampening of dynamic wind effects in tall structures, particularly of circular cross-section. It consists of an impermeable membrane loosely suspended about the structure as a closed envelope and provided with openings allowing the air

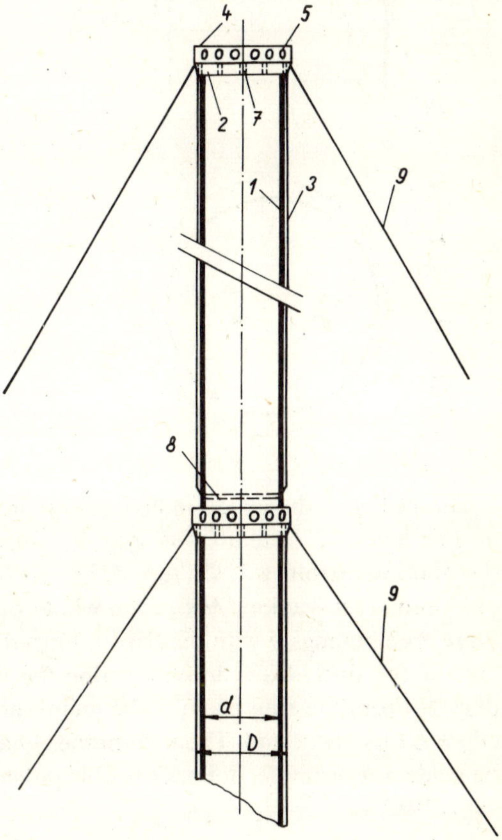

Fig. I-34. Vertical section of a guyded radio mast provided with a device for damping of dynamic wind effects (Czechoslovak patent No. 136 568 [13], [17]).

flow from the outside into the space between the membrane and the jacket of the structure [13], [17].

The divice is illustrated in the vertical section of a guyed radio mast in Fig. I-34. Its function under wind action is shown in detail in the vertical section of the structure in Fig. I-35 and in its cross-section in Fig. I-36.

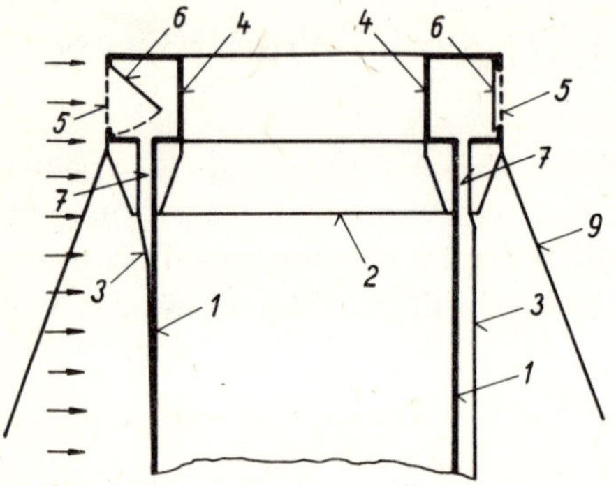

Fig. I-35. Detail of the damping device in the vertical section of the mast [13].

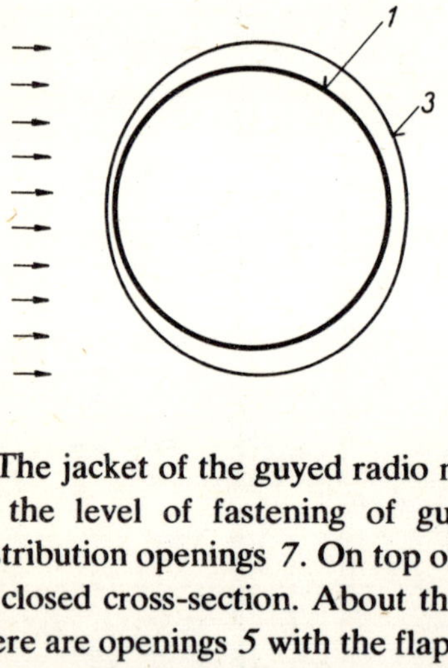

Fig. I-36. Horizontal section of the radio mast with the suspended membrane representing variable cross-section.

The jacket of the guyed radio mast of circular cross-section *1* is provided, at the level of fastening of guys *9*, with a ring beam *2* with vertical distribution openings *7*. On top of the ring beam *2* there is a collector duct *4* of closed cross-section. About the whole outer circumference of the duct *4* there are openings *5* with the flaps *6* hinged on their upper edge to the outer face of the duct *4* and loosely fitting the openings *5* on the inside of the duct *4*. From the ring beam *2* the membrane *3* is suspended, closed by the clip *8* on its lower end. The membrane *3* has the form of a cylinder and can be made, for example, of impermeable polyamide or polyesther fabric coated with PVC.

Fig. I-35 shows how the wind, represented by the arrows, causes the flaps *6* on the windward side to open, while in the places of wind suction the flaps *6* remain closed. The air flows through the openings *5* into the duct *4* and proceeds through the distribution openings *7* into the space between the membrane *3* and the jacket *1* of the structure. In this way internal overpressure originates in that space, which stabilizes the membrane *3* and, together with wind pressure and suction, causes mambrane *3* to take the form of a cylinder of variable cross-section shown in Fig. I-36. The form of the variable cross-section depends on the direction and the velocity of the wind and the smoothness of the membrane surface.

The dynamic wind effects affecting the radio mast are dampend by being distributed by the air cushion of variable thickness about the jacket *1* on all sides of the mast and mutually cancelled.

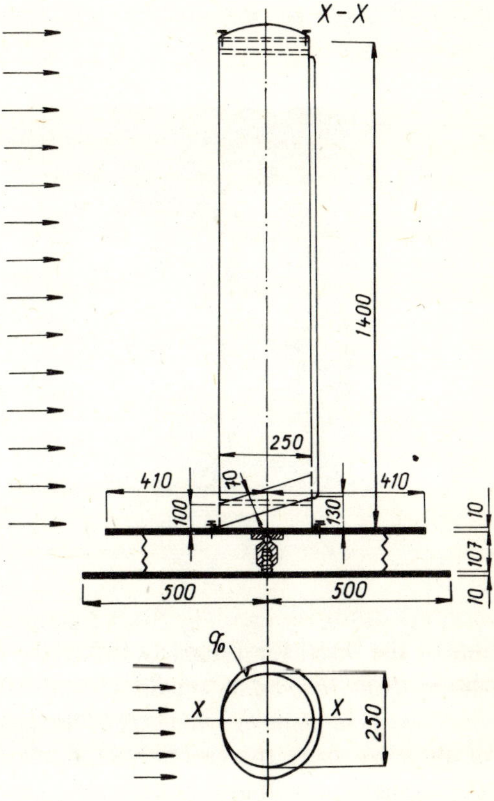

Fig. I-37. Vertical and horizontal sections of the cylinder with a suspended membrane (model of the damping device according to the Czechoslovak patent No. 136 568).

The proposed divice can be modified so that the openings are connected to blowers which are actuated by the wind and automatically blow air into the space between the jacket *1* of the structure and the membrane *3*.

The author has verified experimentally the proposed device in a wind tunnel, using cylinders and suspended cylindrical membranes of various diameters. The cylinder was fastened to a slab, hinged in its centra and elastically supported about its circumference (Fig. I-37).

The air was blown between the membrane and the cylinder through a slanting pipe of diameter 70 mm whose opening is shown in Fig. I-38 together with the mounting of the model. The magnitude of the internal overpressure q_0 in the space between the cylinder and the membrane, and the vibrations of the elastically mounted upper part of the model in two points of the slab, one of which was in the plane of symmetry of the model parallel with the direction of air flow and the other in the plane of symmetry of the model perpendicular to the the direction of air flow, were measured for various wind velocities v_0.

Fig. I-38. Mounting of the model of the damping device.

In Fig. I-39a through Fig. I-39f the upper vibration records (magnified 20 times) are taken for various wind velocities v_0 at a point ***P*** of the upper elastically mounted slab, at a distance of 39 cm from its centre. The point ***P*** is situated in the plane of symmetry of the model perpendicular to the direction of air flow, where the greatest amplitudes of the vibrations of the cylinder with suspended membrane occurred. These records were made for the case

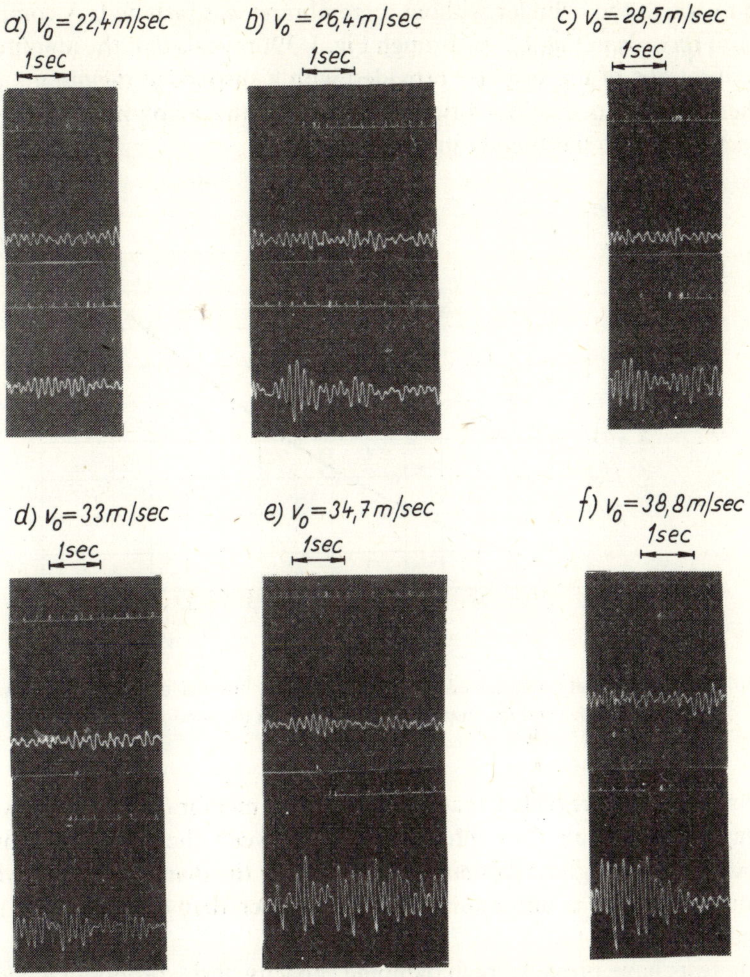

Fig. I-39. Records of the vibrations of the model shown in Figs. I-37 and I-38 in a wind tunnel for various wind velocities. The upper records concern the cylinder with the suspended membrane, the lower the cylinder without a membrane.

when the diameter of the cylindrical membrane was 275 mm. For this particular case the measured dependence of the overpressure q_0 on the velocity v_0 is shown in Fig. I-40. The lower vibration records (also magnified 20 times) in Fig. I-39a through Fig. I-39f refer to the point P of the upper slab to which the cylinder without a membrane was fastened. A comparison of both records in Fig. I-39a through Fig. I-39f reveals that the amplitudes of the vibrations of the cylinder provided with a suspended membrane are for higher wind velocities 3—4 times lower than the amplitudes of the same cylinder without the membrane.

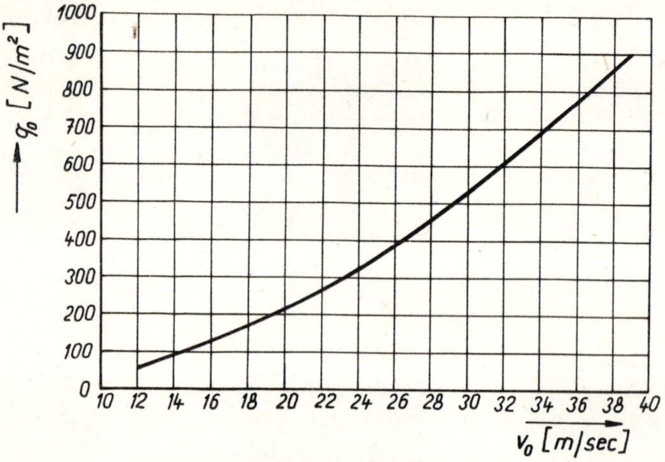

Fig. I-40. Experimentally ascertained dependence of the internal overpressure between the cylinder and the suspended membrane on the wind velocity.

The tests have revealed that the suspended membrane provided with an opening allowing air flow into the space between the membrane and the cylinder has the highest damping capacity when the diameter of the cylindrical membrane D is approximately 10 % larger than the diameter of the cylinder d ($D \approx 1.1d$)

The tests have proved a high damping capacity of the proposed device and a high stability of the suspended membrane achieved by the internal overpressure due to air flow.

Chapter II

STATICS OF AIR-SUPPORTED STRUCTURES

Introduction

In recent times the static problems of air-supported structures have become the subject of intensive research. This is understandably so, due to the growing rate of construction of air-supported structures for the most varied purposes, the application of new forms and designs as well as the application of new construction materials with viscoelastic properties. Apart from the problems of linear mechanics [11], [19], [20], [45], [47], [57] also complex problems of non-linear mechanics [22], [33], [43], [46], [66] and the rheology of membranes have been investigated [24], [41].

Air-supported structures are a special case of membrane structures. Their ability to transfer various loads is due to the prestressing of the construction material by tensile membrane forces which depend on the magnitude of the internal overpressure and the geometric form of the structure. Since the jacket of an air-supported structure behaves as an anisotropic membrane of zero flexural rigidity, only membrane states of stress can be produced in it by external loads. Thus air-supported structures can be investigated by means of the membrane theory of surface structures.

This chapter is concerned primarily with the static structural design of air-supported structures of various forms. Our principal purpose will be the investigation of the membrane forces in the jacket caused by the most frequently occuring types of load and the determination of the necessary internal overpressure. The last part of the chapter deals with some mechanical properties of the fabrics used and the method of assessment of the safety of the designed air-supported structure.

1

Fundamental Geometric Relations of the Theory of Surfaces

1.1 *Definition of a Surface*

In a rectangular system of coordinates x, y, z a surface can be defined by parametric equations (Fig. II-1)

$$x=x(\xi, \eta), \quad y=y(\xi, \eta), \quad z=z(\xi, \eta), \tag{1.1}$$

by an explicit equation

$$z=f(x, y) \tag{1.2}$$

or by an implicit equation

$$F(x, y, z)=0. \tag{1.3}$$

The position of a point on the surface can be unambiguously determined by its curvilinear coordinates ξ and η which represent two independent parameters in the equations (1.1). The first system of coordinate lines on the surface

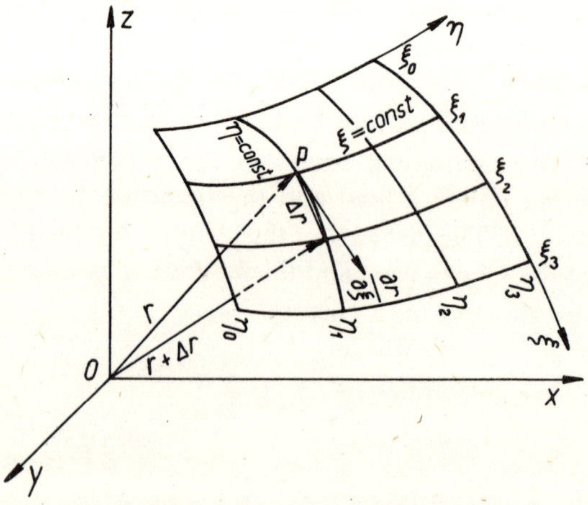

Fig. II-1. Curvilinear coordinates on a surface.

is defined by the equation $\xi = \text{const}$ ($\xi = \xi_0$, $\xi = \xi_1$, $\xi = \xi_2$, etc.). The other system of coordinate lines on the surface comprizes the curves with the equation $\eta = \text{const}$ ($\eta = \eta_0$, $\eta = \eta_1$, $\eta = \eta_2$, etc.). On every coordinate line the positive sense is that in which ξ or η increase.

1.2 *The First Fundamental Form of a Surface*

The parametric equations (1.1) can be expressed more briefly by the vector equation

$$\boldsymbol{r} = \boldsymbol{r}(\xi, \eta), \tag{1.4}$$

where $\boldsymbol{r}$ is the position vector (radius vector) of the point P of the surface relative to the origin O (Fig. II-1).

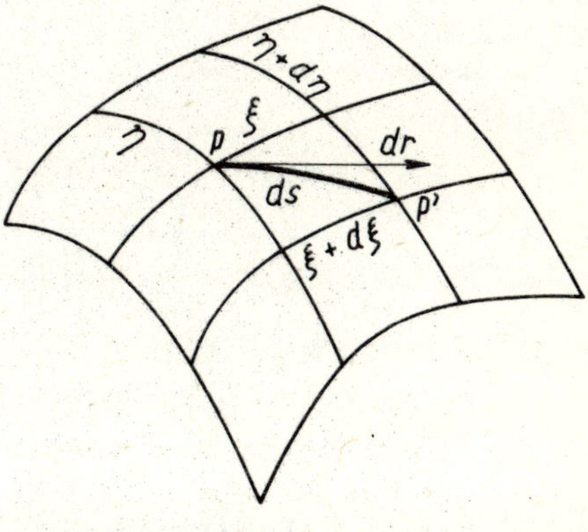

Fig. II-2. Distance of two points on a surface.

The square of the differential $\mathrm{d}s$ of the arc passing through the point P equals to the scalar product (Fig. II-2)

$$\mathrm{d}s^2 = \mathrm{d}\boldsymbol{r} . \mathrm{d}\boldsymbol{r} = \mathrm{d}\boldsymbol{r}^2, \tag{1.5}$$

where

$$\mathrm{d}\boldsymbol{r} = \frac{\partial \boldsymbol{r}}{\partial \xi}\mathrm{d}\xi + \frac{\partial \boldsymbol{r}}{\partial \eta}\mathrm{d}\eta \tag{1.6}$$

is the total differential of the radius vector. After the substitution of (1.6) into (1.5) we obtain the first quadratic differential form of the surface

$$ds^2 = g_{11}\,d\xi^2 + 2g_{12}\,d\xi\,d\eta + g_{22}\,d\eta^2, \tag{1.7}$$

where

$$g_{11} = \left(\frac{\partial \boldsymbol{r}}{\partial \xi}\right)^2, \quad g_{12} = \frac{\partial \boldsymbol{r}}{\partial \xi}\cdot\frac{\partial \boldsymbol{r}}{\partial \eta}, \quad g_{22} = \left(\frac{\partial \boldsymbol{r}}{\partial \eta}\right)^2 \tag{1.8}$$

are the coefficients of the first fundamental form.

If we express the radius vector $\boldsymbol{r}$ by means of Cartesian coordinates x, y, z, we obtain from (1.8)

$$\begin{aligned} g_{11} &= \left(\frac{\partial x}{\partial \xi}\right)^2 + \left(\frac{\partial y}{\partial \xi}\right)^2 + \left(\frac{\partial z}{\partial \xi}\right)^2, \\ g_{12} &= \frac{\partial x}{\partial \xi}\frac{\partial x}{\partial \eta} + \frac{\partial y}{\partial \xi}\frac{\partial y}{\partial \eta} + \frac{\partial z}{\partial \xi}\frac{\partial z}{\partial \eta}, \\ g_{22} &= \left(\frac{\partial x}{\partial \eta}\right)^2 + \left(\frac{\partial y}{\partial \eta}\right)^2 + \left(\frac{\partial z}{\partial \eta}\right)^2. \end{aligned} \tag{1.9}$$

For the angle Λ between the coordinate lines at the point (ξ, η) ve have [26], [64]

$$\begin{aligned} \cos \Lambda &= \frac{\dfrac{\partial \boldsymbol{r}}{\partial \xi}\dfrac{\partial \boldsymbol{r}}{\partial \eta}}{\left|\dfrac{\partial \boldsymbol{r}}{\partial \xi}\right| \left|\dfrac{\partial \boldsymbol{r}}{\partial \eta}\right|} = g_{12}(g_{11}g_{22})^{-\frac{1}{2}}, \\ \sin \Lambda &= \left(\frac{g_{11}g_{22} - g_{12}^2}{g_{11}g_{22}}\right)^{\frac{1}{2}}. \end{aligned} \tag{1.10}$$

For Λ we choose an obtuse angle ($\sin \Lambda > 0$); the square roots in (1.10) are considered positive.

In an orthogonal system ($\Lambda = \pi/2$) it follows from (1.10) that $g_{12} = 0$, and consequently,

$$ds^2 = g_{11}\,d\xi^2 + g_{12}\,d\eta^2. \tag{1.11}$$

1.3 *The Second Fundamental Form of a Surface*

The curvature $\varkappa_n$ of a normal section of a surface, i.e. of the curve in which the surface is intersected by a plane containing the normal to the surface, equals [7], [64]

$$\varkappa_n = \frac{h_{11}\,d\xi^2 + 2h_{12}\,d\xi\,d\eta + h_{22}\,d\eta^2}{g_{11}\,d\xi^2 + 2g_{12}\,d\xi\,d\eta + g_{22}\,d\eta^2}, \tag{1.12}$$

where

$$h_{11} = \frac{\dfrac{\partial^2 \boldsymbol{r}}{\partial \xi^2}\,\dfrac{\partial \boldsymbol{r}}{\partial \xi}\,\dfrac{\partial \boldsymbol{r}}{\partial \eta}}{(g_{11}g_{22} - g_{12}^2)^{\frac{1}{2}}}, \quad h_{12} = \frac{\dfrac{\partial^2 \boldsymbol{r}}{\partial \xi\,\partial \eta}\,\dfrac{\partial \boldsymbol{r}}{\partial \xi}\,\dfrac{\partial \boldsymbol{r}}{\partial \eta}}{(g_{11}g_{22} - g_{12}^2)^{\frac{1}{2}}},$$

$$h_{22} = \frac{\dfrac{\partial^2 \boldsymbol{r}}{\partial \eta^2}\,\dfrac{\partial \boldsymbol{r}}{\partial \xi}\,\dfrac{\partial \boldsymbol{r}}{\partial \eta}}{(g_{11}g_{22} - g_{12}^2)^{\frac{1}{2}}} \tag{1.13}$$

are the coefficients of the second fundamental form. The numerator in (1.12) is the second quadratic differential form of the surface. The curvature $\varkappa_n$ equals thus to the ratio of the two fundamental forms.

If the radius vector is expressed in Cartesian coordinates x, y, z, we obtain from (1.13)

$$h_{11} = (g_{11}g_{22} - q_{12}^2)^{-\frac{1}{2}} \begin{vmatrix} \dfrac{\partial^2 x}{\partial \xi^2} & \dfrac{\partial x}{\partial \xi} & \dfrac{\partial x}{\partial \eta} \\ \dfrac{\partial^2 y}{\partial \xi^2} & \dfrac{\partial y}{\partial \xi} & \dfrac{\partial y}{\partial \eta} \\ \dfrac{\partial^2 z}{\partial \xi^2} & \dfrac{\partial z}{\partial \xi} & \dfrac{\partial z}{\partial \eta} \end{vmatrix},$$

$$h_{12} = (g_{11}g_{22} - g_{12}^2)^{-\frac{1}{2}} \begin{vmatrix} \dfrac{\partial^2 x}{\partial \xi\,\partial \eta} & \dfrac{\partial x}{\partial \xi} & \dfrac{\partial x}{\partial \eta} \\ \dfrac{\partial^2 y}{\partial \xi\,\partial \eta} & \dfrac{\partial y}{\partial \xi} & \dfrac{\partial y}{\partial \eta} \\ \dfrac{\partial^2 z}{\partial \xi\,\partial \eta} & \dfrac{\partial z}{\partial \xi} & \dfrac{\partial z}{\partial \eta} \end{vmatrix}, \tag{1.14}$$

$$h_{22} = (g_{11}g_{22} - g_{12}^2)^{-\frac{1}{2}} \begin{vmatrix} \dfrac{\partial^2 x}{\partial \eta^2} & \dfrac{\partial x}{\partial \xi} & \dfrac{\partial x}{\partial \eta} \\ \dfrac{\partial^2 y}{\partial \eta^2} & \dfrac{\partial y}{\partial \xi} & \dfrac{\partial y}{\partial \eta} \\ \dfrac{\partial^2 z}{\partial \eta^2} & \dfrac{\partial z}{\partial \xi} & \dfrac{\partial z}{\partial \eta} \end{vmatrix}.$$

1.4 *Principal Curvatures of a Surface*

The rotation of the normal section about the normal at the point (ξ, η) is equivalent to the change of the ratio $d\eta/d\xi$. In certain positions of the normal section the curvature $\varkappa_n$ attains its extreme values.

If we write the equation (1.12) in the form

$$h_{11}\,d\xi^2+2h_{12}\,d\xi\,d\eta+h_{22}\,d\eta^2-\varkappa_n(g_{11}\,d\xi^2+2g_{12}\,d\xi\,d\eta+g_{22}\,d\eta^2)=0 \tag{1.15}$$

and differentiate (1.15) first with respect to $d\xi$, while considering $d\eta$ constant, and then with respect to $d\eta$, while considering $d\xi$ constant, we obtain

$$\begin{aligned} h_{11}\,d\xi+h_{12}\,d\eta-\varkappa_n(g_{11}\,d\xi+g_{12}\,d\eta)&=0,\\ h_{12}\,d\xi+h_{22}\,d\eta-\varkappa_n(g_{12}\,d\xi+g_{22}\,d\eta)&=0. \end{aligned} \tag{1.16}$$

If we eliminate the ratio $d\eta/d\xi$ from (1.16), we obtain a quadratic equation with respect to $\varkappa_n$

$$(g_{11}g_{22}-g_{12}^2)\varkappa_n^2-(g_{11}h_{22}+g_{22}h_{11}-2g_{12}h_{12})\varkappa_n+(h_{11}h_{22}-h_{12}^2)=0. \tag{1.17}$$

From (1.17) it is possible to determine two extreme values $\varkappa_1$ and $\varkappa_2$ of $\varkappa_n$ which are called the principal curvatures of the surface at the given point. The respective directions, determined by the ratio $d\eta/d\xi$ according to (1.16), are called the principal directions of the surface.

A principal curve on the surface is such a curve whose tangent at any point lies in a principal direction.

For the Gaussian curvature K and the mean curvature H of the surface we obtain from (1.17)

$$\begin{aligned} K&=\varkappa_1\varkappa_2=\frac{h_{11}h_{22}-h_{12}^2}{g_{11}g_{22}-g_{12}^2},\\ H&=\frac{1}{2}(\varkappa_1+\varkappa_2)=\frac{1}{2}\,\frac{g_{11}h_{22}-2g_{12}h_{12}+g_{22}h_{11}}{g_{11}g_{22}-g_{12}^2}. \end{aligned} \tag{1.18}$$

From the quantities K and H, whose equations for various surfaces are given in the papers dealing with the differential geometry of surfaces, it is possible to determine the principal curvatures

$$\varkappa_{1,2}=H\pm\sqrt{H^2-K}. \tag{1.19}$$

Let us identify now the curvilinear coordinate lines with the principal curves of the surface. If $\varkappa_1$ is the principal curvature on the curve $\eta=\text{const}$ and $\varkappa_2$ the principal curvature of the surface on the curve $\xi=\text{const}$, and if we substitute in the equations (1.16) $d\xi=0$ or $d\eta=0$, we obtain

$$h_{12}-g_{12}\varkappa_1=0, \quad h_{12}-g_{12}\varkappa_2=0. \tag{1.20}$$

For $\varkappa_1 \neq \varkappa_2$ we obtain from (1.20) that $g_{12}=0$ and $h_{12}=0$; consequently, the principal directions are mutually orthogonal. For this case equation (1.17) yields the following simple expressions for the principal curvatures:

$$\varkappa_1=\frac{1}{R_1}=\frac{h_{11}}{g_{11}}, \quad \varkappa_2=\frac{1}{R_2}=\frac{h_{22}}{g_{22}}, \tag{1.21}$$

where R_1 and R_2 are the principal curvature radii.

If the surface is defined explicitely by equation (1.2) and the variables $x, y (x=\xi, y=\eta)$ are selected as parameters, the equations (1.9) and (1.14) yield the following expressions for the coefficients of the first and the second fundamental forms:

$$\begin{aligned} &g_{11}=1+\left(\frac{\partial z}{\partial x}\right)^2, \quad g_{12}=\frac{\partial z}{\partial x}\frac{\partial z}{\partial y}, \quad g_{22}=1+\left(\frac{\partial z}{\partial y}\right)^2, \\ &h_{11}=\frac{1}{\omega}\frac{\partial^2 z}{\partial x^2}, \quad h_{12}=\frac{1}{\omega}\frac{\partial^2 z}{\partial x\,\partial y}, \quad h_{22}=\frac{1}{\omega}\frac{\partial^2 z}{\partial y^2}, \end{aligned} \tag{1.22}$$

where

$$\omega^2=1+\left(\frac{\partial z}{\partial x}\right)^2+\left(\frac{\partial z}{\partial y}\right)^2. \tag{1.23}$$

The surface determined by the explicit equation (1.2) evidently has the parametric equations

$$x=\xi, \quad y=\eta, \quad z=z(\xi, \eta). \tag{1.24}$$

In this case the coefficients of the first and the second fundamental forms are determined by the equations (1.22), where x and y have been replaced by ξ and η.

For a surface of revolution arising by the rotation of a curve (meridian) about the z-axis and defined by the parametric equations (Fig. II-3)

$$x=r(z)\cos\beta, \quad y=r(z)\sin\beta, \quad z=z, \tag{1.25}$$

in which $\xi=z$ and $\eta=\beta$, we obtain, according to (1.9) and (1.14),

$$g_{11}=1+r'^2, \qquad g_{12}=0, \qquad g_{22}=r^2,$$
$$h_{11}=-\frac{r''}{\sqrt{1+r'^2}}, \quad h_{12}=0, \qquad h_{22}=\frac{r}{\sqrt{1+r'^2}}, \tag{1.26}$$

where r' and r'' are derivatives with respect to z (r being the radius of the parallel circle, β the angle of rotation of the meridian about the z-axis which be directed downward in our discussion of surfaces of revolution).

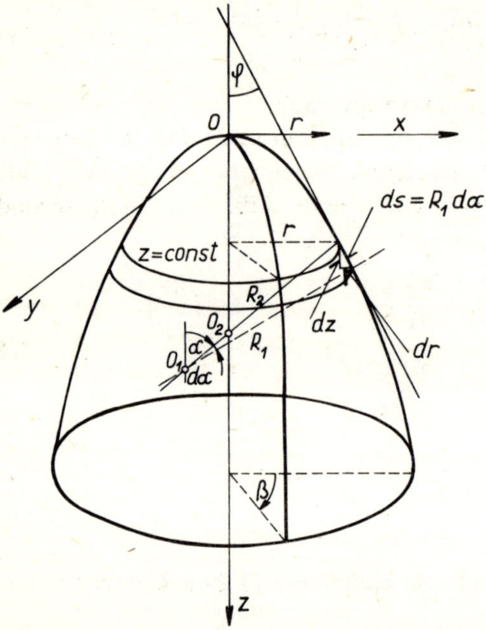

Fig. II-3. Surface of revolution.

According to (1.21) and (1.26) the principal curvatures of a surface of revolution are

$$\varkappa_1=\frac{1}{R_1}=-\frac{r''}{(1+r'^2)^{3/2}},$$
$$\varkappa_2=\frac{1}{R_2}=\frac{1}{r(1+r'^2)^{1/2}}. \tag{1.27}$$

1.5 *Lamé's Parameters and Gauss—Codazzi Equations*

If we replace the quantities g_{11} and g_{22} with Lamé's parameters A_1 and A_2 defined by the relations

$$A_1=\sqrt{g_{11}}, \quad A_2=\sqrt{g_{22}} \tag{1.28}$$

then, according to (1.10),

$$\cos \Lambda = \frac{g_{12}}{A_1 A_2} \tag{1.29}$$

and the formula (1.7) acquires the form

$$ds^2 = A_1^2 \, d\xi^2 + 2A_1A_2 \cos \Lambda \, d\xi \, d\eta + A_2^2 \, d\eta^2. \tag{1.30}$$

In rectangular coordinates $\cos \Lambda = 0$ and, according to (1.30), we obtain

$$ds^2 = A_1^2 \, d\xi^2 + A_2^2 \, d\eta^2. \tag{1.31}$$

The following three Gauss—Codazzi equations relate Lamé's parameters to the principal curvatures of a surface with a rectangular system of coordinate curves $\xi = \text{const}$ and $\eta = \text{const}$, identified with the principal curves of the surface [20], [45]:

$$\begin{aligned} &\frac{\partial}{\partial \eta}(A_1 \varkappa_1) = \varkappa_2 \frac{\partial A_1}{\partial \eta}, \\ &\frac{\partial}{\partial \xi}(A_2 \varkappa_2) = \varkappa_1 \frac{\partial A_2}{\partial \xi}, \end{aligned} \tag{1.32}$$

$$\frac{\partial}{\partial \xi}\left(\frac{1}{A_1}\frac{\partial A_2}{\partial \xi}\right) + \frac{\partial}{\partial \eta}\left(\frac{1}{A_2}\frac{\partial A_1}{\partial \eta}\right) = -\varkappa_1 \varkappa_2 A_1 A_2 .$$

For a surface of revolution the differential of the arc length satisfies the relation (Fig. II-3)

$$ds^2 = R_1^2 \, d\alpha^2 + r^2 \, d\beta^2, \tag{1.33}$$

where α is the angle between the axis of revolution and the normal.

A comparison of (1.31) and (1.33) yields

$$A_1 = R_1(\alpha), \quad A_2 = r(\alpha), \tag{1.34}$$

while

$$\frac{\partial A_1}{\partial \beta}=0, \quad r=R_2 \sin \alpha. \tag{1.35}$$

If we substitute (1.34) and (1.35) into (1.32), we obtain only one Gauss—Codazzi equation for the surface of revolution

$$R_1 \cos \alpha=\frac{\mathrm{d}}{\mathrm{d}\alpha}(R_2 \sin \alpha). \tag{1.36}$$

2

Surfaces of General Form

2.1 *Equations of Equilibrium*

In the structural analysis of air-supported membrane structures we shall use a rectangular system of curvilinear coordinates, assuming that the coordinate curves are the principal curves on the middle surface of the membrane. Thus

$$\cos \Lambda = 0, \quad g_{12} = 0. \tag{2.1}$$

We shall derive presently the equations of equilibrium when forces are applied to a cutaway element of an area $A_1 A_2 \, d\xi \, d\eta$, enclosed between two pairs of arcs of the principal curves (Fig. II-4). The internal normal and tangential (shearing) forces will be referred to a unit arc length in the middle surface. We shall call these internal forces having dimensions N/m, kN/m or N/cm, etc. the unit membrane forces; for the sake of brevity we shall call them either unit forces or membrane forces. We shall denote the components of the external continuous load $q(\xi, \eta)$ in the direction of tangents to the curves ξ, η and the normal n as q_1, q_2, q_n.

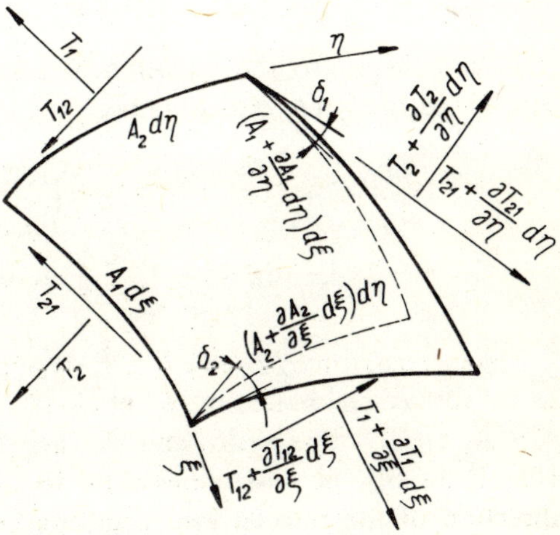

Fig. II-4. Unit membrane forces affecting an element of a surface.

The arcs of lengths $A_1\,\mathrm{d}\xi$ and $A_2\,\mathrm{d}\eta$ in the middle surface are subjected to the effect of normal membrane forces T_2, T_1 and tangential membrane forces T_{21}, T_{12}; the unit lengths of the adjoining arcs of length $[A_1+(\partial A_1/\partial\eta)\,\mathrm{d}\eta]\,\mathrm{d}\xi$ and $[A_2+(\partial A_2/\partial\xi)\,\mathrm{d}\xi]\,\mathrm{d}\eta$ are subjected to the effect of internal normal forces $T_2+(\partial T_2/\partial\eta)\,\mathrm{d}\eta$, $T_1+(\partial T_1/\partial\xi)\,\mathrm{d}\xi$ and internal tangential forces $T_{21}+(\partial T_{21}/\partial\eta)\,\mathrm{d}\eta$, $T_{12}+(\partial T_{12}/\partial\xi)\,\mathrm{d}\xi$. The tangents and the normals to neighbouring arcs of the principal ξ-curves form an angle δ_1, and the tangents to the neighbouring arcs of the principal η-curves form an angle δ_2 (see Fig. II-4)

$$\delta_1=\frac{\dfrac{\partial A_2}{\partial\xi}\,\mathrm{d}\xi\,\mathrm{d}\eta}{\left(A_1+\dfrac{\partial A_1}{\partial\eta}\,\mathrm{d}\eta\right)\mathrm{d}\xi},\quad \delta_2=\frac{\dfrac{\partial A_1}{\partial\eta}\,\mathrm{d}\eta\,\mathrm{d}\xi}{\left(A_2+\dfrac{\partial A_2}{\partial\xi}\,\mathrm{d}\xi\right)\mathrm{d}\eta}. \tag{2.2}$$

If the force $[T_2+(\partial T_2/\partial\eta)\,\mathrm{d}\eta]\times[A_1+(\partial A_1/\partial\eta)\,\mathrm{d}\eta]\,\mathrm{d}\xi$ is multiplied by a small angle δ_1 and the force $[T_{12}+(\partial T_{12}/\partial\xi)\,\mathrm{d}\xi]\times[A_2+(\partial A_2/\partial\xi)\,\mathrm{d}\xi]\,\mathrm{d}\eta$ by a small angle δ_2 and if these projections of the forces on the direction of the tangent to the ξ-curve are added to other forces acting in this direction, the following equation of equilibrium is obtained (neglecting minor quantities of higher order):

$$\begin{aligned}&\left(\frac{\partial T_1}{\partial\xi}A_2+T_1\frac{\partial A_2}{\partial\xi}\right)\mathrm{d}\xi\,\mathrm{d}\eta+\left(\frac{\partial T_{21}}{\partial\eta}A_1+T_{21}\frac{\partial A_1}{\partial\eta}\right)\mathrm{d}\xi\,\mathrm{d}\eta-\\&-T_2\frac{\partial A_2}{\partial\xi}\,\mathrm{d}\xi\,\mathrm{d}\eta+T_{12}\frac{\partial A_1}{\partial\eta}\,\mathrm{d}\xi\,\mathrm{d}\eta+q_1A_1A_2\,\mathrm{d}\xi\,\mathrm{d}\eta=0.\end{aligned} \tag{2.3}$$

Analogously also the equation of equilibrium for the direction of the tangent to the η-curve is obtained, viz.:

$$\begin{aligned}&\left(\frac{\partial T_2}{\partial\eta}A_1+T_2\frac{\partial A_1}{\partial\eta}\right)\mathrm{d}\xi\,\mathrm{d}\eta+\left(\frac{\partial T_{12}}{\partial\xi}A_2+T_{12}\frac{\partial A_2}{\partial\xi}\right)\mathrm{d}\xi\,\mathrm{d}\eta-\\&-T_1\frac{\partial A_1}{\partial\eta}\,\mathrm{d}\xi\,\mathrm{d}\eta+T_{21}\frac{\partial A_2}{\partial\xi}\,\mathrm{d}\xi\,\mathrm{d}\eta+q_2A_1A_2\,\mathrm{d}\xi\,\mathrm{d}\eta=0.\end{aligned} \tag{2.4}$$

The internal forces $T_1+(\partial T_1/\partial\xi)\,\mathrm{d}\xi$ and $T_2+(\partial T_2/\partial\eta)\,\mathrm{d}\eta$ and the normal to the surface element at the point (ξ,η) form the angles $\pi/2-A_1\,\mathrm{d}\xi/R_1$, $\pi/2-A_2\,\mathrm{d}\eta/R_2$, where R_1 and R_2 are the principal radii in this point (Fig. II-5a, b). The projections of the resultants of these internal forces in the direction of the normal are, neglecting small quantities of higher order,

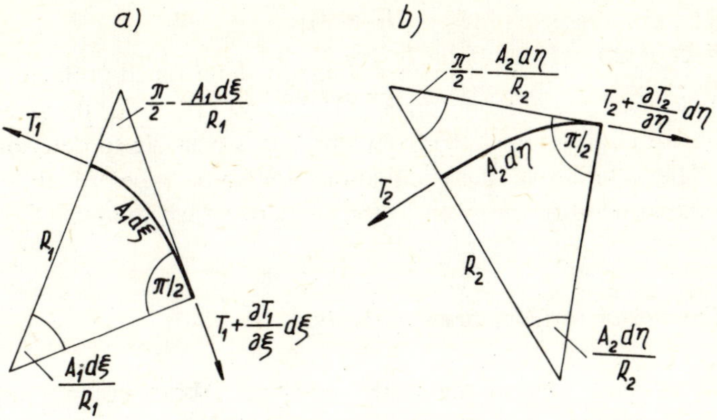

Fig. II-5. a), b) Normal sections of a surface passing through the principal directions of the mebrane forces T_1 and T_2.

$T_1A_1A_2\,\mathrm{d}\xi\,\mathrm{d}\eta/R_1$ and $T_2A_1A_2\,\mathrm{d}\xi\,\mathrm{d}\eta/R_2$, since $\cos(\pi/2-A_1\,\mathrm{d}\xi/R_1)\approx A_1\,\mathrm{d}\xi/R_1$ and $\cos(\pi/2-A_2\,\mathrm{d}\eta/R_2)\approx A_2\,\mathrm{d}\eta/R_2$. If these components of internal forces are added to the external load, the equation of equilibrium in the direction of the normal to the surface is obtained,

$$T_1\frac{A_1A_2}{R_1}\,\mathrm{d}\xi\,\mathrm{d}\eta+T_2\frac{A_1A_2}{R_2}\,\mathrm{d}\xi\,\mathrm{d}\eta+q_nA_1A_2\,\mathrm{d}\xi\,\mathrm{d}\eta=0. \tag{2.5}$$

After dividing the equations (2.3), (2.4) and (2.5) by the product $\mathrm{d}\xi\,\mathrm{d}\eta$ and some rearrangements, we obtain the following equations of equilibrium of the surface element:

$$\begin{aligned}
&\frac{\partial(A_2T_1)}{\partial\xi}-T_2\frac{\partial A_2}{\partial\xi}+\frac{\partial(A_1T_{21})}{\partial\eta}+T_{12}\frac{\partial A_1}{\partial\eta}+q_1A_1A_2=0,\\
&\frac{\partial(A_1T_2)}{\partial\eta}-T_1\frac{\partial A_1}{\partial\eta}+\frac{\partial(A_2T_{12})}{\partial\xi}+T_{21}\frac{\partial A_2}{\partial\xi}+q_2A_1A_2=0,\\
&T_1R_2+T_2R_1=q_nR_1R_2.
\end{aligned} \tag{2.6}$$

The load q_n acting on a surface of positive Gaussian curvature ($K>0$) is considered positive, if it is applied to the surface in the direction opposite to that of the centres of its principal curvatures.

The moment equilibrium condition with regard to the normal to the element yields the relation

$$T_{12} = T_{21}. \tag{2.7}$$

The membrane state of stress is a statically determined problem, since for three unknown membrane forces T_1, T_2 and $T_{12} = T_{21}$ there are three equations of equilibrium (2.6) from which it is possible to determine these forces. The solution of these equations for some types of air-supported membrane structures are given in the sections which follow.

2.2 *Conditions for Membrane Creasing*

Membranes made out of the usual industrial fabrics are not capable of transferring compression and thus they crease when negative membrane forces occur.

The minimum normal unit force at a given point of the surface equals the lower value of the principal unit force (see Subsection 8.3)

$$T_{min} = \frac{1}{2}\{T_1 + T_2 - \sqrt{(T_1 - T_2)^2 + T_{12}^2}\}. \tag{2.8}$$

The membrane does not crease when $T_{min} \geqq 0$ i.e. when

$$T_1 + T_2 \geqq \sqrt{(T_1 - T_2)^2 + 4T_{12}^2}. \tag{2.9}$$

From the inequality (2.9) the following conditions are obtained:

$$T_1 T_2 \geqq T_{12}^2, \quad T_1 \geqq 0, \quad T_2 \geqq 0. \tag{2.10}$$

Particularly the last two inequalities (2.10) will be used in the following sections of this chapter for the determination of the required internal overpressure.

2.3 *Physical Relations for Elastic and Viscoelastic Membranes*

The relative elongations ε_1, ε_2 and the relative displacement γ_{12} at a point (ξ, η) of an orthotropic elastic membrane can be determined from the physical relations

$$\varepsilon_1 = \frac{T_1}{E_1} - \frac{\nu_2}{E_2} T_2, \quad \varepsilon_2 = \frac{T_2}{E_2} - \frac{\nu_1}{E_1} T_1, \quad \gamma_{12} = \frac{T_{12}}{E_{12}}, \tag{2.11}$$

where E_1, E_2 are the tensile moduli of elasticity of dimensions N/cm, kN/m, etc., E_{12} is the shear modulus of elasticity of the same dimension, and ν_1, ν_2 are Poisson's ratios (Chapter II, Subsection 8.2).

According to the Betti's theorem, we have the following relation between these constants:

$$\frac{\nu_1}{E_1}=\frac{\nu_2}{E_2}. \tag{2.12}$$

When analyzing relative deformations at the time t due to a long-term load, it is necessary to replace the relations (2.11) by the physical relations in the operator form

$$\begin{aligned} C(\varepsilon_1)&=A(T_1)-B(T_2),\\ R(\varepsilon_2)&=N(T_2)-P(T_1),\\ D(\gamma_{12})&=M(T_{12}), \end{aligned} \tag{2.13}$$

where

$$\begin{gathered} A=\sum_{i=0}^{a} a_i\frac{\partial^i}{\partial t^i},\quad B=\sum_{i=0}^{b} b_i\frac{\partial^i}{\partial t^i},\quad C=\sum_{i=0}^{c} c_i\frac{\partial^i}{\partial t^i},\\ N=\sum_{i=0}^{n} n_i\frac{\partial^i}{\partial t^i},\quad P=\sum_{i=0}^{p} p_i\frac{\partial^i}{\partial t^i},\quad R=\sum_{i=0}^{r} r_i\frac{\partial^i}{\partial t^i},\\ D=\sum_{i=0}^{d} d_i\frac{\partial^i}{\partial t^i},\quad M=\sum_{i=0}^{m} m_i\frac{\partial^i}{\partial t^i} \end{gathered} \tag{2.14}$$

are linear time differential operators with constant coefficients to be determined experimentally. The constitutive equations (2.13) hold for a viscoelastic orthotropic membrane.

2.4 *Dependence of Relative Deformations on Time*

Under the assumption that the membrane forces T_1, T_2 and T_{12}, determined by the equations of equilibrium (2.6), do not vary with time t, all derivatives with respect to time in the right-hand sides of the constitutive equations (2.13) are equal to zero.

The first relation in (2.13) acquires the form

$$c_0\varepsilon_1+c_1\frac{\partial\varepsilon_1}{\partial t}+c_2\frac{\partial^2\varepsilon_1}{\partial t^2}+\ldots+c_c\frac{\partial^c\varepsilon_1}{\partial t^c}=a_0T_1(\xi,\eta)-b_0T_2(\xi,\eta). \tag{2.15}$$

The relative deformation ε_1 can be expressed in the form of a product of two functions,

$$\varepsilon_1(\xi, \eta, t) = F_1(t)\Phi_1(\xi, \eta), \tag{2.16}$$

where the function $F_1(t)$ depends only on time, and the function $\Phi_1(\xi, \eta)$ only on the coordinates. After substituting (2.16) into (2.15) we obtain

$$c_0F_1 + c_1\dot{F}_1 + c_2\ddot{F}_1 + \ldots + c_cF^{(c)} = \frac{a_0T_1(\xi, \eta) - b_0T_2(\xi, \eta)}{\Phi_1(\xi, \eta)}, \tag{2.17}$$

where the dots denote derivatives of the function F_1 with respect to time. As the left-hand side of (2.17) depends on time t and the right-hand side depends only on the coordinates, the equation can be satisfied only when both sides equal to a constant $\omega_0 \neq 0$.

Thus

$$c_0F_1 + c_1\dot{F}_1 + c_2\ddot{F}_1 + \ldots + c_cF_1^{(c)} = \omega_0 \tag{2.18}$$

and

$$\Phi_1(\xi, \eta) = \frac{1}{\omega_0}\left[a_0T_1(\xi, \eta) - b_0T_2(\xi, \eta)\right]. \tag{2.19}$$

If $\lambda_1, \lambda_2, \ldots, \lambda_c$ are the non-multiple roots of the characteristic equation

$$c_0 + c_1\lambda + c_2\lambda^2 + \ldots + c_c\lambda^c = 0, \tag{2.20}$$

corresponding to equation (2.18) with its right-hand side replaced by 0, the general solution of the non-homogenous differential equation (2.18) has the form

$$F_1(t) = \frac{\omega_0}{c_0} + \sum_{i=1}^{c} K_i e^{\lambda_i t}, \tag{2.21}$$

where K_i, $i = 1, 2, \ldots, c$, are integration constants.

At $t = 0$ we have the initial conditions

$$\begin{gathered} \varepsilon_1(\xi, \eta, 0) = \frac{T_1(\xi, \eta)}{E_{1,0}} - \frac{\nu_{2,0}}{E_{2,0}} T_2(\xi, \eta), \\ \dot{\varepsilon}_1(\xi, \eta, 0) = 0, \quad \ddot{\varepsilon}_1(\xi, \eta, 0) = 0, \quad \ldots, \quad \varepsilon^{(c-1)}(\xi, \eta, 0) = 0, \end{gathered} \tag{2.22}$$

where $E_{1,0}$, $E_{2,0}$ are the initial tensile moduli of an orthotropic viscoelastic membrane, and $\nu_{2,0}$ is the Poisson's ratio at the beginning of the loading of the membrane structure.

If we substitute (2.19) and (2.21) into (2.16), we obtain

$$\varepsilon_1(\xi, \eta, t) = \frac{1}{\omega_0}\left(\frac{\omega_0}{c_0} + \sum_{i=1}^{c} K_i e^{\lambda_i t}\right)[a_0 T_1(\xi, \eta) - b_0 T_2(\xi, \eta)], \qquad (2.23)$$

and after the substitution of (2.23) for $t=0$ into the first equation (2.22) we obtain

$$\sum_{i=1}^{c} K_i = \frac{T_1/E_{1,0} - T_2 \nu_{2,0}/E_{2,0}}{a_0 T_1 - b_0 T_2}\,\omega_0 - \frac{\omega_0}{c_0}. \qquad (2.24)$$

As the left-hand side of equation (2.24) is constant, the equation can be satisfied only when the following conditions have been complied with:

$$a_0 = \frac{1}{E_{1,0}}, \quad b_0 = \frac{\nu_{2,0}}{E_{2,0}}; \qquad (2.25)$$

and then also the right-hand side is constant. The ralations (2.25) determine the physical meaning of the constants a_0 and b_0.

According to (2.24) and (2.25),

$$\sum_{i=1}^{c} K_i = \left(1 - \frac{1}{c_0}\right)\omega_0 \qquad (2.26)$$

and, after the substitution of (2.23) into the last $c-1$ initial conditions (2.22), we obtain

$$\sum_{i=1}^{c} \lambda_i K_i = 0, \quad \sum_{i=1}^{c} \lambda_i^2 K_i = 0, \quad \ldots, \quad \sum_{i=1}^{c} \lambda_i^{c-1} K_i = 0. \qquad (2.27)$$

If we introduce new constants k_i by putting

$$K_i = (1 - 1/c_0)\omega_0 k_i, \qquad (2.28)$$

we obtain, after the substitution of (2.28) into the equations (2.26) and (2.27), a system of c linear equations

$$\begin{gathered} \sum_{i=1}^{c} k_i = 1, \quad \sum_{i=1}^{c} \lambda_i k_i = 0, \\ \sum_{i=1}^{c} \lambda_i^2 k_i = 0, \quad \ldots, \quad \sum_{i=1}^{c} \lambda_i^{c-1} k_i = 0, \end{gathered} \qquad (2.29)$$

whose solution yields all the constants k_i, $i = 1, 2, \ldots, c$.

Combining (2.28) with equation (2.23) we eliminate the unknown quantity ω_0; thus we obtain the dependence of the relative deformation ε_1 on time t

$$\varepsilon_1(\xi, \eta, t) = \frac{1}{c_0}[a_0 T_1(\xi, \eta) - b_0 T_2(\xi, \eta)]\left[1 + (c_0 - 1)\sum_{i=1}^{c} k_i e^{\lambda_i t}\right]. \tag{2.30}$$

If all roots λ_i, $i = 1, 2, \ldots, c$, of the characteristic equation (2.20) are negative, the relative deformation settles asymptotically as $t \to \infty$, according to (2.30), as

$$\varepsilon_1(\xi, \eta, \infty) = \frac{1}{c_0}[a_0 T_1(\xi, \eta) - b_0 T_2(\xi, \eta)]. \tag{2.31}$$

Consequently, the long-term tensile modulus $E_{1,\infty}$ of a viscoelastic membrane in the direction of the force T_1 is

$$E_{1,\infty} = \frac{c_0}{a_0} = c_0 E_{1,0}. \tag{2.32}$$

Further,

$$\frac{\nu_{2,\infty}}{E_{2,\infty}} = \frac{b_0}{c_0},$$

where $E_{2,\infty}$ is the long-term tensile modulus in the direction of the force T_2 and $\nu_{2,\infty}$ is the asymptotical Poisson's ratio for a viscoelastic membrane as $t \to \infty$.

Let us note that the solution of the system of equations (2.29) always exists, as the determinant of this system

$$\Delta = \begin{vmatrix} 1 & 1 & 1 & \ldots & 1 \\ \lambda_1 & \lambda_2 & \lambda_3 & \ldots & \lambda_c \\ \lambda_1^2 & \lambda_2^2 & \lambda_3^2 & \ldots & \lambda_c^2 \\ \ldots & \ldots & \ldots & \ldots & \ldots \\ \lambda_1^{c-1} & \lambda_2^{c-1} & \lambda_3^{c-1} & \ldots & \lambda_c^{c-1} \end{vmatrix} \tag{2.33}$$

is the Vandermond determinant which differs from zero ($\Delta \neq 0$).

By the same method as the relation (2.30) we can also derive the dependence of the relative deformations ε_2 and γ_{12} on time.

We obtain

$$\varepsilon_2(\xi, \eta, t) = \frac{1}{r_0}[n_0 T_2(\xi, \eta) - p_0 T_1(\xi, \eta)]\left[1 + (r_0 - 1)\sum_{i=1}^{r} l_i e^{\alpha_i t}\right], \tag{2.34}$$

where α_i, $i=1, 2, \ldots, r$, are the roots of the equation

$$r_0+r_1\alpha+r_2\alpha^2+\ldots+r_r\alpha^r=0 \tag{2.35}$$

and l_i, $i=1,2,\ldots,r$, are constants which can be determined by the solution of the following system of r linear equations:

$$\begin{gathered}\sum_{i=1}^{r} l_i=1, \quad \sum_{i=1}^{r} \alpha_i l_i=0, \\ \sum_{i=1}^{r} \alpha_i^2 l_i=0, \quad \ldots, \quad \sum_{i=1}^{r} \alpha_i^{r-1} l_i=0.\end{gathered} \tag{2.36}$$

Analogously

$$\gamma_{12}(\xi, \eta, t)=\frac{m_0}{d_0} T_{12}(\xi, \eta)\left[1+(d_0-1)\sum_{i=1}^{m} s_i e^{\beta_i t}\right], \tag{2.37}$$

where β_i, $i=1, 2, \ldots, m$, are the roots of the equation

$$d_0+d_1\beta+d_2\beta^2+\ldots+d_d\beta^d=0 \tag{2.38}$$

and s_i, $i=1,2,\ldots,m$, are constants which can be determined by the solution of the following system of m linear equations:

$$\begin{gathered}\sum_{i=1}^{m} s_i=1, \quad \sum_{i=1}^{m} \beta_i s_i=0, \\ \sum_{i=1}^{m} \beta_i^2 s_i=0, \quad \ldots, \quad \sum_{i=1}^{m} \beta_i^{m-1} s_i=0.\end{gathered} \tag{2.39}$$

As $t\rightarrow\infty$, the relative deformations ε_2 and γ_{12} stabilize at

$$\begin{gathered}\varepsilon_2(\xi, \eta, \infty)=\frac{1}{r_0}\left[n_0 T_2(\xi, \eta)-p_0 T_1(\xi, \eta)\right], \\ \gamma_{12}(\xi, \eta, \infty)=\frac{m_0}{d_0} T_{12}(\xi, \eta).\end{gathered} \tag{2.40}$$

Taking into account (2.12) and (2.25), we conclude that in the formulae (2.34) and (2.37)

$$n_0=\frac{1}{E_{2,0}}, \quad p_0=b_0=\frac{\nu_{1,0}}{E_{1,0}}, \quad m_0=\frac{1}{E_{12,0}},$$

where $E_{12,0}$ is the initial shear modulus, and in (2.40)

$$\frac{r_0}{n_0}=E_{2,\infty},\quad \frac{p_0}{r_0}=\frac{\nu_{1,\infty}}{E_{1,\infty}},\quad \frac{d_0}{m_0}=E_{12,\infty},$$

where $E_{12,\infty}$ is the long-term shear modulus.

2.5 *Geometric Relations and Relations between Displacements and Membrane Forces*

The displacements u and v in the direction of the tangents to principal curves $\xi=\text{const}$, $\eta=\text{const}$, the displacement w in the direction of the normal to the surface, and the relative deformations ε_1, ε_2, γ_{12} satisfy the equations [20]:

$$\varepsilon_1=\frac{1}{A_1}\frac{\partial u}{\partial\xi}+\frac{1}{A_1A_2}\frac{\partial A_1}{\partial\eta}v-\frac{w}{R_1},$$

$$\varepsilon_2=\frac{1}{A_2}\frac{\partial v}{\partial\eta}+\frac{1}{A_1A_2}\frac{\partial A_2}{\partial\xi}u-\frac{w}{R_2},\tag{2.41}$$

$$\gamma_{12}=\frac{A_1}{A_2}\frac{\partial}{\partial\eta}\left(\frac{u}{A_1}\right)+\frac{A_2}{A_1}\frac{\partial}{\partial\xi}\left(\frac{v}{A_2}\right).$$

The positive sense of the displacements u and v coincides with the positive sense of the coordinates ξ and η, and the positive sense of the displacement (deflection) w for the positive Gaussian curvature lies in the direction from the surface to the centres of principal curvatures, i.e. opposite to that of q_n.

The first two equations (2.41) yield, after the elimination of the deflection w,

$$\frac{1}{R_2A_1}\frac{\partial u}{\partial\xi}-\frac{1}{R_1A_1A_2}\frac{\partial A_2}{\partial\xi}u-\frac{1}{R_1A_2}\frac{\partial v}{\partial\eta}+\frac{1}{R_2A_1A_2}\frac{\partial A_1}{\partial\eta}v==\frac{\varepsilon_1}{R_2}-\frac{\varepsilon_2}{R_1}.\tag{2.42}$$

In wiew of (2.11), the equations (2.42) and the last equation (2.41) yield the following two relations between the displacements u and v and the membrane forces:

$$\frac{1}{R_2A_1}\frac{\partial u}{\partial\xi}-\frac{1}{R_1A_1A_2}\frac{\partial A_2}{\partial\xi}u-\frac{1}{R_1A_2}\frac{\partial v}{\partial\eta}+\frac{1}{R_2A_1A_2}\frac{\partial A_1}{\partial\eta}v=$$

$$=\left(\frac{1}{R_2}+\frac{\nu_1}{R_1}\right)\frac{T_1}{E_1}-\left(\frac{1}{R_1}+\frac{\nu_2}{R_2}\right)\frac{T_2}{E_2},\tag{2.43}$$

$$\frac{A_1}{A_2}\frac{\partial}{\partial\eta}\left(\frac{u}{A_1}\right)+\frac{A_2}{A_1}\frac{\partial}{\partial\xi}\left(\frac{v}{A_2}\right)=\frac{T_{12}}{E_{12}}.$$

Having determined the displacements u and v from the system (2.43), we can determine the deflection w from the first or the second equation (2.41)

$$w = -R_1\left(\frac{T_1}{E_1} - \frac{\nu_2}{E_2}T_2\right) + \frac{R_1}{A_1}\frac{\partial u}{\partial \xi} + \frac{R_1}{A_1A_2}\frac{\partial A_1}{\partial \eta}v. \tag{2.44}$$

The equations (2.43) and (2.44) hold for an elastic orthotropic membrane and can be used for the calculations of displacements due to short-term loading.

In the calculations of displacements due to long-term loading the expressions (2.30), (2.34) and (2.37) must be substituted for the relative deformations in the equations (2.41).

As $t \to \infty$, the displacements u and v are determined by the solution of equations

$$\frac{1}{R_2A_1}\frac{\partial u}{\partial \xi} - \frac{1}{R_1A_1A_2}\frac{\partial A_2}{\partial \xi}u - \frac{1}{R_1A_2}\frac{\partial v}{\partial \eta} + \frac{1}{R_2A_1A_2}\frac{\partial A_1}{\partial \eta}v =$$

$$= \left(\frac{a_0}{c_0R_2} + \frac{p_0}{r_0R_1}\right)T_1 - \left(\frac{n_0}{r_0R_1} + \frac{b_0}{c_0R_2}\right)T_2, \tag{2.45}$$

$$\frac{A_1}{A_2}\frac{\partial}{\partial \eta}\left(\frac{u}{A_1}\right) + \frac{A_2}{A_1}\frac{\partial}{\partial \xi}\left(\frac{v}{A_2}\right) = \frac{m_0}{d_0}T_{12},$$

obtained from equation (2.42) and the last equation (2.41), by taking into account (2.31) and (2.40).

The deflection w due to long-term loading, as $t \to \infty$, is determined by the expression

$$w = -\frac{R_1}{c_0}(a_0T_1 - b_0T_2) + \frac{R_1}{A_1}\frac{\partial u}{\partial \xi} + \frac{R_1}{A_1A_2}\frac{\partial A_1}{\partial \eta}v \tag{2.46}$$

which follows from the first equation (2.41) after we take into account (2.31).

The changes of the principal curvatures $\Delta\varkappa_1$, $\Delta\varkappa_2$ and the distorsion of the middle surface χ are determined by the geometric relations [63]

$$\Delta\varkappa_1 = \frac{\partial \varkappa_1}{\partial \xi}\frac{u}{A_1} + \frac{\partial \varkappa_1}{\partial \eta}\frac{v}{A_2} - \varkappa_1^2 w - \frac{1}{A_1}\frac{\partial}{\partial \xi}\left(\frac{1}{A_1}\frac{\partial w}{\partial \xi}\right) - \frac{1}{A_1A_2^2}\frac{\partial A_1}{\partial \eta}\frac{\partial w}{\partial \eta},$$

$$\Delta\varkappa_2 = \frac{\partial \varkappa_2}{\partial \xi}\frac{u}{A_1} + \frac{\partial \varkappa_2}{\partial \eta}\frac{v}{A_2} - \varkappa_2^2 w - \frac{1}{A_2}\frac{\partial}{\partial \eta}\left(\frac{1}{A_2}\frac{\partial w}{\partial \eta}\right) - \frac{1}{A_1A_2^2}\frac{\partial A_2}{\partial \xi}\frac{\partial w}{\partial \xi}, \tag{2.47}$$

$$\chi=(\varkappa_1-\varkappa_2)\left[\frac{A_1}{A_2}\frac{\partial}{\partial\eta}\left(\frac{u}{A_1}\right)-\frac{A_2}{A_1}\frac{\partial}{\partial\xi}\left(\frac{v}{A_2}\right)-\right.$$

$$\left.-\frac{2}{A_1A_2}\left(\frac{\partial^2 w}{\partial\xi\,\partial\eta}-\frac{1}{A_1}\frac{\partial A_1}{\partial\eta}\frac{\partial w}{\partial\xi}-\frac{1}{A_2}\frac{\partial A_2}{\partial\xi}\frac{\partial w}{\partial\eta}\right)\right].$$

If the displacements u and v are small in comparison with the deflection w, the equations (2.47) can be simplified by the omission if the terms comprizing u and v and their derivatives.

Equations (2.41) and (2.42) were derived by assuming the validity of the theory of small deformations. If the displacements u and v are small in comparison with the deflection w, the equations (2.41) can be simplified by the omission of all terms comprizing the desplacements u and v and their derivatives.

Equations (2.6), (2.11) and (2.13) were derived under the same assumption.

2.6 *Influence of Geometric Deviations on the State of Stress of Air-Supported Structures*

In case of large relative deformations of the membrane or large displacements it is necessary to investigate the state of stress and deformation by means of the condition of equilibrium of forces applied to the cutaway element of the deformed membrane. In the theory of large deformations, moreover, the linear differential geometric relations (2.41) and (2.44) are no longer valid, and the validity of the physical relations (2.11) is restricted and dependent on the material of the membrane.

The design of cylindrical membrane structures, when finite displacements are considered, will be dealt with in Subsection 6.4 of this chapter.

In air-supported and suspended membrane structures whose skin and covering are made of the usual fabrics, the relative deformations due to loading are small. Therefore, in approximate calculation of relative deformations of the membrane due to short-term loads the linear relations (2.11) can be used; for more accurate calculations non-linear relations are used, which have to be determined experimentally. For the calculations of relative membrane deformations due to long-term load, e.g. the internal overpressure, one has to use the equations (2.30), (2.34) and (2.37) or the simple equations (2.31) and (2.40).

In this subsection we shall derive differential equations for additional membrane forces originating in the membrane as a result of geometric deviations from the designed shape of the membrane structure. The geometric deviations may be due to the inaccuracies of production of the skin, inaccuracies in the staking out of the outline of the structure during its anchorage, or the displacements originating as a consequence of the deformation of the membrane and/or the changes in boundary conditions caused by the deformations of anchorage elements.

The geometric deviations from the designed shape of the structure can be expressed by the changes of the coefficients characterizing the original principal shape of the surface and by the changes of the principal radii.

For the designed shape of the membrane structure the quantities

$$A_1, \quad A_2, \quad R_1, \quad R_2 \tag{2.48}$$

are known, and for a membrane structure with geometrical deviations we have the quantities

$$A_1^* = A_1 + a_1, \quad A_2^* = A_2 + a_2, \quad R_1^* = R_1 + r_1, \quad R_2^* = R_2 + r_2, \tag{2.49}$$

where $a_i = a_i(\xi, \eta)$, $r_i = r_i(\xi, \eta)$, $i = 1, 2$, are the changes of the quantities (2.48). These changes can be ascertained by the measurements of the shape of the erected structure or by a calculation; both methods can be used simultaneously; the changes can be also partly estimated.

The quantities (2.48) correspond to the membrane forces

$$T_1, \quad T_2, \quad T_{12} = T_{21} \tag{2.50}$$

satisfying the differential equations (2.6); the quantities (2.49) correspond to membrane forces

$$T_1 + t_1, \quad T_2 + t_1, \quad T_{12} + t_{12} = T_{21} + t_{21}, \tag{2.51}$$

where t_1, t_2 and $t_{12} = t_{21}$ are additional membrane forces to be determined.

Provided the loading components q_1, q_2 and q_n are identical for the designed shape of the structure and for the structure with geometric deviations, we obtain, after replacing in the equations of equilibrium (2.6) the quantities (2.48) and (2.50) by the quantities (2.49) and (2.51) respectively, and after some rearrangements, the equations

$$
\begin{aligned}
&\frac{\partial(A_2T_1)}{\partial\xi}+\frac{\partial}{\partial\xi}[(A_2+a_2)t_1]+\frac{\partial(a_2T_1)}{\partial\xi}-T_2\frac{\partial A_2}{\partial\xi}-t_2\frac{\partial}{\partial\xi}(A_2+a_2)-\\
&\quad-T_2\frac{\partial a_2}{\partial\xi}+\frac{\partial(A_1T_{12})}{\partial\eta}+\frac{\partial}{\partial\eta}[(A_1+a_1)t_{12}]+\frac{\partial(a_1T_{12})}{\partial\eta}+T_{12}\frac{\partial A_1}{\partial\eta}+\\
&+t_{12}\frac{\partial}{\partial\eta}(A_1+a_1)+T_{12}\frac{\partial a_1}{\partial\eta}+q_1A_1A_2+q_1(a_1A_2+a_2A_1+a_1a_2)=0,\\
&\frac{\partial(A_1T_2)}{\partial\eta}+\frac{\partial}{\partial\eta}[(A_1+a_1)t_2]+\frac{\partial(a_1T_2)}{\partial\eta}-T_1\frac{\partial A_1}{\partial\eta}-t_1\frac{\partial}{\partial\eta}(A_1+a_1)-\\
&\quad-T_1\frac{\partial a_1}{\partial\eta}+\frac{\partial(A_2T_{12})}{\partial\xi}+\frac{\partial}{\partial\xi}[(A_2+a_2)t_{12}]+\frac{\partial(a_2T_{12})}{\partial\xi}+T_{12}\frac{\partial A_2}{\partial\xi}+\\
&+t_{12}\frac{\partial}{\partial\xi}(A_2+a_2)+T_{12}\frac{\partial a_2}{\partial\xi}+q_2A_1A_2+q_2(a_1A_2+a_2A_1+a_1a_2)=0,\\
&\qquad T_1R_2+t_1(R_2+r_2)+T_1r_2+T_2R_1+t_2(R_1+r_1)+T_2r_1=\\
&\qquad\quad =q_nR_1R_2+q_n(r_1R_2+r_2R_1+r_1r_2).
\end{aligned}
\tag{2.52}
$$

If we subtract the equations (2.6) from the corresponding equations (2.52) and take into account the notation introduced by (2.49), we obtain for the additional membrane forces t_1, t_2 and t_{12} the differential equations

$$
\begin{aligned}
&\frac{\partial(A_2^*t_1)}{\partial\xi}-t_2\frac{\partial A_2^*}{\partial\xi}+\frac{\partial(A_1^*t_{12})}{\partial\eta}+t_{12}\frac{\partial A_1^*}{\partial\eta}+q_1^*A_1^*A_2^*=0,\\
&\frac{\partial(A_1^*t_2)}{\partial\eta}-t_1\frac{\partial A_1^*}{\partial\eta}+\frac{\partial(A_2^*t_{12})}{\partial\xi}+t_{12}\frac{\partial A_2^*}{\partial\xi}+q_2^*A_1^*A_2^*=0,\\
&\qquad t_1R_2^*+t_2R_1^*=q_n^*R_1^*R_2^*,
\end{aligned}
\tag{2.53}
$$

where

$$
\begin{aligned}
&q_1^*A_1^*A_2^*=q_1(a_1A_2+a_2A_1+a_1a_2)+\frac{\partial(a_2T_1)}{\partial\xi}-T_2\frac{\partial a_2}{\partial\xi}+\frac{\partial(a_1T_{12})}{\partial\eta}+T_{12}\frac{\partial a_1}{\partial\eta},\\
&q_2^*A_1^*A_2^*=q_2(a_1A_2+a_2A_1+a_1a_2)+\frac{\partial(a_1T_2)}{\partial\eta}-T_1\frac{\partial a_1}{\partial\eta}+\frac{\partial(a_2T_{12})}{\partial\xi}+T_{12}\frac{\partial a_2}{\partial\eta},\\
&\qquad q_n^*R_1^*R_2^*=q_n(r_1R_2+r_2R_1+r_1r_2)-T_1r_2-T_2r_1.
\end{aligned}
\tag{2.54}
$$

In the investigation of the state of stress of a membrane structure with geometric deviations it is necessary to determine first the membrane forces (2.50) by solving the equations (2.6). Then it is possible to calculate the components of the equivalent load q_1^*, q_2^* and q_n^* from (2.54) and, by solving equations (2.53), determine the additional membrane forces t_1, t_2 and t_{12}, thus determining also the resulting membrane forces (2.51). Equations (2.53) have the same form as equations (2.6).

3

Surfaces of Revolution of General Curvature

3.1 *Equations of Equilibrium and Their Solution under Symmetric Load*

According to (1.26), (1.27) and (1.28) the Lamé's parameters and the principal radii of a surface of revolution, defined by the parametric equations (1.25), are (Fig. II-3)

$$\begin{aligned} A_1 &= (1+r'^2)^{1/2}, \qquad & A_2 &= r, \\ R_1 &= -\frac{1}{r''}(1+r'^2)^{3/2}, \qquad & R_2 &= r(1+r'^2)^{1/2}, \end{aligned} \tag{3.1}$$

where $r=r(z)$ is the radius of a parallel circle in the section $z=\text{const}$, R_1 is the principal radius of the surface of revolution on the curve $\beta=\text{const}$ (meridian), and R_2 is the principal radius of the surface of revolution on the curve $z=\text{const}$ (parallel circle).

After substituting (3.1) into the equations (2.6) we obtain the equations of equilibrium of an element of a surface of revolution

$$\begin{gathered} \frac{\partial}{\partial z}(rT_1) - r'T_2 + (1+r'^2)^{1/2}\frac{\partial T_{21}}{\partial \beta} + r(1+r'^2)^{1/2}q_1 = 0, \\ (1+r'^2)^{1/2}\frac{\partial T_2}{\partial \beta} + \frac{\partial}{\partial z}(rT_{12}) + r'T_{21} + r(1+r'^2)^{1/2}q_2 = 0, \\ -\frac{rr''}{1+r'^2}T_1 + T_2 - r(1+r'^2)^{1/2}q_n = 0, \end{gathered} \tag{3.2}$$

where the unit force T_1 and the load q_1 are tangent to the meridian, the unit force T_2 and the load q_2 are tangent to the parallel circle, and the load q_n has the direction of the normal.

Let us present now the solution of equations (3.2) for the case of a symmetric load

$$q_1 = q_1(z), \quad q_2 = 0, \quad q_n = q_n(z), \tag{3.3}$$

which does not depend on the angle β, i.e. does not vary along the parallel circle. In such a case $T_{12}=0$ and the normal membrane forces T_1 and T_2 are the principal internal forces which depend only on z $[T_1=T_1(z), T_2=T_2(z)]$.

Then the first equation (3.2) takes the form

$$\frac{\mathrm{d}}{\mathrm{d}z}(rT_1)-r'T_2+r(1+r'^2)^{1/2}q_1=0 \tag{3.4}$$

and the third equation (3.2) yields

$$T_2=\frac{rr''}{1+r'^2}T_1+r(1+r'^2)^{1/2}q_n. \tag{3.5}$$

Susbtituting (3.5) into (3.4), we obtain

$$\frac{\mathrm{d}}{\mathrm{d}z}(rT_1)-\frac{rr'r''}{1+r'^2}T_1+r(1+r'^2)^{1/2}q_1-rr'(1+r'^2)^{1/2}q_n=0. \tag{3.6}$$

If we define the function $V_1=V_1(z)$ according to Vlasov [63] by the relation

$$T_1=\frac{1}{r}(1+r'^2)^{1/2}V_1, \tag{3.7}$$

we obtain, substituting (3.7) into (3.6), the equation

$$\frac{\mathrm{d}V_1}{\mathrm{d}z}+rq_1-rr'q_n=0,$$

whose solution is

$$V_1=\int_0^z rr'q_n\,\mathrm{d}z-\int_0^z rq_1\,\mathrm{d}z+C, \tag{3.8}$$

where C is the integration constant.

The substitution of (3.8) into (3.7) yields for the unit force T_1 the expression

$$T_1=\frac{1}{r}(1+r'^2)^{1/2}\left[\int_0^z rr'q_n\,\mathrm{d}z-\int_0^z rq_1\,\mathrm{d}z+C\right]. \tag{3.9}$$

According to Fig. II-3 and the last equation (3.1),

$$\sin \alpha = \frac{r}{R_2} = (1 + r'^2)^{-1/2},$$

$$\cos \alpha = r'(1 + r'^2)^{-1/2} = r' \sin \alpha = \frac{rr'}{R_2}, \tag{3.10}$$

$$\mathrm{d}z = \mathrm{d}s \sin \alpha = R_1 \sin \alpha \, \mathrm{d}\alpha,$$

where α is the angle between the normal to the surface and the z-axis. We shall identify the origin O with the apex of the surface, i.e. with the point of intersection of the meridians with the z-axis.

If we multiply equation (3.9) by 2π and use the relations (3.10), we obtain, after rearrangements,

$$2\pi T_1 r \sin \alpha = P(\alpha) + C, \tag{3.11}$$

where

$$P(\alpha) = \int_0^{\alpha} (q_n \cos \alpha - q_1 \sin \alpha) 2\pi r R_1 \, \mathrm{d}\alpha \tag{3.12}$$

is the resultant of the vertical components of the loads q_n and q_1 applied to the part of the surface of revolution separated by the section $\alpha = \text{const}$. Since the left-hand side of equation (3.11) represents the resultant of the vertical components of internal forces T_1, both resultants have be equal,

$$2\pi T_1 r \sin \alpha = P(\alpha). \tag{3.13}$$

A comparison of the equations (3.11) and (3.13) yields $C=0$. Consequently, equation (3.13) can be obtained directly from the condition of equilibrium in the vertical direction of the part of the surface of revolution separated by the section $\alpha = \text{const}$.

Since $C=0$, we have, according to (3.9) and (3.10), the following expression for T_1:

$$T_1 = \frac{R_2}{r^2} \left[\int_0^z q_n r r' \, \mathrm{d}z - \int_0^z q_1 r \, \mathrm{d}z \right]. \tag{3.14}$$

Substituting (3.14) into (3.5), we obtain, taking into account (3.1), the following expression for the unit force T_2:

$$T_2 = R_2 q_n - \frac{R_2^2}{r^2 R_1} \left[\int_0^z q_n r r' \, \mathrm{d}z - \int_0^z q_1 r \, \mathrm{d}z \right]. \tag{3.15}$$

3.2 *Membrane Forces Due to Internal Overpressure. Equation of the Meridian for* $T_2=0$

If the surface of revolution is affected only by the internal overpressure q_0, then $q_1=0$ and $q_n=q_0$ = const. As $r'\,\mathrm{d}z=\mathrm{d}r$, we obtain for this particular case from the equations (3.14) and (3.15), after integration, the following formulae for the principal membrane forces:

$$T_{1,q}=\frac{1}{2}\,q_0R_2,$$
$$T_{2,q}=q_0R_2\left(1-\frac{R_2}{2R_1}\right)=T_{1,q}\left(2-\frac{R_2}{R_1}\right). \tag{3.16}$$

According to (3.16), the surface of revolution complies with the conditions (2.10) only if the Gaussian curvature is non-negative and if the inequality

$$2R_1\geqq R_2 \tag{3.17}$$

is valid, regardless of the magnitude of the internal overpressure. Inequality (3.17) must be taken care of in the design of air-supported structures consisting of one or several surfaces of revolution (see Chapter III, Subsection 2.3). The equation of the meridian of a surface of revolution, for which the membrane force T_2 at an internal overpressure equals zero in all points of the surface, can be determined from the condition

$$2R_1-R_2=0. \tag{3.18}$$

If we substitute the expressions (3.1) for R_1 and R_2 in (3.18), we obtain, after rearrangements, a non-linear differential equation of the meridian

$$rr''+2r'^2+2=0.$$

The first integral of this equation is

$$r^4r'^2+r^4=C_1. \tag{3.19}$$

According to (3.19),

$$r'=\frac{1}{r^2}\,(C_1-r^4)^{1/2} \tag{3.20}$$

and, consequently, to yield a real value for r', all points of the surface must comply with the inequality

$$C_1\geqq r^4.$$

After the separation of variables in equation (3.20) and integration we obtain an elliptic integral in the form

$$z=\int r^2(C_1-r^4)^{-1/2}\,\mathrm{d}r+C_2, \qquad (3.21)$$

where C_2 is the second arbitrary constant.

We shall present now an approximate evaluation of the elliptic integral (3.21) for $C_1 \gg r^4$. For this purpose we shall write $(C_1-r^4)^{-1/2}$ in the form $C_1^{-1/2}(1-r^4/C_1)^{-1/2}$, which we shall expand into Newton series

$$C_1^{-1/2}\left(1-\frac{r^4}{C_1}\right)^{-1/2}=C_1^{-1/2}\left[1+\binom{-\frac{1}{2}}{1}\frac{r^4}{C_1}+\binom{-\frac{1}{2}}{2}\frac{r^8}{C_1^2}+\binom{-\frac{1}{2}}{3}\frac{r^{12}}{C_1^3}+\dots\right]=$$

$$=C_1^{-1/2}\left(1-\frac{1}{2}\frac{r^4}{C_1}+\frac{3}{8}\frac{r^8}{C_1^2}-\frac{15}{48}\frac{r^{12}}{C_1^3}+\dots\right).$$

Since $r^4/C_1 \ll 1$, the series is fast convergent.

If we substitute this series into (3.21) and confine it to its first four terms, we obtain, after integration, the equation of the meridian

$$z=\frac{r^3}{\sqrt{C_1}}\left[\frac{1}{3}-\frac{1}{14}\frac{r^4}{C_1}+\frac{3}{88}\left(\frac{r^4}{C_1}\right)^2-\frac{1}{48}\left(\frac{r^4}{C_1}\right)^3\right]+C_2. \qquad (3.22)$$

The highest value of the radius of the parallel circle r_{max} is attained for $r'=0$; consequently, according to equation (3.20), $C_1=r_{max}^4$. If the meridian passes through the origin of coordinates ($z=0$, $r=0$), then $C_2=0$.

The design of the shape of this type of surface and the calculation of the state of stress of the membrane skin, taking into account its deformability due to the effect of the internal overpressure, are dealt with in [32], see Chapter III, Subsection 2.3.

3.3 *Membrane Forces Due to Dead Load and Symmetric Snow Load*

The components of the dead weight g of the unit surface of the membrane in the direction of the tangent and the normal to the meridian of revolution are (Fig. II-6)

$$q_1=g\sin\alpha, \quad q_2=0, \quad q_n=-g\cos\alpha \qquad (3.23)$$

and, if we take into account the relation (3.10),

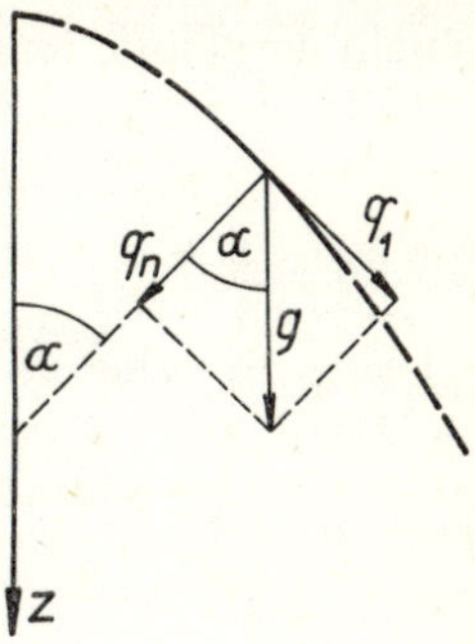

Fig. II-6. Components of load due to dead weight.

$$q_1 = g\frac{r}{R_2}, \quad q_n = -g\frac{rr'}{R_2}. \tag{3.24}$$

If we substitute (3.24) into (3.14), then, using the formula (3.1) for R_2, we obtain, at any point, for the membrane force $T_{1,g}$ caused by the dead weight ($g = \text{const}$)

$$T_{1,g} = -\frac{R_2 g}{r^2}\int_0^z R_2(z)\,\mathrm{d}z. \tag{3.25}$$

The membrane force $T_{2,g}$ due to the dead weight of the membrane can be determined from (3.5) by substituting the formulae (3.24) and (3.25) for q_n and T_1 respectively, or from the third equation (2.6)

$$T_2 = R_2\left(q_n - \frac{T_1}{R_1}\right). \tag{3.26}$$

Combining (3.24) and (3.25) with (3.26), we obtain

$$T_{2,g} = g\left[\frac{R_2^2}{R_1 r^2}\int_0^z R_2(z)\,\mathrm{d}z - rr'\right]. \tag{3.27}$$

The symmetric snow load $\bar{q}$ per unit surface of a horizontal projection of the membrane will be considered in accordance with the German Standard DIN 1055 [47]

$$\bar{q} = q_s \cos\alpha, \quad 0 \leqq \alpha \leqq \pi/2, \tag{3.28}$$

where $q_s = \text{const}$ is the maximum permissible load. The unit surface of the membrane is loaded by the vertical snow load

$$q = \bar{q}\cos\alpha = q_s\cos^2\alpha \tag{3.29}$$

and its components in the directions of the tangent and the normal to the meridian are

$$q_1 = q \sin\alpha = q_s \sin\alpha \cos^2\alpha, \qquad q_n = -q\cos\alpha = -q_s\cos^3\alpha. \tag{3.30}$$

Using the relations (3.10), we obtain

$$q_1 = q_s \frac{r^3 r'^2}{R_2^3}, \quad q_n = q_s \frac{r^3 r'^3}{R_2^3}. \tag{3.31}$$

Combining (3.31) with (3.14) and using formula (3.1) for R_2, we obtain the following expression for the membrane force $T_{1,s}$ due to snow load at a point (z, r):

$$T_{1,s} = \frac{q_s R_2}{r^2} \int_0^z \frac{r^2 r'^2}{R_2(z)} dz. \tag{3.32}$$

Here

$$0 < z \leqq z_{\pi/2}, \tag{3.33}$$

while $z_{\pi/2}$ is the distance from the origin of the coordinates corresponding to $\alpha = \pi/2$.

For the membrane force $T_{2,s}$ we obtain from equation (3.26), by (3.31) and (3.32), the formula

$$T_{2,s} = -q_s \frac{r^3 r'^3}{R_2^2} - \frac{R_2}{R_1} T_{1,s}, \tag{3.34}$$

where z lies within the interval (3.33).

In the case when the angle α takes also values in excess of $\pi/2$, no snow load is applied to the membrane for $z > z_{\pi/2}$ ($q_1 = q_n = 0$). In this part of the surface of revolution the membrane force $T_{1,s}$ is determined from (3.32). After carrying out the integration, let us set $z = z_{\pi/2}$. Then

$$T_{1,s} = -q_s \frac{R_2(z)}{r^2(z)} \left[\int_0^z \frac{r^2 r'^2}{R_2(z)} dz \right]_{z = z_{\pi/2}}. \tag{3.35}$$

The membrane force $R_{2,s}$ for $z > \pi/2$ is, according to (3.26), ($q_n = 0$)

$$T_{2,s} = -\frac{R_2(z)}{R_1(z)} T_{1,s} = q_s \frac{R_2^2(z)}{r^2(z) R_1(z)} \left[\int_0^z \frac{r^2 r'^2}{R_2(z)} dz \right]_{z = z_{\pi/2}}. \tag{3.36}$$

On smooth surfaces made of the usual fabrics the snow slides already when the angle $\alpha = \pi/4$. Therefore, in the expressions (3.35) and (3.36) the value $z_{\pi/2}$ can be replaced by the value $z_{\pi/4}$ which corresponds to this angle.

3.4 *Membrane Forces Due to Wind Load*

In the investigation of the membrane forces due to asymmetric loads we shall use, instead of (1.25), the parametric equations

$$x = r(\alpha)\cos\beta, \quad y = r(\alpha)\sin\beta, \quad z = z(\alpha), \tag{3.37}$$

where the parameters α and β have the meaning shown in Fig. II-3.

For the differential of the arc length on the surface we have the equation

$$ds^2 = R_1^2\,d\alpha^2 + r^2\,d\beta, \tag{3.38}$$

and the Lamé's parameters, according to Subsection 1.5, are

$$A_1 = R_1, \quad A_2 = r \tag{3.39}$$

while

$$\frac{\partial A_1}{\partial \beta} = 0, \quad r = R_2 \sin\alpha. \tag{3.40}$$

According to Fig. II-3,

$$R_1\,d\alpha = \frac{dr}{\cos(\alpha + d\alpha)} = \frac{dr}{\cos\alpha},$$

whence

$$\frac{dr}{d\alpha} = R_1\cos\alpha = \frac{d}{d\alpha}(R_2\sin\alpha), \tag{3.41}$$

which is the Gauss—Codazzi relation (1.36) between the principal radii of a surface of revolution.

If we substitute (3.39) in (2.6) and set $\xi = \alpha$ and $\eta = \beta$, we obtain, by (2.7), (3.10), (3.40) and (3.41), the following equations of equilibrium:

$$\frac{\partial}{\partial\alpha}(rT_1) + \frac{\partial T_{12}}{\partial\beta}R_1 - T_2R_1\cos\alpha + q_1rR_1 = 0,$$

$$\frac{\partial}{\partial \alpha}(rT_{12}) + T_{12}R_1 \cos \alpha + \frac{\partial T_2}{\partial \beta} R_1 + q_2 r R_1 = 0, \tag{3.42}$$

$$T_1 + T_2 R_1 \sin \alpha = q_n r R_1 .$$

The static wind load, i.e. the dynamic pressure and suction, effect the smooth surface in the radial direction only. Therefore

$$q_1 = q_2 = 0. \tag{3.43}$$

We adopt the following law for the distribution of the wind load q_n along a surface of revolution:

$$q_n = q_0^* + q_v \sin \alpha \cos \beta , \tag{3.44}$$

where q_0^* and q_v are constants. The direction of the wind is determined by the plane $\beta = 0$ (Fig. II-3). The quantity q_v is always taken non-positive, the quantity q_0^* is given a positive value, if it represents suction, and a negative value, if it represents pressure.

The membrane force due to the radially symmetric component of load q_0^* is $T_{12} = 0$. The membrane forces T_1 and T_2 are determined by the equations (3.16), where q_0^* is replaced by q_0.

The membrane forces due to the antisymmetric component

$$q_{an} = q_v \sin \alpha \cos \beta \tag{3.45}$$

can be written in the form

$$T_1 = S_1(\alpha) \cos \beta , \quad T_2 = S_2(\alpha) \cos \beta , \quad T_{12} = S_{12}(\alpha) \sin \beta . \tag{3.46}$$

Let us substitute (3.43), (3.45) and (3.46) into (3.42). After the division by $\cos \beta$ or $\sin \beta$ we obtain three ordinary differential equations

$$\frac{d}{d\alpha}(rS_1) + S_{12}R_1 - S_2 R_1 \cos \alpha = 0,$$

$$\frac{d}{d\alpha}(rS_{12}) + S_{12}R_1 \cos \alpha - S_2 R_1 = 0, \tag{3.47}$$

$$S_1 r + S_2 R_1 \sin \alpha = q_v r R_1 \sin \alpha ,$$

for the three unknown functions $S_1(\alpha)$, $S_2(\alpha)$, $S_{12}(\alpha)$.

If we eliminate the function S_2, using the last equation, and using (3.40) and (3.41), we obtain the following two equations:

$$\frac{dS_1}{d\alpha}+\left(1+\frac{R_1}{R_2}\right)\cot\alpha S_1+\frac{R_1}{R_2}\frac{1}{\sin\alpha}S_{12}=q_vR_1\cos\alpha,$$
$$\frac{dS_{12}}{d\alpha}+2\frac{R_1}{R_2}\cot\alpha S_{12}+\frac{1}{\sin\alpha}S_1=q_vR_1. \tag{3.48}$$

The solution of these equations for a spherical surface is given in Subsection 5.2 of Chapter II.

In many cases it is advisable to discuss the surfaces of revolution, defined by parametric equations (3.37), by means of the equations of equilibrium

$$\frac{1}{R_1}\frac{\partial T_1}{\partial\alpha}+\frac{\cot\alpha}{R_2}(T_1-T_2)+\frac{1}{R_2\sin\alpha}\frac{\partial T_{12}}{\partial\beta}+q_1=0,$$
$$\frac{1}{R_1}\frac{\partial T_{12}}{\partial\alpha}+2\frac{\cot\alpha}{R_2}T_{12}+\frac{1}{R_2\sin\alpha}\frac{\partial T_2}{\partial\beta}+q_2=0, \tag{3.49}$$
$$\frac{T_1}{R_1}+\frac{T_2}{R_2}=q_n,$$

which follow from the equations (2.6), in view of (3.39) and (3.40).

If we introduce instead of T_1 and T_{12} new functions U and V according to Novoshilov [45]

$$U=T_1R_2\sin^2\alpha,\quad V=T_{12}R_2\sin^2\alpha, \tag{3.50}$$

we obtain from (3.49) after the elimination of T_2 the following two equations:

$$\frac{R_2^2\sin\alpha}{R_1}\frac{\partial U}{\partial\alpha}+\frac{\partial V}{\partial\beta}=(q_n\cos\alpha-q_1\sin\alpha)R_2^3\sin^2\alpha,$$
$$\frac{\partial V}{\partial\alpha}-\frac{R_2}{\sin\alpha}\frac{\partial U}{\partial\beta}=-\left(q_2\sin\alpha+\frac{\partial q_n}{\partial\beta}\right)R_1R_2^2\sin\alpha. \tag{3.51}$$

The elimination of the function V from the equations (3.51) yields the following second order partial differential equation for the unknown function U:

$$\frac{1}{R_1R_2\sin\alpha}\frac{\partial}{\partial\alpha}\left(\frac{R_2^2\sin\alpha}{R_1}\frac{\partial U}{\partial\alpha}\right)+\frac{1}{R_1\sin^2\alpha}\frac{\partial^2U}{\partial\beta^2}=F(\alpha,\beta), \tag{3.52}$$

where

$$F(\alpha, \beta)=\frac{1}{R_1R_2 \sin \alpha}\frac{\partial}{\partial \alpha}[(q_n \cos \alpha - q_1 \sin \alpha)R_2^3 \sin^2 \alpha]+ + R_2\left(\frac{\partial^2 q_n}{\partial \beta^2}+\frac{\partial q_2}{\partial \beta}\sin \alpha\right). \tag{3.53}$$

For the load components described by

$$q_1=\bar{q}_1 \cos \beta, \quad q_2=\bar{q}_2 \sin \beta, \quad q_n=\bar{q}_n \cos \beta, \tag{3.54}$$

where

$$\bar{q}_1=\bar{q}_1(\alpha), \quad \bar{q}_2=\bar{q}_2(\alpha), \quad \bar{q}_n=\bar{q}_n(\alpha)$$

are the quantities dependent on the angle α only, the function U can be written as

$$U=\bar{U}(\alpha) \cos \beta, \tag{3.55}$$

and equation (3.52) becomes an ordinary differential equation for the function $\bar{U}=\bar{U}(\alpha)$:

$$\frac{1}{R_1R_2 \sin \alpha}\frac{d}{d\alpha}\left(\frac{R_2^2 \sin \alpha}{R_1}\frac{d\bar{U}}{d\alpha}\right)-\frac{1}{R_1 \sin^2 \alpha}\bar{U}=\bar{F}(\alpha), \tag{3.56}$$

where

$$\bar{F}(\alpha)=\frac{1}{R_1R_2 \sin \alpha}\frac{d}{d\alpha}[(\bar{q}_n \cos \alpha - \bar{q}_1 \sin \alpha)R_2^3 \sin^2 \alpha]-R_2(\bar{q}_n-\bar{q}_2 \sin \alpha). \tag{3.57}$$

If we introduce a new function Y by [45]

$$Y=UR_2 \sin \alpha, \tag{3.58}$$

and use the Gauss—Codazzi equation (3.41), we obtain from equation (3.56)

$$\frac{d}{d\alpha}\left(\frac{1}{R_1 \sin \alpha}\frac{dY}{d\alpha}\right)=R_1\bar{F}(\alpha). \tag{3.59}$$

Applying two quadratures to equation (3.59) and integration by parts to its right-hand side, we obtain*)

---

*) In the integrands of all equations of this chapter α is considered variable. Before the integral, however, α has the same meaning as its upper limit.

$$Y = C_2 + C_1 \int_{\alpha'}^{\alpha} R_1 \sin\alpha \, d\alpha + \int_{\alpha'}^{\alpha} \Phi R_1 \sin\alpha \tag{3.60}$$

where C_1 and C_2 are integration constants and

$$\begin{aligned} \Phi = (\bar{q}_n \cos\alpha - \bar{q}_1 \sin\alpha) R_2^2 \sin\alpha - \\ - \int_{\alpha'}^{\alpha} (\bar{q}_n \sin\alpha + \bar{q}_1 \cos\alpha - \bar{q}_2) R_1 R_2 \sin\alpha \, d\alpha . \end{aligned} \tag{3.61}$$

The lower limit α', which can be chosen arbitrarily, will be identified with the upper boundary of the surface.

According to (3.50), (3.55) and the last equation (3.49), we have

$$T_1 = \bar{T}_1 \cos\beta , \quad T_2 = \bar{T}_2 \cos\beta , \tag{3.62}$$

where, by (3.54), (3.58), (3.60) and the last equation (3.49),

$$\bar{T}_1 = \frac{1}{R_2^2 \sin^3\alpha} \left[C_2 + C_1 \int_{\alpha'}^{\alpha} R_1 \sin\alpha \, d\alpha + \int_{\alpha'}^{\alpha} \Phi R_1 \sin\alpha \, d\alpha \right] \tag{3.63}$$

and

$$\bar{T}_2 = \bar{q}_n R_2 - \frac{1}{R_1 R_2 \sin^3\alpha} \left[C_2 + C_1 \int_{\alpha'}^{\alpha} R_1 \sin\alpha \, d\alpha + \int_{\alpha'}^{\alpha} \Phi R_1 \sin\alpha \, d\alpha \right] . \tag{3.64}$$

The tangential mebrane force T_{12} can be written in the form

$$T_{12} = \bar{T}_{12} \sin\beta , \tag{3.65}$$

where the function $\bar{T}_{12} = \bar{T}_{12}(\alpha)$ depends only on the angle α. If we write the function U in the first equation (3.51) as in the first expression in (3.50) and use the first relation (3.62) and (3.63), we obtain, by integration, and using (3.65) and the second relation (3.50), the following formula for the quantity $\bar{T}_{12}$:

$$\bar{T}_{12} = \bar{T}_1 \cos\alpha - \frac{C_1}{R_2 \sin\alpha} + \bar{f}(\alpha) , \tag{3.66}$$

where

$$\bar{f}(\alpha) = \frac{1}{R_2 \sin\alpha} \int_{\alpha'}^{\alpha} (\bar{q}_n \sin\alpha + \bar{q}_1 \cos\alpha - \bar{q}_2) R_1 R_2 \sin\alpha \, d\alpha . \tag{3.67}$$

The integration constants C_1 and C_2 can be determined from the quantities $\bar{T}_1$ and $\bar{T}_{12}$ which correspond to the angle α'. If we set $\alpha=\alpha'$ in the formulae (3.63) and (3.66), we get

$$\begin{aligned} C_2 &= R_2'^2 \sin^3 \alpha' \bar{T}_1', \\ C_1 &= T_1' \bar{R}' \sin \alpha' \cos \alpha' - T'_2 \bar{R}' \sin \alpha' \end{aligned} \tag{3.68}$$

from which we obtain for $\alpha'=0$, i.e. for surfaces with no opening in the apex,

$$C_1 = C_2 = 0. \tag{3.69}$$

The unit forces due to the antisymmetric component (3.45) of the wind load are obtained by taking in (3.63) and (3.64)

$$\bar{q}_1 = \bar{q}_2 = 0, \quad \bar{q}_n = q_v \sin \alpha. \tag{3.70}$$

It follows now from (3.62), (3.65), (3.66) and (3.69) that the unit forces due to the antisymmetric load (3.45) on a surface of revolution closed at the top are given by the formulae

$$\begin{aligned} T_{1,v} &= \left[\frac{1}{R_2^2 \sin^3 \alpha} \int_0^\alpha \Phi_v R_1 \sin \alpha \, d\alpha\right] \cos \beta, \\ T_{2,v} &= \left[q_v R_2 \sin \alpha - \frac{1}{R_1 R_2 \sin^3 \alpha} \int_0^\alpha \Phi_v R_1 \sin \alpha \, d\alpha\right] \cos \beta, \\ T_{12,v} &= \left[\frac{\cos \alpha}{R_2^2 \sin^3 \alpha} \int_0^\alpha \Phi_v R_1 \sin \alpha \, d\alpha + f_v(\alpha)\right] \sin \beta, \end{aligned} \tag{3.71}$$

where

$$\begin{aligned} \Phi_v &= q_v R_2^2 \sin^2 \alpha \cos \alpha - q_v \int_0^\alpha \mathrm{R}_1 \mathrm{R}_2 \sin^3 \alpha \, d\alpha, \\ f_v &= \frac{q_v}{R_2 \sin \alpha} \int_0^\alpha \mathrm{R}_1 \mathrm{R}_2 \sin^3 \alpha \, d\alpha. \end{aligned} \tag{3.72}$$

3.5 *Membrane Forces Due to One-Sided Snow Load*

By (3.30), one-sided snow load can be expressed in the form

$$\begin{aligned} q_1 &= \frac{q_s}{2} \sin \alpha \cos^2 \alpha (1 + \cos \beta), \quad q_2 = 0, \\ q_n &= -\frac{q_s}{2} \cos^3 \alpha (1 + \cos \beta); \qquad 0 \leqq \alpha \leqq \pi/2. \end{aligned} \tag{3.73}$$

Thus

$$q_1 = q^{*}_{1,s} + q^{**}_{1,s}, \quad q_n = q^{*}_{n,s} + q^{**}_{n,s}, \tag{3.74}$$

where

$$q^{*}_{1,s} = \frac{q_s}{2} \sin \alpha \cos^2 \alpha, \quad q^{*}_{n,s} = -\frac{q_s}{2} \cos^3 \alpha \tag{3.75}$$

are the axially symmetric load components, and

$$q^{**}_{1,s} = \frac{q_s}{2} \sin \alpha \cos^2 \alpha \cos \beta, \quad q^{**}_{2,s} = 0,$$
$$q^{**}_{n,s} = -\frac{q_s}{2} \cos^3 \alpha \cos \beta \tag{3.76}$$

are the antisymmetric load components relative to the plane $\beta = 0$. The unit membrane forces $T^{*}_{1,s}$ and $T^{*}_{2,s}$ due to the symmetric load components (3.75) are

$$T^{*}_{1,s} = \frac{T_{1,s}}{2}, \quad T^{*}_{2,s} = \frac{T_{2,s}}{2}, \tag{3.77}$$

where the quantities $T_{1,s}$ and $T_{2,s}$ are determined by (3.32) and (3.34) for surfaces defined by equations (1.25). For a surface of revolution defined by the parametric equations (3.37) the unit forces (3.77) can be determined as follows:

Under symmetric load $T_{12} = 0$, and the membrane forces T_1 and T_2 depend only on the angle α. Therefore, (3.50) implies $\partial V/\partial \beta = 0$ and $\partial U/\partial \beta = 0$; the first equation (3.51) yields

$$\frac{dU}{d\alpha} = (q_n \cos \alpha - q_1 \sin \alpha) R_1 R_2 \sin \alpha, \tag{3.78}$$

whence

$$U = C + \int_{\alpha'}^{\alpha} (q_n \cos \alpha - q_1 \sin \alpha) R_1 R_2 \sin \alpha \, d\alpha, \tag{3.79}$$

where C is the integration constant.

Substituting (3.79) into the first equation (3.50), we obtain

$$T_1 = \frac{C}{R_2 \sin^2 \alpha} + \frac{1}{R_2 \sin^2 \alpha} \int_{\alpha'}^{\alpha} (q_n \cos \alpha - q_1 \sin \alpha) R_1 R_2 \sin \alpha \, d\alpha. \tag{3.80}$$

If the unit force T_1' corresponds to the angle α', it follows from equation (3.80) that

$$C = T_1' R_2 \sin^2 \alpha' . \tag{3.81}$$

In a surface closed at the top $\alpha' = 0$, and according to (3.81), we have $C = 0$, so that the unit force T_1 is

$$T_1 = \frac{1}{R_2 \sin^2 \alpha} \int_0^\alpha (q_n \cos \alpha - q_1 \sin \alpha) R_1 R_2 \sin \alpha \, d\alpha . \tag{3.82}$$

For the unit force T_2 we obtain from the third equation (3.49) the formula

$$\begin{aligned} T_2 &= q_n R_2 - \frac{1}{R_1 \sin^2 \alpha} \int_0^\alpha (q_n \cos \alpha - q_1 \sin \alpha) R_1 R_2 \sin \alpha \, d\alpha = \\ &= q_n R_2 - \frac{R_2}{R_1} T_1 . \end{aligned} \tag{3.83}$$

If we replace q_1 and q_n in formulae (3.82) and (3.83) by the quantities $q^*_{1,s}$ and $q^*_{n,s}$ determined by (3.75), we obtain for the unit forces $T^*_{1,s}$ and $T^*_{2,s}$ the formulae:

$$\begin{aligned} T^*_{1,s} &= -\frac{q_s}{2 R_2 \sin^2 \alpha} \int_0^\alpha R_1 R_2 \cos^2 \alpha \sin \alpha \, d\alpha , \\ T^*_{2,s} &= -\frac{q_s R_2}{2} \cos^3 \alpha + \frac{q_s}{2 R_1 \sin^2 \alpha} \int_0^\alpha R_1 R_2 \cos^2 \alpha \sin \alpha \, d\alpha . \end{aligned} \tag{3.84}$$

The unit forces $T^{**}_{1,s}$, $T^{**}_{2,s}$ and $T^{**}_{12,s}$ due to the antisymmetric load components (3.76) can be obtained from equations (3.62) and (3.66), if in (3.63) and (3.64)

$$\bar{q}_1 = \frac{q_s}{2} \sin \alpha \cos^2 \alpha , \quad \bar{q}_n = -\frac{q_s}{2} \cos^3 \alpha . \tag{3.85}$$

Using (3.85), we obtain from (3.61) and (3.67)

$$\Phi_s = -\frac{q_s R_2^2}{2} \cos^2 \alpha \sin \alpha , \quad \bar{f}(\alpha) \equiv 0 ,$$

and, consequently,

$$T^{**}_{1,s} = -\left[\frac{q_s}{2 R_2^2 \sin^3 \alpha} \int_0^\alpha R_1 R_2^2 \cos^2 \alpha \sin^2 \alpha \, d\alpha \right] \cos \beta ,$$

$$T^{**}_{2,s} = -\left[\frac{q_s R_2}{2}\cos^3\alpha - \frac{q_s}{2R_1R_2\sin^3\alpha}\int_0^\alpha R_1R_2^2\cos^2\alpha\sin^2\alpha\,\mathrm{d}\alpha\right]\cos\beta, \tag{3.86}$$

$$T^{**}_{12,s} = -\left[\frac{q_s\cos\alpha}{2R_2^2\sin^3\alpha}\int_0^\alpha R_1R_2^2\cos^2\alpha\sin^2\alpha\,\mathrm{d}\alpha\right]\sin\beta = T^{**}_{1,s}\cos\alpha\tan\beta.$$

The resulting unit forces $T_{1,s}$, $T_{2,s}$ and $T_{12,s}$ due to one-sided snow load (3.73) for $\alpha \leqq \pi/2$ can be obtained by the superposition of the forces (3.84) and (3.86)

$$T_{1,s} = T^*_{1,s} + T^{**}_{1,s}, \quad T_{2,s} = T^*_{2,s} + T^{**}_{2,s}, \quad T_{12,s} = T^{**}_{12,s}. \tag{3.87}$$

For a membrane structure in which the angle α attains values in excess of $\pi/2$ we have $q_s = 0$ when $\alpha > \pi/2$. In the lower part of the surface the unit forces can be determined, once again, from the formulae (3.84) and (3.86) by substituting the values $\alpha = \pi/2$ for the upper limits of the definite integrals occuring in them and by setting $q_s = 0$ in the first terms of the formulae for $T^*_{2,s}$ and $T^{**}_{2,s}$. In the case of smoth surfaces the upper limit of these integrals can be replaced by the value $\alpha = \pi/4$.

The resulting membrane forces due to the combination of various types of loads applied simultaneously to the structure can be obtained by the superposition of the forces due to the individual types of load. The principle of superposition can be used because the equations of equilibrium, determining the membrane forces, are linear.

3.6 *Deformation of a Membrane Structure of Revolution*

Consider a surface of revolution described by the equations (1.25). It follows from (3.1) that for such a surface the geometric relations (2.41) acquire the form ($\xi = z$, $\eta = \beta$, $R_1 = -A_1^3/r''$, $R_2 = A_1 r$):

$$\varepsilon_1 = \frac{1}{A_1}\frac{\partial u}{\partial z} + \frac{r''}{A_1^3}w,$$

$$\varepsilon_2 = \frac{1}{r}\frac{\partial v}{\partial \beta} + \frac{r'}{A_1 r}u - \frac{1}{A_1 r}w, \tag{3.88}$$

$$\gamma_{12} = \frac{1}{r}\frac{\partial u}{\partial \beta} + \frac{r}{A_1}\frac{\partial}{\partial z}\left(\frac{v}{r}\right),$$

where

$$A_1=(1+r'^2)^{1/2}.$$

If the surface of revolution is defined by parametric equations (3.37), then, according to (3.39), (3.40) and (3.41),

$$A_1=R_1(\alpha),\quad A_2=R_2(\alpha)\sin\alpha,\quad \frac{\partial A_2}{\partial \xi}=R_1\cos\alpha,$$

and the equations (2.41) can be written as follows ($\xi=\alpha$, $\eta=\beta$)

$$\begin{aligned}
\varepsilon_1&=\frac{1}{R_1}\frac{\partial u}{\partial \alpha}-\frac{w}{R_1},\\
\varepsilon_2&=\frac{1}{R_2\sin\alpha}\frac{\partial v}{\partial \beta}+\frac{\cot\alpha}{R_2}u-\frac{w}{R_2},\\
\gamma_{12}&=\frac{1}{R_2\sin\alpha}\frac{\partial u}{\partial \beta}-\frac{1}{R_1}\frac{\partial v}{\partial \alpha}-\frac{\cot\alpha}{R_2}v.
\end{aligned}\tag{3.89}$$

Eliminating the displacement w from the first two equations (3.89) and applying of the physical relations (2.11) or using the equations (2.43), we obtain two equations for the unknown displacements u and v

$$\begin{aligned}
\frac{\partial u}{\partial \alpha}-u\cot\alpha-\frac{\partial v}{\partial \beta}&=(R_1+\nu_1R_2)\frac{T_1}{E_1}-(R_2+\nu_2R_1)\frac{T_2}{E_2},\\
\frac{1}{\sin\alpha}\frac{\partial u}{\partial \beta}+\frac{R_2}{R_1}\frac{\partial v}{\partial \alpha}-v\cot\alpha&=R_2\frac{T_{12}}{E_{12}},
\end{aligned}\tag{3.90}$$

while the deflection w satisfies, according to (2.44),

$$w=-R_1\left(\frac{T_1}{E_1}-\frac{\nu_2}{E_2}T_2\right)+\frac{\partial u}{\partial \alpha}.\tag{3.91}$$

Equations (3.90) and (3.91) are valid only for elastic deformation of an orthotropic membrane. These equations can be used also for the calculation of the initial deformation of a viscoelastic membrane, if the quantities E_1, E_2, ν_1, ν_2 and E_{12} are replaced by the constants $E_{1,0}$, $E_{2,0}$, $\nu_{1,0}$, $\nu_{2,0}$, $E_{12,0}$, corresponding to the viscoelastic membrane at the beginning of loading (at the time $t=0$).

The displacements of a viscoelastic membrane due to long-term loads (as the time $t\to\infty$) can be determined, once again, from equations (3.90) and (3.91), by putting

$$E_1 = E_{1,\infty} = \frac{c_0}{a_0}, \quad \frac{\nu_1}{E_1} = \frac{\nu_{1,\infty}}{E_{1,\infty}} = \frac{p_0}{r_0}, \tag{3.92}$$
$$E_2 = E_{2,\infty} = \frac{r_0}{n_0}, \quad \frac{\nu_2}{E_2} = \frac{\nu_{2,\infty}}{E_{2,\infty}} = \frac{b_0}{c_0}, \quad E_{12} = E_{12,\infty} = \frac{d_0}{m_0},$$

where the quantities at the right-hand sides of these relations are the coefficients of the differential operators (2.14).

In the case of axially symmetric loading we have $T_{12}=0$, $v=0$ and the first equation (3.90) becomes an ordinary differential equation

$$\frac{du}{d\alpha} - u \cot \alpha = (R_1 + \nu_1 R_2)\frac{T_1}{E_1} - (R_2 + \nu_2 R_1)\frac{T_2}{E_2}, \tag{3.93}$$

whose general solution can be obtained by the method of variation of constants.

Thus we obtain

$$u = \sin \alpha \left[\int \frac{f(\alpha)}{\sin \alpha}\, d\alpha + C \right], \tag{3.94}$$

where $f(\alpha)$ is the right-hand side of equation (3.93) and C is the inegration constant which is determined from the condition at the point of fastening of the membrane. If the surface of revolution is fastened to the foundations along a parallel circle $\alpha = \alpha_1$, than $u(\alpha_1) = 0$ and from (3.94) we obtain

$$C = -\left[\int \frac{f(\alpha)}{\sin \alpha}\, d\alpha \right]_{\alpha=\alpha_1}. \tag{3.95}$$

Thus, the inegration constant equals the negative value of the primitive function of the integrand $f(\alpha)/\sin \alpha$ at the point $\alpha = \alpha_1$.

If we consider only the load due to the internal overpressure q_0, then, by (3.16), the function $f(\alpha)$ in equation (3.94) is

$$f(\alpha) = \frac{q_0 R_1 R_2}{E_1 E_2} \left[\frac{E_2}{2}\left(1 + \nu_1 \frac{R_2}{R_1}\right) - E_1 \left(\nu_2 + \frac{R_2}{R_1}\right)\left(1 - \frac{R_2}{2R_1}\right) \right]. \tag{3.96}$$

According to (3.93) and (3.94),

$$\frac{du}{d\alpha} = f(\alpha) + \cos \alpha \left[\int \frac{f(\alpha)}{\sin \alpha}\, d\alpha + C \right] \tag{3.97}$$

and the substitution of (3.16) and (3.97) into (3.91) yields for the displacement w due to the internal overpressure, the formula

$$w = -q_0 R_1 R_2 \left[\frac{1}{2E_1} - \frac{\nu_2}{E_2}\left(1 - \frac{R_2}{2R_1}\right)\right] + f(\alpha) + \cos\alpha \left[\int \frac{f(\alpha)}{\sin\alpha}\,d\alpha + C\right]. \tag{3.98}$$

The resulting displacements due to simultaneous loads, e.g. to internal overpressure, dead weight of the membrane, or wind and snow, are obtained by a superposition of the effects of the individual types of loads. This can be done since the relations (3.90) and (3.91) between the displacements and the internal forces are linear differential equations.

3.7 *Effect of Boundary Conditions on Membrane Forces*

At the place of membrane fastening, i.e. where $\alpha = \alpha_1$, we obtain from (3.98), by (3.95), that the displacement $w(\alpha_1)$ is given by

$$w(\alpha_1) = -q_0 R_1(\alpha_1) R_2(\alpha_1) \left[\frac{1}{2E_1} - \frac{\nu_2}{E_2}\left(1 - \frac{R_2(\alpha_1)}{2R_1(\alpha_1)}\right)\right] + f(\alpha_1), \tag{3.99}$$

where $f(\alpha_1)$ is the value of the function (3.96) along the parallel circle $\alpha = \alpha_1$. Generally $w(\alpha_1) \neq 0$, which means that the internal overpressure produces a radial displacement in the place of membrane fastening. If the structural design of the anchorage does not permit such displacement, the membrane forces in the vicinity of the parallel circle $\alpha = \alpha_1$ change so as to satisfy condition $w(\alpha_1) = 0$. Consequently, in the vicinity of $\alpha = \alpha_1$ the actual membrane forces differ from the membrane forces calculated according to (3.16). Only in the case of an entirely unelongated membrane the formulae (3.16) yield correct values of membrane forces in the vicinity of anchorages; indeed, (3.99) implies $w(\alpha_1) = 0$ for $E_1 \to \infty$ and $E_2 \to \infty$.

In air-supported structures there can appear also deviations from the boundary conditions (3.99) and $u(\alpha_1) = 0$, due to imperfections of the skin material, defects of the staking out of the points of anchorage and in its execution as well as non-uniform deformation of anchorage elements. For this reason the actual membrane forces in some places of skin anchorage can be many times higher than those calculated from the formulae (3.16). Substantial differences between actual and theoretical membrane forces were ascertained experimentally in the case of an air-supported structure of a cylindrical shape with quarter-spherical ends along mobile anchorages [28].

The deviations from the boundary conditions (3.99) and $u(\alpha_1)=0$ can manifest themselves very unfavourably, in some loading stages, in the vicinity of the anchorages, particularly by the appearance of excessive sheraring forces T_{12}. Therefore, the skin of air-supported structures should be strengthened in the vicinity of anchorages, if major inaccuracies in its manufacture or erection can be expected.

According to the Saint-Venant principle the effect of boundary conditions on the membrane state of stress is of a local character. If major geometric deviations in the skin of the structure occur, their effect on the magnitude of membrane forces may be investigated by the method described in Subsection 2.6 of this chapter.

4

Quadric Surfaces of Revolution

4.1 *Equations of the Meridians of Quadric Surfaces of Revolution and Their Principal Radii*

The quadric surfaces of revolution are obtained by rotating quadric curves about the z-axis. With respect to the r, z coordinate system their meridians are determined by an equation (Fig. II-3)

$$r = \frac{a}{b}\sqrt{2bz + mz^2}, \tag{4.1}$$

where a, b are constants different from zero and m is a given parameter. The origin O of the r, z coordinate system is the point of intersection of the meridian with the z-axis (axis of revolution).

When $m = -1$ and $a \neq b$, equation (4.1) yields the equation of an ellipse, when $m = -1$ and $a = b$, the equation yields a circle; when $m = 0$, $a = R_0/\sqrt{2}$, $b = R_0/2$, the meridian is a parabola, and when $m = +1$, the meridian is a hyperbola (see Section 5, Chapter II). The parametric equations of these surfaces have the form (1.25).

According to (4.1),

$$\begin{aligned} r' &= \frac{a}{b}\frac{b + mz}{(2bz + mz^2)^{1/2}}, \\ r'' &= -\frac{ab}{(2bz + mz^2)^{3/2}}. \end{aligned} \tag{4.2}$$

Substituting (4.1) and (4.2) into (3.1), we obtain the following formulae for the principal radii:

$$\begin{aligned} R_1 &= \frac{1}{ab^4}[m(b^2 + ma^2)z^2 + 2b(b^2 + ma^2)z + a^2b^2]^{3/2}, \\ R_2 &= \frac{a}{b^2}[m(b^2 + ma^2)z^2 + 2b(b^2 + ma^2)z + a^2b^2]^{1/2}. \end{aligned} \tag{4.3}$$

According to (4.3) the characteristic curvature of a quadric surface of revolution is given by

$$\frac{R_2^3}{R_1}=\frac{a^4}{b^2}=R_0^2, \tag{4.4}$$

where R_0 is the value of a principal radius at the apex of the surface, i.e. at the point $z=0$, $r=0$ (see Subsection 2.4, Chapter III).

If the quadric surfaces of revolution are defined by parametric equations of the form (3.37), then (see Subsection 2.4, Chapter III)

$$R_1=\frac{R_0}{(1+\gamma\sin^2\alpha)^{3/2}}, \quad R_2=\frac{R_0}{(1+\gamma\sin^2\alpha)^{1/2}}. \tag{4.5}$$

After substituting (4.5) for R_1 into the equation $dz/d\alpha=R_1\sin\alpha$ (see Fig. II-3) we obtain

$$\frac{dz}{d\alpha}=\frac{R_0\sin\alpha}{(1+\gamma\sin^2\alpha)^{3/2}}; \tag{4.6}$$

integrating this we get

$$z=-\frac{R_0\cos\alpha}{(1+\gamma)\sqrt{1+\gamma\sin^2\alpha}}+C, \tag{4.7}$$

where C is an arbitrary constant and $\gamma\neq-1$.

If we identify the apex of the surface with the origin of the coordinate axes, then $z=0$ for $\alpha=0$, and from (4.7) we obtain

$$C=\frac{R_0}{1+\gamma}. \tag{4.8}$$

Writing for R_2 in the second equation (3.40) the expression given in (4.5) and substituting (4.8) into (4.7), we obtain the parametric equations of a meridian of a quadric surface of revolution

$$\begin{aligned} r&=\frac{R_0\sin\alpha}{(1+\gamma\sin^2\alpha)^{1/2}},\\ z&=\frac{R_0}{1+\gamma}\left[1-\frac{\cos\alpha}{(1+\gamma\sin^2\alpha)^{1/2}}\right]. \end{aligned} \tag{4.9}$$

If $\gamma=-1$, the meridian is a parabola and equation (4.6) takes the form

$$\frac{dz}{d\alpha}=\frac{R_0 \sin \alpha}{\cos^3 \alpha}, \tag{4.10}$$

whence

$$z=\frac{R_0}{2 \cos^2 \alpha}+C. \tag{4.11}$$

If $z=0$ for $\alpha=0$, (4.11) yields $C=-R_0/2$ and, consequently, thc meridian of a paraboloid of revolution is defined by the following parametric equations:

$$r=R_0 \tan \alpha, \quad z=\frac{1}{2} R_0 \tan^2 \alpha. \tag{4.12}$$

4.2 *Load Due to Internal Overpressure*

The membrane forces $T_{1,q}$ and $T_{2,q}$ due to internal overpressure q_0 in a quadric surface can be obtained by substituting (4.3) into (3.16). We have ($T_{12}=0$)

$$\begin{aligned} T_{1,q}&=\frac{q_0 a}{2b^2}[m(b^2+ma^2)z^2+2b(b^2+ma^2)z+a^2b^2]^{1/2}, \\ T_{2,q}&=T_{1,q}\{2-a^2b^2[m(b^2+ma^2)z^2+2b(b^2+ma^2)z+a^2b^2]^{-1}\}. \end{aligned} \tag{4.13}$$

If we use (4.5) instead of (4.3), the membrane forces due to internal overpressure are determined by the formulae

$$\begin{aligned} T_{1,q}&=\frac{q_0 R_0}{2}(1+\gamma \sin^2 \alpha)^{-1/2}, \\ T_{2,q}&=\frac{q_0 R_0}{2}\frac{1-\gamma \sin^2 \alpha}{(1+\gamma \sin^2 \alpha)^{1/2}}. \end{aligned} \tag{4.14}$$

4.3 *Load Due to Dead Weight*

For the calculation of the membrane forces $T_{1,g}$ and $T_{2,g}$ due to the dead weight of the membrane g we shall use first the formulae (3.25) and (3.27). If we substitute in the integral appearing in them the expression for R_2 given in (4.3), the integral becomes

$$J_g(z)=\int_0^z R_2(z)\,\mathrm{d}z=\frac{a}{b^2}\int_0^z [m(b^2+ma^2)z^2+2b(b^2+ma^2)z+a^2b^2]^{1/2}\,\mathrm{d}z. \tag{4.15}$$

In this equation the integration constant is absent, as its zero value was determined from the boundary condition in Subsection 3.1, Chapter II. The evaluation of the integral in (4.15) depends on the sign of the discriminant

$$\Delta=-4b^4(b^2+ma^2) \tag{4.16}$$

of the polynomial in square brackets in (4.15), and on the term

$$c=m(b^2+ma^2) \tag{4.17}$$

at z^2 [21].

If $c>0$, we obtain

$$J_g(z)=-\frac{ab^2}{2m}I+\frac{a}{2mb^2}(mz+b)\,[m(b^2+ma^2)z^2+2b(b^2+ma^2)z+a^2b^2]^{1/2}-\frac{a^2}{2m}, \tag{4.18}$$

where

$$I=\frac{a}{b^2}\int_0^z \frac{\mathrm{d}z}{R_2(z)}. \tag{4.19}$$

For $c>0$ and $\Delta>0$ the integration in (4.19) yields

$$I=[m(b^2+ma^2)]^{-1/2}\left\{\operatorname{arcsinh}\left[\frac{mz+b}{b^2}(b^2+ma^2)^{1/2}\right]\right\}_0^z, \tag{4.20}$$

and for $c<0$ and $\Delta<0$ we have

$$I=-[-m(b^2+ma^2)]^{-1/2}\left\{\arcsin\left[\frac{mz+b}{b^2}(b^2+ma^2)^{1/2}\right]\right\}_0^z. \tag{4.21}$$

The substitution of (4.18) for the integral in the formulae (3.25) and (3.27) and an application of (4.1), (4.2) and (4.3) yields the following formulae for the membrane forces $T_{1,g}$ and $T_{2,g}$:

$$T_{1,g}=gR_2\frac{b^2}{2bz+mz^2}\left[\frac{b^2}{2ma}I-(mz+b)\frac{R_2}{2ma^2}+\frac{1}{2m}\right], \tag{4.22}$$

$$T_{2,g}=-gR_2\left(\frac{b^2}{2bz+mz^2}\frac{R_2}{R_1}\left\{\left[\frac{b^2}{2ma}I-(mz+b)\frac{R_2}{2ma^2}\right]+\frac{1}{2m}\right\}+\frac{a^2}{b^2R_2}(mz+b)\right).$$

If $\Delta=0$, then, according to (4.16), $b^2+ma^2=0$. Consequently (4.3) yields

$$R_1=R_2=\frac{a^2}{b}, \tag{4.23}$$

which is the case of a spherical surface whose membrane forces were investigated in Subsection 5.2 of Chapter II.

The membrane forces $T_{1,g}$ and $T_{2,g}$ can be also determined from the formulae (3.82) and (3.83) by the substitution of (4.5) for R_1 and R_2 and (3.23) for q_1 and q_n. Employing (3.23), we can write formulae (3.82) and (3.83) in the form

$$\begin{aligned} T_{1,g} &= -\frac{g}{R_2 \sin^2 \alpha}\int_0^\alpha R_1 R_2 \sin \alpha \, \mathrm{d}\alpha, \\ T_{2,g} &= -gR_2 \cos \alpha + \frac{g}{R_1 \sin^2 \alpha}\int_0^\alpha R_1 R_2 \sin \alpha \, \mathrm{d}\alpha = \\ &= -gR_2 \cos \alpha - \frac{R_2}{R_1} T_{1,g} \end{aligned} \tag{4.24}$$

whence, after substituting (4.5), we obtain

$$\begin{aligned} T_{1,g} &= -\frac{gR_0}{\sin^2 \alpha (1+\gamma \sin^2 \alpha)^{1/2}}\int_0^\alpha \frac{\sin \alpha \, \mathrm{d}\alpha}{(1+\gamma \sin^2 \alpha)^2}, \\ T_{2,g} &= -\frac{gR_0 \cos \alpha}{(1+\gamma \sin^2 \alpha)^{1/2}} - (1+\gamma \sin^2 \alpha) T_{1,g}. \end{aligned} \tag{4.25}$$

The substitution of

$$t=\cos \alpha, \quad \mathrm{d}\alpha = -\frac{\mathrm{d}t}{\sin \alpha}, \quad \sin^2 \alpha = 1-t^2 \tag{4.26}$$

transforms the integral

$$J^* = \int_0^\alpha \frac{\sin \alpha \, \mathrm{d}\alpha}{(1+\gamma \sin^2 \alpha)^2} \tag{4.27}$$

into

$$J^* = -\int_{t_1}^{t_2} \frac{\mathrm{d}t}{(1+\gamma-\gamma t^2)^2}, \tag{4.28}$$

which yields [21]

$$J^* = -\frac{1}{2(1+\gamma)} I^* - \left[\frac{t}{2(1+\gamma)(1+\gamma-\gamma t^2)}\right]_{t_1}^{t_2}, \tag{4.29}$$

where

$$I^* = \int_{t_1}^{t_2} \frac{\mathrm{d}t}{1+\gamma-\gamma t^2}. \tag{4.30}$$

For $\Delta^* < 0$, where

$$\Delta^* = -4\gamma(1+\gamma) \tag{4.31}$$

is the discriminant of the binomial between the brackets in (4.28), we have

$$I^* = -\left[\frac{1}{2\sqrt{\gamma(1+\gamma)}} \operatorname{arctanh} \frac{-\gamma t}{2\sqrt{\gamma(1+\gamma)}}\right]_{t_1}^{t_2} \tag{4.32}$$

and for $\Delta^* > 0$ we have

$$I^* = \left[\frac{1}{2\sqrt{-\gamma(1+\gamma)}} \arctan \frac{-\gamma t}{2\sqrt{-\gamma(1+\gamma)}}\right]_{t_2}^{t_1}. \tag{4.33}$$

Taking into account (4.26), we get for $\Delta^* < 0$

$$J^* = +\frac{1}{2(1+\gamma)} \left[\frac{1}{2\sqrt{\gamma(1+\gamma)}} \operatorname{arctanh} \frac{-\gamma \cos\alpha}{2\sqrt{\gamma(1+\gamma)}} - \frac{\cos\alpha}{1+\gamma\sin^2\alpha}\right]_0^\alpha. \tag{4.34}$$

Substituting (4.33) into (4.29) and applying (4.26), we obtain for $\Delta^* > 0$

$$J^* = -\frac{1}{2(1+\gamma)} \left[\frac{1}{2\sqrt{-\gamma(1+\gamma)}} \arctan \frac{-\gamma \cos\alpha}{2\sqrt{-\gamma(1+\gamma)}} + \frac{\cos\alpha}{1+\gamma\sin^2\alpha}\right]_0^\alpha. \tag{4.35}$$

If we substitute the limits 0 and α into the integrals (4.34) and (4.35), we obtain, according to (4.25), the following formulae for the unit membrane force $T_{1,g}$ due to the dead weight g of the membrane:

a) for $\Delta^* < 0$, i.e. for $\gamma(1+\gamma) > 0$

$$T_{1,g} = -\frac{gR_0}{2(1+\gamma)\sin^2\alpha(1+\sin^2\alpha)^{1/2}} \times \\ + \left[\frac{1}{2\sqrt{\gamma(1+\gamma)}} \left(\operatorname{arctanh} \frac{-\gamma\cos\alpha}{2\sqrt{\gamma(1+\gamma)}} - \operatorname{arctanh} \frac{-\gamma}{2\sqrt{\gamma(1+\gamma)}}\right) - \\ - \frac{\cos\alpha}{1+\gamma\sin^2\alpha} + 1\right], \tag{4.36}$$

b) for $\Delta^*>0$, i.e. for $\gamma(1+\gamma)<0$

$$T_{1,g}=\frac{gR_0}{2(1+\gamma)\sin^2\alpha(1+\gamma\sin^2\alpha)^{1/2}}\times$$
$$\times\left[\frac{1}{2\sqrt{-\gamma(1+\gamma)}}\left(\arctan\frac{-\gamma\cos\alpha}{2\sqrt{-\gamma(1+\gamma)}}-\arctan\frac{-\gamma}{2\sqrt{-\gamma(1+\gamma)}}\right)+\right. \tag{4.37}$$
$$\left.+\frac{\cos\alpha}{1+\gamma\sin^2\alpha}-1\right].$$

The unit membrane force $T_{2,g}$ can be determined from the second formula (4.25).

The case $\Delta=0$, i.e. when $\gamma=0$ or $\gamma=-1$ will be considered in Subsections 5.2c and 5.3c of this chapter. We shall derive there the corresponding formula for $T_{1,g}$.

4.4 *Symmetric Snow Load*

According to (4.1) and (4.2),

$$r^2r'^2=\frac{a^4}{b^4}(b+mz)^2=\frac{a^4}{b^4}(b^2+2mbz+m^2z^2). \tag{4.38}$$

Substituting this into (3.32), we get

$$T_{1,s}=-\frac{q_sR_2a^4}{r^2b^4}(b^2J_1+2mbJ_2+m^2J_3), \tag{4.39}$$

where

$$J_1=\int_0^z\frac{dz}{R_2},\quad J_2=\int_0^z\frac{z\,dz}{R_2},\quad J_3=\int_0^z\frac{z^2\,dz}{R_2}, \tag{4.40}$$

while the principal radius R_2 is defined by the second formula in (4.3).

The integral J_1 equals

$$J_1=\frac{b^2}{a}I, \tag{4.41}$$

where I is determined, for $c>0$ and $\Delta>0$ by (4.20) and for $c<0$ and $\Delta<0$ by (4.21).

For the integral J_2 we obtain [21]

$$J_2 = \frac{b^4(R_2 - a^2/b)}{a^2 m(b^2 + ma^2)} \tag{4.42}$$

and the integral J_3 is given by

$$J_3 = -\frac{ab^4}{2m(b^2+ma^2)} I + \frac{b^4 R_2}{2ma^2(b^2+ma^2)} z, \tag{4.43}$$

where I has the same meaning as in (4.41).

After substituting (4.1), (4.41), (4.42) and (4.43) into (4.39) we obtain

$$\begin{aligned} T_{1,s} &= -q_s R_2 \frac{a^2 b^2}{2bz + mz^2} \times \\ &\times \left[\left(\frac{1}{a} - \frac{a}{2(b^2+ma^2)} \right) I + \frac{2(bR_2 - a^2)}{a^2(b^2+ma^2)} + \frac{mR_2}{2a^2(b^2+ma^2)} z \right]. \end{aligned} \tag{4.44}$$

For $T_{2,s}$ we have, according to (3.34) and taking into account (4.1) and (4.2),

$$T_{2,s} = -q_s R_2 \left[\frac{a^6}{R_2^3 b^6} (b + mz)^3 + \frac{T_{1,s}}{q_s R_1} \right], \tag{4.45}$$

where R_1 and R_2 are defined by the expressions (4.3).

Formulae (4.44) and (4.45) apply to the part of the surface on which there is snow, i.e. for $z \leqq z_s$, where $z_s = \text{const}$ is the horizontal plane passing through the boundary between the snow-covered and snow-free parts of the surface. For $z > z_s$ we have, according to (3.35) and (4.44),

$$\begin{aligned} T_{1,s} &= -q_s R_2(z) \frac{a^2 b^2}{2bz + mz^2} \times \\ &\times \left[\left(\frac{1}{a} - \frac{a}{2(b^2+ma^2)} \right) I(z_s) + \frac{2[bR_2(z_s) - a^2]}{a^2(b^2+ma^2)} + \frac{mR_2(z_s)}{2a^2(b^2+ma^2)} z_s \right], \end{aligned} \tag{4.46}$$

where $R_2(z_s)$ is the value of the principal radius along the parallel circle $z = z_s$ and $I(z_s)$ is the value of (4.20) or (4.21) for $z = z_s$.

For $T_{2,s}$, when $z > z_s$, we have, according to (3.36),

$$T_{2,s} = -\frac{R_2(z)}{R_1(z)} T_{1,s}, \tag{4.47}$$

where $T_{1,s}$ should be replaced by its value according to (4.46). As a rule z_s corresponds to the angle $\alpha = \pi/2$ or $\alpha = \pi/4$, as we already have mentioned in Subsection 3.3 of Chapter II.

In the calculation of membrane forces due to symmetric snow load it is also possible to use the formulae (3.82) and (3.83), replacing q_1 and q_n by the expressions (3.30), and R_1, R_2 by their values (4.5).

In this way we obtain

$$T_{1,s} = -\frac{q_s R_0 (1+\gamma \sin^2 \alpha)^{1/2}}{\sin^2 \alpha} \int_0^\alpha \frac{\cos^2 \alpha \sin \alpha}{(1+\gamma \sin^2 \alpha)^2} \, d\alpha,$$
$$T_{2,s} = -q_s R_0 \frac{\cos^3 \alpha}{(1+\gamma \sin^2 \alpha)^{1/2}} - (1+\gamma \sin^2 \alpha) T_{1,s}. \qquad (4.48)$$

The change of variables (4.26) transforms the integral

$$J_s = \int_0^\alpha \frac{\cos^2 \alpha \sin \alpha}{(1+\gamma \sin^2 \alpha)^2} \, d\alpha \qquad (4.49)$$

in the first equation (4.48) into

$$J_s = -\int_{t_1}^{t_2} \frac{t^2 \, dt}{(1+\gamma-\gamma t^2)^2}. \qquad (4.50)$$

Thus we obtain [21]

$$J_s = \left[\frac{t}{2\gamma(1+\gamma-\gamma t^2)}\right]_{t_1}^{t_2} - \frac{1}{2\gamma} I^*, \qquad (4.51)$$

where I^* is the integral (4.30). If we return to the variable α, we have

$$J_s = \left[\frac{\cos \alpha}{2\gamma(1+\gamma \sin^2 \alpha)}\right]_0^\alpha - \frac{1}{2\gamma} I^*, \qquad (4.52)$$

where for $\gamma(\gamma+1)>0$ we have, according to (4.32) and (4.26),

$$I^* = -\frac{1}{2\sqrt{\gamma(1+\gamma)}} \left[\operatorname{arctanh} \frac{-\gamma \cos \alpha}{2\sqrt{\gamma(1+\gamma)}}\right]_0^\alpha \qquad (4.53)$$

and for $\gamma(1+\gamma)<0$ we have, according to (4.33),

$$I^* = \frac{1}{2\sqrt{-\gamma(1+\gamma)}} \left[\arctan \frac{-\gamma \cos \alpha}{2\sqrt{-\gamma(1+\gamma)}}\right]_0^\alpha. \qquad (4.54)$$

Substituting (4.53) or (4.54) into (4.52) and then into the first relation (4.48), we obtain for $\gamma(1+\gamma)>0$

$$T_{1,s} = -q_s R_0 \frac{(1+\gamma\sin^2\alpha)^{1/2}}{\sin^2\alpha}\left[\frac{\cos\alpha}{2\gamma(1+\gamma\sin^2\alpha)} - \frac{1}{2\gamma} + \frac{1}{4\gamma\sqrt{\gamma(1+\gamma)}}\times\right.$$

$$\left.\times\left(\operatorname{arctanh}\frac{-\gamma\cos\alpha}{2\sqrt{\gamma(1+\gamma)}} - \operatorname{arctanh}\frac{-\gamma}{2\sqrt{\gamma(1+\gamma)}}\right)\right] \tag{4.55}$$

and for $\gamma(1+\gamma)<0$, taking into account (4.54), we have

$$T_{1,s} = -q_s R_0 \frac{(1+\gamma\sin^2\alpha)^{1/2}}{\sin^2\alpha}\left[\frac{\cos\alpha}{2\gamma(1+\gamma\sin^2\alpha)} - \frac{1}{2\gamma} - \frac{1}{4\gamma\sqrt{-\gamma(1+\gamma)}}\times\right.$$

$$\left.\times\left(\arctan\frac{-\gamma\cos\alpha}{2\sqrt{-\gamma(1+\gamma)}} - \arctan\frac{-\gamma}{2\sqrt{-\gamma(1+\gamma)}}\right)\right]. \tag{4.56}$$

If we determine the force $T_{1,s}$ according to (4.55) or (4.56), we can calculate the force $T_{2,s}$ from the second equation (4.48).

Formulae (4.55) and (4.56) hold for $\alpha \leqq \pi/2$, where $q_s>0$. For $\alpha>\pi/2$ we have $q_s=0$ from the plane $\alpha=\pi/2$ downwards, so that the upper limit of the integral (4.49) can be taken $\alpha=\pi/2$. For the force $T_{1,s}$ we have, according to (4.55), for $\gamma(1+\gamma)>0$ and $\alpha>\pi/2$, the following formula:

$$T_{1,s} = -q_s R_0 \frac{(1+\gamma\sin^2\alpha)^{1/2}}{2\gamma\sin^2\alpha}\left[-1 - \frac{1}{2\sqrt{\gamma(1+\gamma)}}\operatorname{arctanh}\frac{-\gamma}{2\sqrt{\gamma(1+\gamma)}}\right]. \tag{4.57}$$

To obtain the formula for $T_{1,s}$ for $\gamma(1+\gamma)<0$ and $\alpha>\pi/2$ we need only to substitute in (4.57) $-\arctan(-\gamma)/\{2\sqrt{-\gamma(1+\gamma)}\}$ for $\operatorname{arctanh}(-\gamma)/$ $/\{2\sqrt{\gamma(1+\gamma)}\}$ and $-\gamma(1+\gamma)$ for $\gamma(1+\gamma)$.

For the force $T_{2,s}$ we have, for $\alpha>\pi/2$,

$$T_{2,s} = -(1+\gamma\sin^2\alpha)T_{1,s}. \tag{4.58}$$

4.5 *One-Side Snow Load*

The forces $T^*_{1,s}$ and $T^*_{2,s}$ which are due to the symmetric components of the one-side snow load (3.75), can be determined from (3.77), where $T_{1,s}$ and $T_{2,s}$ have been calculated according to the formulae in the preceding Subsection.

The unit membrane forces due to antisymmetric components (3.76) of the one-sided snow load can be calculated from (3.86).

Consider the integral

$$J_s(\alpha)=\int_0^{\alpha} R_1 R_2^2 \cos^2 \alpha \sin^2 \alpha \, d\alpha, \tag{4.59}$$

occuring in equation (3.86). By (4.5), this integral can be written as

$$\bar{J}_s(\alpha)=R_0^3 \bar{I}_s(\alpha), \tag{4.60}$$

where

$$\bar{I}_s(\alpha)=\int_0^{\alpha} \frac{\cos^2 \alpha \sin^2 \alpha}{(1+\gamma \sin^2 \alpha)^{5/2}} \, d\alpha. \tag{4.61}$$

Substituting (4.60) into (3.86) and taking into account that holds for $\alpha \leqq \pi/2$, we get

$$\begin{aligned} T^{**}_{1,s} &= -\frac{q_s R_0}{2} \frac{(1+\gamma \sin^2 \alpha)\bar{I}_s(\alpha)}{\sin^3 \alpha} \cos \beta, \\ T^{**}_{2,s} &= -\frac{q_s R_0}{2} \left[\frac{\cos^3 \alpha}{(1+\gamma \sin^2 \alpha)^{1/2}} - \frac{(1+\gamma \sin^2 \alpha)^2 \bar{I}_s(\alpha)}{\sin^3 \alpha}\right] \cos \beta, \\ T^{**}_{12,s} &= T^{**}_{1,s} \cos \alpha \tan \beta. \end{aligned} \tag{4.62}$$

For a spherical surface we have $\gamma=0$ and

$$\bar{I}_s(\alpha)=\int_0^{\alpha} \cos^2 \alpha \sin^2 \alpha \, d\alpha, \tag{4.63}$$

and for a paraboloid of revolution we have $\gamma=-1$ and

$$\bar{I}_s(\alpha)=\int_0^{\alpha} \frac{\sin^2 \alpha}{\cos^3 \alpha} \, d\alpha. \tag{4.64}$$

These two cases are dealt with in Section 5 of Chapter II.

For $\gamma \neq 0$ and $\gamma \neq -1$ it is advisable to evaluate the integral $\bar{I}_s(\alpha)$ numerically, e.g. according to the Simpson rule.

For $\alpha > \pi/2$ the following equations take the place of (4.62):

$$T_{1,\,s}^{**}=-\frac{q_s R_0}{2}\frac{(1+\gamma\sin^2\alpha)\bar{I}_s(\pi/2)}{\sin^3\alpha}\cos\beta,$$

$$T_{2,\,s}^{**}=-\frac{q_s R_0}{2}\frac{(1+\gamma\sin^2\alpha)^2}{\sin^3\alpha}\bar{I}_s(\pi/2)\cos\beta, \qquad (4.65)$$

$$T_{12,s}^{**}=T_{1,\,s}^{**}\cos\alpha\tan\beta,$$

where $\bar{I}_s(\pi/2)$ is the value of the integral (4.61) with the upper limit of $\alpha=\pi/2$.

The resulting membrane forces $T_{1,\,s}$, $T_{2,\,s}$, $T_{12,\,s}$ due to one-side snow load can be determined by a superposition of the components of internal forces, according to (3.87).

4.6 *Wind Load*

According to (3.71), the determination of the unit membrane forces in the quadric surfaces due to the antisymmetric components of the wind load (3.45) requires evaluating the integrals:

$$J_{1,\,v}=\int_0^\alpha R_1R_2\sin^3\alpha\,d\alpha,$$
$$J_{2,\,v}=\int_0^\alpha \Phi_v R_1\sin\alpha\,d\alpha, \qquad (4.66)$$

where Φ_v is the function described by equation (3.72).

Utilizing (4.5), we obtain

$$J_{1,\,v}=R_0^2 I_{1,\,v}(\alpha), \qquad (4.67)$$

where

$$I_{1,\,v}(\alpha)=\int_0^\alpha\frac{\sin^3\alpha\,d\alpha}{(1+\gamma\sin^2\alpha)^2}. \qquad (4.68)$$

The change of variables (4.26) leads to

$$I_{1,\,v}=-\int_{t_1}^{t_2}\frac{(1-t^2)\,dt}{(1+\gamma-\gamma t^2)^2}=J^*-J_s, \qquad (4.69)$$

where J^* and J_s are the integrals (4.28) and (4.50). Substituting (4.29) and (4.52) for these integrals in (4.69) and taking into account (4.26), we obtain

$$I_{1,\mathrm{v}}=\frac{I^*}{2\gamma(1+\gamma)}-\frac{1+2\gamma}{2\gamma(1+\gamma)}\left[\frac{\cos\alpha}{1+\gamma\sin^2\alpha}\right]_0^\alpha, \tag{4.70}$$

where the integral I^* is determined by the expression (4.53) or (4.54).

For $\gamma(1+\gamma)>0$,

$$I_{1,\mathrm{v}}(\alpha)=-\frac{1}{2\gamma^{3/2}(1+\gamma)^{3/2}}\left(\operatorname{arctanh}\frac{-\gamma\cos\alpha}{2\sqrt{\gamma(1+\gamma)}}-\operatorname{arctanh}\frac{-\gamma}{2\sqrt{\gamma(1+\gamma)}}\right)-$$
$$-\frac{1+2\gamma}{2\gamma(1+\gamma)}\left(\frac{\cos\alpha}{1+\gamma\sin^2\alpha}-1\right) \tag{4.71}$$

and for $\gamma(1+\gamma)<0$ we have

$$I_{1,\mathrm{v}}(\alpha)=\frac{1}{4[-\gamma(1+\gamma)]^{3/2}}\left(\arctan\frac{-\gamma\cos\alpha}{2\sqrt{-\gamma(1+\gamma)}}-\arctan\frac{-\gamma}{2\sqrt{-\gamma(1+\gamma)}}\right)-$$
$$-\frac{1+2\gamma}{2\gamma(1+\gamma)}\left(\frac{\cos\alpha}{1+\gamma\sin^2\alpha}-1\right). \tag{4.72}$$

Substituting (4.67) and the formula (3.72) for Φ_v in the second equation of (4.66), and taking account of (4.5), we obtain

$$J_{2,\mathrm{v}}=q_\mathrm{v}R_0^3[I_{2,\mathrm{v}}(\alpha)-I_{3,\mathrm{v}}(\alpha)], \tag{4.73}$$

where

$$I_{2,\mathrm{v}}(\alpha)=\int_0^\alpha\frac{\sin^3\alpha\cos\alpha}{(1+\gamma\sin^2\alpha)^{5/2}}\,\mathrm{d}\alpha,$$
$$I_{3,\mathrm{v}}(\alpha)=\int_0^\alpha\frac{\Psi(\alpha)\sin\alpha}{(1+\gamma\sin^2\alpha)^{5/2}}\,\mathrm{d}\alpha, \tag{4.74}$$

while

$$\Psi(\alpha)=\int\frac{\sin^3\alpha\,\mathrm{d}\alpha}{(1+\gamma\sin^2\alpha)}.$$

By a change of variables

$$t=\sin\alpha,\quad \mathrm{d}\alpha=\frac{\mathrm{d}t}{\cos\alpha} \tag{4.75}$$

we write the integral $I_{2,\mathrm{v}}$ as

$$I_{2,\mathrm{v}}=\int_{t_1}^{t_2}\frac{t^3}{(1+\gamma t^2)^{5/2}}\,\mathrm{d}t, \tag{4.76}$$

which can be expressed [21] as

$$I_{2,\mathrm{v}}=\left[-\frac{1}{\gamma^2(1+\gamma t^2)^{1/2}}+\frac{1}{3\gamma^2(1+\gamma t^2)^{3/2}}\right]_{t_1}^{t_2} \tag{4.77}$$

and thus, due to (4.75),

$$I_{2,\mathrm{v}}=-\frac{1}{\gamma^2(1+\gamma\sin^2\alpha)^{1/2}}+\frac{1}{3\gamma^2(1+\gamma\sin^2\alpha)^{3/2}}+\frac{2}{3\gamma^2}. \tag{4.78}$$

The integral $I_{3,\mathrm{v}}$ for $\gamma=0$ and $\gamma=-1$ can be calculated directly (see Subsection 5.3f); for $\gamma\neq 0$ and $\gamma=-1$ we can use, for example, the Simpson rule.

By substituting (4.67) and (4.73) for the integrals (4.66) in equations (3.71) and using the formulae (4.5) we obtain

$$T_{1,\mathrm{v}}=q_{\mathrm{v}}R_0\frac{1+\gamma\sin^2\alpha}{\sin^3\alpha}[I_{2,\mathrm{v}}(\alpha)-I_{3,\mathrm{v}}(\alpha)]\cos\beta, \tag{4.79}$$

$$T_{2,\mathrm{v}}=q_{\mathrm{v}}R_0\left\{\frac{\sin\alpha}{(1+\gamma\sin^2\alpha)^{1/2}}-\frac{(1+\gamma\sin^2\alpha)^2}{\sin^3\alpha}[I_{2,\mathrm{v}}(\alpha)-I_{3,\mathrm{v}}(\alpha)]\right\}\cos\beta,$$

$$T_{12,\mathrm{v}}=q_{\mathrm{v}}R_0\left\{\frac{(1+\gamma\sin^2\alpha)\cos\alpha}{\sin^3\alpha}[I_{2,\mathrm{v}}(\alpha)-I_{3,\mathrm{v}}(\alpha)]+\right.$$

$$\left.+\frac{I_{1,\mathrm{v}}(\alpha)}{(1+\gamma\sin^2\alpha)^{1/2}\sin\alpha}\right\}\sin\beta.$$

The membrane forces due to the symmetric component q_0^* of the wind load are, according to (3.16) and (4.5),

$$T_{1,\mathrm{v}}^*=\frac{q_0^*R_0}{2(1+\gamma\sin^2\alpha)^{1/2}},\quad T_{12,\mathrm{v}}^*=0,$$

$$T_{2,\mathrm{v}}^*=\frac{q_0^*R_0}{2}\frac{1-\gamma\sin^2\alpha}{(1+\gamma\sin^2\alpha)^{1/2}}. \tag{4.80}$$

The unit membrane forces due to the overall wind load (3.44) can be obtained by the superposition of the components of internal forces (4.79) and (4.80),

$$T_1=T_{1,\mathrm{v}}+T_{1,\mathrm{v}}^*,\quad T_2=T_{2,\mathrm{v}}+T_{2,\mathrm{v}}^*,\quad T_{12}=T_{12,\mathrm{v}}. \tag{4.81}$$

4.7 *Deformation of Quadric Surfaces of Revolution*

The displacements u, v and w of membrane structures in the shape of quadric surfaces of revolution can be determined by solving the differential equations (3.90) and (3.91), where we have to substitute the formulae (4.5) for the principal radii R_1 and R_2 and we have to express the membrane forces T_1, T_2 and T_{12} by the formulae derived in this section for the antisymmetric components of load. The displacement u due to the symmetric load can be determined from equation (3.94), the displacement w from equation (3.91).

Let us deal now with the calculation of the displacements u and w due to the load produced by internal overpressure.

Substituting (4.5) into (3.96), we obtain

$$f(\alpha)=\frac{q_0R_0^2}{E_1}\frac{1}{(1+\gamma\sin^2\alpha)^2}\times$$
$$\times\left\{\frac{1}{2}[1+\nu_1(1+\gamma\sin^2\alpha)]-\frac{E_1}{E_2}[\nu_2+(1+\gamma\sin^2\alpha)]\left[1-\frac{1}{2}(1+\gamma\sin^2\alpha\right]\right\} \tag{4.82}$$

and, by (3.94),

$$u=\frac{q_0R_0^2}{E_1}[F(\alpha)+C]\sin\alpha, \tag{4.83}$$

where C is an integration constant and

$$F(\alpha)=\left(\frac{1}{2}-\frac{E_1}{E_2}\nu_2\right)K_1(\alpha)+\left[\frac{\nu_1}{2}-\frac{E_1}{E_2}\left(1-\frac{\nu_2}{2}\right)\right]K_2(\alpha)+\frac{E_1}{2E_2}K_3(\alpha), \tag{4.84}$$

while

$$K_1(\alpha)=\int\frac{d\alpha}{(1+\gamma\sin^2\alpha)^2\sin\alpha},$$
$$K_2(\alpha)=\int\frac{d\alpha}{(1+\gamma\sin^2\alpha)\sin\alpha}, \tag{4.85}$$
$$K_3(\alpha)=\int\frac{d\alpha}{\sin\alpha}=\ln\tan\frac{\alpha}{2}.$$

Performing the substitution (4.26) we can write the indefinite integrals K_1 and K_2 in the form

$$K_1 = -\int \frac{dt}{(1+\gamma-\gamma t^2)^2(1-t^2)},$$
$$K_2 = -\int \frac{dt}{(1+\gamma-\gamma t^2)(1-t^2)}. \tag{4.86}$$

The integrals (4.86) can be expressed in a closed form for a general γ by a decomposition of their integrands into partial fractions.

In the case of a spherical surface, $\gamma=0$ and, according to (4.85),

$$K_1(\alpha)=K_2(\alpha)=K_3(\alpha). \tag{4.87}$$

In the case of a paraboloid of revolution, $\gamma=-1$ and

$$K_1(\alpha)=\int \frac{d\alpha}{\cos^4\alpha \sin\alpha}=\frac{1}{\cos\alpha}+\frac{1}{3\cos^3\alpha}+\ln\tan\frac{\alpha}{2},$$
$$K_2(\alpha)=\int \frac{d\alpha}{\cos^2\alpha \sin\alpha}=\frac{1}{\cos\alpha}+\ln\tan\frac{\alpha}{2}. \tag{4.88}$$

To obtain the formula for the radial displacement w we substitute (4.5) into (3.98):

$$w=-\frac{q_0R_0^2}{E_1}\left[\frac{1}{2}+\left(\frac{1}{2}-\frac{E_1}{E_2}\nu_2\right)\frac{1}{(1+\gamma\sin^2\alpha)^2}\right]+$$
$$+f(\alpha)+\cos\alpha[F(\alpha)+C]. \tag{4.89}$$

Here $f(\alpha)$ and $F(\alpha)$ are the functions defined by the equations (4.82) and (4.84). The integration constant C, occuring in équations (4.83) and (4.89) is equal to $-F(\alpha_1)$, where α_1 is the angle corresponding to the parallel circle along which the air-supported structure is anchored.

5

Elliptical, Spherical, Parabolic and Hyperbolic Surfaces

5.1 *Ellipsoid of Revolution*

a) *Equations of the Meridian and Principal Radii*

The meridian of an ellipsoid of revolution whose apex (pole) is identical with the origin O of the r, z coordinate system is described by equation (4.1), where $m=-1$.

Thus

$$r^2 = -\frac{a^2}{b^2} z^2 + \frac{2a^2}{b} z \tag{5.1}$$

and the principal radii are, according to (4.3),

$$\begin{aligned} R_1 &= \frac{1}{ab^4} [(a^2-b^2)z^2 - 2b(a^2-b^2)z + a^2b^2]^{3/2}, \\ R_2 &= \frac{a}{b^2} [(a^2-b^2)z^2 - 2b(a^2-b^2)z + a^2b^2]^{1/2}, \end{aligned} \tag{5.2}$$

where a, b are the lengths of the main demi-axes of the ellipsoid (Fig. II-7).

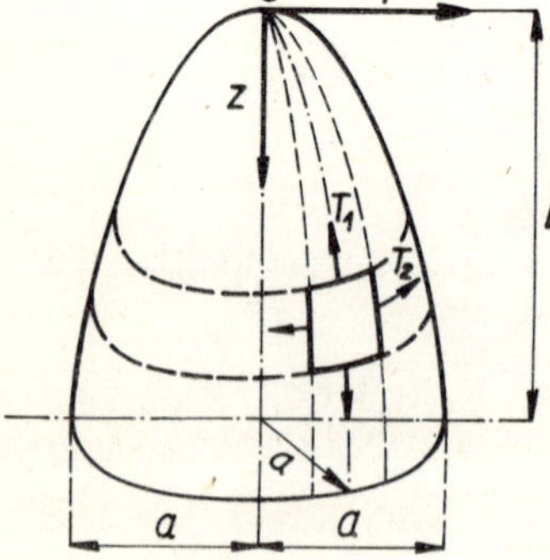

Fig. II-7. Ellipsoid of revolution.

At the apex of the surface, i.e. for $z=0$ we obtain from (5.2)

$$R_1(0)=R_2(0)=R_0=\frac{a^2}{b}. \tag{5.3}$$

The parametric equations of the meridian of the above ellipsoid of revolution, is obtained by a substitution of (5.3) and

$$\gamma=\frac{a^2}{b^2}-1 \tag{5.4}$$

into equations (4.9).

We have

$$\begin{aligned} r&=\frac{a^2}{b}\frac{\sin\alpha}{\left[1+\left(\frac{a^2}{b^2}-1\right)\sin^2\alpha\right]^{1/2}},\\ z&=b\left(1-\frac{\cos\alpha}{\left[1+\left(\frac{a^2}{b^2}-1\right)\sin^2\alpha\right]^{1/2}}\right) \end{aligned} \tag{5.5}$$

and, by (4.5), (5.3) and (5.4), the principal radii are

$$\begin{aligned} R_1&=\frac{a^2}{b\left[1+\left(\frac{a^2}{b^2}-1\right)\sin^2\alpha\right]^{3/2}},\\ R_2&=\frac{a^2}{b\left[1+\left(\frac{a^2}{b^2}-1\right)\sin^2\alpha\right]^{1/2}}. \end{aligned} \tag{5.6}$$

For $0>\gamma>-1$, according to (5.4), we have $a<b$, which means that the horizontal demi-axis is smaller than the vertical demi-axis. For $\gamma>0$ it is the other way round $(a>b)$.

b) *Internal Overpressure*

Setting $m=-1$ in (4.13), we obtain the following formulae for the unit membrane forces due to internal overpressure:

$$\begin{aligned} T_{1,q}&=\frac{q_0 a}{2b^2}\left[(a^2-b^2)z^2-2b(a^2-b^2)z+a^2b^2\right]^{1/2},\\ T_{2,q}&=T_{1,q}\{2-a^2b^2[(a^2-b^2)z^2-2b(a^2-b^2)z+a^2b^2]^{-1}\}. \end{aligned} \tag{5.7}$$

Considering that the meridian of an elliptic surface is determined by the parametric equations (5.5) and substituting the formulae (5.3) and (5.4) into (4.14), we obtain the following formulae for the above forces:

$$T_{1,q}=\frac{q_0a^2}{2b}\left[1+\left(\frac{a^2}{b^2}-1\right)\sin^2\alpha\right]^{-1/2},$$

$$T_{2,q}=\frac{q_0a^2}{2b}\frac{1-\left(\frac{a^2}{b^2}-1\right)\sin^2\alpha}{\left[1+\left(\frac{a^2}{b^2}-1\right)\sin^2\alpha\right]^{1/2}}. \tag{5.8}$$

At the apex of the surface, i.e. for $z=0$ or $\alpha=0$, we have, according to (5.7) or (5.8),

$$T_{1,q}(0)=T_{2,q}(0)=q_0\frac{a^2}{2b} \tag{5.9}$$

and for $z=b$ we obtain

$$T_{1,q}(b)=q_0\frac{a}{2},\quad T_{2,q}(b)=q_0\frac{a(2b^2-a^2)}{2b^2}. \tag{5.10}$$

From the conditions $\mathrm{d}T_{1,q}/\mathrm{d}z=0$ and $\mathrm{d}T_{2,q}/\mathrm{d}z=0$, or $\mathrm{d}T_{1,q}/\mathrm{d}\alpha=0$ and $\mathrm{d}T_{2,q}/\mathrm{d}\alpha=0$, it follows that the extreme values of the forces $T_{1,q}$ and $T_{2,q}$ are obtained for $z=0$ and $z=b$.

When $a<b$, the forces $T_{1,q}$ and $T_{2,q}$ attain their maximum for $z=b$ and their minimum for $z=0$.

An ellipsoid of revolution with $a>b\sqrt{2}$ is not suitable for an air-supported structure; indeed the second equation (5.10) yields in that case a negative membrane force ($T_{2,q}<0$) in the direction of the parallel circle, and this would result in the creasing of the whole membrane. For this reason we shall confine our further considerations to the case $a<b$, which corresponds, according to (5.4), to $0>\gamma>-1$.

c) *Dead Weight*

Setting $m=-1$ in the equations (4.22), we obtain

$$T_{1,g}=-gR_2\frac{b^2}{2bz-z^2}\left[\frac{b^2}{2a}I-(b-z)\frac{R_2}{2a^2}+\frac{1}{2}\right],$$

$$T_{2,g}=gR_2\left\{\frac{b^2}{2bz-z^2}\frac{R_2}{R_1}\left[\frac{b^2}{2a}I-(b-z)\frac{R_2}{2a^2}+\frac{1}{2}\right]-\frac{a^2}{b^2R_2}(b-z)\right\},\tag{5.11}$$

where R_1 and R_2 are determined by the equations (5.2). Here

$$I=-(b^2-a^2)^{-1/2}\left\{\arcsin\left[\frac{b-z}{b^2}(b^2-a^2)^{1/2}\right]-\arcsin\frac{1}{b}(b^2-a^2)^{1/2}\right\},\tag{5.12}$$

according to (4.21), as for $a<b$, by (4.16) and (4.17), we have $\Delta<0$ and $c<0$.

We can also obtain formulae for the same forces by substituting (5.3) and (5.4) in equation (4.37) and the second equation (4.25):

$$T_{1,g}=\frac{gb}{2\sin^2\alpha\left[1-\left(1-\frac{a^2}{b^2}\right)\sin^2\alpha\right]^{1/2}}\times$$
$$\times\left(\frac{b}{2a\left(1-\frac{a^2}{b^2}\right)^{1/2}}\left\{\arctan\left[\frac{b}{2a}\left(1-\frac{a^2}{b^2}\right)^{1/2}\cos\alpha\right]-\right.\right.$$
$$\left.\left.-\arctan\left[\frac{b}{2a}\left(1-\frac{a^2}{b^2}\right)^{1/2}\right]\right\}+\frac{\cos\alpha}{1-\left(1-\frac{a^2}{b^2}\right)\sin^2\alpha}-1\right),\tag{5.13}$$

$$T_{2,g}=-\frac{ga^2\cos\alpha}{b\left[1-\left(1-\frac{a^2}{b^2}\right)\sin^2\alpha\right]^{1/2}}-\left[1-\left(1-\frac{a^2}{b^2}\right)\sin^2\alpha\right]T_{1,g}.$$

d) *Symmetric Snow Load*

If we set $m=-1$ in the equations (4.44) and (4.45), we obtain the membrane forces due to symmetric snow load (3.28) in the part of the surface covered by snow:

$$T_{1,s}=-q_sR_2\frac{a^2b^2}{2bz-z^2}\left[\left(\frac{1}{a}-\frac{a}{2(b^2-a^2)}\right)I+\frac{2(bR_2-a^2)}{a^2(b^2-a^2)}-\frac{R_2}{2a^2(b^2-a^2)}z\right],$$

$$T_{2,s}=-q_sR_2\left[\frac{a^6}{R_2^3b^6}(b-z)^3+\frac{T_{1,s}}{q_sR_1}\right].\tag{5.14}$$

Here R_1 and R_2 are determined by (5.2) and I is given by (5.12).

In the lower part of the surface, which is not covered by snow, i.e. for $z > z_s$ ($\alpha > \pi/2$), we have, according to (4.46) and (4.47),

$$T_{1,s} = -q_s R_2(z) \frac{a^2 b^2}{2bz - z^2} \left[\left(\frac{1}{a} - \frac{a}{2(b^2 - a^2)} \right) I(z_s) + \frac{2[bR_2(z_s) - a^2]}{a^2(b^2 - a^2)} - \right.$$

$$\left. - \frac{R_2(z_s)}{2a^2(b^2 - a^2)} z_s \right],$$

$$T_{2,s} = -\frac{R_2(z)}{R_1(z)} T_{1,s}. \tag{5.15}$$

If we substitute (5.3) and (5.4) in equation (4.56) and the second equation (4.48), we obtain, for $\alpha < \pi/2$,

$$T_{1,s} = q_s \frac{a^2}{b} \frac{\left[1 - \left(1 - \frac{a^2}{b^2}\right) \sin^2 \alpha\right]^{1/2}}{2 \sin^2 \alpha} \times$$

$$\times \left[\frac{\cos \alpha}{\left(1 - \frac{a^2}{b^2}\right) \left[1 - \left(1 - \frac{a^2}{b^2}\right) \sin^2 \alpha\right]} - \frac{1}{1 - \frac{a^2}{b^2}} - \frac{b}{2a\left(1 - \frac{a^2}{b^2}\right)^{3/2}} \times \right.$$

$$\left. \times \left(\arctan \frac{b\left(1 - \frac{a^2}{b^2}\right)^{1/2} \cos \alpha}{2a} - \arctan \frac{b\left(1 - \frac{a^2}{b^2}\right)^{1/2}}{2a} \right) \right], \tag{5.16}$$

$$T_{2,s} = -q_s \frac{a^2}{b} \frac{\cos^3 \alpha}{\left[1 - \left(1 - \frac{a^2}{b^2}\right) \sin^2 \alpha\right]^{1/2}} - \left[1 - \left(1 - \frac{a^2}{b^2}\right) \sin^2 \alpha\right] T_{1,s}$$

and for $\alpha > \pi/2$, according to (4.57) and (4.58),

$$T_{1,s} = q_s \frac{a^2}{b} \frac{\left[1 - \left(1 - \frac{a^2}{b^2}\right) \sin^2 \alpha\right]^{1/2}}{2\left(1 - \frac{a^2}{b^2}\right) \sin^2 \alpha} \times$$

$$\times \left[-1 + \frac{b}{2a\left(1 - \frac{a^2}{b^2}\right)^{1/2}} \arctan \frac{b\left(1 - \frac{a^2}{b^2}\right)^{1/2}}{2a} \right],$$

$$T_{2,s} = -\left[1 - \left(1 - \frac{a^2}{b^2}\right) \sin^2 \alpha\right] T_{1,s}. \tag{5.17}$$

e) *One-Sided Snow Load*

According to the law of superposition the internal forces $\bar{T}_{1,s}$, $\bar{T}_{2,s}$, $\bar{T}_{12,s}$, due to one-sided snow load are

$$\bar{T}_{1,s}=\frac{T_{1,s}}{2}+T^{**}_{1,s},\quad \bar{T}_{2,s}=\frac{T_{2,s}}{2}+T^{**}_{2,s},\quad \bar{T}_{12,s}=T^{**}_{12,s},\tag{5.18}$$

where $T_{1,s}$ and $T_{2,s}$ are the forces obtained from the equations (5.14), (5.16) for $\alpha\leqq\pi/2$, or from the equations (5.15), (5.17) for $\alpha>\pi/2$. For $\alpha\leqq\pi/2$ we can determine the forces $T^{**}_{1,s}$, $T^{**}_{2,s}$ and $T^{**}_{12,s}$ from the equations (4.62), taking into account (5.3) and (5.4),

$$T^{**}_{1,s}=-q_s\frac{a^2}{2b}\,\frac{1-\left(1-\frac{a^2}{b^2}\right)\sin^2\alpha}{\sin^3\alpha}\,\bar{I}_s(\alpha)\cos\beta,$$

$$T^{**}_{2,s}=-q_s\frac{a^2}{2b}\left\{\frac{\cos^3\alpha}{\left[1-\left(1-\frac{a^2}{b^2}\right)\sin^2\alpha\right]^{1/2}}-\frac{\left[1-\left(1-\frac{a^2}{b^2}\right)\sin^2\alpha\right]^2}{\sin^3\alpha}\,\bar{I}_s(\alpha)\right\}\times$$

$$\times\cos\beta,\tag{5.19}$$

$$T^{**}_{12,s}=T^{**}_{1,s}\cos\alpha\tan\beta.$$

Here

$$\bar{I}_s(\alpha)=\int_0^\alpha\frac{\cos^2\alpha\sin^2\alpha\,d\alpha}{\left[1-\left(1-\frac{a^2}{b^2}\right)\sin^2\alpha\right]^{5/2}}.\tag{5.20}$$

For $\alpha>\pi/2$ we have, according to (4.65),

$$T^{**}_{1,s}=-q_s\frac{a^2}{2b}\,\frac{1-\left(1-\frac{a^2}{b^2}\right)\sin^2\alpha}{\sin^3\alpha}\,\bar{I}_s(\pi/2)\cos\beta,$$

$$T^{**}_{2,s}=-q_s\frac{a^2}{2b}\,\frac{\left[1-\left(1-\frac{a^2}{b^2}\right)\sin^2\alpha\right]^2}{\sin^3\alpha}\,\bar{I}_s(\pi/2)\cos\beta,\tag{5.21}$$

$$T^{**}_{12,s}=T^{**}_{1,s}\cos\alpha\tan\beta,$$

where $\bar{I}_s(\pi/2)$ is the value of the integral (5.20) with the upper limit $\alpha=\pi/2$.

f) *Wind Load*

The unit membrane forces T_1, T_2 and T_{12} due to wind load, variable along the surface according to (3.44), can be obtained by the superposition of the magnitudes of the forces determined by the formulae (4.79) and (4.80). For an elliptical surface

$$T_1 = q_0^* \frac{a^2}{2b} \frac{1}{1-\left(1-\frac{a^2}{b^2}\sin^2\alpha\right)^{1/2}} +$$

$$+ q_v \frac{a^2}{b} \frac{1-\left(1-\frac{a^2}{b^2}\sin^2\alpha\right)}{\sin^3\alpha} [I_{2,v}(\alpha) - I_{3,v}(\alpha)] \cos\beta,$$

$$T_2 = q_0^* \frac{a^2}{2b} \frac{1+\left(1-\frac{a^2}{b^2}\right)\sin^2\alpha}{1-\left(1-\frac{a^2}{b^2}\right)\sin^2\alpha)^{1/2}} + q_v \frac{a^2}{b} \left\{ \frac{\sin\alpha}{\left[1-\left(1-\frac{a^2}{b^2}\right)\sin^2\alpha\right]^{1/2}} - \right.$$

$$\left. - \frac{\left[1-\left(1-\frac{a^2}{b^2}\right)\sin^2\alpha\right]^2}{\sin^3\alpha} [I_{2,v}(\alpha) - I_{3,v}(\alpha)] \right\} \cos\beta, \qquad (5.22)$$

$$T_{12} = q_v \frac{a^2}{b} \left\{ \frac{\left[1-\left(1-\frac{a^2}{b^2}\right)\sin^2\alpha\right]\cos\alpha}{\sin^3\alpha} [I_{2,v}(\alpha) - I_{3,v}(\alpha)] + \right.$$

$$\left. + \frac{I_{1,v}(\alpha)}{\left[1-\left(1-\frac{a^2}{b^2}\right)\sin^2\alpha\right]^{1/2}\sin\alpha} \right\} \cos\beta,$$

where

$$I_{1,v}(\alpha) = \frac{b^3}{4a^3\left(1-\frac{a^2}{b^2}\right)^{1/2}} \left\{ \arctan\left[\frac{b}{2a}\left(1-\frac{a^2}{b^2}\right)^{1/2}\cos\alpha\right] - \right.$$

$$\left. - \arctan\left[\frac{b}{2a}\left(1-\frac{a^2}{b^2}\right)^{1/2}\right] \right\} - \frac{b^2}{2a^2} \frac{\cos\alpha}{1-\left(1-\frac{a^2}{b^2}\right)\sin^2\alpha} + \frac{b^2}{2a^2},$$

$$I_{2,v}(\alpha)=\frac{1}{\left(1-\frac{a^2}{b^2}\right)^2}\left[-\frac{1}{\left[1-\left(1-\frac{a^2}{b^2}\right)\sin^2\alpha\right]^{1/2}}+\frac{1}{3\left[1-\left(1-\frac{a^2}{b^2}\right)\sin^2\alpha\right]^{3/2}}+\frac{2}{3}\right], \tag{5.23}$$

$$I_{3,v}(\alpha)=\int_0^\alpha \frac{\Psi(\alpha)\sin\alpha\, d\alpha}{1-\left(1-\frac{a^2}{b^2}\right)\sin^2\alpha},$$

and

$$\Psi(\alpha)=\frac{b^3}{4a^3\left(1-\frac{a^2}{b^2}\right)^{1/2}}\arctan\left[\frac{b}{2a}\left(1-\frac{a^2}{b^2}\right)^{1/2}\cos\alpha\right]-\frac{b^2}{2a^2}\frac{\cos\alpha}{1-\left(1-\frac{a^2}{b^2}\right)\sin^2\alpha}. \tag{5.24}$$

5.2 *Spherical Surface*

a) *Equations of the Meridian*

The basic equations for the spherical surface can be obtained as a special case of the equations deduced for an elliptical surface in the preceding subsection. If $a=b$, we obtain

$$r^2=2az-z^2, \tag{5.25}$$

where a is the radius of the spherical surface, as

$$a=R_1=R_2=R_0 \tag{5.26}$$

according to (5.2) and (5.3).

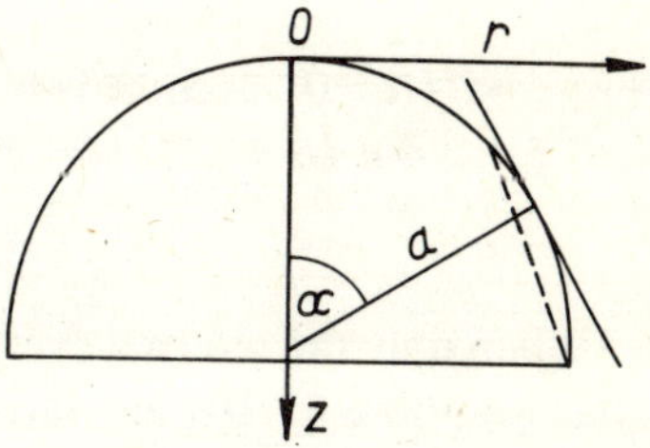

Fig. II-8. Spherical surface.

By (5.4), in the case of a spherical surface, we have $\gamma=0$, and the parametric equations of the meridian are, according to (5.5), (Fig. II-8)

$$r=a\sin\alpha\,,\quad z=a(1-\cos\alpha)\,. \tag{5.27}$$

b) *Internal Overpressure*

The unit membrane forces due to an internal overpressure q_0 are, according to (3.16) or (5.7) or (5.8),

$$T_{1,q}=T_{2,q}=\frac{1}{2}\,q_0 a\,. \tag{5.28}$$

c) *Dead Weight and the Necessary Internal Overpressure*

For $a=b$ an indeterminate form of the type 0/0 occurs in equations (5.11), (5.13); therefore, we deduce the necessary expressions for $T_{1,g}$ and $T_{2,g}$ directly from the equations (3.25), (3.27) or (3.82), (3.83), where we substitute for q_n and q_1 the expressions (3.23) and set $R_1=R_2=a$.

Thus we obtain

$$\begin{aligned} T_{1,g}&=-ga^2\,\frac{z}{2az-z^2}\,,\\ T_{2,g}&=ga\left(\frac{az}{2az-z^2}-1+\frac{z}{a}\right) \end{aligned} \tag{5.29}$$

or

$$\begin{aligned} T_{1,g}&=-ga\,\frac{1-\cos\alpha}{\sin^2\alpha}\,,\\ T_{2,g}&=ga\left(\frac{1-\cos\alpha}{\sin^2\alpha}-\cos\alpha\right). \end{aligned} \tag{5.30}$$

For $z=0$ $(\alpha=0)$, according to (5.29) or (5.30), $T_{1,g}(0)=T_{2,g}(0)=-ga/2$ and for $z=a$ $(\alpha=\pi/2)$ we obtain

$$T_{1,g}(a)=-ga\,,\quad T_{2,g}(a)=ga\,. \tag{5.31}$$

When only the internal overpressure q_0 and the dead weight g are applied, the conditions (2.10) are satisfied for

$$T_1 = T_{1,q} + T_{1,g} \geqq 0, \tag{5.32}$$

whence we obtain, after the substitution of (5.28) and the first equation (5.30),

$$q_0 > 2g \frac{1 - \cos\alpha}{\sin^2\alpha}. \tag{5.33}$$

This condition has to be satisfied by the internal overpressure to ensure that creasing of the membrane does not occur. According to (5.33), in an air-supported structure of hemispherical shape, the internal overpressure must be $q_0 > 2g$, since the expression on the right-hand side of inequality (5.33) attains its maximum $2g$ for $\alpha = \pi/2$. In the case of a structure consisting of a spherical surface anchored along a parallel circle $\alpha = 3/4\pi$ the maximum value $2(2+\sqrt{2})g$ of this expression is found at the anchorage; therefore, we have to ensure that $q_0 > 2(2+\sqrt{2})g \approx 6.83\, g$.

The spherical surface of a radome is usually bordered by a horizontal plane $z = 1.5a$ at the bottom, where the membrane force $T_{1,g}$ satisfies $T_{1,g} = -2ga$, according to (5.32). In this case inequality (5.32) yields for the necessary internal overpressure the condition $q_0 \geqq 4g$. According to the second equation (5.27), the parallel circle in the plane $z = 1.5a$ corresponds to the angle $\alpha = \arccos(-0.5) = 120° = 2/3\pi$.

d) *Symmetric Snow Load and the Necessary Internal Overpressure*

For $m = -1$ and $a = b$, according to (4.38),

$$r^2 r'^2 = (a - z)^2. \tag{5.34}$$

Subsituting (5.25), (5.26) and (5.34) into the formulae (3.32) and (3.34), we obtain for $z \leqq a$ $(\alpha < \pi/2)$ the expressions

$$T_{1,s} = -\frac{q_s}{2az - z^2}\left(a^2 z - az^2 + \frac{1}{3} z^3\right), \tag{5.35}$$

$$T_{2,s} = -q_s \frac{(a-z)^3}{a^2} - T_{1,s} = -q_s\left[\frac{(a-z)^3}{a^2} + \frac{a^2 z - az^2 + z^3/3}{2az - z^2}\right].$$

From the equations (4.48) we obtain for $\alpha \leqq \pi/2$, by setting $\gamma = 0$ and integrating, for the same membrane forces, the formulae

$$T_{1,s} = -\frac{q_s a}{3} \frac{1-\cos^3 \alpha}{\sin^2 \alpha},$$
$$T_{2,s} = -q_s a \cos^3 \alpha - T_{1,s} = \frac{q_s a}{3}\left(\frac{1-\cos^3 \alpha}{\sin^2 \alpha} - 3\cos^3 \alpha\right). \tag{5.36}$$

For $z > a$, i.e. for $\alpha > \pi/2$, we have, according to (3.35) and (3.36),

$$T_{1,s} = -\frac{q_s a^3}{3(2az - z^2)}, \quad T_{2,s} = -T_{1,s} = \frac{q_s a^3}{3(2az - z^2)} \tag{5.37}$$

or, according to (4.48), by taking $a = \pi/2$ for the upper limit of the integral and after the elimination of the first term of the expression for $T_{2,s}$,

$$T_{1,s} = -\frac{q_s a}{3 \sin^2 \alpha}, \quad T_{2,s} = -T_{1,s} = \frac{q_s a}{3 \sin^2 \alpha}. \tag{5.38}$$

For $z = a$ ($\alpha = \pi/2$), according to (5.35) and (5.36) respectively, the unit membrane forces have the magnitudes $T_{1,s} = -q_s a/3$, $T_{2,s} = q_s a/3$ and for $z = 1{,}5a$, according to (5.37), $T_{1,s} = -4/9 q_s a$, $T_{2,s} = 4/9 q_s a$.

The necessary internal overpressure which prevents the creasing of the membrane can be deduced, once again, from the conditions (2.10). If we consider the loads due to internal overpressure, dead weight and symmetric snow load, then $T_{12} = 0$ and, therefore, the inequalities (2.10) are satisfied when

$$T_1 = T_{1,q} + T_{1,g} + T_{1,s} > 0, \quad T_2 = T_{2,q} + T_{2,g} + T_{2,s} > 0. \tag{5.39}$$

As the forces $T_{2,g}$ and $T_{2,s}$ are mostly positive and attain negative values for small angles α only, when they are relatively small in comparison with the negative values of the forces $T_{1,g}$ and $T_{1,s}$, it is the first inequality (5.39) that is decisive for the calculation of the necessary internal overpressure. If we substitute into it the respective expressions (5.28), (5.29) and (5.35) or (5.30) and (5.36), we obtain

$$q_0 \geqq \frac{2}{2az - z^2}\left[ga^2 z + q_s\left(a^2 z - az^2 + \frac{1}{3} z^3\right)\right], \tag{5.40}$$

or

$$q_0 \geqq \frac{2a}{\sin^2 \alpha}\left[g(1-\cos \alpha) + \frac{q_s}{3}(1-\cos^3 \alpha)\right] \tag{5.41}$$

respectively.

The right-hand sides of the inequalities (5.40) and (5.41) acquire their maximum value $g+q_s$ for $z=0$ $(\alpha=0)$, therefore, we obtain for the spherical air-supported structure the necessary internal overpressure

$$q_0 \geqq g+q_s. \tag{5.42}$$

e) *One-Sided Snow Load and the Necessary Internal Overpressure*

According to (3.77) and (5.36), the membrane forces $T^*_{1,s}$ and $T^*_{2,s}$ on a spherical surface due to the symmetric components (3.75) of one-sided snow load are for $\alpha \leqq \pi/2$

$$\begin{aligned} T^*_{1,s} &= -\frac{q_s a}{6}\frac{1-\cos^3\alpha}{\sin^2\alpha}, \\ T^*_{2,s} &= \frac{q_s a}{6}\left(\frac{1-\cos^3\alpha}{\sin^2\alpha}-3\cos^3\alpha\right), \end{aligned} \tag{5.43}$$

and for $\alpha>\pi/2$, by (5.38),

$$T^*_{1,s} = -\frac{q_s a}{6\sin^2\alpha}, \quad T^*_{2,s} = -T^*_{1,s} = \frac{q_s a}{6\sin^2\alpha}. \tag{5.44}$$

The membrane forces $T^{**}_{1,s}$, $T^*_{2,s}$ and $T^{**}_{12,s}$, due to the antisymmetric components (3.76) of one-sided snow load, are determined by the expressions (4.62), where $\gamma=0$, $R_0=a$ and $\bar{I}_s(\alpha)$ is the integral (4.63), given by [21]

$$\bar{I}_s(\alpha) = -\frac{1}{8}\left(\frac{\sin 4\alpha}{4}-\alpha\right) = \frac{1}{32}(4\alpha-\sin 4\alpha). \tag{5.45}$$

According to (3.87), the resulting membrane forces due to one-sided snow load are determined for $\alpha \leqq \pi/2$ by the sum of the expressions (5.43) and (4.62),

$$\begin{aligned} T_{1,s} &= -\frac{q_s a}{2\sin^2\alpha}\left(\frac{1-\cos^3\alpha}{3}+\frac{1}{32}\frac{4\alpha-\sin 4\alpha}{\sin\alpha}\cos\beta\right), \\ T_{2,s} &= \frac{q_s a}{2}\left[\frac{1-\cos^3\alpha}{3\sin^2\alpha}-\cos^3\alpha-\left(\cos^3\alpha-\frac{1}{32}\frac{4\alpha-\sin 4\alpha}{\sin^3\alpha}\right)\cos\beta\right], \\ T_{12,s} &= -\frac{q_s a}{64}\frac{4\alpha-\sin 4\alpha}{\sin^3\alpha}\cos\alpha\sin\beta, \end{aligned} \tag{5.46}$$

and for $\alpha>\pi/2$, they are, by (4.65), (5.44) and (3.87) $(\bar{I}_s(\pi/2)=\pi/16)$,

$$T_{1,s}=-\frac{q_s a}{2\sin^2\alpha}\left(\frac{1}{3}+\frac{\pi}{16\sin\alpha}\cos\beta\right),$$

$$T_{2,s}=-T_{1,s}=\frac{q_s a}{2\sin^2\alpha}\left(\frac{1}{3}+\frac{\pi}{16\sin\alpha}\cos\beta\right), \tag{5.47}$$

$$T_{12,s}=-q_s a\frac{\pi\cos\alpha}{32\sin^3\alpha}\sin\beta.$$

For $\alpha=0$ some terms appearing in the equations (5.46) become indeterminate expressions of the type 0/0. The use of de l'Hospital's rule yields for $\beta=0$ and $\alpha=0$ the following values of the membrane forces:

$$T_{1,s}(0,0)=-\frac{7}{12}q_s a,\quad T_{2,s}(0,0)=-\frac{1}{12}q_s a,\quad T_{12,s}(0,0)=0,$$

and for $\beta=0$ and $\alpha=\pi/2$ we obtain

$$T_{1,s}\left(0,\frac{\pi}{2}\right)\approx-0.36q_s a,\quad T_{2,s}\left(0,\frac{\pi}{2}\right)=-1.14q_s a,\quad T_{12,s}\left(0,\frac{\pi}{2}\right)=0.$$

The highest positive values of membrane forces $T_{1,s}$ and $T_{2,s}$ are obtained for $\beta=\pi$.

The tangential force $T_{12,s}$ is of greatest magnitude in the plane of $\beta=\pi/2$. For $\alpha=0$ and $\beta=\pi/2$ we have $T_{12,s}(0,\pi/2)=-q_s a/96$ and for $\alpha=\pi/2$, $\beta=\pi/2$ we obtain

$$T_{12,s}\left(\frac{\pi}{2},\frac{\pi}{2}\right)=-\frac{\pi}{32}q_s a.$$

The necessary internal overpressure q_0 for a spherical air-supported structure with one-sided snow load can be obtained from the inequalities (2.10). If the very low dead weight of the membrane $g=10\ \mathrm{N/m^2}$ is neglected, the resulting membrane forces T_1, T_2 and T_{12} due to the internal overpressure and one-sided snow load are

$$T_1=T_{1q}+T_{1,s},\quad T_2=T_{2q}+T_{2,s},\quad T_{12}=T_{12,s}. \tag{5.48}$$

If we substitute (5.28) into (5.48) and then into (2.10), we obtain the following inequalities for the necessary internal overpressure:

$$q_0\geqq-\frac{1}{a}(T_{1,s}+T_{2,s})+\frac{1}{a}\sqrt{(T_{1,s}-T_{2,s})^2+4T_{12,s}^2},$$

$$q_0\geqq-\frac{2T_{1,s}}{a},\quad q_0\geqq-\frac{2T_{2,s}}{a} \tag{5.49}$$

From the second inequality (5.49) we obtain, using (5.46) that, for $\alpha \leqq \pi/2$,

$$q_0 \geqq q_s \left(\frac{1-\cos^3 \alpha}{3 \sin^2 \alpha} + \frac{1}{32} \frac{4\alpha - \sin 4\alpha}{\sin^3 \alpha} \cos \beta \right) \tag{5.50}$$

and for $\alpha \geqq \pi/2$, by (5.47),

$$q_0 \geqq q_s \left(\frac{1}{2 \sin^2 \alpha} + \frac{\pi}{16 \sin^3 \alpha} \cos \beta \right). \tag{5.51}$$

The right-hand side of the inequality (5.50) attains its extreme value for $\beta = 0$ and $\alpha = 0$. The application of de l'Hospital's rule yields

$$q_0 \geqq \frac{7}{6} q_s \tag{5.52}$$

and, according to (5.51), for $\beta = 0$, we have

$$q_0 \geqq q_s \left(\frac{1}{3 \sin^2 \alpha} + \frac{\pi}{16 \sin^3 \alpha} \right). \tag{5.53}$$

For $\alpha < 3/4\pi$ the inequality (5.52) is the one which decides about the magnitude of the necessary internal overpressure. If the spherical membrane is anchored along a parallel where $\alpha > 3/4\pi$, then inequality (5.53) must be used for the calculation of the necessary internal overpressure.

For the internal overpressure q_0, which satisfies both inequalities (5.52) and (5.53), all conditions of (5.49) are complied with; consequently, the membrane does not crease in any part of the spherical surface.

If we consider, apart from the one-sided snow load, also the dead weight of the membrane, we obtain instead of (5.52) and (5.53), using (5.30), the following approximate criteria for the magnitude of the necessary internal overpressure

$$q > \frac{7}{6} q_s + g,$$

$$q_0 > q_s \left(\frac{1}{3 \sin^2 \alpha} + \frac{\pi}{16 \sin^3 \alpha} \right) + 2g \frac{1-\cos \alpha}{\sin^2 \alpha}. \tag{5.54}$$

The inequality which yields the higher internal overpressure is always decisive.

In the case of a spherical membrane anchored along the parallel circle

$\alpha = 3/4\pi$ the second inequality (5.54) yields the condition for the necessary internal overpressure

$$q_0 > 1.22 q_s + 6.83\ g\,.$$

f) *Wind Load and the Necessary Internal Overpressure*

The membrane forces due to the antisymmetric component (3.45) of the wind load are determined by the formulae (3.46), where the quantities S_1 and S_{12} can be determined by solving the equations (3.48) which, for the spherical surface, acquire the form $(R_1 = R_2 = a)$

$$\begin{aligned} &\frac{\mathrm{d}S_1}{\mathrm{d}\alpha} + 2\cot\alpha\, S_1 + \frac{1}{\sin\alpha}\, S_{12} = q_v a \cos\alpha\,, \\ &\frac{\mathrm{d}S_{12}}{\mathrm{d}\alpha} + 2\cot\alpha\, S_{12} + \frac{1}{\sin\alpha}\, S_1 = q_v a\,. \end{aligned} \tag{5.55}$$

If we take the sum and the difference of the equations (5.55) and introduce the symbols [57]

$$Z_1 = S_1 + S_{12}\,, \quad Z_2 = S_1 - S_{12}\,, \tag{5.56}$$

we obtain for the functions Z_1 and Z_2 the following differential equations:

$$\begin{aligned} &\frac{\mathrm{d}Z_1}{\mathrm{d}\alpha} + \left(2\cot\alpha + \frac{1}{\sin\alpha}\right) Z_1 = q_v a (1 + \cos\alpha)\,, \\ &\frac{\mathrm{d}Z_2}{\mathrm{d}\alpha} + \left(2\cot\alpha - \frac{1}{\sin\alpha}\right) Z_2 = -q_v a (1 - \cos\alpha)\,. \end{aligned} \tag{5.57}$$

If we denote

$$\begin{aligned} &P_1 = 2\cot\alpha + \frac{1}{\sin\alpha}\,, \qquad && Q_1 = q_v a (1 + \cos\alpha)\,, \\ &P_2 = 2\cot\alpha - \frac{1}{\sin\alpha}\,, \qquad && Q_2 = -q_v a (1 - \cos\alpha)\,, \end{aligned} \tag{5.58}$$

the general solutions of equations (5.57) have the form [54]

$$Z_i = \mathrm{e}^{-\int P_i \,\mathrm{d}\alpha} \left(C_i + \int Q_i \mathrm{e}^{\int P_i \,\mathrm{d}\alpha}\right), \qquad i = 1, 2\,, \tag{5.59}$$

where C_i, $i = 1, 2$, are integration constants.

Since for $0<\alpha<\pi$

$$\int P_1\,\mathrm{d}\alpha = 2\ln\sin\alpha + \ln\tan\frac{\alpha}{2} = \ln[\sin\alpha(1-\cos\alpha)],$$
$$\int P_2\,\mathrm{d}\alpha = \ln[\sin\alpha(1+\cos\alpha)] \tag{5.60}$$

and

$$\int Q_1 e^{\int P_1\,\mathrm{d}\alpha} = q_v a\int\sin^3\alpha\,\mathrm{d}\alpha - q_v a\cos\alpha\left(1-\frac{\cos^2\alpha}{3}\right),$$
$$\int Q_2 e^{\int P_2\,\mathrm{d}\alpha} = -q_v a\int\sin^3\alpha\,\mathrm{d}\alpha = q_v a\cos\alpha\left(1-\frac{\cos^2\alpha}{3}\right), \tag{5.61}$$

we obtain, substituting (5.60) and (5.61) into (5.59), the following equalities:

$$Z_1 = \frac{1+\cos\alpha}{\sin^3\alpha}\left[C_1 - q_v a\left(\cos\alpha - \frac{1}{3}\cos^3\alpha\right)\right],$$
$$Z_2 = \frac{1-\cos\alpha}{\sin^3\alpha}\left[C_2 + q_v a\left(\cos\alpha - \frac{1}{3}\cos^3\alpha\right)\right]. \tag{5.62}$$

If we substitute (5.62) into (5.56), we obtain, using (3.62) and (3.65), the membrane forces T_1 and T_{12} due to the antisymmetric component of the load,

$$T_{1,v} = \frac{\cos\beta}{\sin^3\alpha}\left[\frac{C_1+C_2}{2} + \frac{C_1-C_2}{2}\cos\alpha - q_v a\left(\cos^2\alpha - \frac{1}{3}\cos^4\alpha\right)\right],$$
$$T_{12,v} = \frac{\sin\beta}{\sin^3\alpha}\left[\frac{C_1-C_2}{2} + \frac{C_1+C_2}{2}\cos\alpha - q_v a\left(\cos^2\alpha - \frac{1}{3}\cos^4\alpha\right)\right]. \tag{5.63}$$

The integration constants C_1 and C_2 can be determined from the conditions of the equilibrium of forces applied to a part of the spherical surface, e.g. a hemispherical surface. If $\alpha=\pi/2$, then, according to (5.63),

$$T_{1,v}(\pi/2,\beta) = \frac{C_1+C_2}{2}\cos\beta,\quad T_{12,v}(\pi/2,\beta) = \frac{C_1-C_2}{2}\sin\beta. \tag{5.64}$$

The equation of equilibrium of moments of forces relative to the line of intersection of the planes $\alpha=\pi/2$ and $\beta=\pi/2$, i.e. relative to the diameter perpendicular to the wind direction, has the form

$$\int_0^{2\pi} T_{1,v}(\pi/2,\beta)a^2\cos\beta\,\mathrm{d}\beta = a^2\frac{C_1+C_2}{2}\int_0^{2\pi}\cos^2\beta\,\mathrm{d}\beta = 0, \tag{5.65}$$

as the moment due to the radial wind load with regard to this intersection line equals zero.

The condition of equilibrium of the components of internal forces and external load (3.45) in the wind direction, i.e. in the direction parallel to the plane $\beta=0$, has the form

$$\int_0^{2\pi} T_{12,v}(\pi/2, \beta) a \sin\beta \, d\beta - \int_{(F)} q_{an} \sin\alpha \cos\beta \, dF = 0, \qquad (5.66)$$

where

$$dF = r\, d\beta . a\, d\alpha = a^2 \sin\alpha \, d\alpha \, d\beta .$$

Substitution of (3.45) and the second formula in (5.64) into (5.66) yields

$$a\pi \frac{C_1 - C_2}{2} = \int_0^{2\pi} d\beta \int_0^{\pi/2} q_v a^2 \sin^3\alpha \cos^2\beta \, d\alpha = -\frac{2}{3}\pi q_v a^2 . \qquad (5.67)$$

From equation (5.65) we obtain

$$C_1 = -C_2 \qquad (5.68)$$

whence, by (5.67),

$$C_1 = \frac{2}{3} q_v a, \quad C_2 = -\frac{2}{3} q_v a . \qquad (5.69)$$

If we substitute (5.69) into (5.63) then, using the third equation (3.47), the formula (3.46) for T_2 and the formula (5.27) for r, we obtain

$$T_{1,v} = \frac{1}{3} q_v a \frac{\cos\alpha}{\sin^3\alpha} (2 - 3\cos\alpha + \cos^3\alpha) \cos\beta ,$$

$$T_{2,v} = -\frac{q_v a}{3\sin^3\alpha} (2\cos\alpha - 3\sin^2\alpha - 2\cos^4\alpha) \cos\beta , \qquad (5.70)$$

$$T_{12,v} = \frac{q_v a}{3\sin^3\alpha} (2 - 3\cos\alpha + \cos^3\alpha) \sin\beta .$$

The above formulae (5.70) can be deduced in an easier way from the equations (3.71).

The membrane forces due to the symmetric component q_0^* of the wind load are, according to (3.16),

$$T_1^* = T_2^* = \frac{1}{2} q_0^* a, \quad T_{12}^* = 0. \qquad (5.71)$$

The resulting membrane forces $T_{1,R}$, $T_{2,R}$, $T_{12,R}$ due to wind load, approximated by the expression (3.44), are obtained by the superposition of the mutually corresponding quantities of (5.70) and (5.71),

$$T_{1,R}=T_{1,v}+T_1^*, \quad T_{2,R}=T_{2,v}+T_2^*, \quad T_{12,R}=T_{12,v}. \tag{5.72}$$

Let us note that the quantity q_0^* is positive, if it represents suction, and negative, if it represents pressure. The quantity q_v is negative, as it always represents the maximum value of the antisymmetric components of wind pressure applied to the windward side of the surface.

The conditions (2.10) for the necessary internal overpressure under the static wind load applied to the spherical surface, with regard to (5.28), (5.71) and (5.72), have the form

$$\left(\frac{1}{2}q_0a+\frac{1}{2}q_0^*a+T_{1,v}\right)\left(\frac{1}{2}q_0a+\frac{1}{2}q_0^*a+T_{2,v}\right)\geqq T_{12,v}^2,$$
$$q_0\geqq -q_0^*-2\frac{T_{1,v}}{a}, \quad q_0\geqq -q_0^*-2\frac{T_{2,v}}{a}. \tag{5.73}$$

If α is in the interval $0\leqq\alpha\leqq\pi/2$, the third inequality (5.73), taken for $\beta=0$, is decisive for the calculation of the magnitude of the internal overpressure [47]. If we substitute in it the formula (5.70) for $T_{2,v}$ with $\beta=0$, we obtain the following inequality for the minimum required internal overpressure

$$q_0\geqq q_0^*+\frac{2q_v}{3\sin^3\alpha}(2\cos\alpha-3\sin^2\alpha-2\cos^4\alpha). \tag{5.74}$$

For $\alpha>\pi/2$ inequality (5.74) can be used as an approximate criterion for the calculation of the necessary internal overpressure.

For example, if

$$q_0^*=-0.3q_v, \quad q_v\approx 0.613v_0^2, \tag{5.75}$$

where v_0 is the wind velocity in m/sec and q_v is the dynamic wind pressure in N/m^2, we obtain, applying (5.74) with $\alpha=\pi/2$, i.e. for a hemispherical surface,

$$q_0\geqq 1.7q_v\approx 1.04v_0^2. \tag{5.76}$$

For an air-supported structure anchored along the parallel circle at $\alpha=2/3\pi=120°$ the necessary internal overpressure, according to (5.74) and (5.75), has to satisfy

$$q_0 \geqq -3.3 q_v \approx 2.02 v_0^2. \tag{5.77}$$

For a spherical surface anchored along the parallel circle at $\alpha = \pi/3 = 60°$ we obtain

$$q_0 \geqq 1.1 q_v \approx 0.67 v_0^2. \tag{5.78}$$

For $v_0 = 30$ m/sec ($v_0 = 108$ km/h) we have, according to (5.76), ($\alpha = \pi/2$) the inequality $q_0 \geqq 936$ N/m^2, according to (5.77), ($\alpha = 2/3\pi$) the condition $q_0 \geqq 1\,818$ N/m^2, and according to (5.78), ($\alpha = \pi/3$) the inequality $q_0 \geqq 603$ N/m^2.

The above criteria for the magnitude of the necessary internal overpressure are excessively strict, as the membrane does not crease in any place, if they are observed. However, an air-supported structure may be geometrically as wel as aerodynamically stable, if the creasing of the membrane is of local character only. If we permit, for example, in the case of a hemispherical surface, creasing of the membrane on the windward side such that the deformed surface is part of an inclined plane showed by dashed lines in Fig. II-8, the necessary internal overpressure q_0 equals the dynamic wind pressure whose magnitude on the windward side near the anchorage (at the point $\alpha = \pi/2$, $\beta = 0$) is

$$q(\pi/2, 0) = 0.7 \cdot 0.613 v_0^2 = 0.43 v_0^2.$$

Thus the necessary internal overpressure is

$$q_0 \geqq q(\pi/2, 0) = 0.43 v_0^2 \tag{5.79}$$

This yields for $v_0 = 30$ m/sec the inequality $q_0 \geqq 387$ N/m^2, and this value is more than 2.4 times smaller than value obtained from condition (5.76). The criterion (5.79) was verified experimentally.

According to the Japanese standard [30] the wind pressure q is calculated from the equation

$$q = 0.613 v_{max}^2 \sqrt[4]{\frac{h}{10}}, \tag{5.80}$$

where q is in N/m^2, v_{max} (m/sec) is the maximum design wind velocity and h (m) is the height of the structure above ground level. For a hemispherical surface the necessary internal overpressure is stipulated as $q_0 \geqq 0.7q$ and for a three-quarter spherical shape as $q_0 \geqq q$ or $q_0 \geqq 0.8q$, if major membrane deformations are permitted.

5.3 *Paraboloid of Revolution*

a) *Equations of the Meridian and Principal Radii*

If we have in (4.1) $m=0$, $a=R_0/\sqrt{2}$ and $b=R_0/2$, we obtain in the coordinates r, z the equation of a quadratic parabola (Fig. II-9)

$$r^2=2R_0z$$

and the principal radii are, according to (4.3),

$$R_1=R_0\left(1+2\frac{z}{p}\right)^{3/2}, \quad R_2=R_0\left(1+2\frac{z}{p}\right)^{1/2}. \tag{5.81}$$

At the apex of the surface, i.e. for $z=0$, according to (5.81), $R_1(0)=$ $=R_2(0)=R_0=p$.

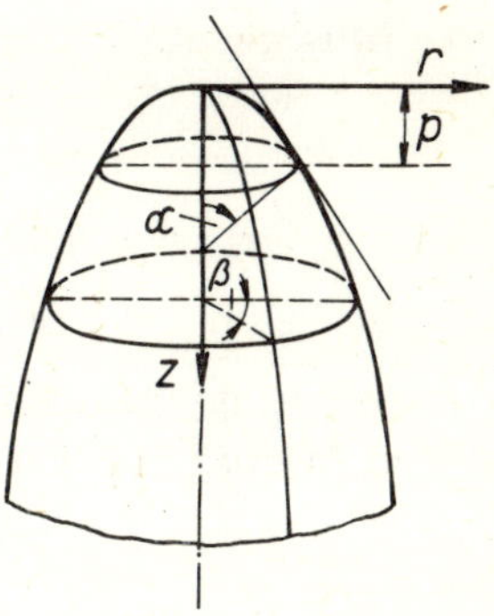

Fig. II-9. Paraboloid of revolution.

The parametric equations of the meridian of a paraboloid of revolution have the form (4.12)

$$r=R_0\tan\alpha, \quad z=\frac{1}{2}R_0\tan^2\alpha \tag{5.82}$$

and the principal radii, according to (4.5) for $\gamma=-1$, are determined by the following functions of α:

$$R_1=\frac{R_0}{\cos^3\alpha}, \quad R_2=\frac{R_0}{\cos\alpha}. \tag{5.83}$$

b) *Internal Overpressure*

Substituting (5.81) into (3.16), or using (4.13), we obtain

$$T_{1,q}=\frac{1}{2}q_0R_0(1+2z/p)^{1/2},\quad T_{2,q}=\frac{1}{2}q_0R_0\frac{1+4z/p}{(1+2z/p)^{1/2}}. \tag{5.84}$$

Applying the formulae (5.83), we obtain

$$T_{1,q}=\frac{q_0R_0}{2\cos\alpha},\quad T_{2,q}=\frac{1}{2}q_0R_0\frac{1+\sin^2\alpha}{\cos\alpha}. \tag{5.85}$$

c) *Dead Weight*

The integral in the first formula (4.25) where $\gamma=-1$ equals

$$J^*=\int_0^\alpha \frac{\sin\alpha}{\cos^4\alpha}\,d\alpha=\frac{1}{3\cos^3\alpha}-\frac{1}{3},$$

and, consequently,

$$T_{1,g}=-gR_0\frac{1-\cos^3\alpha}{3\sin^2\alpha\cos^4\alpha},$$
$$T_{2,g}=-gR_0\left(1-\frac{1-\cos^3\alpha}{3\sin^2\alpha\cos^2\alpha}\right). \tag{5.86}$$

Applying de l'Hospital's rule at the apex of the surface, i.e. for $\alpha=0$, we obtain from (5.86) that $T_{1,g}(0) = T_{2,g}(0) = -gR_0/2$.

d) *Symmetric Snow Load*

The integral in the first formula (4.48) for $\gamma=-1$ equals

$$J_s=\int_0^\alpha \frac{\sin\alpha}{\cos^2\alpha}\,d\alpha=\frac{1}{\cos\alpha}-1$$

and, consequently,

$$T_{1,s}=-q_sR_0\frac{1-\cos\alpha}{\sin^2\alpha},$$
$$T_{2,s}=-q_sR_0\cos^2\alpha\left(1-\frac{1-\cos\alpha}{\sin^2\alpha}\right). \tag{5.87}$$

These formulae are valid for any part of the surface, as in the case of a paraboloid of revolution we allways have $\alpha<\pi/2$.

e) *One-Sided Snow Load*

The integral (4.64) is given by [21]

$$\bar{I}_s(\alpha)=\frac{\sin\alpha}{2\cos^2\alpha}-\frac{1}{2}\ln\tan\left(\frac{\pi}{4}-\frac{\alpha}{2}\right). \tag{5.88}$$

If we substitute $\gamma=-1$ and (5.88) into (4.62), we obtain the following membrane forces due to the antisymmetric load component:

$$T^{**}_{1,\,s}=-\frac{q_s R_0}{4\sin^2\alpha}\left[1-\frac{\cos^2\alpha}{\sin\alpha}\ln\tan\left(\frac{\pi}{4}+\frac{\alpha}{2}\right)\right]\cos\beta,$$

$$T^{**}_{2,\,s}=-\frac{q_s R_0}{2}\cos^2\alpha\left\{1-\frac{1}{2\sin^2\alpha}\left[1-\frac{\cos^2\alpha}{\sin\alpha}\ln\tan\left(\frac{\pi}{4}+\frac{\alpha}{2}\right)\right]\right\}\cos\beta, \tag{5.89}$$

$$T_{12,\,s}=T^{**}_{1,\,s}\cos\alpha\tan\beta.$$

The membrane forces due to the overall one-sided snow load can be obtained from equations (5.18) by substituting there (5.87) and (5.89).

f) *Wind Load*

For $\gamma=-1$ the integrals (4.68) and (4.74) are given by

$$I_{1,\,v}=I_{2,\,v}=\int_0^\alpha\frac{\sin^3\alpha}{\cos^4\alpha}\,d\alpha=-\frac{1}{\cos\alpha}+\frac{1}{3\cos^3\alpha}+\frac{2}{3},$$

$$I_{3,\,v}=-\int_0^\alpha\frac{\sin\alpha}{\cos^4\alpha}\,d\alpha+\frac{1}{3}\int_0^\alpha\frac{\sin\alpha}{\cos^6\alpha}\,d\alpha=-\frac{1}{3\cos^3\alpha}+\frac{1}{15\cos^5\alpha}+\frac{4}{15}. \tag{5.90}$$

Substituting $\gamma=-1$ and (5.90) into (4.79), we obtain

$$T_{1,\,v}=q_v R_0\frac{\cos^2\alpha}{\sin^3\alpha}\left(\frac{6}{15}-\frac{1}{\cos\alpha}+\frac{2}{3\cos^3\alpha}-\frac{1}{15\cos^5\alpha}\right)\cos\beta,$$

$$T_{2,v} = q_v R_0 \left[\tan\alpha - \frac{\cos^4\alpha}{\sin^3\alpha}\left(\frac{6}{15} - \frac{1}{\cos\alpha} + \frac{2}{3\cos^3\alpha} - \frac{1}{15\cos^5\alpha}\right)\right]\cos\beta,$$

$$T_{12,v} = q_v R_0 \left[\cot^3\alpha\left(\frac{6}{15} - \frac{1}{\cos\alpha} + \frac{2}{3\cos^3\alpha} - \frac{1}{15\cos^5\alpha}\right) + \right. \tag{5.91}$$

$$\left. + \frac{1}{\sin\alpha\cos\alpha}\left(\frac{2}{3} - \frac{1}{\cos\alpha} + \frac{1}{3\cos^3\alpha}\right)\right]\sin\beta.$$

According to (4.80),

$$T^*_{1,v} = \frac{q^*_0 R_0}{2\cos\alpha},$$

$$T^*_{2,v} = \frac{q^*_0 R_0}{2}\frac{1+\sin^2\alpha}{\cos\alpha}, \tag{5.92}$$

$$T^*_{12,v} = 0.$$

The membrane forces due to the overall wind load can be determined according to the equations (4.81), combined with (5.91) and (5.92).

5.4 *Two-Sheet Hyperboloid of Revolution*

a) *Eguations of the Meridian and Principal Radii*

If $m=1$, we obtain from (4.1) the equation of the meridian of a two-sheet hyperboloid of revolution (Fig. II-10)

$$r^2 = \frac{a^2}{b^2}(2bz + z^2). \tag{5.93}$$

According to (4.3), its principal radii are

$$R_1 = \frac{1}{ab^4}\left[(a^2+b^2)z^2 + 2b(a^2+b^2)z + a^2b^2\right]^{3/2},$$

$$R_2 = \frac{a}{b^4}\left[(a^2+b^2)z^2 + 2b(a^2+b^2)z + a^2b^2\right]^{1/2}, \tag{5.94}$$

where a, b are the parameters of the hyperbola. At the apex of the surface, i.e. for $z=0$, we obtain from (5.94) for the principal radii the expression (5.3).

Fig. II-10. A two-sheet hyperboloid of revolution.

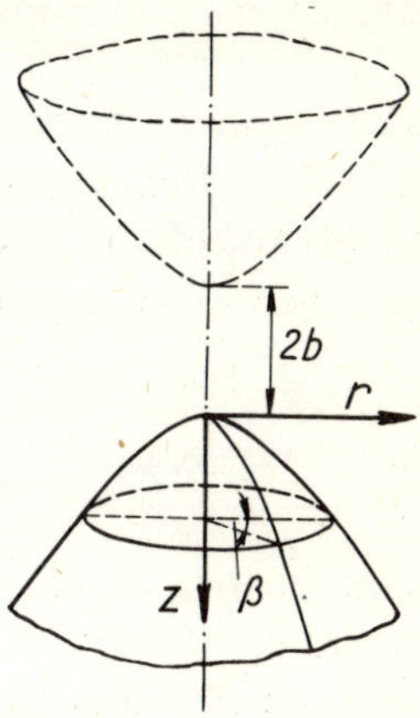

The parametric equations of the meridian have the form (4.9), where $\gamma < -1$ and the principal radii depend on the angle α, according to (4.5).

b) *Internal Overpressure*

Formulae (4.13) yield for $m=1$

$$T_{1,q}=\frac{q_0 a}{2b^2}[(a^2+b^2)z^2+2b(a^2+b^2)z+a^2b^2]^{1/2},$$
$$T_{2,q}=T_{1,q}\{2-a^2b^2[(a^2+b^2)z^2+2b(a^2+b^2)z+a^2b^2]^{-1}\}. \quad (5.95)$$

For $z=0$ we obtain

$$T_{1,q}(0)=T_{2,q}(0)=\frac{a^2}{2b}q_0.$$

Instead of (5.95) we can use the equations (4.14) for the calculation of the membrane forces due to the internal overpressure q_0.

c) *Dead Weight*

The membrane forces due to the dead weight can be determined according to (4.36) and from the second equation (4.25) or from equation (4.22), where the integral I has to be evaluated as in (4.20).

d) *Symmetric Snow Load*

The membrane force $T_{1,s}$ can be determined from the formula (4.55) and the force $T_{2,s}$ from the second equation (4.48); both can be calculated from (4.44) and (4.45), where I has to be evaluated as in (4.20). Let us note that for this surface we allways have $\alpha<\pi/2$.

e) *One-Sided Snow Load*

The resulting membrane forces due to this load can be determined from the equations (5.18), where for $T_{1,s}$ and $T_{2,s}$ we have to use the formulae determined in accordance with the preceding paragraph d), and $T_{1,s}^{**}$, $T_{2,s}^{**}$ and $T_{12,s}^{**}$ have to be evaluated as in (4.62).

f) *Wind Load*

Internal forces due to wind load are calculated according to the equations (4.81). The integrals in (4.79) are replaced by the expressions (4.71), (4.78) and formula (4.74) is used for $I_{3,v}(\alpha)$, where

$$\Psi(\alpha)=\frac{1}{4\gamma^{3/2}(1+\gamma)^{3/2}}\operatorname{arctanh}\frac{-\gamma\cos\alpha}{2\sqrt{\gamma(1+\gamma)}}-\frac{1+2\gamma}{2\gamma(1+\gamma)}\frac{\cos\alpha}{1+\gamma\sin^2\alpha}. \tag{5.96}$$

6

Surfaces of Revolution with Zero Gaussian Curvature

6.1 *Conical Surface*

a) *Equations of the Meridian and Principal Radii*

The meridian of a conical surface is a straight line described by the equation

$$r = z \cot \alpha , \tag{6.1}$$

where $\alpha = \text{const}$ (Fig. II-11).

The principal radii are

$$R_1 = \infty , \quad R_2 = \frac{r}{\sin \alpha} = z \frac{\cos \alpha}{\sin^2 \alpha} . \tag{6.2}$$

b) *Internal Overpressure*

Substitution of (6.2) into (3.16) yields

$$T_{1,q} = q_0 z \frac{\cos \alpha}{2 \sin^2 \alpha} , \quad T_{2,q} = q_0 z \frac{\cos \alpha}{\sin^2 \alpha} . \tag{6.3}$$

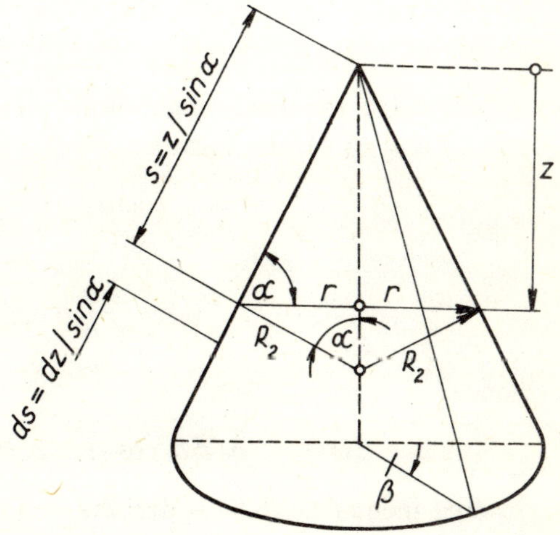

Fig. II-11. Conical surface.

c) *Dead Weight and the Necessary Internal Overpressure*

Substituting (6.1) and (6.2) into (3.25) and (3.27), we obtain

$$T_{1,g} = -gz\frac{1}{2\sin^2\alpha}, \quad T_{2,g} = -gz\cot^2\alpha. \tag{6.4}$$

By (6.3) and (6.4), conditions (2.10) take the form

$$\begin{aligned} q_0 z\frac{\cos\alpha}{2\sin^2\alpha} - gz\frac{1}{2\sin^2\alpha} &\geqq 0, \\ q_0 z\frac{\cos\alpha}{\sin^2\alpha} - gz\cot^2\alpha &\geqq 0, \end{aligned} \tag{6.5}$$

whence we obtain for the necessary internal overpressure

$$q_0 \geqq \frac{g}{\cos\alpha}, \quad q_0 \geqq g\cos\alpha. \tag{6.6}$$

The first inequality (6.6) is decisive as it yields a higher internal overpressure.

d) *Symetric Snow Load and the Necessary Internal Overpressure*

Substituting (6.1) and (6.2) in the formulae (3.32) and (3.34), we get

$$T_{1,s} = -q_s z\frac{\cot^2\alpha}{2}, \quad T_{2,s} = -q_s z\frac{\cos^4\alpha}{\sin^2\alpha}. \tag{6.7}$$

If we neglect the dead weight of the membrane, conditions (2.10) acquire, by (6.3) and (6.7), the form

$$\begin{aligned} q_0 z\frac{\cos\alpha}{2\sin^2\alpha} - q_s z\frac{\cot^2\alpha}{2} &\geqq 0, \\ q_0 z\frac{\cos\alpha}{\sin^2\alpha} - q_s z\frac{\cos^4\alpha}{\sin^2\alpha} &\geqq 0, \end{aligned} \tag{6.8}$$

whence

$$q_0 \geqq q_s\cos\alpha, \quad q_0 \geqq q_s\cos^3\alpha. \tag{6.9}$$

The first inequality (6.9) is decisive.

If we consider, apart from the snow load q_s, also the dead weight of the membrane g, we obtain the following condition for the necessary internal overpressure:

$$q_0 \geqq q_s \cos\alpha + \frac{g}{\cos\alpha}. \tag{6.10}$$

e) *One-Sided Snow Load and the Necessary Internal Overpressure*

According to Fig. II-11,

$$ds = R_1\, d\alpha = \frac{dz}{\sin\alpha}. \tag{6.11}$$

The integral in the formulae (3.86), can be evaluated with the use of (6.2) and (6.11),

$$\int_0^\alpha R_1 R_2^2 \cos^2\alpha \sin^2\alpha\, d\alpha = \frac{\cos^4\alpha}{\sin^3\alpha}\int_0^z z^2\, dz = \frac{\cos^4\alpha}{\sin^3\alpha}\frac{z^3}{3}. \tag{6.12}$$

Substituting (6.2) and (6.12) into (3.86), we obtain

$$\begin{aligned} T^{**}_{1,s} &= -q_s z \frac{\cot^2\alpha}{6}\cos\beta, \\ T^{**}_{2,s} &= -q_s z \frac{\cos^4\alpha}{2\sin^2\alpha}\cos\beta, \\ T^{**}_{12,s} &= -q_s z \frac{\cos^3\alpha}{6\sin^2\alpha}\sin\beta, \end{aligned} \tag{6.13}$$

The symmetric load component, according to (3.77) and (6.7), produces the forces

$$T^{*}_{1,s} = -q_s z \frac{\cot^2\alpha}{4}, \quad T^{*}_{2,s} = -q_s z \frac{\cos^4\alpha}{2\sin^2\alpha}, \tag{6.14}$$

whence the resulting membrane forces due to one-sided snow load are, according to (3.87),

$$\begin{aligned} T_{1,s} &= -q_s z \frac{\cot^2\alpha}{2}\left(\frac{1}{2} + \frac{1}{3}\cos\beta\right), \\ T_{2,s} &= -q_s z \frac{\cos^4\alpha}{2\sin^2\alpha}(1 + \cos\beta), \\ T_{12,s} &= T^{**}_{12,s}. \end{aligned} \tag{6.15}$$

If we consider the internal overpressure and one-sided snow load without the dead weight of the membrane, the resulting membrane forces T_1, T_2 are determined by the sum of the forces (6.3) and (6.15), and $T_{12} = T_{12,s}$.

It is greatest practical importance that the second inequality (2.10) should hold $\beta = 0$, i.e.

$$q_0 z \frac{\cos \alpha}{2 \sin^2 \alpha} \geqq \frac{5}{12} q_s z \cot^2 \alpha,$$

whence we obtain the condition

$$q_0 \geqq \frac{5}{6} \cos \alpha \tag{6.16}$$

for the necessary internal overpressure.

f) *Wind Load and the Necessary Internal Overpressure*

The membrane forces due to the symmetric component q_0^* are determined by the expressions of the form (6.3),

$$T_{1,v}^* = q_0^* z \frac{\cos \alpha}{2 \sin^2 \alpha}, \quad T_{2,v}^* = q_0^* z \frac{\cos \alpha}{\sin^2 \alpha}. \tag{6.17}$$

The membrane forces due to the antisymmetric component (3.45) are determined from the equations (3.71).

Substituting (6.2) and (6.11) into (3.72), we obtain

$$\Phi_v = q_v z^2 \frac{\cos^3 \alpha}{\sin^2 \alpha} - q_s \sin \alpha \int_0^z z \, dz = q_v z^2 \left(\frac{\cos^3 \alpha}{\sin^2 \alpha} - \frac{\sin \alpha}{2} \right),$$
$$f_v = q_v z \frac{\sin^2 \alpha}{\cos \alpha}. \tag{6.18}$$

If we substitute (6.11) and (6.18) into (3.71), we get

$$T_{1,v} = \frac{q_v z}{6} (2 \cot \alpha - \tan \alpha) \cos \beta,$$
$$T_{2,v} = q_v z \cot \alpha \cos \beta, \tag{6.19}$$
$$T_{12,v} = q_v z \frac{\sin \beta}{3 \sin \alpha}.$$

The equations for the resulting membrane forces due to wind load can be obtained by the superposition of (6.17) and (6.19), according to (4.81).

If we substitute (6.3), (6.17) and the formulae (6.19) for $T_{1,v}$ and $T_{2,v}$ with $\beta=0$ into the last two inequalities (2.10), we obtain

$$q_0 z \frac{\cos\alpha}{2\sin^2\alpha} + q_0^* z \frac{\cos\alpha}{2\sin^2\alpha} + \frac{q_v z}{6}(2\cot\alpha - \tan\alpha) \geqq 0,$$
$$q_0 z \frac{\cos\alpha}{\sin^2\alpha} + q_0^* z \frac{\cos\alpha}{\sin^2\alpha} + q_v z \cot\alpha \geqq 0, \quad (6.20)$$

which implies the following conditions for the necessary internal overpressure:

$$q_0 \geqq -q_0^* - \frac{q_v}{3}\left(2\sin\alpha - \frac{\sin^3\alpha}{\cos^2\alpha}\right),$$
$$q_0 \leqq -q_0^* - q_v \sin\alpha\,. \quad (6.21)$$

While it may happen that the internal overpressure of an air-supported structure satisfies both conditions (6.21), it need not satisfy, for certain angles α, the first condition (2.10). But this does not mean that the air-supported structure does not satisfy the requirements of shape and aerodynamic stability.

An analogous method can be used also for the determination of the necessary internal overpressure for any combination of simultaneously acting loads, e.g. for wind load and symmetric or one-sided snow load, taking into account the dead weight of the membrane, if required.

g) *The Necessary Internal Overpressure for Combined Loads*

The membrane forces T_1 and T_2 due to one-sided snow load, the dead weight of the membrane and internal overpressure satisfy

$$T_1 = T_{1,s} + T_{1,g} + T_{1,q}, \quad T_2 = T_{2,s} + T_{2,g} + T_{2,q}. \quad (6.22)$$

The necessary internal overpressure can be determined by means of the inequality $T_1 \geqq 0$ only, which, by (6.22), has the form

$$T_{1,s} + T_{1,g} + T_{1,q} \geqq 0. \quad (6.23)$$

Substituting (6.3), (6.4) and the first expression (6.15) with $\beta=0$ into (6.23), we obtain

$$-q_s z \frac{5 \cot^2 \alpha}{12} - gz \frac{1}{2 \sin^2 \alpha} + q_0 z \frac{\cos \alpha}{2 \sin^2 \alpha} \geqq 0, \tag{6.24}$$

whence we deduce for the necessary internal overpressure

$$q_0 \geqq \frac{g}{\cos \alpha} + \frac{5}{6} q_s \cos \alpha . \tag{6.25}$$

Inequality (6.25) implies a larger value for q_0 than the condition $T_2 \geqq 0$.

A comparison of (6.10) and (6.25) yields the conclusion that for the same intensity of snow load q_s the symmetric snow load is decisive for the magnitude of the internal overpressure.

According to (6.10) we obtain, for example for $\alpha=30°$,

$$q_0 \geqq 2g + q_s/2 . \tag{6.26}$$

Under simultaneous action of the internal overpressure, the dead weight of the membrane, symmetric snow load and wind load, the membrane force T_1 is

$$T_1 = T_{1,q} + T_{1,g} + T_{1,s} + T_{1,v}^* + T_{1,v} . \tag{6.27}$$

Substituting (6.3), (6.4), (6.7), (6.17) and the expression (6.19) with $\beta=0$ into (6.27), we can write the inequality $T_1 \geqq 0$ in the form

$$q_0 z \frac{\cos \alpha}{2 \sin^2 \alpha} - gz \frac{1}{2 \sin^2 \alpha} - q_s z \frac{\cot^2 \alpha}{2} + q_0^* z \frac{\cos \alpha}{2 \sin^2 \alpha} + \\ + \frac{q_v z}{6} (2 \cot \alpha - \tan \alpha) \geqq 0, \tag{6.28}$$

whence we obtain for the necessary internal overpressure

$$q_0 \geqq \frac{g}{\cos \alpha} + q_s \cos \alpha - q_0^* - \frac{q_v}{3} \left(2 \sin \alpha - \frac{\sin^3 \alpha}{\cos^2 \alpha} \right), \tag{6.29}$$

where the quantity q_v has to be negative, as q_v is the dynamic wind pressure acting in the direction from the surface towards its centres of curvature (see Subsection 3.4).

6.2 *Cylindrical Surface with a Vertical Axis*

a) *Internal Overpressure and Dead Weight*

In the case of a cylindrical surface with a vertical axis, terminating in a rigid slab at its upper end (Fig. II-12a), we have

$$R_1=\infty, \quad r=R_2=R=\text{const}, \quad \alpha=\pi/2. \tag{6.30}$$

The vertical part of the cylindrical surface limited by the section $z=\text{const}$ is loaded by the internal overpressure q_0, the dead weight of the membrane g and the dead weight of the rigid slab g_d (in N/m^2, for example). From the equation of equilibrium of internal and external forces applied to that part of the surface we obtain

$$T_1=\frac{1}{2}q_0R-\frac{1}{2}g_dR-gz \tag{6.31}$$

and, according to (3.16), by (6.30),

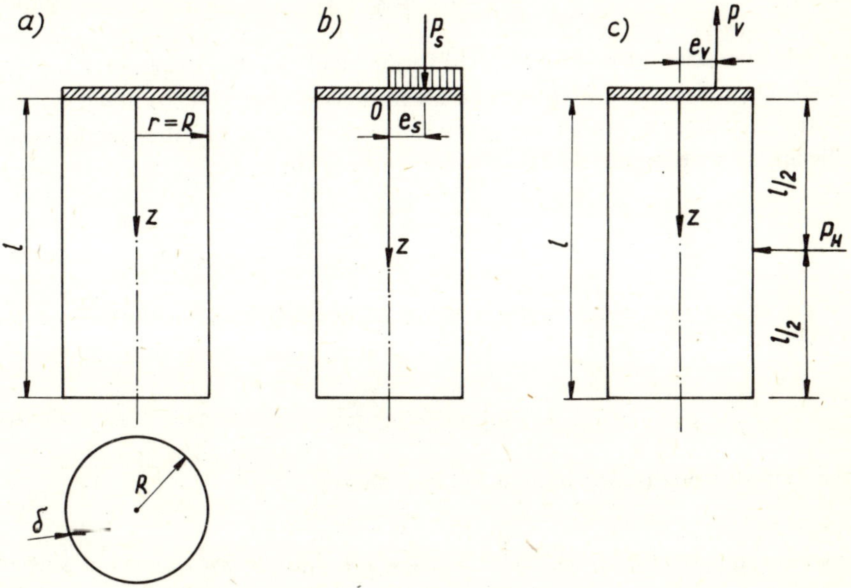

Fig. II-12. a) Cylindrical surface with a vertical axis. b) One-sided snow load. c) Vertical and horizontal resultants of wind load.

$$T_2 = q_0 R. \tag{6.32}$$

If we substitute (6.31) into the inequality $T_1 \geqq 0$ and set $z = l$, we obtain the following condition for the necessary internal overpressure:

$$q_0 \geqq g_d + 2g \frac{l}{R},$$

where l is the height of the cylindrical surface.

b) *Snow Load*

For the snow load q_s, uniformly distributed along the surface of the rigid slab,

$$T_{1,s} = -\frac{1}{2} q_s R, \quad T_{2,s} = 0 \tag{6.33}$$

and, in wiew of (6.31), the necessary internal overpressure satisfies

$$q_0 \geqq g_d + 2g \frac{l}{R} + q_s. \tag{6.34}$$

For one-sided snow load applied to the rigid slab the resultant P_s of this load acts in the plane $\beta = 0$ with an eccentricity e_s (Fig. II-12b). If we assume that the cylindrical surface consists of an elastic membrane, the magnitude of the largest membrane force on the side of the point of application of the force P_s is

$$\min T_{1,s} = -\frac{P_s}{2\pi R} + \min \sigma_s \delta, \tag{6.35}$$

where σ_s is the stress due to the bending moment $P_s e_s$ and δ is the membrane thickness. If δ is small in comparison with R, then

$$\min \sigma_s = -\frac{P_s e_s}{\pi R^2 \delta} \tag{6.36}$$

and, substituting (6.36) into (6.35), we obtain

$$\min T_{1,s} = -\frac{P_s}{\pi R}\left(\frac{1}{2} + \frac{e_s}{R}\right) \tag{6.37}$$

and

$$T_{2,s} = 0.$$

If we consider a uniformly distributed one-sided snow load q_s applied to one half of the circular rigid slab, then

$$P_s = \frac{q_s}{2} \pi R^2, \quad e_s = \frac{3}{4\pi} R \tag{6.38}$$

and, substituting (6.38) into (6.37), we obtain

$$\min T_{1,s} \approx -0{,}4622 q_s R. \tag{6.39}$$

The necessary internal overpressure with one-sided snow load is determined from the inequality

$$q_0 \geqq g_d + 2g \frac{l}{R} + 0{,}9244 q_s. \tag{6.40}$$

c) *Wind Load*

Setting $R_1 = \infty$, $R_2 = R$ and substituting the expression (3.44) with $\alpha = \pi/2$ into the last equation (2.6), we obtain

$$T_2 = q_\delta^* R + q_v R \cos \beta. \tag{6.41}$$

The largest negative membrane force T_1 on the leeward side at a point of the fastening of the membrane can be determined approximately from the bending stress of a cantilever affected by the vertical resultant P_v and the horizontal resultant P_H of wind pressure and wind suction. According to Fig. II-12c, the highest membrane compressive force is

$$\min T_1 = \min \sigma_v \delta = -\frac{P_v e_v + P_H l/2}{\pi R^2} + \frac{P_v}{2\pi R}, \tag{6.42}$$

where

$$P_H = 2q_\delta^* Rl \int_0^\pi \cos \beta \, d\beta - 2q_v R \int_0^\pi \cos^2 \beta \, d\beta = -\pi q_v Rl. \tag{6.43}$$

To obtain the inequalities for the necessary internal overpressure we have to substitute the sum of (6.31) and (6.42) with $z = l$ and the sum of (6.32) and (6.41) with $\beta = 0$ into the last two conditions (2.10), and next apply (6.43):

$$\begin{aligned} q_0 &\geqq g_d + 2g \frac{l}{R} - q_v \frac{l^2}{R^2} + \frac{2P_v e_v}{\pi R^3} - \frac{P_v}{\pi R^2}, \\ q_0 &\geqq -q_\delta^* - q_v. \end{aligned} \tag{6.44}$$

6.3 *Cylindrical Surface with a Horizontal Axis*

a) *Equations of Equilibrium and Their Solution*

In the coordinates x, ϑ the element of length ds on a circular cylindrical surface is given by (Fig. II-13)

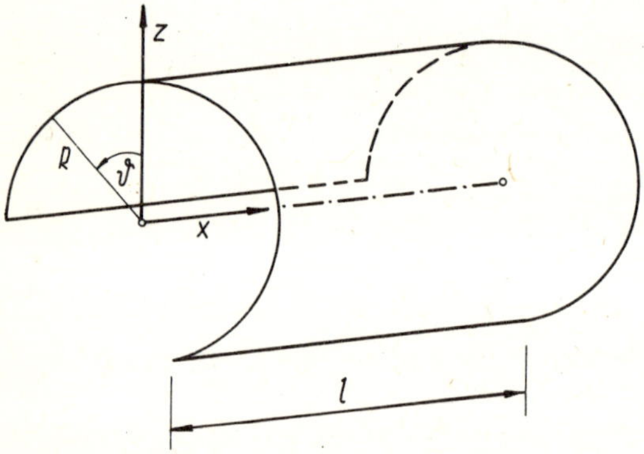

Fig. II-13. Cylindrical surface with a horizontal axis.

$$ds^2 = dx^2 + R^2\, d\vartheta^2. \tag{6.45}$$

Lamé's parameters and the principal radii are

$$\begin{aligned} A_1 &= 1, \qquad & A_2 &= R, \\ R_1 &= \infty, \qquad & R_2 &= R. \end{aligned} \tag{6.46}$$

If we substitute (6.46) into (2.6) and set $\xi = x$, $\eta = \vartheta$, we obtain the equations of equilibrium

$$\begin{aligned} R\frac{\partial T_1}{\partial x} + \frac{\partial T_{12}}{\partial \vartheta} + q_1 R &= 0, \\ R\frac{\partial T_{12}}{\partial x} + \frac{\partial T_2}{\partial \vartheta} + q_2 R &= 0, \\ T_2 &= q_n R, \end{aligned} \tag{6.47}$$

where q_1 is the external load component in the x-direction, q_2 is its component in the direction of a tangent, and q_n is the radial load.

From the last two equations we obtain

$$T_{12}=C_1(\vartheta)-\int\left(q_2+\frac{\partial q_n}{\partial\vartheta}\right)\mathrm{d}x. \tag{6.48}$$

Substituting (6.48) into the first equation (6.47) and integrating, we get

$$T_1=C_2(\vartheta)-\frac{x}{R}\frac{\mathrm{d}C_1}{\mathrm{d}\vartheta}+\int\left[-q_1+\frac{1}{R}\int\left(\frac{\partial q_2}{\partial\vartheta}+\frac{\partial^2 q_n}{\partial\vartheta^2}\right)\mathrm{d}x\right]\mathrm{d}x, \tag{6.49}$$

where $C_1(\vartheta)$ and $C_2(\vartheta)$ are arbitrary functions of the angle ϑ.

In the case when the load components q_2 and q_n are independent of x and $q_1=0$, we have

$$\begin{aligned} T_2&=q_nR,\\ T_{12}&=C_1(\vartheta)-\left(q_2+\frac{\mathrm{d}q_n}{\mathrm{d}\vartheta}\right)x,\\ T_1&=C_2(\vartheta)-\frac{x}{R}\frac{\mathrm{d}C_1}{\mathrm{d}\vartheta}+\left(\frac{\mathrm{d}q_2}{\mathrm{d}\vartheta}+\frac{\mathrm{d}^2q_n}{\mathrm{d}\vartheta^2}\right)\frac{x}{2R}. \end{aligned} \tag{6.50}$$

For boundary conditions symmetric with respect to the plane $x=l/2$, where l is the length of the cylindrical membrane, we have $T_{12}(l/2)=0$ and, consequently,

$$C_1(\vartheta)=\left(q_2+\frac{\mathrm{d}q_n}{\mathrm{d}\vartheta}\right)\frac{l}{2}. \tag{6.51}$$

Substituting (6.51) into (6.50), we get

$$\begin{aligned} T_2&=q_nR,\\ T_{12}&=\left(q_2+\frac{\mathrm{d}q_n}{\mathrm{d}\vartheta}\right)\left(1-2\frac{x}{l}\right)\frac{l}{2},\\ T_1&=C_2(\vartheta)-\left(\frac{\mathrm{d}q_2}{\mathrm{d}\vartheta}+\frac{\mathrm{d}^2q_n}{\mathrm{d}\vartheta^2}\right)\left[\frac{x}{l}-\left(\frac{x}{l}\right)^2\right]\frac{l^2}{2R}. \end{aligned} \tag{6.52}$$

The equations for the function $C_2(\vartheta)$ for various load types are given in Subsection 6.3g.

b) *Internal Overpressure*

If we set $q_n=q_0$, $q_2=0$ in the equations (6.52), we obtain

$$T_{2,q}=q_0R,\quad T_{12}=0,\quad T_{1,q}=C_{2,q}(\vartheta). \tag{6.53}$$

c) *Dead Weight*

The load components due to the dead weight of the membrane g are

$$q_1=0, \quad q_2=g \sin \vartheta, \quad q_n=-g \cos \vartheta. \tag{6.54}$$

Substitution of (6.54) into (6.52) yields

$$\begin{aligned} T_{2,g} &= -gR \cos \vartheta, \\ T_{12,g} &= gl\left(1-2\frac{x}{l}\right) \sin \vartheta, \\ T_{1,g} &= C_{2,g}(\vartheta)-\frac{gl^2}{R}\left[\frac{x}{l}-\left(\frac{x}{l}\right)^2\right] \cos \vartheta. \end{aligned} \tag{6.55}$$

d) *Symmetric Snow Load*

By (3.30), we have

$$q_1=0, \quad q_2=q_s \sin \vartheta \cos^2 \vartheta, \quad q_n=-q_s \cos^3 \vartheta. \tag{6.56}$$

Substitution of (6.56) into (6.52) yields for $\vartheta \leqq \pi/2$

$$\begin{aligned} T_{2,s} &= -q_s R \cos^3 \vartheta, \\ T_{12,s} &= 2q_s l\left(1-2\frac{x}{l}\right) \sin \vartheta \cos^2 \vartheta, \\ T_{1,s} &= C_{2,s}(\vartheta)-2q_s\frac{l^2}{R}\left[\frac{x}{l}-\left(\frac{x}{l}\right)^2\right] \cos \vartheta(\cos^2 \vartheta-2 \sin^2 \vartheta). \end{aligned} \tag{6.57}$$

For $\vartheta>\pi/2$, we have $q_2=q_n=0$ and, consequently,

$$T_{2,s}=T_{12,s}=0, \quad T_{1,s}=C_{2,s}(\vartheta). \tag{6.58}$$

a) *One-Sided Snow Load*

If we write the components of this load in the form

$$q_1=0,$$
$$q_2=\frac{q_s}{2}\sin\vartheta\cos^2\vartheta(1+\sin\vartheta), \qquad (6.59)$$
$$q_n=-\frac{q_s}{2}\cos^3\vartheta(1+\sin\vartheta),$$

we obtain from the equations (6.52) the following membrane forces for $\vartheta\leqq\pi/2$:

$$T_{2,s}=-\frac{q_sR}{2}\cos^3\vartheta(1+\sin\vartheta),$$
$$T_{12,s}=q_sl\left(1-2\frac{x}{l}\right)\cos^2\vartheta\left[\sin\vartheta(1+\sin\vartheta)-\frac{1}{4}\cos^2\vartheta\right], \qquad (6.60)$$
$$T_{1,s}=C^*_{2,s}(\vartheta)-\frac{q_sl^2}{R}\left[\frac{x}{l}-\left(\frac{x}{l}\right)^2\right]\cos\vartheta\times$$
$$\times[(1+\sin\vartheta)(\cos^2\vartheta-2\sin^2\vartheta)+2\sin\vartheta\cos^2\vartheta].$$

For $\vartheta>\pi/2$, we have

$$T_{2,s}=T_{12,s}=0,\quad T_{1,s}=C^*_{2,s}(\vartheta). \qquad (6.61)$$

f) *Wind Load*

The distribution of wind pressure and wind suction along a smooth cylindrical membrane has been dealt in Subsection 6.4. Here we assume, in the first approach, for the wind action perpendicular to the longitudinal axis of the cylindrical membrane, that

$$q_1=q_2=0,\quad q_n=q^*_0+q_v\sin\vartheta, \qquad (6.62)$$

where the quantity q_v must be negative, as on the windward side, i.e. for $\vartheta>0$, the wind creates pressure.

Substitution of (6.62) into (6.52) yields

$$T_{2,R}=q^*_0R+q_vR\sin\vartheta,$$
$$T_{12,R}=\frac{1}{2}q_vl\left(1-2\frac{x}{l}\right)\cos\vartheta, \qquad (6.63)$$
$$T_{1,R}=C_{2,v}(\vartheta)+\frac{q_vl^2}{2R}\left[\frac{x}{l}-\left(\frac{x}{l}\right)^2\right]\sin\vartheta.$$

g) *Formula for the Function $C_2(\vartheta)$*

From the formulae for the membrane force T_1 for various loads, given in Subsections b), c), d), e) and f) it follows that for $x=0$ and $x=l$

$$T_1(0, \vartheta) = T_1(l, \vartheta) = C_2(\vartheta). \tag{6.64}$$

This means that the function $C_2(\vartheta)$ represents a membrane force T_1 at the ends of the cylindrical surface which must equal the membrane force $T_2^{(K)}(\vartheta, \pi/2)$ at the head terminating the cylindrical surface, i.e.

$$C_2(\vartheta) = T_2^{(K)}\left(\vartheta, \frac{\pi}{2}\right). \tag{6.65}$$

For the case when the cylindrical surface terminates in a horizontal plane $\vartheta = \pm\pi/2$ and is closed off at the ends by spherical surfaces (quarters of a sphere) Trostel [47] deduced from condition (6.65) the following equations for the function $C_2(\vartheta)$:

$$C_{2,q}(\vartheta) = \frac{1}{2} q_0 R,$$

$$C_{2,g}(\vartheta) = gR\left(\frac{1}{1+\cos\vartheta} - \cos\vartheta\right),$$

$$C_{2,s}(\vartheta) = \frac{q_s R}{3}\left(\frac{1-\cos^3\vartheta}{\sin^2\vartheta} - 3\cos^3\vartheta\right),$$

$$C^*_{2,s}(\vartheta) = \frac{q_s R}{2}\left(\frac{1-\cos^3\vartheta}{3\sin^2\vartheta} - 2\cos^3\vartheta + \frac{4\vartheta - \sin\vartheta}{32\sin^3\vartheta}\right) \quad \text{for } \vartheta \geqq 0, \tag{6.66}$$

$$C^*_{2,s}(\vartheta) = \frac{q_s R}{2}\left(\frac{1-\cos^3\vartheta}{3\sin^2\vartheta} - \frac{4\vartheta - \sin 4\vartheta}{32\sin^3\vartheta}\right) \quad \text{for } \vartheta \leqq 0,$$

$$C_{2,v}(\vartheta) = q_0^* \frac{R}{2} - \frac{q_v R}{\sin^3\vartheta}\left(\frac{2}{3}\cos\vartheta - \sin^2\vartheta - \frac{2}{3}\cos^4\vartheta\right) \quad \text{for } \vartheta \geqq 0,$$

$$C_{2,v}(\vartheta) = q_0^* \frac{R}{2} + \frac{q_v R}{\sin^3\vartheta}\left(\frac{2}{3}\cos\vartheta - \sin^2\vartheta - \frac{2}{3}\cos^4\vartheta\right) \quad \text{for } \vartheta \leqq 0.$$

If we substitute (6.66) into the formulae (6.53), (6.55), (6.57), (6.60) and (6.63) for T_1, we obtain all membrane forces for a semi-cylindrical membrane with quarter-spherical ends.

h) *The Necessary Internal Overpressure*

In this subsection we give the conditions for the necessary internal overpessure in a semi-cylindrical membrane with quarter-spherical shaped ends.

If the membrane is affected, apart from the internal overpressure q_0, by its dead weight g only, then, by (6.53), (6.55) and (6.66), the last two conditions (2.10) take the form

$$\begin{aligned} \frac{1}{2} q_0 R + gR\left(\frac{1}{1+\cos\vartheta} - \cos\vartheta\right) - \frac{gl^2}{R}\left[\frac{x}{l} - \left(\frac{x}{l}\right)^2\right] \cos\vartheta \geqq 0, \\ q_0 R - gR \cos\vartheta \geqq 0. \end{aligned} \tag{6.67}$$

If we set $x = l/2$ and $\vartheta = 0$ in the inequalities (6.67), we obtain the following conditions for the necessary internal overpressure:

$$q_0 \geqq g\left(1 + \frac{l^2}{2R^2}\right), \quad q_0 \geqq g. \tag{6.68}$$

For $l/R = 3$ and $g = 10\ \mathrm{N/m^2}$, for example, we obtain, according to the decisive first inequality (6.68), $q_0 \geqq 55\ \mathrm{N/m^2}$.

If we consider, apart from the dead weight, also a symmetric snow load, the condition

$$T_1 = T_{1,q} + T_{1,g} + T_{1,s} \geqq 0 \tag{6.69}$$

is decisive for the determination of the internal overpressure. If we substitute (6.53), (6.55) and (6.57) into (6.69), use the expressions (6.66), and put $x = l/2$ and $\vartheta = 0$, we obtain

$$q_0 \geqq g\left(1 + \frac{l^2}{2R^2}\right) + q_s\left(1 + \frac{l^2}{R^2}\right). \tag{6.70}$$

According to (6.70) the magnitude of the necessary internal overpressure depends not only on the loads g and q_s, but also on the square of the ratio of the length of the semi-cylindrical membrane to its redius. For example, for $g = 10\ \mathrm{N/m^2}$, $q_s = 100\ \mathrm{N/m^2}$ and $l/R = 2$ we obtain from (6.70) that $q_0 \geqq 530\ \mathrm{N/m^2}$. If we neglect g in comparison with q_s, we obtain, according to (6.70), $q_0 > 5q_s$ for $l/R = 2$, and $q_0 > 10q_s$ for $l/R = 3$.

Assuming that wind load is applied to the semi-cylindrical membrane, we obtain, after substituting (6.53) and (6.63) and $x = l/2$, $\vartheta = \pi/2$ into the inequality $T_{1,q} + T_{1,R} \geqq 0$, the following condition for the magnitude of the internal overpressure:

$$q_0 \geqq -q_0^* - q_v\left(2+\frac{l^2}{2R^2}\right). \tag{6.71}$$

The inequality $T_{2,q}+T_{2,v} \geqq 0$ yields

$$q_0 \geqq -q_0^* - q_v. \tag{6.72}$$

The criteria (6.71) and (6.72) are approximate only, as their deduction was based on the assumption that wind pressure and wind suction are distributed along the semi-cylindrical surface according to (6.62). This distribution differs considerably from actual wind load conditions (see the following Subsection 6.4a). In air-supported structures whose shape is a part of a cylinder, smaller than a semi-cylinder the internal overpressure may be taken lower than in the case of a semi-cylindrical membrane.

According to [30], the necessary internal overpressure of a semi-cylindrical air-supported structure with quarter-spherical ends is determined by the condition $q_0=0.6q$, where q is the wind velocity pressure calculated from equation (5.80).

i) *Deformation of a Cylindrical Membrane*

The substitution of (6.46) and $\xi=x$, $\eta=\vartheta$ into (2.41) and (2.47) yields the following geometric relations ($\varkappa_1=0$, $\varkappa_2=1/R$):

$$\varepsilon_1=\frac{\partial u}{\partial x}, \quad \varepsilon_2=\frac{1}{R}\left(\frac{\partial v}{\partial \vartheta}-w\right), \quad \gamma=\frac{1}{R}\frac{\partial u}{\partial \vartheta}+\frac{\partial v}{\partial x},$$

$$\Delta\varkappa_1=\frac{\partial^2 w}{\partial x^2}, \quad \Delta\varkappa_2=\frac{1}{R^2}\left(w-\frac{\partial^2 w}{\partial \vartheta^2}\right), \tag{6.73}$$

$$\chi=-\frac{1}{R^2}\left(\frac{\partial u}{\partial \vartheta}-R\frac{\partial v}{\partial x}-2\frac{\partial^2 w}{\partial x\,\partial \vartheta}\right).$$

If we disregard the elongation of the membrane, $\varepsilon_2=0$, we obtain from the second equation (6.73) the equality

$$w=\frac{\partial v}{\partial \vartheta} \tag{6.74}$$

which we shall use in Chapters IV and V in the investigation of natural and forced vibrations of air-supported structures.

Substituting (2.11) into (6.73), we obtain

$$\frac{\partial u}{\partial x}=\frac{T_1}{E_1}-\frac{\nu_2}{E_2}T_2,\quad \frac{1}{R}\left(\frac{\partial v}{\partial \vartheta}-w\right)=\frac{T_2}{E_2}-\frac{\nu_1}{E_1}T_1,$$
$$\frac{1}{R}\frac{\partial u}{\partial \vartheta}+\frac{\partial v}{\partial x}=\frac{T_{12}}{E_{12}}. \tag{6.74a}$$

The displacements u, v and w can be determined by solving the equations (6.74a),where in the right-hand sides we have to substitute the formulae for the membrane forces for the respective type of load.

6.4 *Design of a Cylindrical Membrane for Finite Deflections*

a) *Wind Pressure and Suction*

The surface exposed to air flow is affected by static load $q_v(\vartheta)$ of the magnitude

$$q_v(\vartheta)=-\frac{1}{2}c(\vartheta)\varrho v_0^2, \tag{6.75}$$

where ϱ is air density, v_0 is the velocity of its flow, and $c(\vartheta)$ is a dimensionless shape coefficient. For a smooth cylindrical surface and wind direction perpendicular to the longitudinal cylinder axis then function $c(\vartheta)$ is shown in Fig. II-14, where the position angle ϑ is measured from the

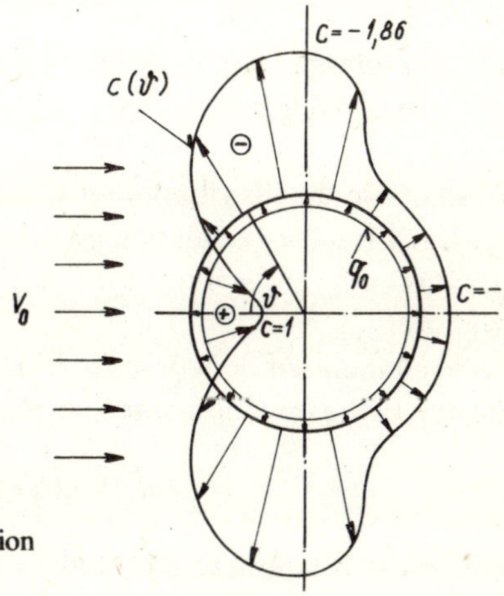

Fig. II-14. Wind pressure and suction distribution about a smooth cylindrical surface.

horizontal plane. For $c(\vartheta)>0$ the surface is affected by pressure, for $c(\vartheta)<0$ by suction. The pressure and the suction act at every point of the surface in the direction of its normal. If we substitute $\varrho=1.247\ \mathrm{kg/m^3}$ in (6.75), which pertains to the temperature of 10 °C and the barometric pressure of 760 mm Hg, we obtain

$$q_v(\vartheta)=-0.613c(\vartheta)v_0^2, \tag{6.76}$$

where v_0 is the wind velocity in m/sec and q_v is in $\mathrm{N/m^2}$.

Formula (6.76) holds for a cylinder with a smooth surface, provided the condition

$$2Rv_0 \geqq 30\ \mathrm{m^2/sec} \tag{6.77}$$

is satisfied. For cylinders with considerably rough surfaces formula (6.76) applies already for $2Rv_0>2\mathrm{m^2/sec}$ however, the curve characterizing the values of the shape coefficient $c(\vartheta)$ is considerably different from that shown in Fig. II-14 [49].

The dependence of the shape coefficient c on the angle ϑ, determined experimentally on a model of an air-supported structure in a wind tunnel, is shown in Fig. II-16 according to [44]. The solid line is valid for the deformable membrane surface of a semi-cylindrical shape for Reynolds number $\mathrm{Re}=1.45\,.\,10^6$ and the ratio $q_0/q_v(0)=0.62$, $q_v(0)$ being the dynamic air pressure calculated according to the formula (6.75) or (6.76) for $c(0)=1$. The dashed line holds for a rigid cylinder.

b) *Membrane Force T_2 Due to Wind Load, Internal Overpressure and Dead Weight*

We shall use the distribution of wind pressure and wind suction according to Fig. II-14 for the calculation of the membrane force T_2 of an air-supported structure of a semi-cylindrical shape, neglecting the effect of the ends of the structure.

An element of length $\mathrm{d}s=R\,\mathrm{d}\vartheta$ of the membrane strip 1 m wide is affected in the direction of the normal by a force $\mathrm{d}N_v$ satisfying (Fig. II-15)

$$\mathrm{d}N_v=q_v(\vartheta)\,\mathrm{d}s=q_v(\vartheta)R\,\mathrm{d}\vartheta. \tag{6.78}$$

Its vertical component $\mathrm{d}V_v$ and its horizontal component $\mathrm{d}H_v$ are

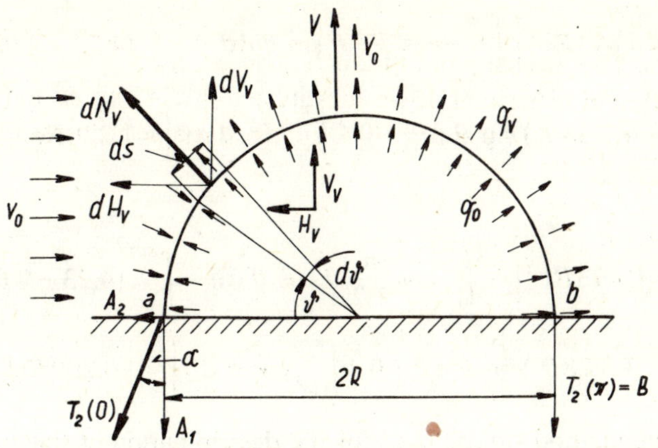

Fig. II-15. Internal overpressure and wind load affecting a semi-cylindrical membrane.

$$dV_v = dN_v \sin\vartheta, \quad dH_v = dN_v \cos\vartheta. \tag{6.79}$$

Substituting (6.78) into (6.79), we obtain, by (6.76),

$$dV_v = -0.613R^2 v_0 c(\vartheta) \sin\vartheta \, d\vartheta, \quad dH_v = -0.613R^2 v_0 c(\vartheta) \cos\vartheta \, d\vartheta. \tag{6.80}$$

The vertical component V_v and the horizontal component H_v of the resultant of the wind load are given by

$$V_v = -0.613Rv_0^2 \int_0^\pi c(\vartheta) \sin\vartheta \, d\vartheta, \quad H_v = -0.613Rv_0^2 \int_0^\pi c(\vartheta) \cos\vartheta \, d\vartheta. \tag{6.81}$$

The positive sense of the forces V_v and H_v is shown in Fig. II-15.

For the calculation of the integrals appearing in the equations (6.81) the following approximate analytical expressions of the shape coefficient $c(\vartheta)$ may be used:

$$\begin{aligned} c(\vartheta) &= \cos\frac{\pi}{1.2}\vartheta && \text{for } 0 \leqq \vartheta \leqq 0.6, \\ c(\vartheta) &= 4.23 - 9.02\vartheta + 3.28\vartheta^2 && \text{for } 0{,}6 \leqq \vartheta \leqq 2.07, \\ c(\vartheta) &= -0.4 && \text{for } 2.07 \leqq \vartheta \leqq \pi. \end{aligned} \tag{6.82}$$

Substitution of (6.82) into (6.81) yields

$$V_v = 0.613Rv_0^2\left[\int_0^{0.6} -\cos\frac{\pi}{1.2}\,\vartheta \sin\vartheta\, d\vartheta - \int_{0.6}^{2.07}(4.23 - 9.02\vartheta + \right.$$

$$\left. + 3.28\vartheta^2)\sin\vartheta\, d\vartheta + 0.4\int_{2.07}^{\pi}\sin\vartheta\, d\vartheta\right] \approx 1.226Rv_0^2 \qquad (6.83)$$

and

$$H_v = 0.613Rv_0^2\left[-\int_0^{0.6}\cos\frac{\pi}{1.2}\,\vartheta\cos\vartheta\, d\vartheta - \int_{0.6}^{2.07}(4.23 - 9.02\vartheta + \right.$$

$$\left. + 3.28\vartheta^2)\cos\vartheta\, d\vartheta + 0.4\int_{2.07}^{\pi}\cos\vartheta\, d\vartheta\right] \approx -0.209Rv_0^2. \qquad (6.84)$$

The same method can be used for the determination of the resultants V_v and H_v due to the load q_v, taking into account the shape coefficient c according to Fig. II-16, whose magnitude can be expressed, for example, by the Lagrange polynomial.

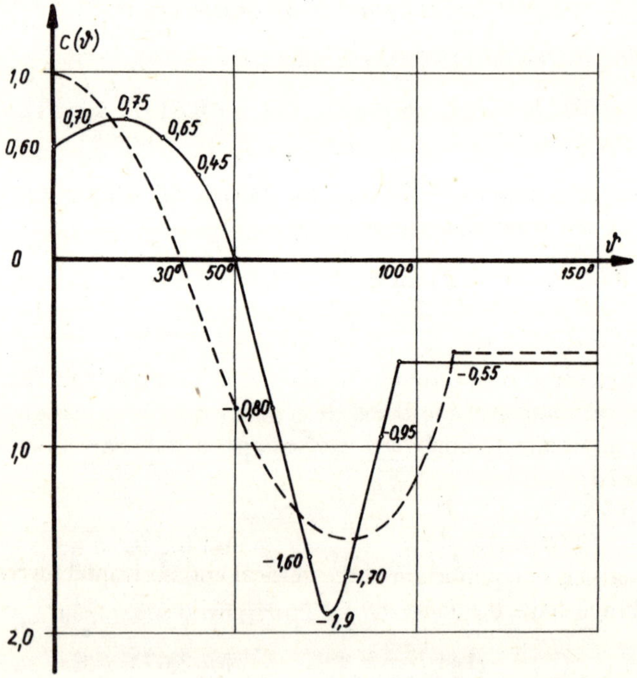

Fig. II-16. Dependence of the coefficient of form $c(\vartheta)$ on the angle ϑ in an undeformed cylindrical surface (dashed line) and in a deformed semi-cylindrical membrane (solid line) [44].

The vertical component of the resultant V_0 due to the internal overpressure is directed upward and satisfies

$$V_0 = q_0 R \int_0^{\pi} \sin \vartheta \, d\vartheta = 2q_0 R, \tag{6.85}$$

while the horizontal component H_0 of the resultant of the internal overpressure equals zero, i.e.

$$H_0 = q_0 R \int_0^{\pi} \cos \vartheta \, d\vartheta = 0. \tag{6.86}$$

The resultant V_g of the dead weight is dierected downward and satisfies

$$V_g = -\pi g R. \tag{6.87}$$

The overall vertical force V due to the loads q_0, q_v and g is directed upward and, according to (6.83), (6.85) and (6.87), it satisfies

$$V = R(2q_0 - \pi g + 1.226 v_0^2). \tag{6.88}$$

The forces V and H_v must be transferred into the foundations by means of the anchorage. For an approximate calculation of reactions we assume that the tangent to the membrane at the place of anchorage on the leeward side is vertical, provided the internal overpressure is sufficiently large. This means, at the place there is only the vertical reaction B, while on the windward side the membrane is affected both by the vertical component A_1 and by the horizontal component A_2 of the reaction. The justification of this assumption was verified experimentally [44].

According to Fig. II-15, we have

$$T_2(\pi) = B, \quad T_2^2(0) = A_1^2 + A_2^2, \tag{6.89}$$

while $T_2(0) = T_2(\pi)$, as the membrane force due to the normal loads q_0 and q_v is constant along the whole length of the arc; when also the dead weight g is taken into account, we have by (6.55),

$$T_2 = T_2(0) - gR \cos(\pi/2 - \vartheta). \tag{6.90}$$

The equations of equilibrium of the vertical and horizontal forces affecting the membrane have the form

$$A_1 + B - V = 0, \quad A_2 + H_v = 0. \tag{6.91}$$

From the first equation (6.89) and from (6.91) we obtain

$$A_1 = V - T_2(0), \tag{6.92}$$

and substituting (6.92) into the second equation (6.89), we get

$$T_2(0) = \frac{1}{2}\left(V + \frac{H_v^2}{V}\right). \tag{6.93}$$

Substitution of (6.88) and (6.84) into equation (6.93) yields the following formula for the tensile force $T_2(0)$:

$$T_2(0) = q_0 R\left(1 - \frac{\pi}{2}\frac{g}{q_0} + 0.613\frac{v_0^2}{q_0} + \frac{0.0011}{1 - \dfrac{\pi g}{2q_0} + 0.613\dfrac{v_0^2}{q_0}}\frac{v_0^4}{q_0^2}\right). \tag{6.94}$$

In formula (6.94) the internal overpressure q_0 and the weight g should be expressed in N/m^2, the wind velocity v_0 in m/sec and the radius R in m. The tensile force $T_2(0)$ is applied to a membrane strip 1 m wide and is of dimension N/m.

For the angle α, between the force $T_2(0)$ and the vertical at the point of anchorage on the windward side, we have, by Fig. II-15,

$$\tan\alpha = \frac{A_2}{A_1}. \tag{6.95}$$

It follows from the equations (6.92) and (6.93) that

$$A_1 = \frac{1}{2}\left(V - \frac{H_v^2}{V}\right). \tag{6.96}$$

After the substitution of (6.96) and $A_2 = -H_v$ into (6.95) we obtain

$$\tan\alpha = -\frac{2VH_v}{V^2 - H_v^2}. \tag{6.97}$$

The membrane force $T_2(0)$ can be determined with sufficient accouracy from the following simple formula [11]:

$$T_2 = R_0(q_0 + 0.613v_0^2), \tag{6.98}$$

which is obtained from formula (6.94) by neglecting the small second and fourth terms. According to (6.98), the wind effect, in the case of a semi-cylindrical surface, can be replaced by the equivalent internal overpressure $\bar{q}_0 = 0{,}613v_0^2$.

In Section 6 of Chapter IV we give another simple formula for the calculation of the membrane force T_2.

c) *Snow Load*

In this subsection we shall deal with the design of an air-supported cylindrical structure affected by symmetric snow load. Similarly as in the preceding two subsections we shall not consider the effect of the ends of the cylindrical structure. We shall also neglect the dead weight and the expansivity of the membrane and we shall assume that the snow load q_s is constant and is applied perpendicularly to the deformed centre line of the arc, representing the cross-section of the air-supported structure. Under these assumptions the cross-section of the deformed cylindrical membrane consists of three circular arcs.

According to Fig. II-17, we have

$$R_0\alpha_s = R_2\alpha_2, \quad 2R_1\alpha_1 + R_2\alpha_2 = R_0\alpha_0 \tag{6.99}$$

and for $\alpha_0 \leqq \pi$,

$$\alpha_0 = 2\alpha_1 + \alpha_2 - \chi_1 \tag{6.100}$$

holds.

The membrane force T_2 on that part of the arc which is loaded by snow is

$$T_2 = (q_0 - q_s)R_2. \tag{6.101}$$

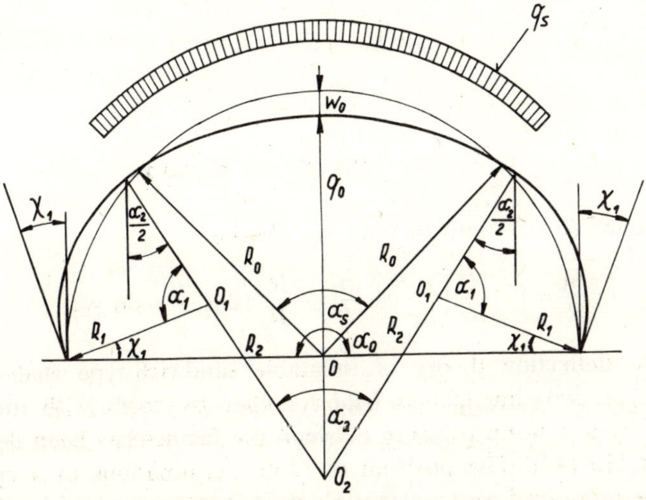

Fig. II-17. Deformation of a cylindrical surface due to the internal overpressure and symmetrical snow load.

Since the loads q_0 and q_s are applied only perpendicularly to the deformed centre line of the arc, the force T_2 is constant along the whole arc length. In that part of the arc, where only the internal overpressure acts, we have

$$T_2 = q_0 R_1 . \tag{6.102}$$

A comparison of (6.101) and (6.102) yields

$$\frac{R_1}{R_2} = 1 - \frac{q_s}{q_0} . \tag{6.103}$$

Moreover, according to Fig. II-17,

$$R_0 \sin \frac{\alpha_0}{2} = R_2 \sin \frac{\alpha_2}{2} + R_1 \sin \left(\frac{\alpha_0}{2} - \chi_1 \right) . \tag{6.104}$$

The five unknown quantities

$$\alpha_1, \alpha_2, \chi_1, R_1, R_2 \tag{6.105}$$

are mutually related by the five independent equations (6.99), (6.100), (6.103) and (6.104) from which all unknown quantities can be determined. After the calculation of the radius R_2 or R_1 we determine the membrane force T_2 from (6.101) or (6.102) respectively.

The deflection w_0 at the arc apex is given by

$$w_0 = R_0 - (R_2 - \overline{OO_2}), \tag{6.106}$$

where

$$\overline{OO_2} = (R_2 - R_1) \cos \frac{\alpha_2}{2} - R_1 \sin \chi_1 . \tag{6.107}$$

Substituting (6.107) into (6.106), we obtain

$$\frac{w_0}{R_0} = 1 - \frac{R_2}{R_0} \left(1 - \cos \frac{\alpha_2}{2} \right) - \frac{R_1}{R_0} \left(\cos \frac{\alpha_2}{2} + \sin \chi_1 \right) . \tag{6.108}$$

A finite deflection theory of inflatable sandwich-type shallow shells consisting of two thin facings held together by cords with the system stabilized by a uniform pressure between the facings has been derived by Nash and Ho [43]. The problem of finite deformations of a cylindrical membrane reinforced with inextensible cords has been solved by Kydoniefs [33].

7

Transcendental Surfaces of Revolution

7.1 *Cycloid of Revolution with a Vertical Axis*

a) *Equations of the Meridian and Principal Radii*

In Subsection 2.3 of Chapter III the following parametric equations of the meridian of a cycloid of revolution are deduced (Figs. II-18a, b):

$$r = C \sin^2 \alpha, \quad z = C\left(\alpha - \frac{\sin 2\alpha}{2}\right). \tag{7.1}$$

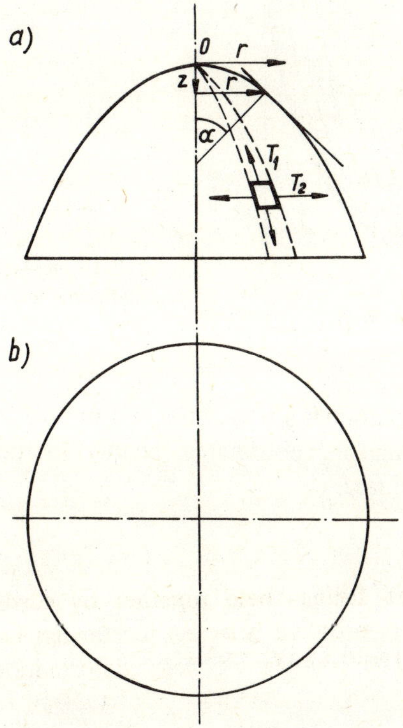

Fig. II-18. Cycloid of revolution with a vertical axis: a) vertical section; b) plan.

For the principal radii we have $R_1 = 2R_2$, $R_2 = r/\sin\alpha$ and hence

$$R_1 = 2C \sin\alpha, \quad R_2 = C \sin\alpha, \tag{7.2}$$

where $C = r_{max}$ (Fig. II-19).

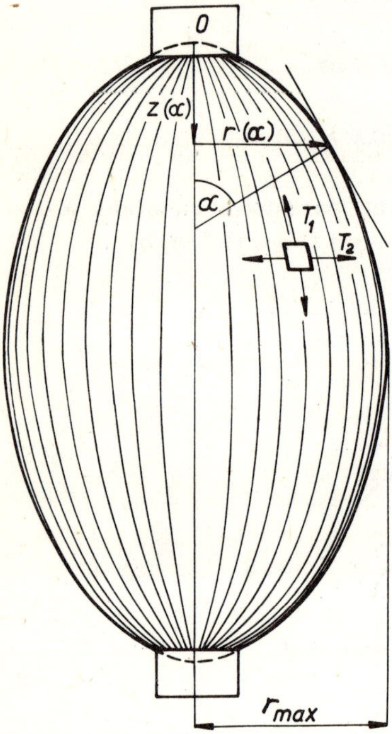

Fig. II-19. Air-supported structure with the jacket in the form of a cycloid of revolution with a horizontal axis (plan).

The cycloid of revolution is a transcendental surface, as its meridian can be expressed, in rectangular coordinates, simply by means of elementary transcendental functions.

b) *Internal Overpressure*

Substitution of (7.2) into (3.16) yields

$$T_{1,q} = \frac{1}{2} q_0 C \sin\alpha, \quad T_{2,q} = \frac{3}{4} q_0 C \sin\alpha. \tag{7.3}$$

c) *Dead Weight*

If we substitute (3.23) and (7.2) into (3.82) and (3.83), we obtain

$$T_{1,g} = -\frac{2gC}{\sin^3\alpha}\int_0^\alpha \sin^3\alpha \, d\alpha = -\frac{2gC}{\sin^3\alpha}\left(\frac{\cos^3\alpha}{3} - \cos\alpha + \frac{2}{3}\right),$$

$$T_{2,g} = -gC\sin\alpha\cos\alpha - \frac{1}{2}T_{1,g} = \tag{7.4}$$

$$= -gC\left[\sin\alpha\cos\alpha + \frac{1}{\sin^3\alpha}\left(\frac{\cos^3\alpha}{3} - \cos\alpha + \frac{2}{3}\right)\right].$$

d) *Symmetric Snow Load*

If we substitute (3.30) and (7.2) into (3.82) and (3.83), we obtain for $\alpha \leqq \pi/2$.

$$T_{1,s} = -\frac{2q_sC}{\sin^3\alpha}\int_0^\alpha \sin^3\alpha\cos^2\alpha \, d\alpha = -\frac{2q_sC}{\sin^3\alpha}\left(\frac{\cos^5\alpha}{5} - \frac{\cos^3\alpha}{3} + \frac{2}{15}\right),$$

$$T_{2,s} = -q_sC\sin\alpha\cos^3\alpha - \frac{1}{2}T_{1,s} = \tag{7.5}$$

$$= -q_sC\left[\sin\alpha\cos^3\alpha - \frac{1}{\sin^3\alpha}\left(\frac{\cos^5\alpha}{5} - \frac{\cos^3\alpha}{3} + \frac{2}{15}\right)\right].$$

e) *One-Sided Snow Load*

The symmetric load components (3.75) procedure, according to (3.77),

$$T^*_{1,s} = \frac{1}{2}T_{1,s} = -\frac{q_sC}{\sin^3\alpha}\left(\frac{\cos^5\alpha}{5} - \frac{\cos^3\alpha}{3} + \frac{2}{15}\right),$$

$$T^*_{2,s} = \frac{1}{2}T_{2,s} = -\frac{1}{2}q_sC\left[\sin\alpha\cos^3\alpha - \right. \tag{7.6}$$

$$\left. - \frac{1}{\sin^3\alpha}\left(\frac{\cos^5\alpha}{5} - \frac{\cos^3\alpha}{3} + \frac{2}{15}\right)\right].$$

Substituting (7.2) into (3.86), we obtain the following membrane forces due to the antisymmetric load components (3.76) for $\alpha \leqq \pi/2$:

$$T^{**}_{1,s} = -\left[\frac{q_s C}{2\sin^5\alpha}\int_0^\alpha \sin^5\alpha\cos^2\alpha\,d\alpha\right]\cos\beta =$$

$$= -\frac{q_s C}{2\sin^5\alpha}\left[\frac{8}{105} - \frac{\cos^3\alpha}{7}\left(\frac{4}{3} + \sin^4\alpha - \frac{4}{5}\cos^2\alpha\right)\right]\cos\beta,$$

$$T^{**}_{2,s} = \left[-\frac{q_s C}{2}\sin\alpha\cos^3\alpha - \frac{1}{2\cos\beta}T^{**}_{1,s}\right]\cos\beta = \tag{7.7}$$

$$= -\frac{q_s C}{2}\left\{\sin\alpha\cos^3\alpha - \frac{1}{2\sin^5\alpha}\left[\frac{8}{105} - \frac{\cos^3\alpha}{7}\left(\frac{4}{3} + \sin^4\alpha - \right.\right.\right.$$

$$\left.\left.\left. - \frac{4}{5}\cos^2\alpha\right)\right]\right\}\cos\beta;$$

$$T^{**}_{12,s} = T^{*}_{1,s}\cos\alpha\tan\beta.$$

The resulting membrane forces are determined according to (3.87), i.e. by the superposition of (7.6) and (7.7).

f) *Wind Load*

The substitution of (7.2) into (3.72) yields

$$\Phi_v = q_v C^2 \sin^3\alpha\cos\alpha - 2q_v C^2\int_0^\alpha \sin^5\alpha\,d\alpha =$$

$$= q_v C^2\left(\sin^3\alpha\cos\alpha + \frac{2}{5}\sin^4\alpha\cos\alpha - \frac{8}{15}\cos^3\alpha + \frac{8}{5}\cos\alpha - \frac{16}{15}\right),$$

$$f_v = \frac{2q_v C}{\sin^2\alpha}\int_0^\alpha \sin^5\alpha\,d\alpha = \tag{7.8}$$

$$= \frac{4q_v C}{\sin^2\alpha}\left(\frac{8}{15} - \frac{1}{5}\sin^4\alpha\cos\alpha + \frac{4}{15}\cos^3\alpha - \frac{4}{15}\cos\alpha\right)$$

and

$$\int_0^\alpha \Phi_v R_1 \sin\alpha\,d\alpha = 2q_v C^3\int_0^\alpha\left(\sin^5\alpha\cos\alpha + \frac{2}{5}\sin^6\alpha\cos\alpha - \right.$$

$$\left. - \frac{8}{15}\sin^2\alpha\cos^3\alpha + \frac{8}{5}\sin^2\alpha\cos\alpha - \frac{16}{15}\sin^2\alpha\right)d\alpha = \tag{7.9}$$

$$=2q_vC^3\left(\frac{1}{6}\sin^6\alpha+\frac{2}{35}\sin^7\alpha+\frac{8}{75}\sin^5\alpha+\frac{16}{45}\sin^3\alpha+\right.$$

$$\left.+\frac{8}{15}\sin\alpha\cos\alpha-\frac{8}{15}\alpha\right).$$

According to (3.71), using (7.2), (7.8) and (7.9), we obtain

$$T_{1,v}=2q_vC\left(\frac{2}{35}\sin^2\alpha+\frac{\sin\alpha}{6}+\frac{8}{75}+\frac{16}{45\sin^2\alpha}+\frac{8}{15}\frac{\cos\alpha}{\sin^4\alpha}-\right.$$

$$\left.-\frac{8}{15}\frac{\alpha}{\sin^5\alpha}\right)\cos\beta,$$

$$T_{2,v}=\frac{q_vC}{\sin^5\alpha}\left(\frac{33}{35}\sin^7\alpha-\frac{1}{6}\sin^6\alpha-\frac{8}{75}\sin^5\alpha-\frac{16}{45}\sin^3\alpha-\right.$$

$$\left.-\frac{8}{15}\sin\alpha\cos\alpha+\frac{8}{15}\alpha\right)\cos\beta, \tag{7.10}$$

$$T_{12,v}=\frac{2q_vC}{\sin^2\alpha}\left[\frac{\cos\alpha}{\sin^3\alpha}\left(\frac{2}{35}\sin^7\alpha+\frac{1}{6}\sin^6\alpha+\frac{8}{75}\sin^5\alpha+\frac{16}{45}\sin^3\alpha+\right.\right.$$

$$+\frac{8}{15}\sin\alpha\cos\alpha-\frac{8}{15}\alpha\right)+2\left(\frac{8}{15}-\frac{1}{5}\sin^4\alpha\cos\alpha+\right.$$

$$\left.\left.+\frac{4}{15}\cos^3\alpha-\frac{4}{5}\cos\alpha\right)\right]\sin\beta.$$

The symmetric wind load component q_v^* produces, according to (7.3),

$$T^*_{1,v}=\frac{1}{2}\,q_v^*C\sin\alpha,\quad T^*_{2,v}=\frac{3}{4}\,q_v^*C\sin\alpha,\quad T^*_{12,v}=0. \tag{7.11}$$

Substitution of (7.10) and (7.11) into (4.81) yields the membrane forces due to the overall wind load (3.44).

g) *The Necessary Internal Overpressure*

If we consider only the load due to the internal overpressure q_0 and the dead weight g, we obtain, according to (2.10),

$$T_{1,q}+T_{1,g}\geqq 0,\quad T_{2,q}+T_{2,g}\geqq 0. \tag{7.12}$$

Substituting (7.3) and (7.4) into (7.12), we get

$$q_0 \geqq \frac{4g}{\sin^4\alpha}\left(\frac{\cos^3\alpha}{3} - \cos\alpha + \frac{2}{3}\right), \tag{7.13}$$

$$q_0 \geqq \frac{4}{3}g\left[\cos\alpha - \frac{1}{\sin^4\alpha}\left(\frac{\cos^3\alpha}{3} - \cos\alpha + \frac{2}{3}\right)\right].$$

To determine the necessary magnitude of the internal ovepressure we have to use the first inequality (7.13) which yields higher values of q_0 than the second inequality. For $\alpha = \pi/2$ we have $q_0 \geqq 8/3g$.

If symmetric snow load of an intensity q_s is applied, apart from q_0 and g, we have to calculate the necessary magnitude of the internal overpressure by means of the second inequality (2.10), i.e.

$$T_{1,q} + T_{1,g} + T_{1,s} \geqq 0. \tag{7.14}$$

Substituting (7.3), (7.4) and (7.5) into (7.14), we obtain

$$q_0 \geqq \frac{4}{\sin^4\alpha}\left[g\left(\frac{\cos^3\alpha}{3} - \cos\alpha + \frac{2}{3}\right) + q_s\left(\frac{\cos^5\alpha}{5} - \frac{\cos^3\alpha}{3} + \frac{2}{15}\right)\right], \tag{7.15}$$

whence, for $\alpha = \pi/2$, we deduce $q_0 \geqq 8(g/3 + q_s/15)$.

For one-sided snow load we substitute $T_{1,s} = T_{1,s}^{*} + T_{1,s}^{**}$ into (7.14), the quantities $T_{1,s}^{*}$ and $T_{1,s}^{**}$ being determined by the first formulae in (7.6) and (7.7). For $\beta = 0$ we have

$$q_0 \geqq \frac{1}{\sin^4\alpha}\left\{4g\left(\frac{\cos^3\alpha}{3} - \cos\alpha + \frac{2}{3}\right) + \right.$$
$$\left. + q_s\left[\frac{12}{35} - \frac{\cos^3\alpha}{7}(6 + \sin^4\alpha) + \frac{18}{35}\cos^5\alpha\right]\right\}, \tag{7.16}$$

and for $\alpha = \pi/2$ we have $q_0 \geqq 8/3g + 12/35q_s$. Of decisive importance, therefore, is the symmetric snow load.

If the air-supported structure is subjected to simultaneous effect of symmetric snow load and wind load, the necessary internal overpressure q_0 is determined by the inequality

$$T_{1,q} + T_{1,g} + T_{1,s} + T_{1,v} + T_{1,v}^{*} \geqq 0, \tag{7.17}$$

where $T_{1,v}$ has to be replaced by the first expression (7.10) taken with $\beta = \pi$, and $T_{1,v}^{*}$ with the first expression (7.11). Thus we obtain

$$q_0 \geqq \frac{4}{\sin^4 \alpha} \left[g\left(\frac{\cos^3 \alpha}{3} - \cos \alpha + \frac{2}{3} \right) + q_s \left(\frac{\cos^5 \alpha}{5} - \frac{\cos^3 \alpha}{3} + \frac{2}{15} \right) \right] +$$

$$+ \frac{4}{\sin^2 \alpha} q_v \left(\frac{2}{35} \sin^2 \alpha + \frac{\sin \alpha}{6} + \frac{8}{75} + \frac{16}{45 \sin^2 \alpha} + \frac{8}{15} \frac{\cos \alpha}{\sin^4 \alpha} - \frac{8}{15} \frac{\alpha}{\sin^5 \alpha} \right) - q_0^*, \tag{7.18}$$

whence we get, for $\alpha = \pi/2$,

$$q_0 \geqq \frac{8}{3} g + \frac{8}{15} q_s - 0{,}607 q_v - q_0^*.$$

If the air-supported structure is fastened to the foundations along the parallel circle with $\alpha < \pi/2$, we substitute in the above inequalities the value of α pertaining to the respective point of fastening of the membrane.

7.2 *Cycloid of Revolution With a Horizontal Axis*

a) *Shape and Loading of the Structure*

A special shape of air-supported structure is obtained from a part of a cycloid of revolution by cutting it by a plane parallel to the axis of revolution. The plan of such a structure with two access chambers is shown in Fig. II-19, its cross-section in Fig II-20, and longitudinal section in Fig. II-21, the strips in these drawings representing the stripes of fabric. Among the advantages of

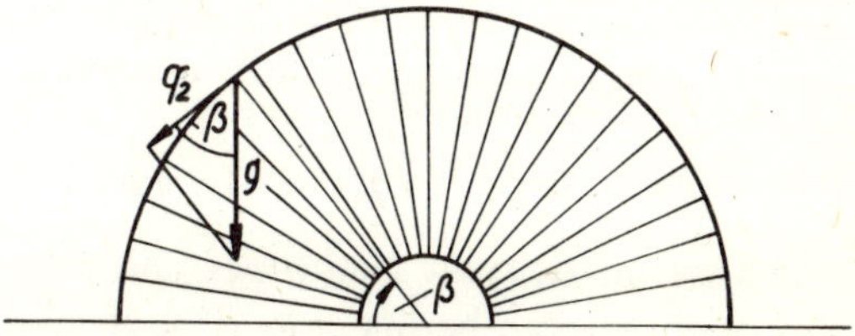

Fig.. II-20. Cross-section of the structure shown in Fig. II-19.

such design, apart from the new shape and a suitable prestress of the membrane by tensile forces, when loaded by internal overpressure, is the fact that the access chambers join directly to the structure in longitudinal direction, not influencing negatively the state of stress of the membrane in their vicinity and enabling their connection to be made without any connecting collars and any special reinforcement.

The load components q_1, q_2, due to the dead weight of the membrane in

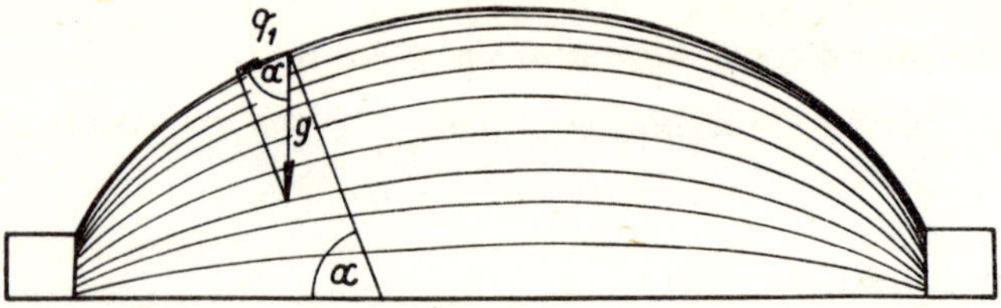

Fig. II-21. Longitudinal section of the structure shown in Fig. II-19.

the direction of the tangent to the meridian and to the parallel circle are, according to Fig. II-20 and Fig. II-21,

$$q_1 = g \cos\alpha, \quad q_2 = g \cos\beta. \tag{7.19}$$

The load component q_n in the direction of the normal to the surface is given by

$$q_n = g \cos\gamma, \tag{7.20}$$

where γ is the angle between the normal and the vertical load g. Since the load components q_1, q_2 and q_n are mutually perpendicular, their direction cosines satisfy

$$\cos^2\alpha + \cos^2\beta + \cos^2\gamma = 1, \tag{7.21}$$

whence

$$\cos\gamma = \sqrt{1 - \cos^2\alpha - \cos^2\beta}. \tag{7.22}$$

Substituting (7.22) into (7.20), we obtain

$$q_n = g\sqrt{1 - \cos^2\alpha - \cos^2\beta}. \tag{7.23}$$

The symmetric snow load $\bar{q}$ per unit area of the horizontal projection of the membrane can be taken as

$$\bar{q} = q_s \cos\alpha \sin\beta, \tag{7.24}$$

where $q_s = \text{const}$, $0 \leqq \alpha \leqq \pi/2$, $0 \leqq \beta \leqq \pi$. The vertical snow load per unit surface of the membrane is

$$q = \bar{q} \cos\alpha \sin\beta = q_s \cos^2\alpha \sin^2\beta, \tag{7.25}$$

and its components in the principal directions of the surface and the normal are (compare with the formulae (7.19) and (7.23))

$$\begin{aligned} q_1 &= q \cos\alpha = q_s \cos^3\alpha \sin^2\beta, \\ q_2 &= q \cos\beta = q_s \cos^2\alpha \sin^2\beta \cos\beta, \\ q_n &= q \cos\gamma = q_s \cos^2\alpha \sin^2\beta\sqrt{1-\cos^2\alpha-\cos^2\beta}. \end{aligned} \tag{7.26}$$

The components of a one-sided snow load are determined for $0 \leqq \beta \leqq \pi/2$ by the formulae (7.26) and for $\beta \geqq \pi/2$ they are $q_1 = q_2 = q_n = 0$.

Static wind load (pressure and suction) affects the surface only in the direction of its normal. If the wind acts in the direction perpendicular to the longitudinal axis of the air-supported structure, so that $q_1 = q_2 = 0$, then an approximate distribution of the load q_n along the surface is given by the formula (see (3.44))

$$q_n = q_0^* + q_v \sin\alpha \cos\beta. \tag{7.27}$$

b) *The Necessary Internal Overpressure under Wind Load*

The membrane forces of a structure as on Fig. II-19, Fig. II-20 and Fig. II-21 due to the internal overpressure are given by the formulae (7.3), and those due to wind loads, by the expressions (7.10) and (7.11). If we neglect the dead weight of the membrane g and do not consider the snow load, we can determine the necessary internal overpressure from the inequality (7.18), where we put $g = 0$, $q_s = 0$ and substitute for α the angle, for which the right-hand side of the inequality attains its maximum.

7.3 General Transcendental Surfaces with Vertical Axis of Revolution

a) *Equations of the Meridian and Principal Radii*

The equation of the meridian of general transcendental surfaces of revolution, of which the cycloid of revolution is a special case, are deduce by the author in Chapter III, Subsection 2.3. According to the second equation (2.39) in Chapter III, the principal radii of these surfaces satisfy ($R_1 = aR_2$, $R_2 = r/\sin\alpha$)

$$R_1 = Ca \sin^{a-1}\alpha, \quad R_2 = C \sin^{a-1}\alpha. \tag{7.28}$$

b) *Internal Overpressure*

Substituting (7.28) into the formulae (3.16), we obtain the following expressions for the membrane forces:

$$T_{1,q}=\frac{1}{2}q_0C\sin^{a-1}\alpha,$$
$$T_{2,q}=q_0C\left(1-\frac{1}{2a}\right)\sin^{a-1}\alpha. \tag{7.29}$$

The ratio of these forces is constant, $T_{1,q}/T_{2,q}=\dfrac{a}{2a-1}$.

c) *Dead Weight*

If we substitute (3.23) and (7.28) into (3.82) and (3.83), we obtain

$$T_{1,g}=-gCa\sin^{1-a}\alpha\int_0^\alpha \sin^{2a-1}\alpha\,d\alpha,$$
$$T_{2,g}=-gC\cos\alpha\sin^{a-1}\alpha+gC\sin^{1-a}\alpha\int_0^\alpha \sin^{2a-1}\alpha\,d\alpha. \tag{7.30}$$

For certain values of the parameter a the integral appearing in the equations (7.30) can be calculated in terms of trigonometric functions. For example, for $a=4$, we obtain from (7.30) $(2a-1=7)$

$$T_{1,g}=-\frac{4gC}{\sin^5\alpha}\left(\frac{16}{35}-\frac{1}{7}\sin^6\alpha\cos\alpha-\frac{6}{35}\sin^4\alpha\cos\alpha+\right.$$
$$\left.+\frac{8}{35}\cos^3\alpha-\frac{24}{35}\cos\alpha\right), \tag{7.31}$$
$$T_{2,g}=-\frac{gC\cos\alpha}{\sin^3\alpha}-\frac{1}{4}T_{1,g}.$$

d) *Symmetric Snow Load*

Substituting (3.30) and (7.28) into (3.82) and (3.83), we obtain

$$T_{1,s} = -q_s Ca \sin^{1-a} \alpha \int_0^{\alpha} (1 - \sin^4 \alpha) \sin^{2a-1} \alpha \, d\alpha,$$
$$T_{2,s} = -q_s C \cos^3 \alpha \sin^{a-1} \alpha - \frac{1}{a} T_{1,s}. \tag{7.32}$$

The membrane forces due to one-sided snow load and wind load can be expressed analogously.

8

Mechanical Properties and Deformation Laws of Industrial Fabrics. Safety of Air-Supported Structures

8.1 *Experimental Results*

Mechanical properties of industrial fabrics depend primarily on the way the stress is applied to the fabric.

In the Institute of Teoretical and Applied Mechanics at the Czechoslovak Academy of Sciences, Prague, where some theoretical and experimental research of membrane structures has been carried out since 1971, there are two loading machines for short-term and long-term tests of industrial fabrics in biaxial tension. This situation is encountred in the skins of air-supported structures and the roofing materials of suspended membrane roofs

Fig. II-22. Test of an industrial fabric in biaxial tension using a cylindrical sample of diameter 6 cm, loaded by an internal overpressure, produced by a liquid, and axial tension produced by type Testatron (Wolpert) testing machine.

(Fig. II-22, Figs. II-23a, b). We shall present some results of tests performed on polyamide fabric coated with PVC, produced by the national corporation of Technolen, Czechoslovakia, under the denomination of 12 824/550/743 (further on referred to as the 12 824 fabric). The tests were made on a machine of the type Testatron (produced by Messrs Wolpert) at the temperature $T = 20°C \pm 2$ °C, and described in detail in [39].

Fig. II-24 shows the working diagrams (relation between the tensile force T and the relative elongation ε) of samples of the 12 824 fabric, of size 25 × 200 mm, loaded by uniaxial tension in the direction of the warp for various loading speeds. The working diagrams under uniaxial tension in the direction of the woof (filling) are shown in Fig. II-25. It can be seen that for the relative deformation speed within the range of $0.02\ \text{sec}^{-1} < \dot{\varepsilon} < 2.84\ \text{sec}^{-1}$ the loading speed has no effect on the strength and the limit deformation of the fabric in the warp direction. However, in the woof direction, for $\dot{\varepsilon} < 1.0\ \text{sec}^{-1}$ the fabric strength is considerably lower than for

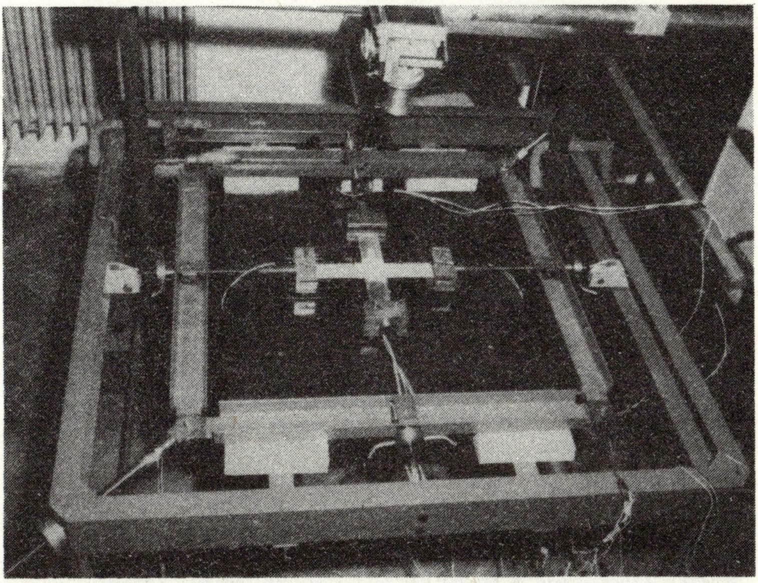

Fig. II-23a. Test of an industrial fabric in biaxial tension using a cross-shaped sample whose deformation is measured by the stereophotogrammetric method. A homogeneous deformation field was obtained by producing the tensile stresses measured in the investigated central square part of the sample by pulling parallel, coating-free fibres in the directions of the warp and woof respectively [2].

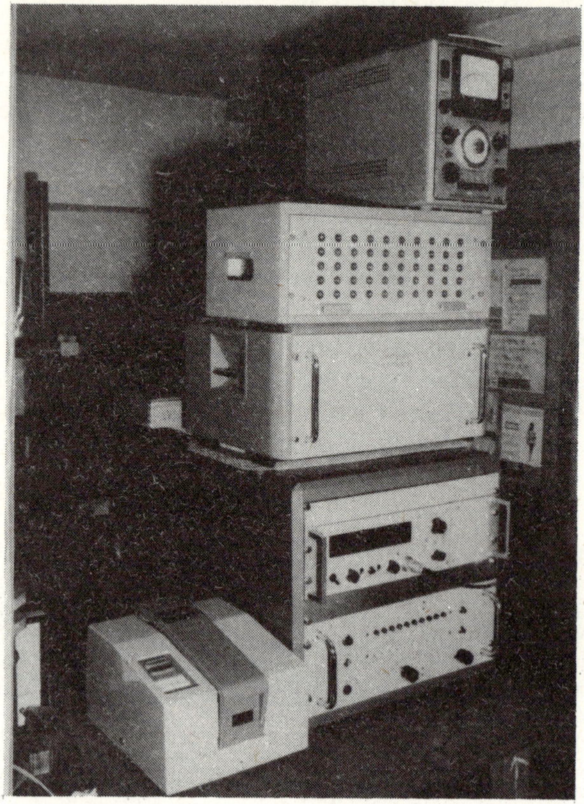

Fig. II-23b. Overall view of the measuring apparatuses for the tests of cross-shaped samples of industrial fabrics.

higher loading speeds at which the strength is ascertained in accordance with the Czechoslovak standard ČSN 80 0815. Also for $\dot{\varepsilon} < 1.0\ \text{sec}^{-1}$ there was observed a considerable dispersal of test results.

Fig. II-26 shows averaged diagrams of membrane force versus deformation under stress relaxation and creep in short-term uniaxial tension in the directions of the warp and of the filling, respectively, for 12 824 fabric (with duration $0 \leqq t < 20$ min). Relaxation and creep tests have proved that the fabric can be considered as a linear viscoelastic material practically in the whole range of the unit membrane force T in warp direction and for $T > 200$ N/cm in the woof direction. The short-term modulus for $T < 100$ N/

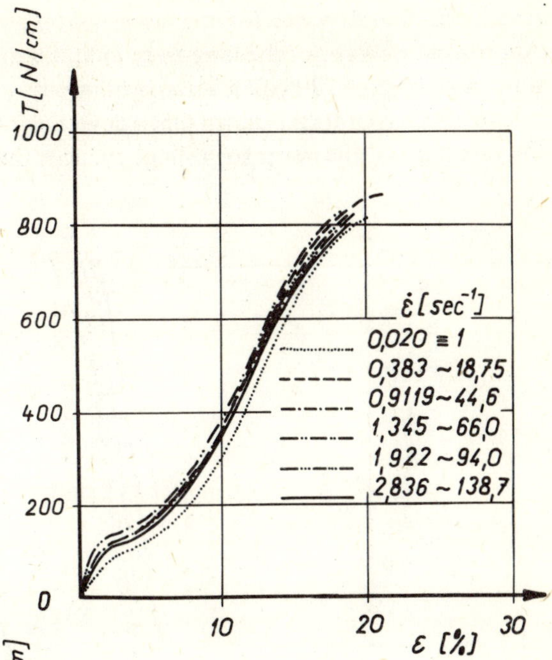

Fig. II-24. Stress-strain diagrams of type 12 824 polyamide fabric coated with PVC under uniaxial tension in the warp direction for various relative deformation speeds [39]. The numbers in the second column give the ratio $\dot{\varepsilon}/\dot{\varepsilon}_0$, where $\dot{\varepsilon}_0 = 0.02\ \text{sec}^{-1}$

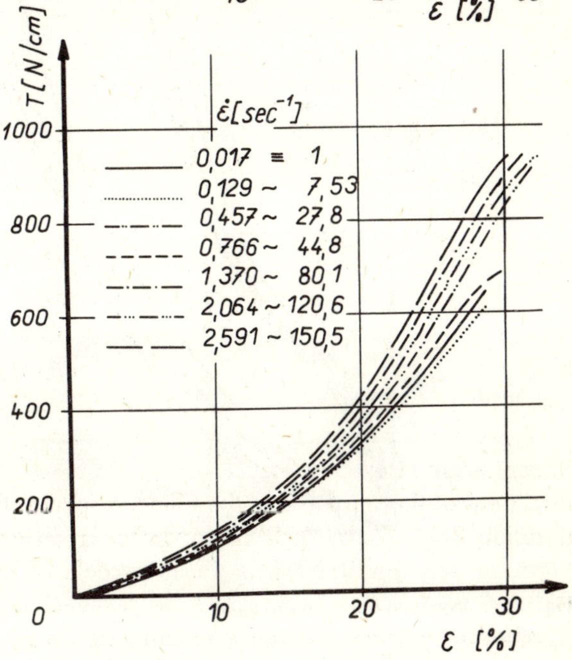

Fig. II-25. Stress-strain diagrams of type 12 824 polyamide fabric coated with PVC under uniaxial tension in the woof direction for various relative deformation speeds [39]. The numbers in the second column give the ratio $\dot{\varepsilon}/\dot{\varepsilon}_0$, where $\dot{\varepsilon}_0 = 0.017\ \textit{sec}^{-1}$.

/cm is $E_w = 3850$ N/cm for the warp and $E_f = 1250$ N/cm for the filling. Approximate values of the long-term modulus are $E_{w,\infty} = 1300$ N/cm for the warp and $E_{f,\infty} = 1130$ N/cm for the filling.

The tests have further shown that a deviation of the tensile force by 5° from the direction of the warp (or filling) reduces the strength to only 68 % (or

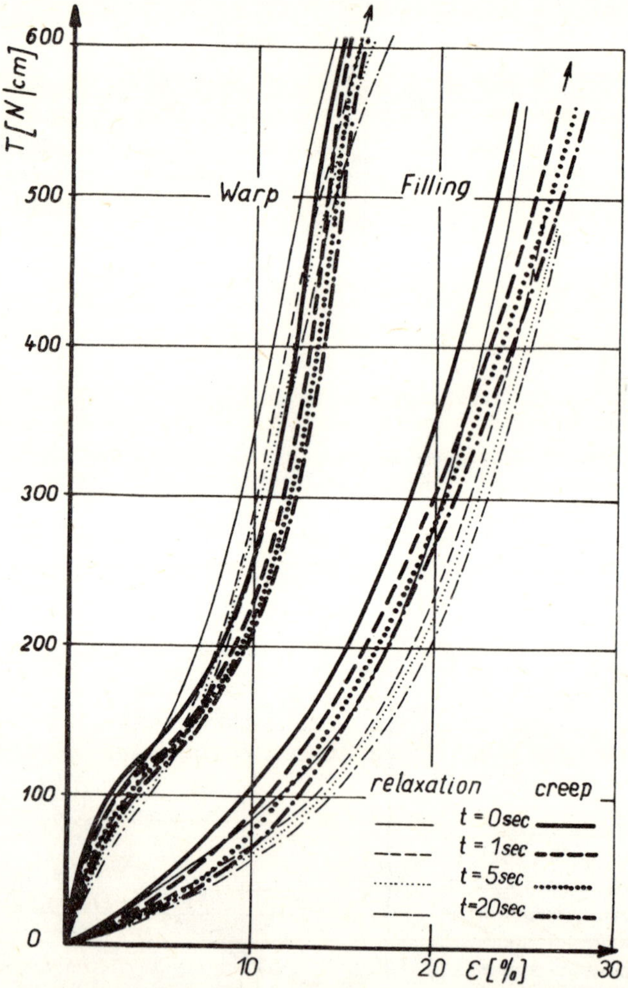

Fig. II-26. Averaged stress-strain diagrams of type 12 824 polyamide fabric coated with PVC under stress relaxation and creep in uniaxial tension in the warp and woof directions [39].

64 % respectively) of its standard value. These results show that the test samples must be very carefully prepared.

The tests of a sample of industrial fabrics in biaxial tension have shown that the shape of the stress-strain diagram depends strongly on the ratio of the stress in the direction of the warp to the stress in the direction of the woof. According to [52], these diagrams for a polyamide fabric are shown in Fig. II-27 and for a polyesther fabric in Fig. II-28. The investigated fabrics are of U.S. origin.

Table I shows uniaxial tensile trength T_s, ultimate relative elongation ε_s and the following impact strength P in the direction of the warp and woof for two types of Trevira polyesther fabrics produced in the Federal Republic of Germany, two types of polyamide fabrics produced by Polydrom, Sweden, two types of synthetic fabrics produced by the Swedish firm of Barracudaverken and the Czechoslovak-made type 12 824 polyamide fabric. All of these PVC coated fabrics are resistant to higher temperatures up to 70—80 °C and can withstand frost of −25 °C to −30°C.

Paper [53] discusses the various materials which are suitable for air-supported structures.

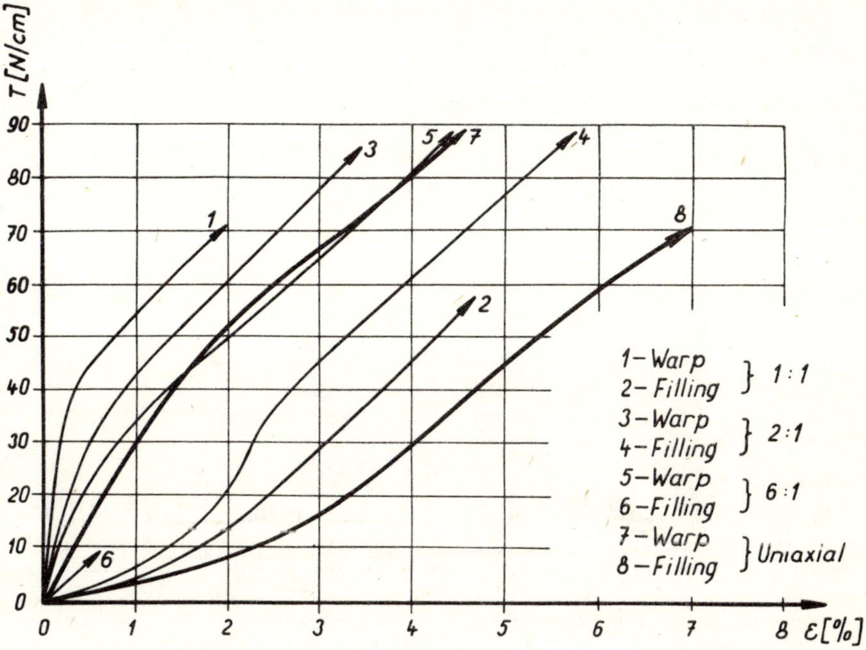

Fig. II-27. Stress-strain diagrams of the polyamide fabric under biaxial tension [52].

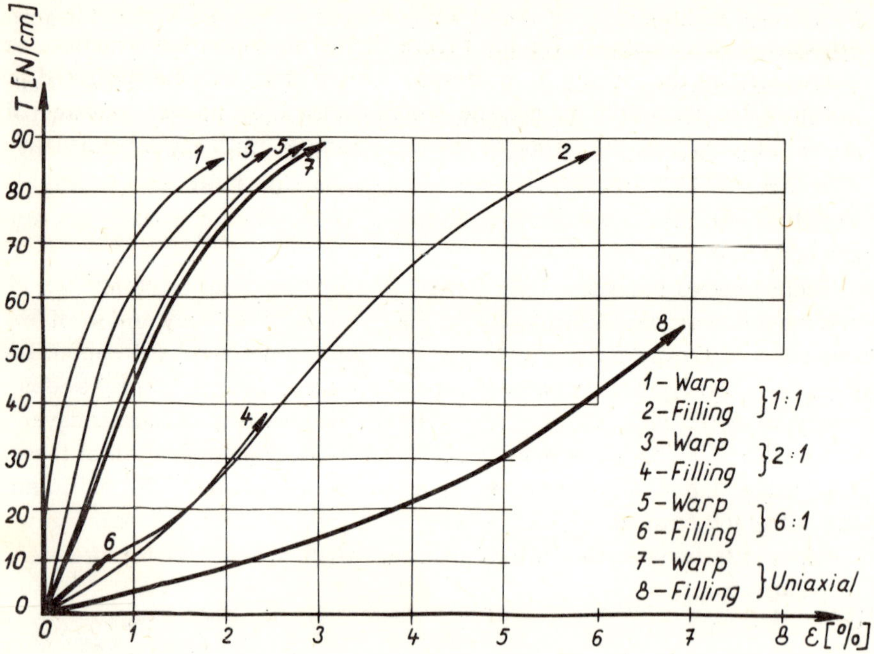

Fig. II-28. Stress-strain diagrams of polyesther fabric under biaxial tension [52].

Table I. Mechanical Properties of Some Industrial Fabrics

Fabric type	Warp			Filling		
	T_s [N/cm]	ε_s [%]	P [N]	T_s [N/cm]	ε_s [%]	P [N]
Trevira with plain weave	600	14	300	560	15	300
Trevira with basket weave	1360	18	700	1200	18	700
Polydrom C 121250	800 to 900	25	650 to 1000	800 to 900	35	650 to 1000
Polydrom C 151570	1160 to 1400	25	1500 to 2000	1160 to 1400	30	1500 to 2000
Barracuda 196	600	—	350	520	—	300
Barracuda 199	980	—	600	820	—	600
12 824/550/743 Fabric	820	19	630	680	30	560

Let us note that the fabric strength in further impact and not tensile strength is often decisive for the failure of the air-supported structure, as particularly in the vicinity of anchorage the structure may be subjected to considerable tangential membrane forces, caused by uneven transfer of tensile forces from the structure into its foundations. This is particularly manifest in the case of mobile anchorage, where screw piles are used. It is advisable, therefore, to reinforce the structure in the vicinity of anchorage with a continuous strip of fabric.

At present, polyesther fabrics are mostly used for air-supported and suspended structures; they are not susceptible to photodegradation and therefore need not be provided, with a light barrier as polyamide fabrics.

8.2 *Constitutive Equations for Elastic Orthotropic Membranes*

Industrial fabrics are governed by deformation laws of orthotropic membranes, because they have different mechanical properties in the direction of the warp and woof (filling). After several loading cycles the elasticity characteristics of a fabric, within a certain stress range under short-term loads, can be considered as constant quantities. These are: the tensile moduli of elasticity E_w and E_f in the direction of the warp and woof, the shear modulus of elasticity G and Poisson's ratios ν_w and ν_f.

If we identify the direction of the tensile unit membrane force T_1 with the direction of the warp and the direction of the tensile unit force T_2 with the direction of the woof, the following relations hold for an elastic orthotropic membrane:

$$\varepsilon_w = \frac{T_1}{E_w} - \frac{\nu_f}{E_f} T_2, \quad \varepsilon_f = \frac{T_2}{E_f} - \frac{\nu_w}{E_w} T_1, \quad \gamma_{w,f} = \frac{T_{12}}{G}, \tag{8.1}$$

where $\varepsilon_w (\varepsilon_f)$ — relative elongation in the warp (the filling) direction,
$\gamma_{w,f}$ — relative displacement (shear strain),
T_{12} — unit shearing force.

According to Betti's theorem we have

$$\frac{\nu_w}{E_w} = \frac{\nu_f}{E_f}. \tag{8.2}$$

If an industrial fabric is subjected to a tensile unit membrane force T_1, forming an angle φ with the warp, also to a unit tensile force $T_2 \perp T_1$ and

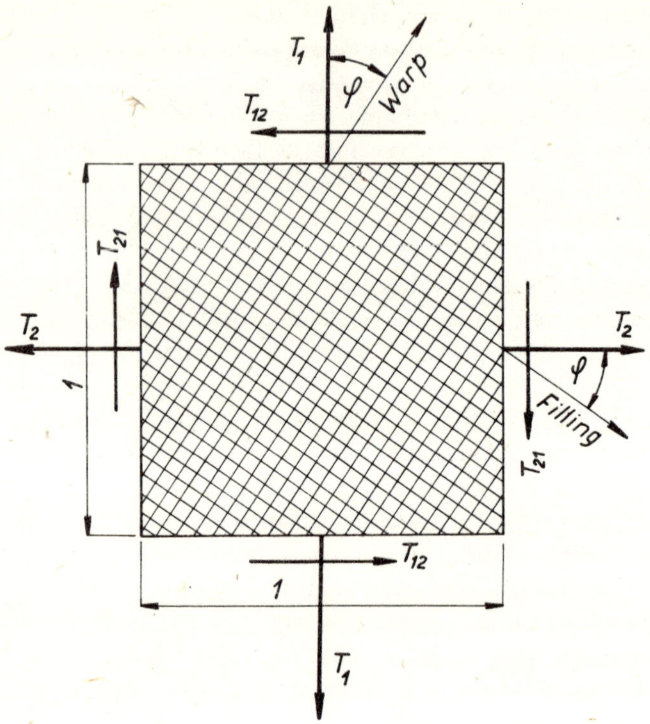

Fig. II-29. Effect of unit membrane forces on a unit surface element of an orthotropic construction fabric.

a shearing force $T_{12}=T_{21}$ per unit length of the fabric (Fig. II-29), the constitutive equations take the form [34]

$$\varepsilon_1=\frac{T_1}{E_1}-\frac{\nu_2}{E_2}T_2+\eta_1 T_{12},$$

$$\varepsilon_2=\frac{T_2}{E_2}-\frac{\nu_1}{E_1}T_1+\eta_2 T_{12}, \qquad (8.3)$$

$$\gamma=\frac{T_{12}}{E_{12}}+\eta_1 T_1+\eta_2 T_2,$$

where

$$\begin{aligned}
&\frac{1}{E_1}=\frac{\cos^4\varphi}{E_w}+\left(\frac{1}{G}-\frac{2\nu_w}{E_w}\right)\sin^2\varphi\cos^2\varphi+\frac{\sin^4\varphi}{E_f},\\
&\frac{1}{E_2}=\frac{\sin^4\varphi}{E_w}+\left(\frac{1}{G}-\frac{2\nu_f}{E_f}\right)\sin^2\varphi\cos^2\varphi+\frac{\cos^4\varphi}{E_f},\\
&\frac{\nu_1}{E_1}=\frac{\nu_2}{E_2}=\frac{1}{4}\left(\frac{1+\nu_w}{E_w}+\frac{1+\nu_f}{E_f}-\frac{1}{G}\right)\sin^2 2\varphi-\frac{\nu_w}{E_w},\\
&\frac{1}{E_{12}}=\left(\frac{1+\nu_w}{E_w}+\frac{1+\nu_f}{E_f}-\frac{1}{G}\right)\sin^2 2\varphi+\frac{1}{G},\\
&\eta_1=\left[2\left(\frac{\sin^2\varphi}{E_f}-\frac{\cos^2\varphi}{E_w}\right)+\left(\frac{1}{G}-\frac{2\nu_w}{E_w}\right)(\cos^2\varphi-\sin^2\varphi)\right]\sin\varphi\cos\varphi,\\
&\eta_2=\left[2\left(\frac{\cos^2\varphi}{E_f}-\frac{\sin^2\varphi}{E_w}\right)-\left(\frac{1}{G}-\frac{2\nu_f}{E_f}\right)(\cos^2\varphi-\sin^2\varphi)\right]\sin\varphi\cos\varphi.
\end{aligned} \tag{8.4}$$

Moreover the following invariant relations are valid:

$$\begin{aligned}
&\frac{1}{E_1}+\frac{1}{E_2}-\frac{2\nu_1}{E_1}=\frac{1}{E_w}+\frac{1}{E_f}-\frac{2\nu_w}{E_w},\\
&\frac{1}{E_{12}}+\frac{4\nu_1}{E_1}=\frac{1}{G}+\frac{4\nu_w}{E_w}.
\end{aligned} \tag{8.5}$$

In the relations (8.3) it is often possible to neglect the terms containing the coefficients η_1 and η_2, as their values are usually small in comparison with other terms. Therefore, it is possible to use the following approximate constitutive equations:

$$\varepsilon_1=\frac{T_1}{E_1}-\frac{\nu_2}{E_2}T_2,\quad \varepsilon_2=\frac{T_2}{E_2}-\frac{\nu_1}{E_1}T_1,\quad \gamma_{12}=\frac{T_{12}}{E_{12}}. \tag{8.6}$$

For the 12 824 polyamide fabric, short-term sample tests stressed in uniaxial tension yielded the following approximate mean values for the Poisson's ratios at the force per unit length $T \leqq 250$ N/cm:

$$\nu_w=0.4,\quad \nu_f=0.14. \tag{8.7}$$

From the Poisson's ratios (8.7) and the values of the tension moduli of elasticity

$$E_w=3\,850 \text{ N/cm},\quad E_f=1\,250 \text{ N/cm}, \tag{8.8}$$

which nearly satisfy the relation (8.2), it is possible to determine aproximately the modulus of elasticity in shear G for the 12 824 fabric under low loads ($T \leqq 100$ N/cm).

Following the method used for the determination of the modulus of elasticity in shear of an orthotropic concrete slab reinforced in two mutually perpendicular direction [57], we obtain

$$\bar{G}=\frac{E_w}{2(1-\nu_w\nu_f)}\left(\sqrt{\frac{E_f}{E_w}}-\nu_f\right), \tag{8.9}$$

or

$$\bar{\bar{G}}=\frac{E_f}{2(1-\nu_w\nu_f)}\left(\sqrt{\frac{E_w}{E_f}}-\nu_w\right). \tag{8.10}$$

On account of (8.2), the formulae (8.9) and (8.10) take the form

$$\bar{G}=\frac{E_w}{2(1+\sqrt{\nu_w\nu_f})}\sqrt{\frac{\nu_f}{\nu_w}} \tag{8.11}$$

and

$$\bar{\bar{G}}=\frac{E_f}{2(1+\sqrt{\nu_w\nu_f})} \tag{8.12}$$

We propose to use the geometric mean of the quantities (8.11) and (8.12) to determine approximately the value of the modulus of elasticity in shear G of an industrial fabric, i.e. we suggest the formula

$$G\approx\sqrt{(\bar{G}\bar{\bar{G}})}=\frac{\sqrt{E_wE_f}}{2(1+\sqrt{\nu_w\nu_f})}, \tag{8.13}$$

which, recalling (8.2), can be written in the form

$$G\approx\frac{E_f}{2(1+\sqrt{\nu_w\nu_f})}\sqrt{\frac{\nu_w}{\nu_f}}, \tag{8.14}$$

or

$$G\approx\frac{E_w}{2(1+\sqrt{\nu_w\nu_f})}\sqrt{\frac{\nu_f}{\nu_w}}. \tag{8.15}$$

A particular case of the relations (8.13), (8.14) and (8.15) is the well known relation

$$G = \frac{E}{2(1+\nu)}, \tag{8.16}$$

applicable to isotropic material.

Sustitution of (8.7) and (8.8) into (8.13) yields, for the 12 824 polyamide fabric under low loads,

$$G \approx 900 \text{ N/cm}. \tag{8.17}$$

Actually the modulus G is not a constant quantity, but depends, among other things, on the state of stress of the fabric.

Using the values (8.7), (8.8) and (8.17), we can determine, in accordance with the relations (8.4) the elasticity constants of the 12 824 fabric occuring in the constitutive equations (8.3) or (8.6).

The deformation laws of viscoelastic orthotropic membranes are dealt with in Subsections 2.3 and 2.4 of this chapter.

8.3 *Assessment of the Safety of Air-Supported Structures*

To design a method of anchorage of an air-supported structure, we have to evaluate the magnitude of the membrane forces along the edges of the structure due to the least favourable load combination.

For the assessment of the safety of an air-supported structure, i.e. of the suitability of the fabric used for the selected shape and size of the structure, under the least favourable load combination, we recommend the use of the inequality

$$T_{max} \leqq T_d. \tag{8.18}$$

Here T_{max} is the highest value of the principal tensile membrane force in the structure and T_d is the design stress of the fabric in the direction of the force T_{max} in biaxial tension, characterized by the ratio T_{min}/T_{max}.. We also recall that [19]

$$\begin{aligned} T_{max} &= \frac{1}{2}\left[T_1 + T_2 + \sqrt{(T_1 - T_2)^2 + 4T_{12}^2}\right], \\ T_{min} &= \frac{1}{2}\left[T_1 + T_2 - \sqrt{(T_1 - T_2)^2 + 4T_{12}^2}\right] \end{aligned} \tag{8.19}$$

are the principal tensile membrane forces which are mutually perpendicular. The angle φ_0 between the forces T_{max} and T_1 satisfies [19]

$$\tan 2\varphi_0 = -\frac{2T_{12}}{T_1 - T_2} \pm \pi. \tag{8.20}$$

Due to the incompleteness of existing experimental data on the strength of the fabrics in various directions in biaxial tension with various ratios of tensile forces, the fabric has been assessed, so far, on the basis of the inequalities

$$T_w \leqq T_{w,d}, \quad T_f \leqq T_{f,d}, \tag{8.21}$$

where T_w and T_f are the tensile membrane forces in the warp and woof directions respectively, generated by the least favourable loads considered, and $T_{w,d}$, $T_{f,d}$ are the design stresses of the fabric in these directions in uniaxial tension.

The forces T_w and T_f are determined from the relations [19]

$$\begin{aligned} T_w &= \frac{1}{2}(T_1 + T_2) + \frac{1}{2}(T_1 - T_2)\cos 2\varphi - T_{12}\sin 2\varphi, \\ T_f &= T_1 \sin^2\varphi + T_2\cos^2\varphi + T_{12}\sin 2\varphi, \end{aligned} \tag{8.22}$$

where φ is the acute angle between the warp and the force T_1 (Fig. II-29).

The shearing membrane force $T_{w,f}$ given by

$$T_{w,f} = \frac{1}{2}(T_1 - T_2)\sin 2\varphi + T_{12}\cos 2\varphi \tag{8.23}$$

exerts a considerable influence on the strength of the fabric; however, as yet, this influence has not been taken into account.

The design stresses of the fabric can be determined from its strength parameters. In this process the reduction of the strength in time must be taken into account. When using the criteria (8.21) it is also necessary to take into account the reduction of the fabric strength due to biaxial loading.

Chapter III

FORMFINDING OF AIR-SUPPORTED STRUCTURES

Introduction

The design of air-supported structures of a stable form must respect the geometric and static conditions of continuity in the whole membrane under the influence of internal overpressure, representing a permanent load. The tensile mebrane forces in the structure due to the internal overpressure ensure the stability of form of the air-supported structure even under the effect of other load types, such as wind and snow loads.

This chapter contains the author's latest results, both theoretical and experimental, in desining forms of air-supported structures. Attention is concentrated chiefly on the surfaces of revolution, skew surfaces over rectangular or square plans and the design of the form of entrance openings.

The closing off of a surface of revolution by a plane perpendicular to the axis of revolution yields a cupola with a circular plan. Further interesting forms of air-supported structures with oval plans can be obtained by slicing the surfaces of revolution by planes not perpendicular to the axis of revolution. In a special case such limiting planes can be parallel to the axis of revolution. This case is possible when the meridians of the surfaces are closed curves, as the limiting plane together with the surface of revolution must always fully enclosed the interior of the air-supported structure. For example, the slicing of a cycloid of revolution by a plane parallel to the axis of revolution yields an air-supported structure in the form similar to that of an inverted boat, which is capable not only to withstand a static load, but also to cope with aerodynamic forces (see Chapter II, Subsection 7.2).

The general principles of the design of air-supported structures are described in [1], [3], [6], [47] and [60].

1

Surfaces of Characteristic Curvature

1.1 *Definition and Fundamental Relations*

By a surface of characteristic curvature we shall mean a surface whose principal curvatures $\varkappa_1$ and $\varkappa_2$ satisfy a relation

$$F(\varkappa_1, \varkappa_2)=0, \tag{1.1}$$

where F is a chosen function. The relation (1.1) shall be called the characteristic curvature of the surface [11].

For a surface with a rectangular system of curvilinear coordinates ξ and η, determined by the principal directions, there are six functions

$$\varkappa_1, \varkappa_2, g_{11}, g_{22}, h_{11}, h_{22} \tag{1.2}$$

of the two variables ξ and η. They satisfy the five independent equations (1.21) and (1.32) of Chapter II. If the surface is of characteristic curvature, equation (1.1) represents the sixth independent relation for the six quantities (1.2), thus defining the form of the surface.

Let us note that the determination of a surface of characteristic curvature is based on the given relation (1.1) and not on given functions g_{ij} and h_{ij} which determine explicitely the form and the size of the surface (Bonnet Theorem).

In a particular case, when the characteristic curvature (1.1) has the form

$$\frac{\varkappa_1}{\varkappa_2}=k, \tag{1.3}$$

where k is a constant (a real number), we obtain, as a result of substituting the relation (1.19) of Chapter II into (1.3), and applying the formulae (1.18) of Chapter II, the following equation of the surface:

$$\begin{aligned}&(g_{11}g_{22}-g_{12}^2)(h_{11}h_{22}-h_{12}^2)-\\&-\frac{k}{1+k^2}(g_{11}h_{22}-2g_{12}h_{12}+g_{22}h_{11})^2=0.\end{aligned} \tag{1.4}$$

Consider a surface of characteristic curvature (1.3), described in Cartesian coordinates x, y, z by an explicit equation $z=f(x, y)$. Substituting the formulae (1.22) from Chapter II into equation (1.4), we obtain the non-linear partial differential equation

$$\begin{gathered}(1+f_x^2+f_y^2)\,(f_{xx}f_{yy}-f_{xy}^2)-\\ -\frac{k}{(1+k)^2}\,[(1+f_x^2)f_{yy}-2f_xf_yf_{xy}+(1+f_y^2)f_{xx}]^2=0.\end{gathered} \tag{1.5}$$

Developable ruled surfaces, in particular cylindrical and conical surfaces, have a zero Gaussian curvature ($K=0$). According to (1.3) for these surfaces $k=0$ and equation (1.5) is simplified to acquire the form

$$f_{xx}f_{yy}-f_{xy}^2=0. \tag{1.6}$$

1.2 *Some Properties of Surfaces of Characteristic Curvature*

Let us discuss now some properties of surfaces satisfying relation (1.3). The ratio of the principal curvature and the normal curvature in a direction deviating from the principal direction of the surface by the same angle is constant at all points of the surface [11].

To prove this property we shall use Euler's theorem, according to which the normal curvature $\varkappa$ of the normal section, whose tangent forms an angle φ with the principal tangent, equals [7]

$$\varkappa=\varkappa_1\cos^2\varphi+\varkappa_2\sin^2\varphi. \tag{1.7}$$

This identity and (1.3) yield for a selected point P on the surface

$${}^P\varkappa={}^P\varkappa_2(k\cos^2\varphi+\sin^2\varphi) \tag{1.8}$$

and, for another point R on the same surface,

$${}^R\varkappa={}^R\varkappa_2(k\cos^2\varphi+\sin^2\varphi). \tag{1.9}$$

Equations (1.8) and (1.9) imply the relation

$$\frac{{}^P\varkappa_2}{{}^P\varkappa}=\frac{{}^R\varkappa_2}{{}^R\varkappa} \tag{1.10}$$

which proves the above mentioned property.

Equation (1.10) can be written equivalently as

$$\frac{^{P}\varkappa}{^{R}\varkappa}=\frac{^{P}\varkappa_2}{^{R}\varkappa_2}, \tag{1.11}$$

whence we get the following property of the surface:

At any two points of a surface of characteristic curvature (1.3) the ratio of the normal curvatures of normal sections of the surface is constant in directions deviating from the principal directions of the surface by the same angle φ.

From (1.3) and (1.18) of Chapter II we obtain

$$K=k\varkappa_2^2, \quad H=\frac{1}{2}(1+k)\varkappa_2. \tag{1.12}$$

Squaring the second equation and dividing both equations by one another, we get

$$\frac{K}{H^2}=\frac{4k}{(1+k)^2}. \tag{1.13}$$

Consequently, at all points of the surface the ratio of the Gaussian curvature to the square of the mean curvature is constant and equals $4k/(1+k)^2$.

2

Survaces of Revolution of Characteristic Curvature

2.1 *Differential Equation of the Meridian*

In this subsection we shall deduce the differential equation of the meridian of a surface of revolution satisfying (1.1) in the form

$$\frac{\varkappa_1^m}{\varkappa_2^n} = k, \tag{2.1}$$

where m, n, k are constants.

For the meridian (directrix) of a surface of revolution defined by the parametric equations (1.25) in Chapter II we obtain, by substituting the expressions (1.27) of Chapter II into (2.1) and rearranging the result, the following ordinary differential equation of second order:

$$r^n r''^m - (-1)^m k(1 + r'^2)^{(3m-n)/2} = 0. \tag{2.2}$$

In the case that $m = n$ and $(-1)^m = -1$, (2.2) can be written as

$$rr'' + \bar{a}r'^2 + \bar{a} = 0, \tag{2.3}$$

where

$$\bar{a} = k^{1/n}. \tag{2.4}$$

If $m = n$ and $(-1)^m = 1$, then $\bar{a} = -k^{1/n}$.

For $n = 3m$ we obtain from (2.2) the following differential equation of the meridian:

$$r^3 r'' = -k^{1/m}. \tag{2.5}$$

If $m = 0$, then (2.2) becomes a first order differential equation:

$$r^2 r'^2 + r^2 = k^{2/n}. \tag{2.6}$$

For $n = 0$, (2.2) acquires the form

$$r''^2 + k^{2/m}(1 + r'^2)^3 = 0, \tag{2.7}$$

and for $k=0$ we obtain $(r \neq 0)$

$$r''=0 \tag{2.8}$$

which is the differential equation of the meridian of conical and cylindrical surfaces $(r=C_1 z+C_2)$.

2.2 *Solution of the Equations (2.3), (2.5) and (2.6)*

Let us deduce now some equations of the meridians with respect to the rectangular coordinates (Fig. II-3).

The first integral of equation (2.3) is determined by the introduction of a new function $y=y(r)$ which is defined by the relations [31]

$$y^2=p^2, \quad p(r)=r'(z). \tag{2.9}$$

According to (2.9) we have

$$\begin{gathered} r'^2(z)=y, \\ r''=p'(r)r' = p'p=\frac{1}{2}\frac{\mathrm{d}}{\mathrm{d}r}[p^2(r)]=\frac{1}{2}y', \end{gathered} \tag{2.10}$$

where

$$p'=\mathrm{d}p/\mathrm{d}r, \quad r'=\mathrm{d}r/\mathrm{d}z, \quad y'=\mathrm{d}y/\mathrm{d}r.$$

By substituting (2.10) into (2.3) we obtain for the function y a linear equation of Euler type,

$$ry'+2\bar{a}y+2\bar{a}=0, \tag{2.11}$$

whose solution has the form

$$y=C_1 r^{-2\bar{a}}-1, \tag{2.12}$$

where C_1 is the integration constant.

By substituting the first expression of (2.10) into (2.12) we obtain the following first integral of equation (2.3):

$$r^{2\bar{a}}r'^2+r^{2\bar{a}}=C_1. \tag{2.13}$$

Separation of variables in equation (2.13) and integration yield the general solution of equation (2.3)

$$z=\int C_1r^{-2\bar{a}}-1)^{-1/2}\,\mathrm{d}r+C_2, \tag{2.14}$$

where C_2 is the second integration constant.

If $k=-1$ and n is an odd number, then according to (2.4), $\bar{a}=-1$ and from (2.14) we obtain the following equation of the meridian

$$z=\frac{1}{\sqrt{C_1}}\ln(r\sqrt{C_1}+\sqrt{C_1r^2-1}) \tag{2.15}$$

whose revolution about the axis z yields a catenoid.

If $k=1$, then according to (2.4), $\bar{a}=1$ and from (2.14) we obtain

$$r^2+(z-C_2)^2=C_1 \tag{2.16}$$

which is the equation of a semicircle of radius $R=C_2=\sqrt{C_1}$.

For $\bar{a}=-1/2$ ($k=(-1/2)^n$) we obtain from (2.14) the equation

$$z=\frac{2}{C_1}\sqrt{C_1r-1}+C_2\,, \tag{2.17}$$

and for $\bar{a}=1/2$ we obtain

$$z=-2\left(\frac{1}{5}r^2-\frac{2}{3}C_1r+C_2^2\right)(c_1-r)^{1/2} \tag{2.18}$$

which is the equation of a cycloid.

Let us solve now equation (2.5). We perform the substitution

$$p(r)=r',\quad r''=p'r'=p'p=\frac{1}{2}\frac{\mathrm{d}}{\mathrm{d}r}(p^2). \tag{2.19}$$

Then (2.5) yields, after separation of variables and integration,

$$\frac{1}{r^2}k^{1/m}=p^2+C_1. \tag{2.20}$$

Applying the first relation (2.19), we arrive at

$$r'^2=\frac{1}{r^2}k^{1/m}-C_1, \tag{2.21}$$

and thus

$$r'=\frac{1}{r}(k^{1/m}-C_1r^2)^{1/2}. \tag{2.22}$$

Hence, separating the variables and integrating, we obtain the general solution of equation (2.5) in the form

$$(C_1 z + C_2)^2 + C_1 r^2 - k^{1/m} = 0. \tag{2.23}$$

If in equation (2.21) we put $C_1 = 0$, then

$$\frac{1}{2}\frac{\mathrm{d}}{\mathrm{d}z}(r^2) = k^{\frac{1}{2m}}, \tag{2.24}$$

whence we obtain the following solution of equation (2.5):

$$r^2 = 2k^{\frac{1}{2m}} z + C. \tag{2.25}$$

The solution of equation (2.6) has the form (2.16), i.e.

$$r^2 + (z - C)^2 = k^{1/n}. \tag{2.26}$$

2.3 *Parametric Equations of the Meridian*

If the characteristic curvature (1.1) has the form

$$R_1/R_2 = g(\alpha), \tag{2.27}$$

where $g(\alpha)$ is a given function of the angle α (Fig. II-3), then equation (3.17) of Chapter II implies that we must have $g(\alpha) \geqq 1/2$ to prevent the creasing of the membrane of the air-supported structure.

If we substitute (2.27) into the Gauss—Codazzi equation (1.36) in Chapter II, we obtain

$$R_2 g(\alpha) \cos \alpha = \frac{\mathrm{d}}{\mathrm{d}\alpha}(R_2 \sin \alpha), \tag{2.28}$$

whence

$$\frac{\mathrm{d}(R_2 \sin \alpha)}{R_2 \sin \alpha} = g(\alpha) \cot \alpha \, \mathrm{d}\alpha \, . \tag{2.29}$$

The integration of (2.29) yields for the function

$$r = R_2 \sin \alpha \, , \tag{2.30}$$

the formula

$$r(\alpha) = C_1 \mathrm{e}^{\int g(\alpha) \cot \alpha \, \mathrm{d}\alpha}, \tag{2.31}$$

where C_1 is an integration constant and e is the base of natural logarithms.

If we substitute (2.27) into equation (see Fig. II-3)

$$\frac{dz}{d\alpha} = R_1 \sin \alpha, \tag{2.32}$$

we obtain, with the help of (2.30),

$$\frac{dz}{d\alpha} = r(\alpha) g(\alpha). \tag{2.33}$$

Hence

$$z = \int r(\alpha) g(\alpha) \, d\alpha + C_2, \tag{2.34}$$

where C_2 is an integration constant. This, together with (2.31), yields

$$z = C_1 \int g(\alpha) e^{\int g(\alpha) \cot \alpha \, d\alpha} \, d\alpha + C_2. \tag{2.35}$$

(2.31) and (2.35) are the parametric equations of a meridian of a surface of revolution deduced by the author in [16]. The selection of the function $g(\alpha) > 1/2$ yields various shapes of the surfaces of revolution. These are suitable for air-supported structures, as the internal overpressure produces only tensile membrane forces at every point of the surface.

If we put, for example,

$$g(\alpha) = a + b \frac{\tan \alpha}{\alpha}, \tag{2.36}$$

where a, b are constants, then the equations (2.35) and (2.31) become

$$\begin{aligned} z(\alpha) &= C_1 \int (a + b \tan \alpha / \alpha) \alpha^b \sin^a \alpha \, d\alpha + C_2, \\ r(\alpha) &= C_1 \alpha^b \sin^a \alpha. \end{aligned} \tag{2.37}$$

The relation between the integration constants C_1 and C_2 can be determined from the condition that the origin of the z, r axes should be identical with the apex of the surface of revolution, i.e.

$$z(0) = 0. \tag{2.38}$$

In a particular case, when $b = 0$, equations (2.37) take the form

$$\begin{aligned} z(\alpha) &= C_1 a \int \sin^a \alpha \, d\alpha + C_2, \\ r(\alpha) &= C_1 \sin^a \alpha. \end{aligned} \tag{2.39}$$

For $a = 1$, $b = 0$ we obtain from the first equation (2.39)

$$z(\alpha) = -C_1 \cos \alpha + C_2. \tag{2.40}$$

Substituting this into (2.38), we get $C_2 = C_1$.

In this case we obtain the parametric equations of the circle

$$\begin{aligned} z(\alpha) &= C_1(1 - \cos \alpha), \\ r(\alpha) &= C_1 \sin \alpha, \end{aligned} \tag{2.41}$$

where the integration constant C_1 is the radius of the circle, thus determining the magnitude of the meridian.

For $a = 2$, $b = 0$ we obtain the parametric equations of the cycloid,

$$\begin{aligned} z(\alpha) &= C_1\left(\alpha - \frac{1}{2}\sin 2\alpha\right), \\ r(\alpha) &= C_1 \sin^2 \alpha. \end{aligned} \tag{2.42}$$

For $a = 3$, $b = 0$ the meridian of the surface of revolution is given by the equations

$$\begin{aligned} z(\alpha) &= C_1(2 + \cos^3 \alpha - 3\cos \alpha), \\ r(\alpha) &= C_1 \sin^3 \alpha \end{aligned} \tag{2.43}$$

and for $a = 4$, $b = 0$,

$$\begin{aligned} z(\alpha) &= C_1\left(\frac{3}{2}\alpha - \sin 2\alpha + \frac{1}{8}\sin 4\alpha\right), \\ r(\alpha) &= C_1 \sin^4 \alpha. \end{aligned} \tag{2.44}$$

If generally $a = 2l$ or $a = 2l + 1$, where l is a natural number, and $b = 0$, the integral in (2.39) can be expressed in terms of trigonometric functions of the angle α [21].

For $a = 2l$, we have, according to (2.39),

$$\begin{aligned} z(\alpha) &= 2C_1 l\left[\frac{1}{2^{2l}}\binom{2l}{l}\alpha + \frac{(-1)^l}{2^{2l-1}}\sum_{j=0}^{l-1}(-1)^j\binom{2l}{j}\frac{\sin(2l-2j)\alpha}{2l-2j}\right] + C_2, \\ r(\alpha) &= C_1 \sin^{2l} \alpha \end{aligned} \tag{2.45}$$

and for $a = 2l + 1$, we have

$$\begin{aligned} z(\alpha) &= C_1(2l+1)\frac{1}{2^l}(-1)^{l+1}\sum_{j=0}^{l}(-1)^j\binom{2l+1}{j}\frac{\cos(2l+1-2j)\alpha}{2l+1-2j} + C_2, \\ r(\alpha) &= C_1 \sin^{2l+1}\alpha. \end{aligned} \tag{2.46}$$

If $a=0$, $b=1$, we obtain from (2.37)

$$\begin{aligned} z(\alpha) &= -C_1 \ln \cos \alpha + C_2, \\ r(\alpha) &= C_1 \alpha \end{aligned} \tag{2.47}$$

and for $a=1$, $b=1$ we obtain from (2.37), after integration, the following parametric equations of the meridian

$$\begin{aligned} z(\alpha) &= C_1\left[\ln \tan\left(\frac{\pi}{4}+\frac{\alpha}{2}\right) - \alpha \cos \alpha\right] + C_2, \\ r(\alpha) &= C_1 \alpha \sin \alpha, \end{aligned} \tag{2.48}$$

the condition (2.38) being satisfied for $C_2=0$.

For $a=2$, $b=1$ the parametric equations satisfying the condition (2.38) have the form

$$\begin{aligned} z(\alpha) &= C_1\left(\frac{\alpha^2}{2} - \frac{\alpha}{2}\sin 2\alpha - \ln \cos \alpha\right), \\ r(\alpha) &= C_1 \alpha \sin^2 \alpha \end{aligned} \tag{2.49}$$

and for $a=3$, $b=1$ the form

$$\begin{aligned} z(\alpha) &= C_1\left[2 + \cos^3 \alpha - 3\cos \alpha - \frac{\sin^3 \alpha}{3} - \sin \alpha + \ln \tan\left(\frac{\pi}{4}+\frac{\alpha}{2}\right)\right], \\ r(\alpha) &= C_1 \alpha \sin^3 \alpha. \end{aligned} \tag{2.50}$$

If we select the function $g(\alpha) > 1/2$ as

$$g(\alpha) = c + \frac{d}{\cos^2 \alpha}, \tag{2.51}$$

where c, d are constants, we obtain from (2.35) and (2.31)

$$\begin{aligned} z(\alpha) &= C_1\left[c\int \frac{\sin^{c+d} \alpha}{\cos^d \alpha}\, d\alpha + d\int \frac{\sin^{c+d} \alpha}{\cos^{d+2} \alpha}\, d\alpha\right] + C_2, \\ r(\alpha) &= C_1(\sin^c \alpha + \tan^d \alpha). \end{aligned} \tag{2.52}$$

For $c=0$, $d=1$ we have, according to (2.52),

$$\begin{aligned} z(\alpha) &= C_1 \tan^2 \alpha + C_2, \\ r(\alpha) &= C_1 \tan \alpha; \end{aligned} \tag{2.53}$$

for $c=0$, $d=2$ we obtain

$$z(\alpha)=\frac{2}{3}C_1\tan^3\alpha+C_2,$$
$$r(\alpha)=C_1\tan^2\alpha, \tag{2.54}$$

and (2.38) yields $C_2=0$.

If $c=1$, $d=1$, we obtain from (2.52)

$$z(\alpha)=C_1\left[-\sin\alpha+\frac{1}{2}\ln\tan\left(\frac{\pi}{4}+\frac{\alpha}{2}\right)+\frac{\sin\alpha}{2\cos\alpha}\right]+C_2,$$
$$r(\alpha)=C_1\frac{\sin^2\alpha}{\cos\alpha}, \tag{2.55}$$

where $C_2=0$.

For $c=1$, $d=2$ we deduce, with the aid of condition (2.38),

$$z(\alpha)=C_1\left(\frac{2}{3}+\cos\alpha-\frac{1}{\cos\alpha}+\frac{2}{3\cos^3\alpha}\right),$$
$$r(\alpha)=C_1\frac{\sin^3\alpha}{\cos\alpha}, \tag{2.56}$$

and for $c=2$, $d=2$ we obtain

$$z(\alpha)=C_1\left(-\alpha+\sin\alpha\cos\alpha+\frac{2}{3}\tan^3\alpha\right),$$
$$r(\alpha)=C_1\frac{\sin^4\alpha}{\cos^2\alpha}. \tag{2.57}$$

Next let us consider the case when

$$g(\alpha)=k+h\frac{\sin^2\alpha}{\cos^2\alpha}, \tag{2.58}$$

where k and h are constants. Substituting (2.58) into (2.35) and (2.31), we obtain

$$z(\alpha)=C_1\int\left(k+h\frac{\sin^2\alpha}{\cos^2\alpha}\right)\frac{\sin^k\alpha}{\cos^h\alpha}\,d\alpha+C_2,$$
$$r(\alpha)=C_1\frac{\sin^k\alpha}{\sin^h\alpha}. \tag{2.59}$$

If we put $k=0$, $h=1$, we obtain, according to (2.59) and with the help of (2.38),

$$z(\alpha)=C_1\frac{1}{2}\left(\frac{\sin\alpha}{\cos^2\alpha}-\ln\frac{1+\sin\alpha}{\cos\alpha}\right),$$
$$r(\alpha)=C_1\frac{1}{\cos\alpha}, \tag{2.60}$$

and for $k=0$, $h=2$ we obtain

$$z(\alpha)=C_1\frac{2}{3}\tan^3\alpha, \quad r(\alpha)=C_1\frac{1}{\cos^2\alpha}. \tag{2.61}$$

For $k=1$, $h=1$ the equations (2.59) and (2.38) yield

$$z(\alpha)=C_1\frac{1}{2}\left(\frac{1}{\cos^2\alpha}-1\right), \quad r(\alpha)=C_1\tan\alpha, \tag{2.62}$$

and for $k=1$, $h=2$ we have

$$z(\alpha)=C_1\left(\frac{1}{3}+\frac{2}{3\cos^3\alpha}-\frac{1}{\cos\alpha}\right),$$
$$r(\alpha)=C_1\frac{\sin\alpha}{\cos^2\alpha}. \tag{2.63}$$

If $k=2$, $h=1$, the parametric equations (2.59) of the meridian take the form

$$z(\alpha)=C_1\left[-\sin\alpha+\frac{1}{2}\frac{\sin\alpha}{\cos^2\alpha}+\frac{1}{2}\ln\tan\left(\frac{\pi}{4}+\frac{\alpha}{2}\right)\right],$$
$$r(\alpha)=C_1\frac{\sin^2\alpha}{\cos\alpha}, \tag{2.64}$$

and for $k=2$, $h=2$ they are

$$z(\alpha)=C_1\frac{2}{3}\tan^3\alpha, \quad r(\alpha)=C_1\tan^2\alpha. \tag{2.65}$$

If the function $g(\alpha)$ is of the form

$$g(\alpha)=p+q\frac{\sin\alpha}{\cos^2\alpha}, \tag{2.66}$$

where p, q are constants, we obtain, as a result of substituting (2.66) into (2.35) and (2.31),

$$z(\alpha)=C_1\left[p\int \sin^p \alpha \frac{(1+\sin \alpha)^q}{\cos^q \alpha} d\alpha + q\int \frac{\sin^{p+1} \alpha(1+\sin \alpha)^q}{\cos^{q+2} \alpha} d\alpha\right]+C_2,$$
$$r(\alpha)=C_1 \sin^p \alpha \frac{(1+\sin \alpha)^q}{\cos^q \alpha}. \tag{2.67}$$

For $p=0$, $q=1$ we get from (2.67) and (2.38)

$$z(\alpha)=C_1 \frac{1}{2}\left[\frac{1}{1-\sin \alpha} - \ln \tan\left(\frac{\pi}{4}+\frac{\alpha}{2}\right)-1\right],$$
$$r(\alpha)=C_1 \frac{1+\sin \alpha}{\cos \alpha}. \tag{2.68}$$

If $p=0$, $q=2$, we obtain, according to (2.67) and (2.38),

$$z(\alpha)=2C_1\left(\frac{2}{3 \cos^3 \alpha}-\frac{1}{\cos \alpha}+\frac{2}{3}\tan^3 \alpha+\frac{1}{3}\right),$$
$$r(\alpha)=C_1 \frac{(1+\sin \alpha)^2}{\cos^2 \alpha}. \tag{2.69}$$

For $p=1$, $q=1$ we have

$$z(\alpha)=C_1\left(\frac{1+\sin \alpha}{2 \cos^2 \alpha}-\sin \alpha+\frac{1}{2}\ln \frac{1+\sin \alpha}{\cos \alpha}\right),$$
$$r(\alpha)=C_1(1+\sin \alpha)\tan \alpha. \tag{2.70}$$

For $p=1$, $q=2$ we deduce

$$z(\alpha)=C_1\left(\cos \alpha-\frac{2}{\cos \alpha}+\frac{4}{3}\frac{1+\sin^3 \alpha}{\cos^3 \alpha}-\frac{1}{3}\right),$$
$$r(\alpha)=C_1 \frac{\sin \alpha}{\cos^2 \alpha}(1+\sin \alpha)^2. \tag{2.71}$$

For $p=2$, $q=1$,

$$z(\alpha)==C_1\left(\frac{1+\sin \alpha}{2 \cos^2 \alpha}-\sin \alpha-\sin^2 \alpha+\frac{1}{2}\ln \frac{1+\sin \alpha}{\cos \alpha}-\frac{1}{2}\right),$$
$$r(\alpha)=C_1(1+\sin \alpha)\frac{\sin^2 \alpha}{\cos \alpha}. \tag{2.72}$$

When $p=2$, $q=2$,

$$z(\alpha)=C_1\left[\left(\frac{1}{2\cos^2\alpha}-2\right)\sin\alpha-\frac{3}{2}\sin^2\alpha-\sin^3\alpha-\right.$$

$$\left.-3\ln\cos\alpha+\frac{3}{2}\ln\tan\left(\frac{\pi}{4}+\frac{\alpha}{2}\right)\right], \tag{2.73}$$

$$r(\alpha)=C_1(1+\sin\alpha)^2\tan^2\alpha.$$

If the function $g(\alpha)$ has the form

$$q(\alpha)=m+\frac{n}{\cos\alpha}, \tag{2.74}$$

we obtain from (2.35) and (2.31) the following parametric equations of the meridian:

$$z(\alpha)=C_1\left[m\int\frac{\sin^{m+n}\alpha}{(1+\cos\alpha)^n}\,\mathrm{d}\alpha+n\int\frac{\sin^{m+n}\alpha}{\cos\alpha(1+\cos\alpha)^n}\,\mathrm{d}\alpha\right]+C_2,$$

$$r(\alpha)=C_1\frac{\sin^{m+n}\alpha}{(1+\cos\alpha)^n}. \tag{2.75}$$

These imply, for $m=0$, $n=1$,

$$z(\alpha)=C_1\ln\frac{1+\cos\alpha}{2\cos\alpha},\quad r(\alpha)=C_1\frac{\sin\alpha}{1+\cos\alpha}. \tag{2.76}$$

If $m=1$, $n=1$, we have

$$z(\alpha)=C_1\left(\ln\frac{1+\sin\alpha}{\cos\alpha}-\sin\alpha\right),\quad r(\alpha)=C_1\frac{\sin^2\alpha}{1+\cos\alpha} \tag{2.77}$$

and for $m=2$, $n=1$ we obtain

$$z(\alpha)=C_1(\cos^2\alpha-\cos\alpha-\ln\cos\alpha),\quad r(\alpha)=C_1\frac{\sin^3\alpha}{1+\cos\alpha}. \tag{2.78}$$

If in the vicinity of the point $\alpha=0$ we have $g(\alpha)<1/2$, then in this area the deduced surfaces of revolution must be replaced by other surfaces, e.g. a conical surface or a quadric surface.

If $g(\alpha)=1/2$, then by (2.39), the parametric equations of the meridian take the form $(a=1/2)$

$$z(\alpha)=\frac{1}{2}C_1\int\sqrt{\sin\alpha}\,\mathrm{d}\alpha+C_2,\quad r(\alpha)=C_1\sqrt{\sin\alpha}. \tag{2.79}$$

and the principal radii of the surface of revolution are, according to (2.27) and (2.30),

$$R_1=\frac{C_1}{2\sqrt{\sin\alpha}},\quad R_2=\frac{C_1}{\sqrt{\sin\alpha}}. \tag{2.80}$$

If we replace the integrand $\sqrt{\sin\alpha}$ in the first equation (2.79) by the first twelve terms of its Maclaurin series, we obtain, after integration,

$$z(\alpha)=C_1\left(\frac{1}{4}\frac{\alpha^3}{3!}-\frac{\alpha^5}{5!}+4\frac{\alpha^7}{7!}-16\frac{\alpha^9}{9!}+64\frac{\alpha^{11}}{11!}\right)+C_2. \tag{2.81}$$

Equation (2.81) is sufficiently accurate for $\alpha\leqq 1(\alpha\leqq 57°)$. (2.38) implies that $C_2=0$, and, according to the second equation (2.79), $C_1=r_{max}$, where r_{max} is the maximum value of the radius of the parallel circle for $\alpha=\pi/2$.

The coordinate $z(\alpha)$ for $\alpha>1$ can be determined from the expression

$$z=C_1\left[\frac{1}{2}\alpha-\frac{1}{4\times 3!}\left(\alpha-\frac{\pi}{2}\right)^3+\frac{2}{5!}\left(\alpha-\frac{\pi}{2}\right)^5-\frac{4}{7!}\left(\alpha-\frac{\pi}{2}\right)^7+\right. \\ \left.+\frac{16}{9!}\left(\alpha-\frac{\pi}{2}\right)^9-\frac{64}{11!}\left(\alpha-\frac{\pi}{2}\right)^{11}\right] \tag{2.82}$$

which is obtained from the first equation (2.79) by expanding the integrand $\sqrt{\sin\alpha}$ into its Taylor series about the point $\alpha=\pi/2$. (2.82) holds for $|\alpha-\pi/2|\leqq 1$.

Kawaguchi [32] derived interesting forms of surfaces of revolution for which the principal membrane forces satisfy a relation

$$T_2=T_1e^{-k\alpha}, \tag{2.83}$$

where k is a constant. Substituting the expressions (3.16) of Chapter II into (2.83), we obtain

$$R_2/R_1=2-e^{-k\alpha} \tag{2.84}$$

which is the case, when the function $g(\alpha)$ is given by $g(\alpha)=1/(2-e^{-k\alpha})$. For $0<k<\infty$ the meridians satisfying the geometric condition (2.84) are intermediate curves between the circle ($k=0$) and the curve determined by the parametric equations (2.79).

2.4 *Principal Radii of Quadric Surfaces*

As follows from equation (4.4) in Chapter II, the quadric surfaces have the characteristic curvature

$$\frac{R_2^3}{R_1} = R_0^2, \tag{2.85}$$

where $R_0 = R_1(0) = R_2(0)$.

In this subsection we shall deduce the formulae for the principal radii, as functions of the parameter α (Fig. II-3).

Substitution of (2.85) into the Gauss—Codazzi equation (1.36) in Chapter II yields

$$\frac{dR_2}{d\alpha} \sin \alpha + R_2\left(1 - \frac{R_2^2}{R_0^2}\right) \cos \alpha = 0, \tag{2.86}$$

whence we obtain, separating the variables,

$$\frac{dR_2}{R_2(1 - R_2^2/R_0^2)} = -\frac{d(\sin \alpha)}{\sin \alpha}. \tag{2.87}$$

The integration of this equation yields

$$\frac{1}{2} \ln \frac{R_2^2}{|1 - R_2^2/R_0^2|} = -\ln |\sin \alpha| + \ln C, \tag{2.88}$$

whence

$$\frac{R_2^2}{1 - R_2^2/R_0^2} = \pm \frac{C^2}{\sin^2 \alpha}, \tag{2.89}$$

where C is the integration constant.

If we set

$$\gamma = \pm R_0^2/C^2, \tag{2.90}$$

we obtain from (2.89) the following formula for the principal radius R_2:

$$R_2 = \frac{R_0}{(1 + \gamma \sin^2 \alpha)^{1/2}}. \tag{2.91}$$

From the relations (2.85) and (2.91) we obtain the formula for the principal radius R_1,

$$R_1 = \frac{R_0}{(1+\gamma \sin^2 \alpha)^{3/2}}. \qquad (2.92)$$

The zero value of the parameter γ corresponds to a spherical surface, $\gamma = -1$ to a paraboloid, $\gamma > -1$ to an ellipsoid and $\gamma < -1$ to a hyperboloid.

According to (2.85), we have

$$\frac{R_1}{R_2} = \frac{R_2^2}{R_0^2}. \qquad (2.93)$$

Condition (3.17) in Chapter II is complied with, provided that

$$R_2 \geqq R_0/\sqrt{2}. \qquad (2.94)$$

Thus the quadric surfaces of revolution are suitable forms for air-supported structures, if they satisfy inequality (2.94).

2.5 *Calculation of the Meridian Length*

The preparation of the fabric jackets of air-supported structures of revolution necessitates the determination of the lengths of their meridians.

If the form of the jacket has been deduced from the assumed relation (2.27) between the principal radii of the surface of revolution, then, according to (2.27) and (2.30), we have

$$R_1 = R_2 g(\alpha) = r(\alpha)\frac{g(\alpha)}{\sin \alpha}. \qquad (2.95)$$

As seen Fig. II-3 the differential of the meridian length is given by

$$\mathrm{d}s = R_1 \,\mathrm{d}\alpha. \qquad (2.96)$$

If we substitute (2.95) into (2.96), we arrive at

$$\mathrm{d}s = r(\alpha)\frac{g(\alpha)}{\sin \alpha}\,\mathrm{d}\alpha, \qquad (2.97)$$

whence we obtain for the meridian length, measured from the apex of the surface to the parallel circle determined by the angle $\alpha = \alpha_r$, the formula

$$s(\alpha) = \int_0^{\alpha_r} r(\alpha)\frac{g(\alpha)}{\sin \alpha}\,\mathrm{d}\alpha. \qquad (2.98)$$

In the examples that follow we shall write α instead of α_r.

If the given function $g(\alpha)$ has the form (2.36), with $a=1$, $b=0$, we obtain, replacing $g(\alpha)$ by 1, and substituting the second expression (2.41) into (2.98) and integrating,

$$s(\alpha)=C_1\alpha. \tag{2.99}$$

For $a=2$, $b=0$ we obtain, with the aid of the second equation (2.42), the following formula for the meridian length:

$$s(\alpha)=2C_1(1-\cos\alpha). \tag{2.100}$$

If $a=3$, $b=0$, we obtain from (2.98), using the second equation (2.43),

$$s(\alpha)=\frac{3}{2}C_1(\alpha-\sin\alpha\cos\alpha), \tag{2.101}$$

and for $a=4$, $b=0$ we have

$$s(\alpha)=4C_1\left(\frac{2}{3}+\frac{1}{3}\cos^3\alpha-\cos\alpha\right). \tag{2.102}$$

For $a=1$, $b=1$ the radius $r(\alpha)$ of the parallel circle is determined by the second equation (2.48); substituting it into (2.98), we obtain

$$s(\alpha)=C_1\left(\frac{\alpha^2}{2}-\ln\cos\alpha\right). \tag{2.103}$$

If $a=2$, $b=1$, we obtain, substituting (2.49) for $r(\alpha)$ in (2.98),

$$s(\alpha)=C_1\left(2\ln\frac{1+\sin\alpha}{\cos\alpha}-\sin\alpha-\alpha\cos\alpha\right). \tag{2.104}$$

and for $a=3$, $b=1$ we have, due to the second equation (2.50),

$$s(\alpha)=C_1\left[\frac{3}{4}\left(\alpha^2-\alpha\sin 2\alpha-\frac{1}{2}\cos 2\alpha\right)+\frac{1}{2}\cos^2\alpha-\ln\cos\alpha-\frac{1}{8}\right]. \tag{2.105}$$

If the function $g(\alpha)$ has the form (2.51), we obtain in a similar way for $c=0$, $d=1$

$$s(\alpha)=\frac{1}{2}C_1\left(\frac{\sin\alpha}{\cos^2\alpha}+\ln\frac{1+\sin\alpha}{\cos\alpha}\right), \tag{2.106}$$

and for $c=0$, $d=2$ we have

$$s(\alpha)=\frac{2}{3}C_1\left(\frac{1}{\cos^3\alpha}-1\right). \tag{2.107}$$

If $c=1$, $d=1$, substitution of (2.55) for $r(\alpha)$ in (2.98) yields

$$s(\alpha)=C_1\left(\frac{1}{2}\tan^2\alpha-\ln\cos\alpha\right). \tag{2.108}$$

For $c=1$, $d=2$ we have

$$s(\alpha)=C_1\sin\alpha\left(\frac{1}{\cos^2\alpha}-1\right), \tag{2.109}$$

and for $c=2$, $d=2$ we obtain

$$s(\alpha)=2C_1\left(\cos\alpha+\frac{1}{3\cos^3\alpha}-\frac{4}{3}\right). \tag{2.110}$$

If the function $g(\alpha)$ has the form (2.58) and $k=0$, $h=1$, we obtain, substituting (2.60) into (2.98),

$$s(\alpha)=\frac{1}{2}C_1\left(\frac{1}{\cos^2\alpha}-1\right). \tag{2.111}$$

Analogously for $k=0$, $h=2$ we have

$$s(\alpha)=\frac{2}{3}C_1\left(\frac{1}{\cos^3\alpha}-1\right). \tag{2.112}$$

If $k=1$, $h=1$, we get

$$s(\alpha)=\frac{1}{2}C_1\left(\ln\frac{1+\sin\alpha}{\cos\alpha}+\frac{\sin\alpha}{2\cos^2\alpha}\right) \tag{2.113}$$

and for $k=1$, $h=2$,

$$s(\alpha)=C_1\left(\tan\alpha+\frac{2}{3}\tan^3\alpha\right). \tag{2.114}$$

For $k=2$, $h=1$,

$$s(\alpha)=C_1\left(\frac{1}{2\cos^2\alpha}-\ln\cos\alpha-\frac{1}{2}\right) \tag{2.115}$$

and for $k=2$, $h=2$,

$$s(\alpha)=\frac{2}{3}C_1\left(\frac{1}{\cos^3\alpha}-1\right). \tag{2.116}$$

For a quadric surface, substitution of (2.92) into (2.96) leads to

$$ds = R_0 \frac{d\alpha}{(1+\gamma \sin^2 \alpha)^{3/2}}, \tag{2.117}$$

whence

$$s(\alpha) = R_0 \int_0^\alpha \frac{d\alpha}{(1+\gamma \sin^2 \alpha)^{3/2}}. \tag{2.118}$$

If $\gamma < 0$, we obtain, according to (2.118),

$$s(\alpha) = R_0 \left\{ \frac{1}{1+\gamma} [E(\gamma, \alpha) - E(\gamma, 0)] - \frac{\gamma}{1+\gamma} \frac{\sin \alpha \cos \alpha}{\sqrt{1+\gamma \sin^2 \alpha}} \right\},$$

where

$$E(\gamma, \alpha) = \int \sqrt{1+\gamma \sin^2 \alpha}\, d\alpha$$

is the elliptic integral of the first order in Legendre normal form.

3

Air-Supported Membrane Structures over a Rectangular Plan

3.1 *Air-Supported Structure with Central Cylindrical Part*

For convenience of production the cylindrical surface is the most suitable form for the central part of the jacket of an air-supported structure over a rectangular plan. The ends, connected to the cylindrical surface, must be so designed as to satisfy continuously both the geometric and the static conditions along the junction of both surfaces. For methodological reasons it is difficult to consider both of these conditions simultaneously, as the

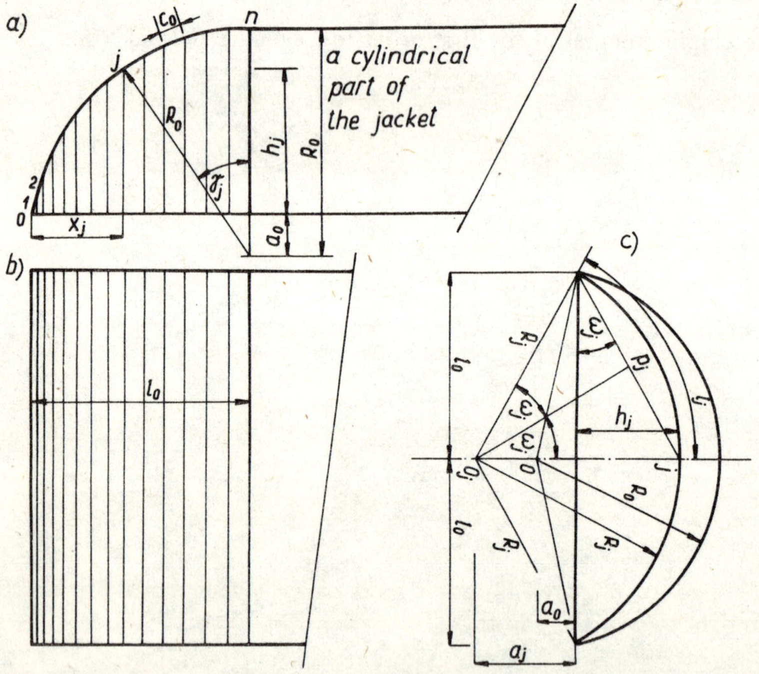

Fig. III-1. a) Longitudinal section of the jacket of an air-supported structure with a cylindrical middle part and ends in the form of a doubly circular warped surface. b) Plan of the air-supported structure. c) Cross-section of the end part.

membrane forces in the jacket depend not only on the type and magnitude of load (e.g. internal overpressure), but also on the geometric form which must be determined for the ends. Therefore, we suggest calculating the form of the ends so that along their joints with the cylindrical surface there are satisfied continuously, to begin with, only the geometrical conditions. The continuity of the static conditions, i.e. the equality of the membrane forces of both joining surfaces at their common points under the effect of the internal overpressure may be verified subsequently theoretically by calculation or experimentally on a model, as these static conditions influence the stability of form of the whole air-supported structure.

We design the ends of the air-supported structure in the circularly circular wedge form belonging among skewed surfaces [7]. This surface is generated by the displacement, along a circle of a radius R_0 in the direction of the longitudinal axis of the air-supported structure, of a circle which continuously intersects the two parallel longer sides of the rectangular plan (Fig. III-1), the radius of the displaced circle changing continuously from $R_n = R_0$ to infinity. For this form the geometric conditions along the connection of the ends with the central cylindrical surface of a radius R_0 are continuously satisfied.

According to Figs. III-1a, b, c the following relations hold for this surface:

$$\gamma_j = \frac{c_0}{R_n}(n-j), \qquad j = 1, 2, \ldots, n,$$

$$h_j = R_n \cos \gamma_i - a_0, \qquad \tan \omega_j = h_j / l_0, \tag{3.1}$$

$$R_j = \frac{\sqrt{l_0^2 + h_j}}{2 \sin \omega_j}, \qquad l_j = 2R_j\omega_j$$

and

$$x_j = l_0 - R_n \sin \gamma_j, \tag{3.2}$$

where $R_n = R_0$.

In the preparation of the fabric for the ends of the air-supported structure it is necessary to use bands of variable width. One side of the band is straight, the other an arc of a circle of radius r_j, its lenght l_j being equal the length of the straight side of the adjoining band (Fig. III-2).

For a small angle τ_j we have, according to Fig III-2, that

$$\sin \tau_j \approx \frac{3}{4} \frac{l_j - l_{j-1}}{\Delta_j}, \tag{3.3}$$

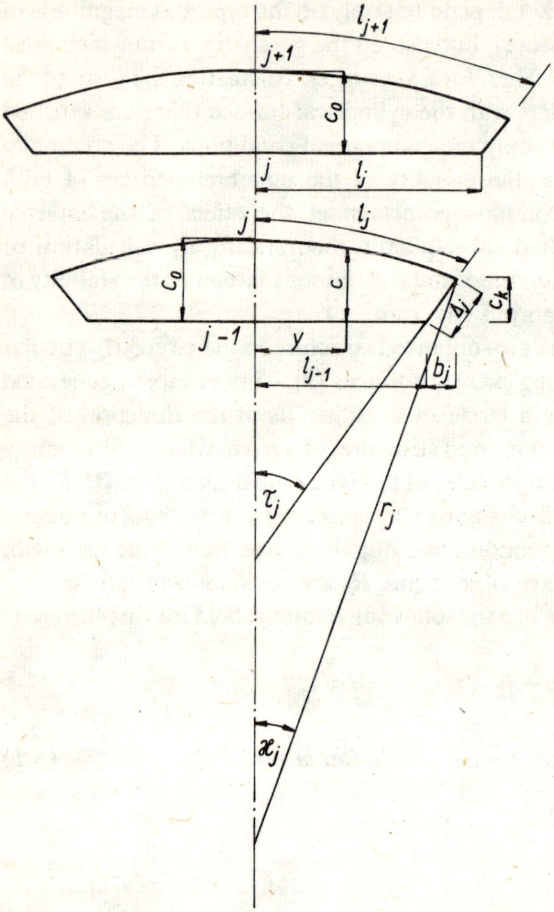

Fig. III-2. Form of the stripes of industrial fabric for the assembly of the end parts of the air-supported structure.

where

$$\Delta_j = x_j - x_{j-1}. \tag{3.4}$$

For the radius r_j the author deduced the formula

$$r_j = \frac{l_j^2}{2(c_0 - \Delta_j \cos \tau_j)}. \tag{3.5}$$

Further it can be shown that

$$\varkappa_j = \frac{l_j}{r_j}, \quad b_j = r_j \sin \varkappa_j - l_{j-1}, \tag{3.6}$$

and

$$c = c_0 - r_j\left[1 - \sqrt{1 - \frac{y^2}{r_j^2}}\right]. \tag{3.7}$$

The slanting part at the end of the band has the length

$$d_j = \sqrt{c_k^2 + b_j^2}\,, \tag{3.8}$$

where c_k corresponds to $y = r_j \sin \varkappa_j$, according to (3.7).

Some of the above mentioned formulae are approximate; therefore $d_j \neq \Delta_j$; however, the differences are negligible.

The accurate lenght D of the plan contour of the end-part of the structure is the sum of all values d_j:

$$D = \sum_{j=1}^{n} d_j\,. \tag{3.9}$$

The method of the design of the form of the jacket of an air-supported structure over a rectangular plan, described above, was used for the preparation of a model over a plan 240 × 500 cm (Fig. III-3). The model was used also for the varification of the static function of the air cushions within the jacket with thermally insulating properties (see Subsection 3.3).

The model showed major creases of the jacket in the corners only, due primarily to the elongation of the membrane of the cylindrical part in the longitudinal direction. Since this elongation is variable along the cross-section of the structure, being zero at the anchorage and maximum at the apex, it is necessary to ascertain experimentally its magnitude and history and eliminate its influence during the preparation of the jacket by continuously reducing the widths of the bands from the anchorage to the apex of the structure. Another method of design of a jacket of stable form consists in changing the geometric form, while taking care not to violate the geometric and the static conditions of continuity, following from the ascertained changes of form of the membrane jacket (see Subsection 3.2).

If the rectangular plan approaches a square plan, the cylindrical part of the jacket is small and the elongation of the membrane in this part has no practical influence on the stability of form of the jacket. This has been varified by the stability of form of an air-supported structure of cushion type (Fig. I-1f) designed by the author in 1978 for a grandstand in Hradec Králové

Fig. III-3. A 1 : 10 model of an air-supported structure with air cushions over a rectangular plan with a jacket consisting of a cylindrical central part and warped surface ends.

(ČSSR) using the formulae given in this section. The structure consisted of 15 air-supported cushions 120 cm thick with a rectangular plan of 3.35 × 3.52 m each.

3.2 *Air-Supported Structure Consisting of Three Skew Surfaces*

The influence of the elongation of the jacket in the longitudinal direction on its form can be totally eliminated by using a form based on three doubly circular wedge-shaped surfaces. The surface is generated by the displacement, in the direction of the longitudinal axis of the structure, along a curve composed of three arcs of circles of the radii R_1, R_2, R_3, of a circle arc of variable diameter which always intersects the two longer parallel sides of the rectangular plan. Fig. III-4 shows a longitudinal section of a jacket of this form. For a 1:10 model of an air-supported structure over a rectangular plan 24 × 50 m the author used the ratio

$$R_3 : R_2 : R_1 = 1:3:9$$

and the actual values $R_1 = 795$ cm, $R_2 = 265$ cm, $R_3 = 88.33$ cm, to approximate the longitudinal section of the jacket deformed by internal overpressure, whose designed form was described in the Subsection 3.1.

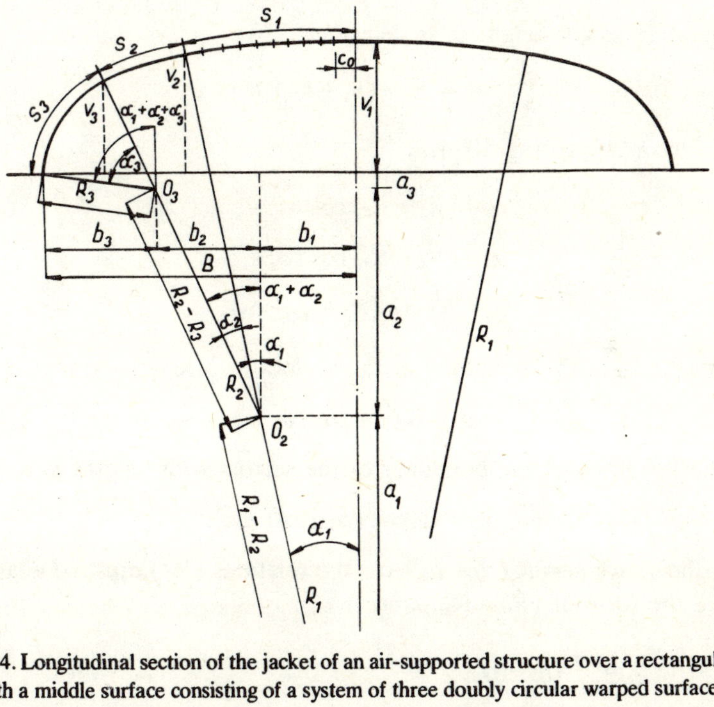

Fig. III-4. Longitudinal section of the jacket of an air-supported structure over a rectangular plan with a middle surface consisting of a system of three doubly circular warped surfaces.

According to Fig. III-4, we have for the central angles α_1, α_2, α_3 of the circular arcs of lengths s_1, s_2, s_3 with the radii R_1, R_2, R_3 the equalities

$$\alpha_1 = n_1 \frac{c_0}{R_1}, \quad \alpha_2 = n_2 \frac{c_0}{R_2}, \quad \alpha_3 = n_3 \frac{c_0}{R_3}, \tag{3.10}$$

where c_0 is the width of the membrane strip at the apices of the jacket cross-sections, and n_1, n_2, n_3 are the number of membrane strips in the individual sectors of circular arcs.

One half of the length of the air-supported structure is

$$B = b_1 + b_2 + b_3, \tag{3.11}$$

where

$$\begin{aligned} b_1 &= (R_1 - R_2) \sin \alpha_1, \\ b &= (R_2 - R_3) \sin (\alpha_1 + \alpha_2), \\ b_3 &= R_3 \sin (\alpha_1 + \alpha_2 + \alpha_3). \end{aligned} \tag{3.12}$$

The greatest jacket height v_1 is determined by

$$v_1 = R_1 - (h_1 + h_2 + h_3), \tag{3.13}$$

where, according to Fig. III-4,

$$\begin{aligned} a_1 &= (R_1 - R_2) \cos \alpha_1, \\ a_2 &= (R_2 - R_3) \cos (\alpha_1 + \alpha_2), \\ a_3 &= R_3 \cos (\alpha_1 + \alpha_2 + \alpha_3). \end{aligned} \tag{3.14}$$

The height v_2 of the boundary of the sectors with lengths s_1 and s_2 is

$$v_2 = R_2 \cos \alpha_1 - a_2 - a_3 \tag{3.15}$$

and the height v_3 of the boundary of the sectors with lengths s_2 and s_3 is

$$v_3 = R_3 \cos (\alpha_1 + \alpha_2) - a_3. \tag{3.16}$$

For the sector $s_3 = R_3(\alpha_1 + \alpha_2 + \alpha_3)$ the relations (3.1) must be adapted to acquire the form of ($R_n = R_3$, $a_0 = a_3$)

$$\begin{aligned} &\bar{\gamma}_j = \frac{c_0}{R_3}(n_3 - j), && j = 1, 2, \ldots, n_3, \\ &h_j = R_3 \cos (\bar{\gamma}_j + \alpha_1 + \alpha_2) - a_3, && \tan \omega_j = \frac{h_j}{l_0}, \\ &R_j = \frac{\sqrt{l_0^2 + h_j^2}}{2 \sin \omega_j}, && l_j = 2R_j\omega_j, \end{aligned} \tag{3.17}$$

where l_0 is the span of the structure and the angle $\bar{\gamma}_j$ is measured from the end point of the sector (see Fig. III-5).

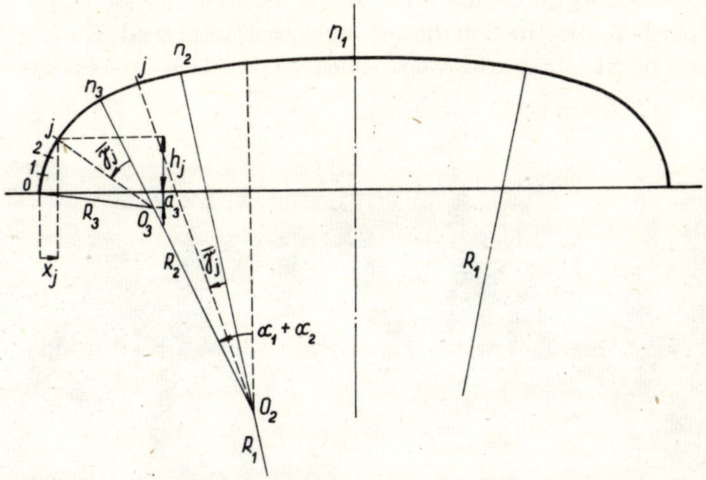

Fig. III-5. Marking of parts and angles in the longitudinal section of the jacket.

For the sector $s_2 = R_2(\alpha_1 + \alpha_2)$, instead of the first two relations (3.17) we have

$$\begin{aligned} \bar{\gamma}_j &= \frac{c_0}{R_2}(n_2 - j), \qquad 0 \leqq j \leqq n_2, \\ h_j &= R_2 \cos(\bar{\gamma}_j + \alpha_1) - a_3 - a_2 \end{aligned} \tag{3.18}$$

and equation (3.2) acquires the form

$$x_j = b_3 + b_2 - R_2 \sin(\bar{\gamma}_j + \alpha_1). \tag{3.19}$$

For the sector $s_1 = R_1\alpha_1$ the first two relations (3.17) take the form (Fig. III-5)

$$\begin{aligned} \bar{\gamma}_j &= \frac{c_0}{R_1}(n_1 - j), \qquad 0 \leqq j \leqq n_1, \\ h_j &= R_1 \cos \bar{\gamma}_j - a_3 - a_2 - a_1 \end{aligned} \tag{3.20}$$

and instead of equation (3.2) we have

$$x_j = b_3 + b_2 + b_1 - R_1 \sin \bar{\gamma}_j. \tag{3.21}$$

The other relations (3.17) and the equations (3.3) to (3.8) hold for all the three sectors considered.

Fig. III-6 shows an overall view of the model of an air-supported hall jacket, in whose construction the above relations were used. The lengths and the widths of the fabric strips were calculated on a Hewlett—Packard digital computers.

Fig. III-6. A 1 : 10 model of an air-supported structure over a rectangular plan in the form of three doubly circular warped surfaces.

The results of the measurements have shown that the structure will be of stable form, if its jacket has circular cross-sections in the directions radial to the longitudinal section of the structure [18].

The form of the jacket, has been theoretically designed so as to comply with all the geometric conditions of continuity on the boundaries of circular skew surfaces, and the form of the directrix of the longitudinal section of the jacket has been suitably corrected. Thus we arrive at the conclusion that by respecting the deformability conditions of the membrane it is also possible to comply with the static conditions of continuity. This fact confirms the autor's theoretical and experimental method of formfinding of an air-supported structure, according to which it is possible to comply, apart from geometric conditions, also with the static conditions of continuity at all places of the membrane by creating the surface of the jacket repeatedly in accordance with a suitably selected geometric theorem, while making corrections of the directrices and generatrices on the basis of their deformability. The tests of two models have proved that the convergence of the selected geometry of the jacket to its definite form is very rapid and that in actual practical cases it is usually sufficient to carry out one correction of the directrix only. This means that it is generally sufficient to carry out the measurements of the form of the directrix of the deformed jacket surface on one model only. If the initial geometric form of the surface has not been properly selected, it is necessary to make the correction of not only the directrix, but also the generatrix, so that corrected form may agree with the deformed membrane of the jacket as much as possible.

The described theoretical and experimental method makes it possible to take into account in the formfinding of an air-supported structure also the viscoelastic properties of the membrane, if the model is made of the same material as the jacket of the actual structure and if the same membrane forces are produced in the membrane of the model and in the jacket of the actual structure under the operational internal overpressure q_0. This is achieved in the $1:n$ model, if its membrane is affected by the internal overpressure q_M given by

$$q_M = nq_0. \tag{3.22}$$

Relation (3.22) follows from the linearity of the equations of equilibrium (2.6) in Chapter II.

3.3 *Formfinding of Air-Supported Structures with Thermally Insulating Jacket and Interactions between the Outer and the Inner Membranes*

According to the Czechoslovak patents 150 309 and 155 310 [17] thermally insulating properties of a jacket of an air-supported structure can be ensured by the provision of air cushions which fill up automatically due to the internal overpressure in a jacket consisting of two membranes mutually connected by seams and provided with openings in the inner membrane (see Chapter I, Section 4).

In a multipurpose air-supported structure over a rectangular plan 24 × 50 m (Fig. I-1e), which was erected in Prague in 1978, the strips of the inner and outer membranes are oriented transversely to the structure (Fig. III-7). In summer excessive heating of the interior of the structure is prevented by a controlled air flow in the air cushions and its expulsion from the interior of the structure, which is made possible by the connection of one row of openings in the air cushions to a pipe whose end, provided with control valves, discharges into the open air. In winter the control valves are closed and the air in the air cushions fulfils its thermally insulating function. A partial opening of the control valves in the winter is effected only from time to time for the sake of ventilation. This process ensures a suitable microclimate inside the hall during the whole year.

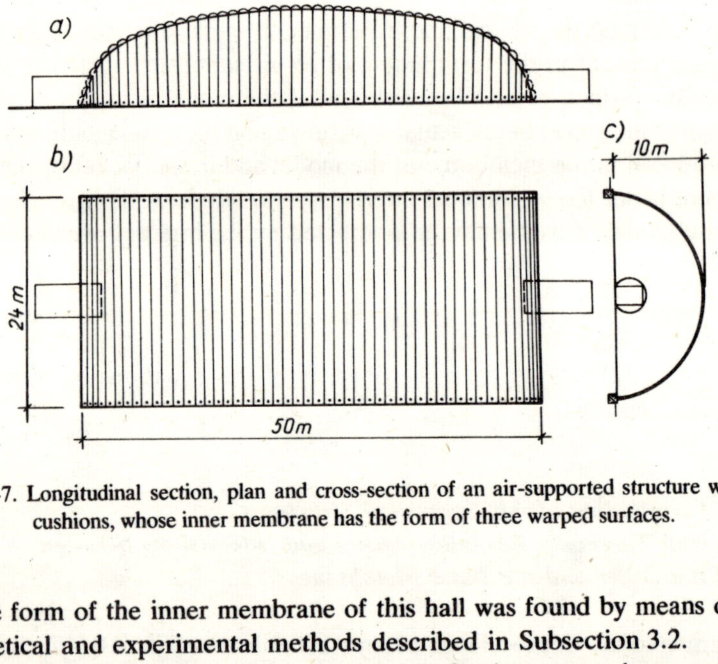

Fig. III-7. Longitudinal section, plan and cross-section of an air-supported structure with air cushions, whose inner membrane has the form of three warped surfaces.

The form of the inner membrane of this hall was found by means of the theoretical and experimental methods described in Subsection 3.2.

The interaction between the outer and the inner membrane will be investigated in the central part of the structure, which we shall approximate by a cylindrical surface of diameter R_0. The form of the outer membrane is generating by the rotation of a circle of a radius r_0 about the cylinder axis x, thus it is part of a torus surface (Fig. III-8).

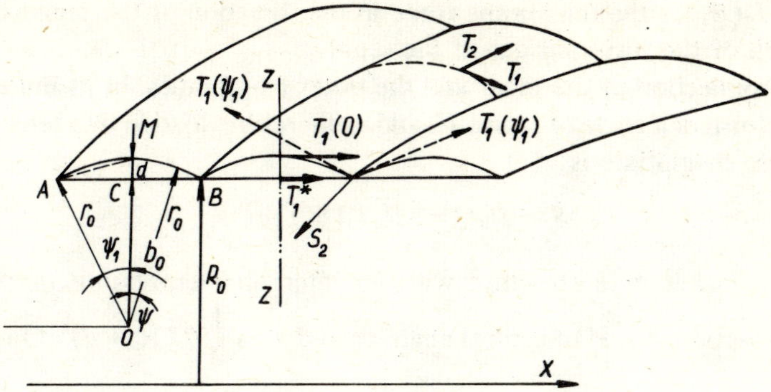

Fig. III-8. Longitudinal section of an air-supported structure with air cushions.

The length of the circular arc AB satisfies

$$s_0 \approx 2p_0 + \frac{2}{3}(p_0 - L) = \frac{2}{3}(4p_0 - L), \tag{3.23}$$

where $p_0 = \overline{AM}$ and $L = \overline{AC} = \overline{BC}$ (Fig. III-8). The length p_0 is determined from the selected values of half the length of the chord L and the height d of the circular arc,

$$p_0 = \sqrt{L^2 - d^2}\,. \tag{3.24}$$

From the relations

$$r_0^2 = L^2 + b_0^2, \quad b = r_0 - d \tag{3.25}$$

we obtain

$$r_0 = \frac{L^2 + d^2}{2d}. \tag{3.26}$$

We have further that

$$\sin \psi_1 = \frac{L}{r_0}, \quad \cos \psi_1 = \frac{L}{r_0}. \tag{3.27}$$

At the place of connection of the outer and the inner membranes the seam is loaded by a radial load Q given by

$$Q = 2T_1(\psi_1) \sin \psi_1, \tag{3.28}$$

where $T_1(\psi_1)$ is the membrane force in the direction of the tangent to the meridian of the torus surface at the seam.

The connection of the inner and the outer membranes, in its initial state, has the form of a circle of radius R_0 and is, therefore, loaded by a tensile force S_2 whose magnitude is

$$S_2 = R_0 Q = 2R_0 T_1(\psi_1) \sin \psi_1 . \tag{3.29}$$

For a jacket of a structure with an internal overpressure q_0 consider a cross-section $Z-Z$ of central angle ϑ_0 and area $\frac{1}{2}\vartheta_0(R_0+d)^2$. One of the parts thus separated is loaded by the resulting force $\frac{1}{2}q_0\vartheta_0(R_0+d)^2$ which is in the state of equilibrium with the resultant of the membrane forces in the inner and the outer membranes. Since the membrane force $T_1(0)$ in the outer membrane at $\psi=0$ is applied along the length $(R_0+d)\vartheta_0$ and the membrane force $T_1^\dagger$ in the inner membrane is applied along the length $R_0\vartheta_0$, we have

$$\vartheta_0[T_1(0)(R_0+d)+T_1^\dagger R_0] = \frac{1}{2}q_0\vartheta_0(R_0+d)^2, \tag{3.30}$$

whence

$$T_1^\dagger = \frac{1}{2}q_0R_0\left(1+\frac{d}{R_0}\right)^2 - T_1(0)\left(1+\frac{d}{R_0}\right). \tag{3.31}$$

The membrane forces in the outer membrane which has the shape of a torus surface, caused by the internal overpressure, are given by the following formulae [47]:

$$T_1 = \frac{1}{2}q_0r_0\frac{2+\frac{r_0}{R_0}\cos\psi}{1+\frac{r_0}{R_0}\cos\psi}, \quad T_2 = \frac{1}{2}q_0r_0 . \tag{3.32}$$

If we substitute the first formula (3.32) into (3.31) and take $\psi=0$, we obtain, for $r_0/R_0 \ll 1$,

$$T_1^\dagger = \frac{1}{2}q_0R_0\left(1+\frac{d}{R_0}\right)^2 - q_0r_0\left(1+\frac{d}{R_0}\right). \tag{3.33}$$

The substitution of the same formula into (3.29), with $\psi=\psi_1$ yields

$$S_2 = q_0 r_0 R_0 \frac{2 + \dfrac{r_0}{R_0} \cos \psi_1}{1 + \dfrac{r_0}{R_0} \cos \psi_1} \sin \psi_1 \tag{3.34}$$

and for $r_0/R_0 \ll 1$,

$$S_2 = 2 q_0 r_0 R_0 \sin \psi_1 . \tag{3.35}$$

The author suggested the use of steel ropes in the central part of the multipurpose hall shown in Fig. III-7 to intercept the force S_2.

4

Formfinding of Entrance Openings

4.1 *Fundamental equations*

We shall consider a form of the entrance opening as suitable if it does not disturb the membrane state of stress of the jacket of an air-supported structure in the vicinity of the opening, under the effect of the internal overpressure which is its permanent loading state. The membrane forces at points of the circumference of the opening will be counteracted by a rope which is generally loaded by a variable force S (Fig. III-9).

In this subsection we shall deduce the differential equation and the formulae for the axial force in the rope and the radius of curvature of the circumference of the entrance opening and, in the following subsection we shall obtain the parametric equations of the curve of the circumference. We shall make the derivation for the cylindrical, cycloidal and spherical surfaces under the assumption that the opening is small in comparison with the surface of the structure. This means that we shall consider the circumference of the opening as a plane curve affected by the membrane forces of the jacket in its plane [16].

The arc length differential ds of a curve defining the entrance opening is affected by the force T_2 dy in the direction of the x-axis and by the force

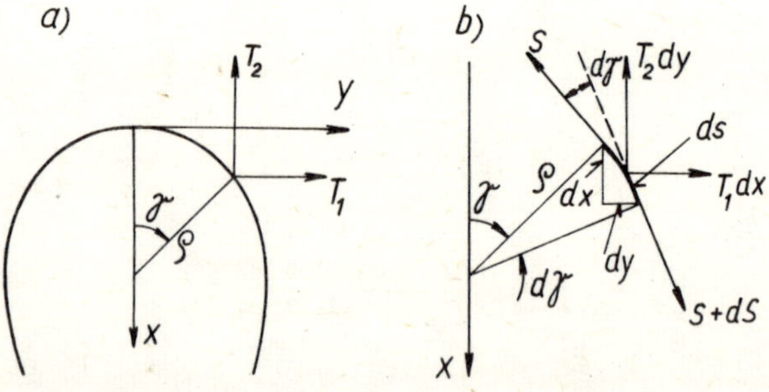

Fig. III-9. a) Entrance opening in the membrane jacket of an air-supported structure. b) Forces affecting the element of rope whose axis is identical with the edge of the entrance opening.

$T_1\,dx$ in the horizontal direction, where T_1 and T_2 are the principal membrane forces in the jacket (Fig. III-9). If the position of the left-hand end of the arc length differential is determined by the normal forming an angle γ with the x-axis, then, according to Fig. III-9b, the following two equations of equilibrium hold in the directions of the tangent and the normal to the curve ($0 \leqq \gamma \leqq \pi/2$):

$$\begin{aligned}(S+\mathrm{d}S)\cos(\mathrm{d}\gamma)-S-T_2\,\mathrm{d}y\sin\gamma+T_1\,\mathrm{d}x\cos\gamma&=0,\\ (S+\mathrm{d}S)\sin(\mathrm{d}\gamma)-T_2\,\mathrm{d}y\cos\gamma-T_1\,\mathrm{d}x\sin\gamma&=0.\end{aligned} \tag{4.1}$$

Since

$$\begin{gathered}\mathrm{d}x=\mathrm{d}s\sin\gamma,\quad \mathrm{d}y=\mathrm{d}s\cos\gamma,\\ \mathrm{d}s=\varrho\,\mathrm{d}\gamma,\quad \cos(\mathrm{d}\gamma)\approx 1,\quad \sin(\mathrm{d}\gamma)\approx\mathrm{d}\gamma,\end{gathered} \tag{4.2}$$

where ϱ is the radius of the curve, we obtain, substituting (4.2) into (4.1) and the neglecting the small quantity of a higher order $\mathrm{d}S\,\mathrm{d}\gamma$, the following relations:

$$\begin{aligned}\mathrm{d}S/\mathrm{d}\gamma&=\varrho(T_2-T_1)\cos\gamma\sin\gamma,\\ S&=\varrho(T_2\cos^2\gamma+T_1\sin^2\gamma).\end{aligned} \tag{4.3}$$

The elimination of ϱ from the relations (4.3) yields for the unknown force S in the rope the differential equation

$$\frac{\mathrm{d}S}{S}=\frac{(T_2-T_1)\cos\gamma\sin\gamma}{T_2\cos^2\gamma+T_1\sin^2\gamma}\,\mathrm{d}\gamma, \tag{4.4}$$

whose integration leads to

$$\ln S=\int f(\gamma)\,\mathrm{d}\gamma-\ln C,$$

whence

$$S=C\,\mathrm{e}^{\int f(\gamma)\,\mathrm{d}\gamma}, \tag{4.5}$$

where e is the base of the natural logarithms, C is an integration constant and

$$f(\gamma)=\frac{(T_2-T_1)\cos\gamma\sin\gamma}{T_2\cos^2\gamma+T_1\sin^2\gamma}. \tag{4.6}$$

If the membrane forces T_1 and T_2 can be expressed as functions of the angle γ and if it is possible to determine the integral $\int f(\gamma)\,\mathrm{d}\gamma$, we can determine the force S from (4.5).

If $F(\gamma)$ is the primitive function of the integrand $f(\gamma)$, then, according to (4.3) and (4.5), for $\gamma=0$ one has

$$S_0 = T_2(0)\varrho_0, \quad S_0 = Ce^{F(0)}, \tag{4.7}$$

where $S_0 = S(0)$, $\varrho_0 = \varrho(0)$ are the values pertaining to $\gamma=0$. From the relations (4.7) it follows that the integration constant satisfies

$$C = T_2(0)\varrho_0/e^{F(0)}. \tag{4.8}$$

Substituting (4.8) into (4.5), we obtain the following equation for the axial force in the rope:

$$S = T_2(0)\varrho_0 e^{F(\gamma)}/e^{F(0)}. \tag{4.9}$$

We can rearrange (4.9), using the first relation (4.7), to obtain

$$S/S_0 = e^{F(\gamma)}/e^{F(0)}, \tag{4.10}$$

where

$$F(\gamma) = \int f(\gamma)\, d\gamma. \tag{4.11}$$

The second relation (4.3) and (4.9) imply the formula for the radius of the curve:

$$\frac{\varrho}{\varrho_0} = \frac{T_2(0)}{e^{F(0)}} \frac{e^{F(\gamma)}}{T_2 \cos^2\gamma + T_1 \sin^2\gamma}. \tag{4.12}$$

The relations (4.10) and (4.12) permit us to determine the variations of the axial force and the radius ϱ accompanying the charges of the angle γ, for the selected value of ϱ_0.

If the principal membrane forces T_1 and T_2 are constant, then according to (4.6) and (4.11),

$$F(\gamma) = -\frac{1}{2} \ln |T_2 \cos^2\gamma + T_1 \sin^2\gamma|, \tag{4.13}$$

and

$$F(0) = -\frac{1}{2} \ln T_2. \tag{4.14}$$

If we substitute (4.13) and (4.14) into (4.10) and (4.12), we obtain the formulae

$$\frac{S}{S_0}=\left(\frac{T_2}{T_2\cos^2\gamma+T_1\sin^2\gamma}\right)^{1/2} \tag{4.15}$$

and

$$\frac{\varrho}{\varrho_0}=\left(\frac{T_2}{T_2\cos^2\gamma+T_1\sin^2\gamma}\right)^{3/2}, \tag{4.16}$$

which are directly useful in engineering practice [18]. Let us note that the formulae (4.15) and (4.16) are valid for $0\leqq\gamma\leqq\pi/2$. For $\gamma<0$ and $\gamma>\pi/2$ it is possible to determine the shape of the whole opening on the basis of symmetry.

A comparison of the formulae (4.15) and (4.16) affords the following interesting simple relation:

$$\varrho/\varrho_0=(S/S_0)^3. \tag{4.17}$$

Let us consider in the cylindrical part of the jacket of an air-supported structure the principal membrane forces T_1 and T_2, where T_1 has the direction of the cylinder axis and T_2 is tangent to the cross-section, due to the internal overpressure q_0. According to Subsection 6.3b in Chapter II,

$$T_1=\frac{1}{2}q_0R_0, \quad T_2=q_0R_0, \tag{4.18}$$

where R_0 is the radius of the cross-section of the cylindrical surface. Substitution of (4.18) into (4.15) and (4.16) yields

$$\frac{S}{S_0}=\left(\frac{2}{2\cos^2\gamma+\sin^2\gamma}\right)^{1/2}, \tag{4.19}$$

$$\frac{\varrho}{\varrho_0}=\left(\frac{2}{2\cos^2\gamma+\sin^2\gamma}\right)^{3/2}, \tag{4.20}$$

where, according to (4.7) and (4.18),

$$S_0=q_0R_0\varrho_0. \tag{4.21}$$

Under the effect of wind of velocity v_0, perpendicular to the direction of the longitudinal axis of the jacket, in the form of a semi-cylinder, the magnitude of the membrane force is determined by the formula (Chapter II, Subsection 6.4b)

$$T_2=R_0(q_0+0.613v_0^2). \tag{4.22}$$

In (4.22) v_0 is in m/sec, R_0 in m, q_0 in N/m^2, and the membrane force T_2 has the dimension of N/m. The wind effect manifests itself, consequently, by creating a force which could equivalently arise from an internal overpressure $\bar{q}_0$ given by

$$\bar{q}_0 = 0.613 v_0^2,$$

where v_0 is in m/sec and $\bar{q}_0$ in N/m^2. Thus the membrane forces due to the internal overpressure q_0 and the effect of the wind in the transverse direction of the cylindrical surface are

$$T_1 = \frac{1}{2} q_0 R_0, \quad T_2 = R_0(q_0 + \bar{q}_0). \tag{4.23}$$

If we substitute (4.23) into (4.15) and (4.16), we obtain

$$\frac{S}{S_0} = \left[\frac{2(1+\bar{q}_0/q_0)}{2(1+\bar{q}_0/q_0)\cos^2\gamma + \sin^2\gamma}\right]^{1/2}, \tag{4.24}$$

$$\frac{\varrho}{\varrho_0} = \left[\frac{2(1+\bar{q}_0/q_0)}{2(1+\bar{q}_0/q_0)\cos^2\gamma + \sin^2\gamma}\right]^{3/2}. \tag{4.25}$$

According to (4.25), the radius of the opening curve depends only on the ratio of $\bar{q}_0/q_0$ and not on the absolute values of the quantities q_0 and $\bar{q}_0$. From the first relation (4.7), with the aid of (4.23), we obtain

$$S_0 = (q_0 + \bar{q}_0) R_0 \varrho_0. \tag{4.26}$$

For $T_1/T_2 = 0$ the formulae (4.15) and (4.16) acquire the form

$$\frac{S}{S_0} = \frac{1}{\cos\gamma}, \quad \frac{\varrho}{\varrho_0} = \frac{1}{\cos^3\gamma}. \tag{4.27}$$

In the cycloidal surface of revolution, whose meridian is a cycloid, the membrane forces due to the internal overpressure are (Chapter II, Subsection 7.1b)

$$T_1 = \frac{1}{2} q_0 r_{max} \sin\alpha, \qquad T_2 = \frac{3}{4} q_0 r_{max} \sin\alpha, \tag{4.28}$$

where α is the angle between the axis of revolution (z-axis) and the normal to the surface (Fig. II-18, Fig. II-19).

If we place the opening in the centre of the surface, where $\alpha = \pi/2$, then according to (4.28) we have

$$T_1 = \frac{1}{2} q_0 r_{\max}, \quad T_2 = \frac{3}{4} q_0 r_{\max}. \tag{4.29}$$

Provided the opening is small, we can consider the forces T_1 and T_2 as constant along the whole of its circumference, and their magnitudes as determined by the romulae (4.29).

If we substitute (4.29) into (4.15) and (4.16), we arrive at

$$\begin{aligned} \frac{S}{S_0} &= \left(\frac{3}{3\cos^2\gamma + 2\sin^2\gamma}\right)^{1/2}, \\ \frac{\varrho}{\varrho_0} &= \left(\frac{3}{3\cos^2\gamma + 2\sin^2\gamma}\right)^{3/2}. \end{aligned} \tag{4.30}$$

In the case of a spherical surface of radius R_0 affected by the internal overpressure q_0 only, the equation of equilibrium of forces in the direction of the normal to the circumference of the opening implies

$$S = T_0 \varrho, \tag{4.31}$$

where (Chapter II, Subsection 5.2b)

$$T_0 = \frac{1}{2} q_0 R_0. \tag{4.32}$$

From the equation of equilibrium in the direction tangent to the circumference of the opening it follows that $\mathrm{d}S = 0$, whence $S = \text{const}$; according to (4.31) also $\varrho = \text{const}$. Thus the opening has the form of a circle of radius ϱ and the axial force in the rope around its circumference is

$$S = \frac{1}{2} q_0 R_0 \varrho. \tag{4.33}$$

The relation (4.33) follows also from the equations (4.7) and (4.15). If we put $T_2 = T_1 = T_0$, then (4.15) implies $S = S_0$ and, substituting (4.32) in the first equation (4.7), we obtain formula (4.33).

4.2 *Parametric Equations of the Entrance Opening Curve*

The form of the curve forming the circumference of the opening in the membrane, which does not disturb the membrane state of stress in the jacket of the air-supported structure, can be calculated directly from the parametric equations which we shall derive now for the case that T_1 and T_2 are both constant.

According to Fig. III-9b,

$$\frac{dx}{d\gamma}=\varrho \sin \gamma, \quad \frac{dy}{d\gamma}=\varrho \cos \gamma, \tag{4.34}$$

where x, y are Cartesian coordinates. If we substitute (4.16) in (4.34) and integrate, we get

$$\begin{aligned} x(\gamma)&=\varrho_0 T_2^{3/2}\int\frac{\sin \gamma}{(T_2 \cos^2 \gamma+T_1 \sin^2 \gamma)^{3/2}}\,d\gamma+C_1, \\ y(\gamma)&=\varrho_0 T_2^{3/2}\int\frac{\cos \gamma}{(T_2 \cos^2 \gamma+T_1 \sin^2 \gamma)^{3/2}}\,d\gamma+C_2, \end{aligned} \tag{4.35}$$

where C_1 and C_2 are integration constants.

Calculating the indefinite integrals in the equations (4.35), we obtain [21]

$$\begin{aligned} x(\gamma)&=-\varrho_0\frac{T_2}{T_1}\frac{\cos \gamma}{[1-(1-T_1/T_2)\sin^2 \gamma]^{1/2}}+C_1, \\ y(\gamma)&=\varrho_0\frac{\cos \gamma}{[1-(1-T_1/T_2)\sin^2 \gamma]^{1/2}}+C_2 \end{aligned} \tag{4.36}$$

The integration constants C_1, C_2 can be determined from the condition that the origin O of the orthogonal coordinate system is situated in the apex of the opening. That means that for $\gamma=0$ we have

$$x(0)=0, \quad y(0)=0. \tag{4.37}$$

Substituting (4.36) into (4.37), we obtain

$$C_1=\varrho_0 T_2/T_1, \quad C_2=0. \tag{4.38}$$

For $T_1/T_2>0$, the equations (4.36) are the parametric equations of an ellipse with principal demiaxes $a_0=\varrho_0 T_2/T_1$, $b_0=\varrho_0\sqrt{T_2/T_1}$; for $T_1/T_2=0$ we obtain from equation (4.35) the parabola

$$x(\gamma)=\frac{\varrho_0}{2}\tan^2 \gamma, \quad y(\gamma)=\varrho_0 \tan \gamma. \tag{4.39}$$

The parametric equations for the circumference of an opening in a cylindrical air-supported surface, affected only by the internal overpressure q_0, can be obtained by the substitution of the formulae (4.18) into (4.36). Recalling (4.38), we obtain

$$x(\gamma)=2\varrho_0\left[1-\frac{2\cos\gamma}{(2-\sin^2\gamma)^{1/2}}\right],$$
$$y(\gamma)=2\varrho_0\frac{\sin\gamma}{(2-\sin^2\gamma)^{1/2}}. \tag{4.40}$$

If the jacket has the form of a semi-cylindrical surface, affected, apart from the internal overpressure, also by cross wind, we substitute the formulae (4.23) into the equations (4.36).

Then, by (4.38), we obtain the equations

$$x(\gamma)=2\varrho_0\left(1+\frac{\bar{q}_0}{q_0}\right)\left(1-\frac{\cos\gamma}{\left\{1-\left[1-\frac{1}{2(1+\bar{q}_0/q_0)}\right]\sin^2\gamma\right\}^{1/2}}\right),$$
$$y(\gamma)=\varrho_0\frac{\sin\gamma}{\left\{1-\left[1-\frac{1}{2(1+\bar{q}_0/q_0)}\right]\sin^2\gamma\right\}^{1/2}} \tag{4.41}$$

which permit us to study the variability of the opening curve under wind load.

The parametric equations of the opening circumference in the middle of a cycloidal surface can be obtained by the substitution of the formulae (4.29) into the equations (4.36). Then by (4.38), we have

$$x(\gamma)=\frac{3}{2}\varrho_0\left[1-\frac{3\cos\gamma}{(3-\sin^2\gamma)^{1/2}}\right],$$
$$y(\gamma)=\varrho_0\frac{3\sin\gamma}{(3-\sin^2\gamma)^{1/2}}. \tag{4.42}$$

In view of the the fact that for a cycloidal surface the forces T_1 are not parallel as is the case of a cylindrical surface, but are tangent to the meridians, the above solution is approximate. However, near the parallel circle $\alpha=\pi/2$ the cycloidal surface can be approximated by cylindrical surface with sufficient accuracy.

According to (4.36), the parametric equations of a circular opening in a spherical surface have the form $(T_1/T_2=1)$

$$x(\gamma) = \varrho_0(1 - \cos\gamma),$$
$$y(\gamma) = \varrho_0 \sin\gamma. \tag{4.43}$$

In the case of a general change of the magnitude and the direction of the principal membrane forces it is advisable to use the grafic method [18].

Chapter IV

NATURAL VIBRATIONS OF AIR-SUPPORTED STRUCTURES

Introduction

The investigation of the dynamic effects of wind is an essential pre-requisite of a correct design of membrane structures dealt with in Chapter I, ensuring the stability of form as well as aerodynamic stability.

In this chapter we present the methodology of investigating the natural frequencies and the respective modes of natural vibrations of air-supported structures of cylindrical form which are most widely used. A suspended cylindrical membrane, which behaves analogous to the air-supported structure, forms also the principal part of equipment intended for the dampening of dynamic wind effects in tall structures of cylindrical cross-section, such as relay towers, etc. (Chapter I, Section 9). At the end of this chapter we shall give some results of the linear theory of inflatable plates.

In considering to dynamic wind effects, the natural vibrations of the air-supported structure are of fundamental importance, as under the wind load every structure vibrates in some of its natural modes or in a mode comprizing several natural modes. The knowledge of the natural frequencies and the respective natural modes of vibrations enables us to investigate the forced vibrations due primarily to dynamic wind effects, using the modal analysis method (see Chapter V). Also the design of an effective facility for the removal of snow from the surface of air-supported structures by the resonance effect, according to the Czechoslovak patent No. 136 153, requires the determination of natural frequencies of these structures (see Chapter I, Section 1).

The determination of natural frequencies of convex surfaces of revolution, according to the membrane theory, is dealt with in [9] and [42].

Natural vibrations of air-supported structures of cylindrical form are investigated under the following assumptions:

1. The weight of the medium (i.e. air) in which the membrane structure vibrates is neglected.

2. The influence of the inner and outer dampening of the structure on natural vibrations is neglected.

3. The deflections of the cylindrical membrane and its elongation due to static load are considered as of arbitrary magnitude.

4. The form of the cylindrical membrane deformed by static load is replaced in the cross-section of the structure by a system of circular arcs of various radii and straight sectors.

5. Static load is considered in the direction radial to the deformed membrane only.

6. In every circular arc the static load is replaced by a constant mean value.

7. The elongation of the membrane during its vibrations is neglected.

8. The dynamic deflections of the membrane are considered small.

9. The influence of the ends of the air-supported structure is neglected.

Due to assumption 9 we obtain lower values of natural frequencies than the actual ones, while the assumptions 1, 2, 7 and 8 affect the result in the opposite direction. Assumption 3 extends the applicability of the proposed dynamic analysis of air-supported structures. The solution of numerical examples has lead the author to the conclusion that the actual dynamic behaviour of an air-supported structure is more markedly influenced by the magnitude of the axial force in the arc due to static load than by the details of the form of its cross-section. This justifies the assumptions 4, 5 and 6. Also assumption 1 does not spoil noticeably the accuracy of the result [11].

1

Differential Equations of a Freely Vibrating Cylindrical Membrane

In free vibrations of a cylindrical membrane whose cross-section consists of circular arcs, the element of a length ds, cut out of a circular arc of a radius R, is affected by the static radial load q, the static axial force qR, the force of inertia μ ds $\partial^2\bar{v}/\partial t^2$ in the direction of the x-axis (tangent to the arc) in the direction opposite to that of the displacement $\bar{v}$, the force of inertia μ ds $\partial^2\bar{w}/\partial t^2$ in the direction of the y-axis, opposite to that of the radial displacement $\bar{w}$, and the dynamic axial forces $\bar{S}$ and $\bar{S}+(\partial\bar{S}/\partial\vartheta)\,\mathrm{d}\vartheta$. The directions of the displacements and axial forces are shown in Fig. IV-1. The circular arc represents a membrane strip of unit width, e.g. of 1 m.

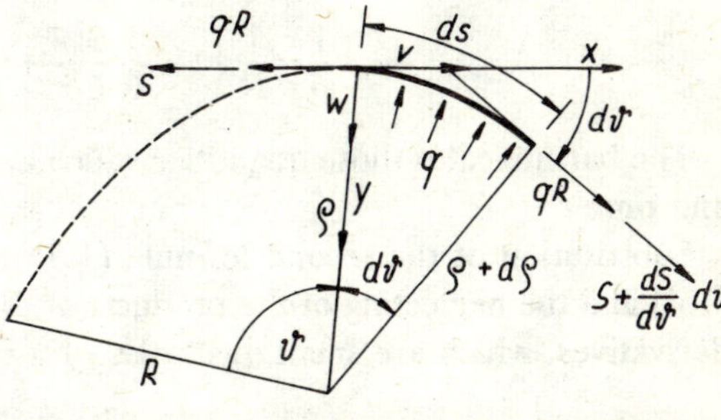

Fig. IV-1. Marking of the positive sense of amplitudes of displacements and forces affecting an element of a circular arc of the membrane.

For the deformed element of the vibrating circular arc of a radius R there hold two equations of dynamic equilibrium, in the direction of the axis x,

$$-\bar{S}-qR+\left(\bar{S}+\frac{\partial\bar{S}}{\partial\vartheta}\,\mathrm{d}\vartheta+qR\right)\cos\mathrm{d}\vartheta-\mu\,\mathrm{d}s\,\frac{\partial^2\bar{v}}{\partial t^2}=0, \tag{1.1}$$

and in the direction of the axis y,

$$(\bar{S}+qR)\cos\left(\frac{\pi}{2}-\mathrm{d}\vartheta\right)-q\,\mathrm{d}s-\mu\,\mathrm{d}s\,\frac{\partial^2\bar{w}}{\partial t^2}=0. \tag{1.2}$$

As we consider small dynamic deflections, we have

$$\mathrm{d}s \approx \bar{\varrho}\,\mathrm{d}\vartheta, \quad \cos \mathrm{d}\vartheta \approx 1, \quad \cos\left(\frac{\pi}{2} - \mathrm{d}\vartheta\right) = \sin \mathrm{d}\vartheta \approx \mathrm{d}\vartheta \tag{1.3}$$

and the equations of equilibrium (1.1) and (1.2) acquire the form

$$\begin{aligned} \frac{1}{\bar{\varrho}} \frac{\partial \bar{S}}{\partial \vartheta} - \mu \frac{\partial^2 \bar{v}}{\partial t^2} &= 0, \\ \frac{1}{\bar{\varrho}} \bar{S} + q\left(\frac{R}{\bar{\varrho}} - 1\right) - \mu \frac{\partial^2 \bar{w}}{\partial t^2} &= 0. \end{aligned} \tag{1.4}$$

If we neglect the elongation during vibrations, the following geometric relations hold [10] for a circular arc:

$$\begin{aligned} \frac{1}{\varrho} = \left(1 + \frac{\partial \bar{\varphi}}{\partial \vartheta}\right), \quad \bar{w} = \frac{\partial \bar{v}}{\partial \vartheta}, \\ \bar{\varphi} = \frac{1}{R}\left(\frac{\partial \bar{w}}{\partial \vartheta} + \int \bar{w}\,\mathrm{d}\vartheta\right). \end{aligned} \tag{1.5}$$

Hence

$$\frac{1}{\bar{\varrho}} = \frac{1}{R}\left(1 + \frac{1}{R} \frac{\partial^2 \bar{w}}{\partial \vartheta^2} + \frac{1}{R}\,\bar{w}\right). \tag{1.6}$$

The bar indicates those quantities which are functions of both the angle ϑ and time t.

Substitution of the second formula (1.5) and of (1.6) into the equations (1.4) and the neglecting of the products of the quantities $\bar{S}$ and $\bar{w}$ and their derivatives, which are small quantities of a higher order, yield

$$\begin{aligned} \frac{1}{R} \frac{\partial^2 \bar{S}}{\partial \vartheta^2} - \mu \frac{\partial^2 \bar{w}}{\partial t^2} &= 0, \\ \frac{1}{R} \bar{S} + \frac{q}{R}\left(\frac{\partial^2 \bar{w}}{\partial \vartheta^2} + \bar{w}\right) - \mu \frac{\partial^2 \bar{w}}{\partial t^2} &= 0, \end{aligned} \tag{1.7}$$

whence we obtain, after the elimination of the quantity $\bar{S}$, the following linear partial differential equation for the radial deflection of the freely vibrating membrane:

$$\frac{\partial^4 \bar{w}}{\partial \vartheta^4} + \frac{\partial^2}{\partial \vartheta^2}\left(\bar{w} - \frac{\mu R}{q} \frac{\partial^2 \bar{w}}{\partial t^2}\right) + \frac{\mu R}{q} \frac{\partial^2 \bar{w}}{\partial t^2} = 0. \tag{1.8}$$

Equation (1.8) is satisfied by the harmonic vibrations

$$\bar{w} = w \sin \omega t, \tag{1.9}$$

where w is the amplitude of the radial deflection and ω is the circular frequency.

Substitution of (1.9) into (1.8) yields the ordinary differential equation of fourth order for the amplitude w of the radial displacement (deflection):

$$\frac{d^4 w}{d\vartheta^4} + (1+\lambda)\frac{d^2 w}{d\vartheta^2} - \lambda w = 0, \tag{1.10}$$

where

$$\lambda = \frac{\mu \omega^2 R}{q}. \tag{1.11}$$

Assuming harmonic vibrations

$$\bar{S} = S \sin \omega t, \tag{1.12}$$

and substituting (1.9) and (1.12) into the second equation (1.7), we obtain the following expression for the amplitude of the dynamic axial force S:

$$S = -q\left[\frac{d^2 w}{d\vartheta^2} + (1+\lambda) w\right]. \tag{1.13}$$

2

General Solution of the Differential Equation and Its Application

The characteristic equation

$$\beta^4+(1+\lambda)\beta^2-\lambda=0, \tag{2.1}$$

pertaining to the differential equation (1.10), has two imaginary and two real roots

$$\begin{aligned} \beta_{1,2}&=\pm\frac{i}{\sqrt{2}}\sqrt{1+\lambda+\sqrt{1+6\lambda+\lambda^2}}\,, \\ \beta_{3,4}&=\pm\frac{1}{\sqrt{2}}\sqrt{-1-\lambda+\sqrt{1+6\lambda+\lambda^2}}\,, \end{aligned} \tag{2.2}$$

where $i=\sqrt{-1}$ and the parameter λ is determined by (1.11).

In terms of

$$\begin{aligned} a&=\frac{\alpha}{\sqrt{2}}\sqrt{1+\lambda+\sqrt{1+6\lambda+\lambda^2}}\,, \\ b&=\frac{\alpha}{\sqrt{2}}\sqrt{-1-\lambda+\sqrt{1+6\lambda+\lambda^2}}\,, \end{aligned} \tag{2.3}$$

the general solution of equation (1.10) may be written as

$$w=C_1\sin a\frac{\vartheta}{\alpha}+C_2\cos a\frac{\vartheta}{\alpha}+C_3\sinh b\frac{\vartheta}{\alpha}+C_4\cosh b\frac{\vartheta}{\alpha}, \tag{2.4}$$

where C_i, $i=1, 2, 3, 4$, are integration constatns, and α is the central angle of the circular arc under consideration, forming part of the cross-section of the freely vibrating membrane (Fig. IV-2).

According to the second formula (1.5) and (2.4), the amplitude of the displacement v in the direction of the tangent is given by the formula

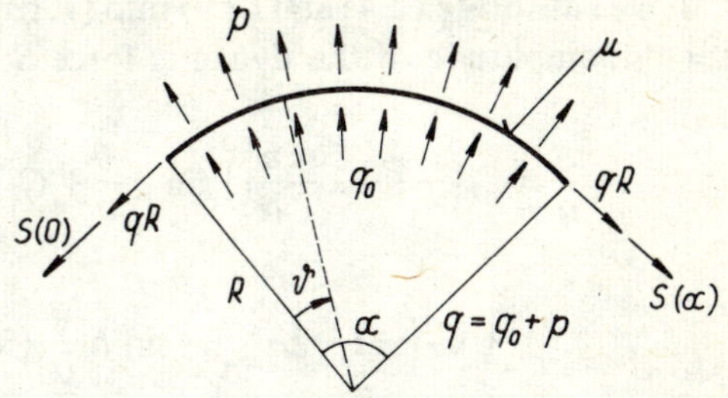

Fig. IV-2. Circular arc of the membrane forming part of the cross-section of a cylindrical air-supported structure.

$$v = -C_1\frac{\alpha}{a}\cos a\frac{\vartheta}{\alpha} + C_2\frac{\alpha}{a}\sin a\frac{\vartheta}{\alpha} + C_3\frac{\alpha}{b}\cosh b\frac{\vartheta}{\alpha} + \\ + C_4\frac{\alpha}{b}\sinh b\frac{\vartheta}{\alpha} + C_5 . \tag{2.5}$$

The integration constant C_5 represents the displacement of the circular arc due to the rotation about its centre. In the case of air-supported structures, which are anchored at the foundations along their whole circumference, such rotation cannot occur during vibrations; therefore

$$C_5 = 0 . \tag{2.6}$$

According to (2.4), we have

$$\frac{dw}{d\vartheta} = C_1\frac{a}{\alpha}\cos a\frac{\vartheta}{\alpha} - C_2\frac{a}{\alpha}\sin a\frac{\vartheta}{\alpha} + C_3\frac{b}{\alpha}\cosh b\frac{\vartheta}{\alpha} + C_4\frac{b}{\alpha}\sinh b\frac{\vartheta}{\alpha} \tag{2.7}$$

and for the amplitude of the angle of rotation of the cross-section φ of the circular arc we obtain (see (1.5))

$$R\varphi = \frac{dw}{d\vartheta} + v = C_1\left(\frac{a}{\alpha} - \frac{\alpha}{a}\right)\cos a\frac{\vartheta}{\alpha} - C_2\left(\frac{a}{\alpha} - \frac{\alpha}{a}\right)\sin a\frac{\vartheta}{\alpha} + \\ + C_3\left(\frac{b}{\alpha} + \frac{\alpha}{b}\right)\cosh b\frac{\vartheta}{\alpha} + C_4\left(\frac{b}{\alpha} + \frac{\alpha}{b}\right)\sinh b\frac{\vartheta}{\alpha} . \tag{2.8}$$

Differentiating (2.7) again with respect to ϑ, we obtain

$$\frac{d^2w}{d\vartheta^2} = -C_1\frac{a^2}{\alpha^2}\sin a\frac{\vartheta}{\alpha} - C_2\frac{a^2}{\alpha^2}\cos a\frac{\vartheta}{\alpha} + C_3\frac{b^2}{\alpha^2}\sinh b\frac{\vartheta}{\alpha} + C_4\frac{b^2}{\alpha^2}\cosh b\frac{\vartheta}{\alpha} . \tag{2.9}$$

If we substitute (2.4) and (2.9) into (1.13), we obtain the following formula for the amplitude of the dynamic force S:

$$\frac{S}{q}=C_1\left(-1-\lambda+\frac{a^2}{\alpha^2}\right)\sin a\frac{\vartheta}{\alpha}+C_2\left(-1-\lambda+\frac{a^2}{\alpha_2}\right)\cos a\frac{\vartheta}{\alpha}+$$
$$+C_3\left(-1-\lambda-\frac{b^2}{\alpha^2}\right)\sin b\frac{\vartheta}{\alpha}+C_4\left(-1-\lambda-\frac{b^2}{\alpha^2}\right)\cosh b\frac{\vartheta}{\alpha}. \tag{2.10}$$

Equations (2.4), (2.5), (2.8) and (2.10) can be used for the investigation of the natural vibrations of an air-supported structure, whose cross-section under static load consists of one circular arc or of several circular arcs of different radii. First we present the case when the air-supported structure is loaded by internal overpressure only (Fig. IV-3). This means that we consider only the mass of the membrane μ, forming the jacket of the air-supported structure, and neglect its weight which is insignificantly low as compared to the magnitude of the internal overpressure.

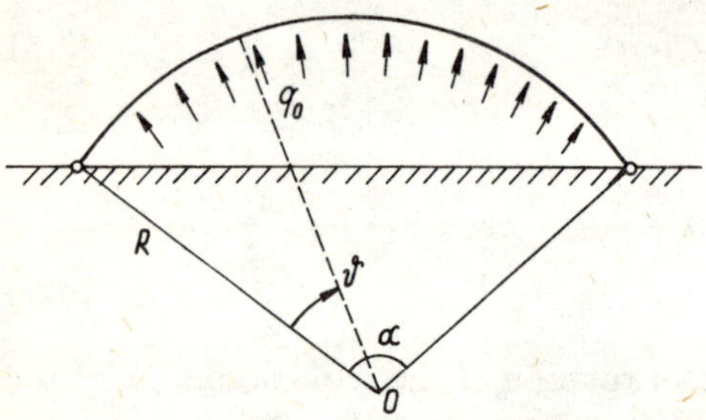

Fig. IV-3. Cross-section of an air-supported membrane structure loaded by internal overpressure.

According to Fig. IV-3, the membrane of the air-supported structure satisfies the following boundary conditions:

$$w(0)=0, \qquad w(\alpha)=0,$$
$$v(0)=0, \qquad v(\alpha)=0. \tag{2.11}$$

Substituting (2.4) and (2.5) with $\vartheta=0$ and $\vartheta=\alpha$ into the equations (2.11), and using (2.6), we obtain the system of four homogeneous algebraic equations ($C_5=0$):

$$\begin{aligned} C_2+C_4&=0, \\ C_1 \sin a+C_2 \cos a+C_3 \sinh b+C_4 \cosh b&=0, \\ -C_1\frac{1}{a}+C_3\frac{1}{b}&=0, \\ -C_1\frac{1}{a}\cos a+C_2\frac{1}{a}\sin a+C_3\frac{1}{b}\cosh b+C_4\frac{1}{b}\sinh b&=0. \end{aligned} \tag{2.12}$$

These equations are linear with respect to the integration constants C_i, $i=1, 2, 3, 4$, and a sufficient condition for the existence of a non-trivial solution is the vanishing of their determinant, whose elements are the coefficients of the integration constants.

This is the condition:

$$\begin{vmatrix} 0 & 1 & 0 & 1 \\ \sin a & \cos a & \sinh b & \cosh b \\ -\frac{1}{a} & 0 & \frac{1}{b} & 0 \\ -\frac{\cos a}{a} & \frac{\sin a}{a} & \frac{\cosh b}{b} & \frac{\sinh b}{b} \end{vmatrix}=0. \tag{2.13}$$

It is possible to determine, by trial and error, the characteristic numbers $\lambda_{01}, \lambda_{02}, \ldots$, for which (2.13) holds.

The principal natural frequency f_1, i.e. the first natural circular frequency ω_1 of the investigated structure, corresponds to the lowest characteristic number which we shall denote by λ_{01}.

The natural circular frequency ω_j can be determined from formula (1.11), if we put $q=q_0$:

$$\omega_j=\sqrt{\frac{q_0}{\mu R}\lambda_{0j}}, \qquad j=1, 2, \ldots. \tag{2.14}$$

The natural frequency is

$$f_j=\frac{\omega_j}{2\pi}, \qquad j=1, 2, \ldots. \tag{2.15}$$

To determine the values of a_{01}, b_{01}, we have to substitute the characteristic number λ_{01} into the formulae (2.3). Substituting those values into the equations (2.4) and (2.5), we obtain the following formulae for the first natural mode of vibrations of the investigated membrane structure ($C_1\neq 0$):

$$w_{01}=C_1\left(\sin a_{01}\frac{\vartheta}{\alpha}+\frac{C_2}{C_1}\cos a_{01}\frac{\vartheta}{\alpha}+\frac{C_3}{C_1}\sinh b_{01}\frac{\vartheta}{\alpha}+\frac{C_4}{C_1}\cosh b_{01}\frac{\vartheta}{\alpha}\right),$$

$$v_{01}=C_1\left(-\frac{\alpha}{a_{01}}\cos a_{01}\frac{\vartheta}{\alpha}+\frac{C_2}{C_1}\frac{\alpha}{a_{01}}\sin a_{01}\frac{\vartheta}{\alpha}+\right. \tag{2.16}$$

$$\left.+\frac{C_3}{C_1}\frac{\alpha}{b_{01}}\cosh b_{01}\frac{\vartheta}{\alpha}+\frac{C_4}{C_1}\frac{\alpha}{b_{01}}\sinh b_{01}\frac{\vartheta}{\alpha}\right).$$

The values of the ratios C_i/C_1, $i=2, 3, 4$, occuring in these formulae are determined by solving of any three algebraic equations of the homogeneous system (2.12) for $a=a_{01}$, $b=b_{01}$.

The amplitudes of deflection w_{01} in the radial direction and the amplitudes of the displacement v_{01} in the direction tandent to the circular membrane are the multiples of any number C_1 and, consequently, have indefinite values.

Thus we come to the conclusion that if we neglect the deformation (elongation) of the membrane, then the natural vibrations of an air-supported structure of cylindrical form are of the same character as the natural vibrations of elastic structures.

If in the equations (2.8) and (2.10) we put $a=a_{01}$, $b=b_{01}$, we obtain formulae for the amplitudes of the angle of rotation φ_1 and the dynamic axial forces S_1 pertaining to the first natural frequency of the investigated membrane structure.

In the same way it is possible to investigate higher natural modes of vibrations of the air-supported structure corresponding to the characteristic numbers λ_{02}, λ_{03}, ... for which equation (2.13) holds.

Instead of (2.13) it is possible to use the characteristic equation, which we obtain by the successive elimination of the integration constants from the equations (2.12). From the first and the third equations (2.12) we obtain

$$C_4=-C_2,\quad C_3=\frac{b}{a}C_1. \tag{2.17}$$

Substituting this into the second and the fourth equations (2.12), we have

$$\left(\sin a-\frac{b}{a}\sinh b\right)C_1+(\cos a-\cosh b)C_2=0,$$

$$\left(-\frac{\cos a}{a}+\frac{1}{a}\cosh b\right)C_1+\left(\frac{\sin a}{a}-\frac{\sinh b}{b}\right)C_2=0, \tag{2.18}$$

whence, after some rearrangements, we obtain the characteristic equation

$$2(1-\cos a \cosh b)+\frac{b^2-a^2}{ab}\sin a \sinh b=0. \tag{2.19}$$

For the air-supported structure in the form of a semi-cylinder, i.e. for $\alpha=\pi$, the first four characteristic numbers for which (2.19) holds have the following values:

$$\lambda_{01}=1.71, \quad \lambda_{02}=5.97, \quad \lambda_{03}=13.07, \quad \lambda_{04}=21.86. \tag{2.20}$$

If

$$q_0=150\ \text{N/m}^2, \quad R=10\ \text{m}, \quad \mu=1\ \text{Nm}^{-3}\ \text{sec}^2=1\ \text{kg/m}^2,$$

we obtain, according to (2.14),

$$\omega_1=5.06\ \text{sec}^{-1}, \quad \omega_2=9.46\ \text{sec}^{-1}, \quad \omega_3=13.99\ \text{sec}^{-1}, \quad \omega_4=18.12\ \text{sec}^{-1}$$

and the first four natural frequencies of the investigated membrane structure are, according to (2.15),

$$f_1=0.81\ \text{sec}^{-1}, \quad f_2=1.51\ \text{sec}^{-1}, \quad f_3=2.22\ \text{sec}^{-1}, \quad f_4=2.89\ \text{sec}^{-1}.$$

Now let us construct by way of an example a system of homogeneous equations for a freely vibrating air-supported structure affected by the internal overpressure q_0 and static wind or snow load.

The distribution of static wind pressure and wind suction over the smooth cylindrical membrane was discussed in Chapter II, Subsection 6.4a. The static wind load can be approximately replaced with p_1, p_2 and p_3, according to Fig. IV-4a [11]. Since the static wind load and the internal overpressure are applied in the direction of the normal to the deformed membrane, the cross-section of the deformed air-supported structure consists of three circular arcs of radii $R_{1,2}$, $R_{2,3}$ and $R_{3,4}$ (Fig. IV-4b) affected by the three constant magnitudes of static load ($q_{1,2}=q_1$, $q_{2,3}=q_2$, $q_{3,4}=q_3$)

$$q_1=q_0+p_1, \quad q_2=q_0+p_2, \quad q_3=q_0+p_3, \tag{2.21}$$

where $p_1<0$ is the mean value of wind pressure in the sector *1—2* and $p_2>0$, $p_3>0$ are the mean values of wind suction in the sectors *2—3* and *3—4*.

For example, for a strip 1 m wide of a semi-cylindrical membrane we have the values [11]

$$p_1=-0.7p_v, \quad p_2=1.2p_v, \quad p_3=0.4p_v, \tag{2.22}$$

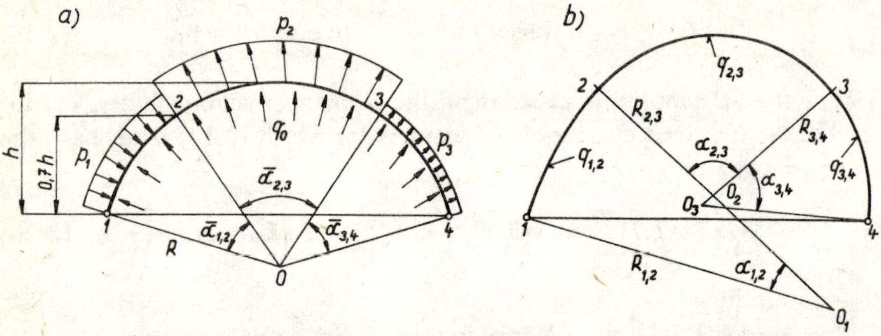

Fig. IV-4. Cross-section of a cylindrical membrane: a) Equivalent loading diagram to replace the internal overpressure and the wind load in the direction perpendicular to the longitudinal axis of the air-supported structure. b) Deformation of the cross-section of the air-supported structure.

where $p_v \approx 0.613 v_0^2$, v_0 being the wind velocity in m/sec and p_v the dynamic wind pressure in N/m^2 on the plane perpendicular to the wind direction.

Three circular arcs of different radii can also replace, approximately, the cross-section of a deformed air-supported structure of cylindrical form affected by symmetric and asymmetric snow load, see Chapter II, Subsection 6.4c.

For a freely vibrating membrane strip consisting of three circular arcs (Fig. IV-4b) we have the following boundary conditions:

$$\begin{aligned} w_1(0)&=0, & v_1(0)&=0,\\ w_1(\alpha_1)&=w_2(0), & v_1(\alpha_1)&=v_2(0),\\ \varphi_1(\alpha_1)&=\varphi_2(0), & S_1(\alpha_1)&=S_2(0),\\ w_2(\alpha_2)&=w_3(0), & v_2(\alpha_2)&=v_3(0),\\ \varphi_2(\alpha_2)&=\varphi_3(0), & S_2(\alpha_2)&=S_3(0),\\ w_3(\alpha_3)&=0, & v_3(\alpha_3)&=0, \end{aligned} \tag{2.23}$$

where the indices signify that the above quantities refer to the sectors *1—2*, *2—3* and *3—4* ($\varphi_i=\varphi_{i,\,i+1}$, $S_i=S_{i,\,i+1}$, $\alpha_i=\alpha_{i,\,i+1}$, $i=1, 2, 3$).

According to (2.4), (2.5), (2.8) and (2.10), we have

$$w_1(0)={}^{(1,2)}C_2+{}^{(1,2)}C_4,$$

$$v_1(0)=-{}^{(1,2)}C_1\frac{\alpha_1}{a_1}+{}^{(1,2)}C_3\frac{\alpha_1}{b_1},$$

$$w_1(\alpha_1)={}^{(1,2)}C_1\sin a_1+{}^{(1,2)}C_2\cos a_1+{}^{(1,2)}C_3\sinh b_1+{}^{(1,2)}C_4\cosh b_1,$$

$$w_2(0)={}^{(2,3)}C_2+{}^{(2,3)}C_4,$$

$$v_2(0)=-{}^{(2,3)}C_1\frac{\alpha_2}{a_2}+{}^{(2,3)}C_3\frac{\alpha_2}{b_2},$$

$$v_1(\alpha_1)=-{}^{(1,2)}C_1\frac{\alpha_1}{a_1}\cos a_1+{}^{(1,2)}C_2\frac{\alpha_1}{a_1}\sin a_1+{}^{(1,2)}C_3\frac{\alpha_1}{b_1}\cosh b_1+$$

$$+{}^{(1,2)}C_4\frac{\alpha_1}{b_1}\sinh b_1,$$

$$R_1\varphi_1(\alpha_1)={}^{(1,2)}C_1\left(\frac{a_1}{\alpha}+\frac{\alpha_1}{a_1}\right)\cos a_1-{}^{(1,2)}C_2\left(\frac{a_1}{\alpha_1}-\frac{\alpha_1}{a_1}\right)\sin a_1+$$

$$+{}^{(1,2)}C_3\left(\frac{b_1}{\alpha_1}+\frac{\alpha_1}{b_1}\right)\cosh b_1+{}^{(1,2)}C_4\left(\frac{b_1}{\alpha_1}+\frac{\alpha_1}{b_1}\right)\sinh b_1\,,$$

$$R_2\varphi_2(0)={}^{(2,3)}C_1\left(\frac{a_2}{\alpha_2}-\frac{\alpha_2}{a_2}\right)+{}^{(2,3)}C_3\left(\frac{b_2}{\alpha_2}+\frac{\alpha_2}{b_2}\right),$$

$$\frac{S_1(\alpha_1)}{q_1}={}^{(1,2)}C_1\left(-1-\lambda_1+\frac{a_1^2}{\alpha_1^2}\right)\sin a_1+{}^{(1,2)}C_2\left(-1-\lambda_1+\frac{a_1^2}{\alpha_1^2}\right)\cos a_1+$$

$$+{}^{(1,2)}C_3\left(-1-\lambda_1-\frac{b_1^2}{\alpha_1^2}\right)\sinh b_1+{}^{(1,2)}C_4\left(-1-\lambda_1-\frac{b_1^2}{\alpha_1^2}\right)\cosh b_1,$$

$$w_2(\alpha_2)={}^{(2,3)}C_1\sin a_2+{}^{(2,3)}C_2\cos a_2+{}^{(2,3)}C_3\sinh b_2+{}^{(2,3)}C_4\cosh b_2,$$

$$\frac{S_2(0)}{q_2}={}^{(2,3)}C_2\left(-1-\lambda_2+\frac{a_2^2}{\alpha_2^2}\right)+{}^{(2,3)}C_4\left(-1-\lambda_2-\frac{b_2^2}{\alpha_2^2}\right),$$

$$w_3(0)={}^{(3,4)}C_2+{}^{(3,4)}C_4,$$

$$v_2(\alpha_2)=-{}^{(2,3)}C_1\frac{\alpha_2}{a_2}\cos a_2+{}^{(2,3)}C_2\frac{\alpha_2}{a_2}\sin a_2+{}^{(2,3)}C_3\frac{\alpha_2}{b_2}\cosh b_2+$$

$$+{}^{(2,3)}C_4\frac{\alpha_2}{b_2}\sinh b_2,$$

$$v_3(0)=-{}^{(3,4)}C_1\frac{\alpha_3}{a_3}+{}^{(3,4)}C_3\frac{\alpha_3}{b_3},\tag{2.24}$$

$$R_2\varphi_2(\alpha_2)={}^{(2,3)}C_1\left(\frac{a_2}{\alpha_2}-\frac{\alpha_2}{a_2}\right)\cos a_2-{}^{(2,3)}C_2\left(\frac{a_2}{\alpha_2}-\frac{\alpha_2}{a_2}\right)\sin a_2+$$

$$+{}^{(2,3)}C_3\left(\frac{b_2}{\alpha_2}+\frac{\alpha_2}{b_2}\right)\cos b_2+{}^{(2,3)}C_4\left(\frac{b_2}{\alpha_2}+\frac{\alpha_2}{b_2}\right)\sinh b_2,$$

$$R_3\varphi_3(0) = {}^{(3,4)}C_1\left(\frac{a_3}{\alpha_3} - \frac{\alpha_3}{a_3}\right) + {}^{(3,4)}C_3\left(\frac{b_3}{\alpha_3} + \frac{\alpha_3}{b_3}\right),$$

$$\frac{S_2(\alpha_2)}{q_2} = {}^{(2,3)}C_1\left(-1 - \lambda_2 + \frac{a_2^2}{\alpha_2^2}\right)\sin a_2 + {}^{(2,3)}C_2\left(-1 - \lambda_2 + \frac{a_2^2}{\alpha_2^2}\right)\cos a_2 +$$

$$+ {}^{(2,3)}C_3\left(-1 - \lambda_2 - \frac{b_2^2}{\alpha_2^2}\right)\sinh b_2 + {}^{(2,3)}C_4\left(-1 - \lambda_2 - \frac{b_2^2}{\alpha_2^2}\right)\cosh b_2,$$

$$\frac{S_3(0)}{q_3} = {}^{(3,4)}C_2\left(-1 - \lambda_3 + \frac{a_3^2}{\alpha_3^2}\right) + {}^{(3,4)}C_4\left(-1 - \lambda_3 - \frac{b_3^2}{\alpha_3^2}\right),$$

$$w_3(\alpha_3) = {}^{(3,4)}C_1 \sin a_3 + {}^{(3,4)}C_2 \cos a_3 + {}^{(3,4)}C_3 \sinh b_3 + {}^{(3,4)}C_4 \cosh b_3,$$

$$v_3(\alpha_3) = -{}^{(3,4)}C_1 \frac{\alpha_3}{a_3}\cos a_3 + {}^{(3,4)}C_2 \frac{\alpha_3}{a_3}\sin a_3 + {}^{(3,4)}C_3 \frac{\alpha_3}{b_3}\cosh b_3 +$$

$${}^{(3,4)}C_4 \frac{\alpha_3}{b_3}\sinh b_3,$$

where ($R_i = R_{i,j}$, $i = 1, 2, 3$)

$$\lambda_i = \frac{\mu_i \omega^2 R_i}{q_i}, \qquad i = 1, 2, 3, \tag{2.25}$$

and the quantities a_i, b_i, $i = 1, 2, 3$, can be determined from the expressions (2.3) by substituting λ_i for λ.

Substitution of the formulae (2.24) into the boundary conditions (2.23) yields the system of twelve homogeneous algebraic equations shown in Table II. These equations are linear with regard to the twelve integration constants.

The solution of this system, for example by the methods presented by the author in [14], yields the characteristic numbers and the corresponding ratios of integration constants which determine the respective natural mode of vibrations of the investigated air-supported structure.

Equations (2.25) yield the following relations between the parameters λ_i:

$$\lambda_2 = \frac{q_1 \mu_2 R_2}{q_2 \mu_1 R_1}\lambda_1, \quad \lambda_3 = \frac{q_1 \mu_3 R_3}{q_3 \mu_1 R_1}\lambda_1, \tag{2.26}$$

Thus the solution of the homogeneous equations in Table II yields the characteristic numbers λ_{0j}, $j = 1, 2, \ldots$, pertaining to the sector *1—2*. From the values of λ_{0j} we calculate, according to (2.14) and (2.15), the natural circular frequencies and the natural frequencies.

The amplitudes of deflection w, displacement v, angle of rotation φ and the axial dynamic force S in the individual sectors are determined from the equations (2.4), (2.5), (2.8) and (2.10) into which we substitute the values of a, b and the corresponding rations of integration constants satisfying the homogeneous equations given in Table II.

The number of resulting homogeneous algebraic equations obtained from the general solution of the differential equation (1.10) is four times larger than the number of circular arcs replacing the cross-section of the membrane.

A substantial reduction of the number of resulting equations can be attained by the application of the semi-deformation method which we shall presently discuss.

3

Deduction of Fundamental Equations of the Semi-Deformation Method

In the semi-deformation method, elaborated by the author in [11], we introduce as unknown quantities the amplitudes of displacements

$$w_i, \; v_i, \; w_j, \; v_j \tag{3.1}$$

of the ends i and j of a circular arc. The positive sense of the end deformations and of the end forces $S_{i,j}$, $S_{j,i}$ is shown in Fig. IV-5.

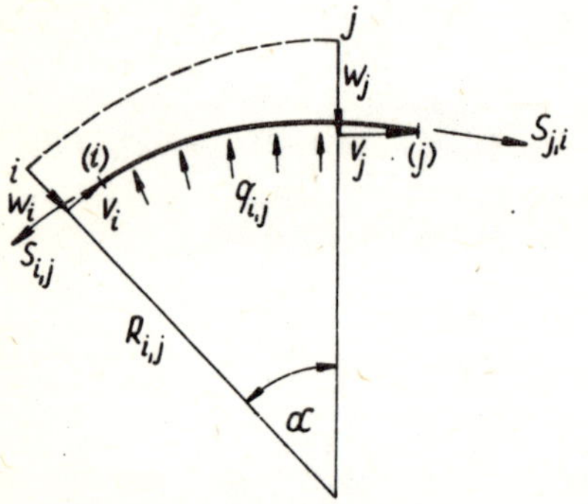

Fig. IV-5. Marking of the positive sense of the amplitudes of the end displacements and end axial forces in a membrane circular arc.

The fundamental equations of the semi-deformation method can be deduced by the elimination of the integration constants C_i, $i = 1, 2, 3, 4$, occuring in the equations (2.4), (2.5), (2.8) and (2.10), and by expressing the quantities $\varphi_{i,j}$, $\varphi_{j,i}$ and $S_{i,j}$, $S_{j,i}$ by means of displacements (3.1).

If we put $\vartheta = 0$ and $\vartheta = \alpha$ in (2.4) and (2.5), we get

$$\begin{aligned} w_i &= C_2 + C_4, \quad v_i = -C_1 \frac{\alpha}{a} + C_3 \frac{\alpha}{b}, \\ w_j &= C_1 \sin a + C_2 \cos a + C_3 \sinh b + C_4 \cosh b, \\ v_j &= -C_1 \frac{\alpha}{a} \cos a + C_2 \frac{\alpha}{a} \sin a + C_3 \frac{\alpha}{b} \cosh b + C_4 \frac{\alpha}{b} \sinh b, \end{aligned} \tag{3.2}$$

whence we obtain for the integration constants the expressions

$$\begin{aligned} C_1 &= V_1 w_i + V_2 w_j + V_3 v_i + V_4 v_j, \\ C_2 &= Z_1 w_i + Z_2 w_j + Z_3 v_i + Z_4 v_j, \\ C_3 &= \frac{b}{a} V_1 w_i + \frac{b}{a} V_2 w_j + \left(\frac{b}{a} V_3 + \frac{b}{\alpha}\right) v_i + \frac{b}{a} V_4 v_j, \\ C_4 &= (1 - Z_1) w_i - Z_2 w_j - Z_3 v_i - Z_4 v_j, \end{aligned} \tag{3.3}$$

where

$$\begin{aligned} V_1 &= \frac{1}{\Delta}\left(\frac{\alpha}{b}\cos a \sinh b - \frac{\alpha}{a}\sin a \cosh b\right), \\ V_2 &= \frac{1}{\Delta}\left(\frac{\alpha}{a}\sin a - \frac{\alpha}{b}\sinh b\right), \\ V_3 &= \frac{1}{\Delta}\left(\cos a \cosh b - \frac{b}{a}\sin a \sinh b - 1\right), \\ V_4 &= \frac{1}{\Delta}(\cosh b - \cos a), \\ Z_1 &= \frac{1}{\Delta}\frac{\alpha}{a}\left(1 - \cos a \cosh b - \frac{a}{b}\sin a \sinh b\right), \\ Z_2 &= \frac{1}{\Delta}\frac{\alpha}{a}(\cos a - \cosh b), \\ Z_3 &= -\frac{1}{\Delta}\left(\sin a \cosh b + \frac{b}{a}\sinh b \cos a\right), \\ Z_4 &= \frac{1}{\Delta}\left(\sin a + \frac{b}{a}\sinh b\right), \end{aligned} \tag{3.4}$$

and

$$\Delta = \frac{\alpha}{a}\left[2(1 - \cos a \cosh b) + \left(\frac{b}{a} - \frac{a}{b}\right)\sin a \sinh b\right]. \tag{3.5}$$

If we substitute the formulae (3.3) into the equations (2.8) and (2.10), and put $\vartheta = 0$ and $\vartheta = \alpha$, we obtain the fundamental equations of the semi-deformation method

$$\begin{aligned}
\varphi_{i,j} &= \frac{1}{R_{i,j}}(-Aw_i - Bw_j - Cv_i + Dv_j),\\
\varphi_{j,i} &= \frac{1}{R_{i,j}}(Bw_i + Aw_j + Dv_i - Cv_j),\\
S_{i,j} &= q_{i,j}(-Ew_i - Fw_j - Gv_i + Hv_j),\\
S_{j,i} &= q_{i,j}(-Fw_i - Ew_j - Hv_i + Gv_j),
\end{aligned} \tag{3.6}$$

where

$$\begin{aligned}
A &= \frac{1}{\Delta}\left(\frac{a}{b}+\frac{b}{a}\right)\left(\frac{b}{a}\sin a \cosh b - \cos a \sinh b\right),\\
B &= \frac{1}{\Delta}\left(\frac{a}{b}+\frac{b}{a}\right)\left(\sinh b - \frac{b}{a}\sin a\right),\\
C &= \frac{1}{\Delta}\left[\left(\frac{\alpha}{b}+2\frac{b}{\alpha}-\frac{b}{a}\frac{\alpha}{a}\right)\sin a \sinh b +\right.\\
&\quad \left.+\left(\frac{a}{\alpha}-2\frac{\alpha}{a}-\frac{b}{a}\frac{b}{\alpha}\right)(1-\cos a \cosh b)\right],\\
D &= \frac{1}{\Delta}\frac{b}{\alpha}\left(\frac{a}{b}+\frac{b}{a}\right)(\cosh b - \cos a),\\
E &= -\frac{1}{\Delta}\left\{\left[\frac{b}{\alpha}\left(\frac{a}{b}-\frac{b}{a}\right)-2\frac{\alpha}{a}(\lambda+1)(1-\cos a \cosh b)+\right.\right.\\
&\quad \left.\left.+\left[\frac{\alpha}{a}\left(\frac{a}{b}-\frac{b}{a}\right)(\lambda+1)-\frac{a}{\alpha}\left(\frac{a}{b}+\frac{b^3}{a^3}\right)\right]\sin a \sinh b\right\},\right.\\
F &= \frac{1}{\Delta}\frac{b}{\alpha}\left(\frac{a}{b}+\frac{b}{a}\right)(\cosh b - \cos a),\\
G &= \frac{1}{\Delta}\left(\frac{a^2}{\alpha^2}+\frac{b^2}{\alpha^2}\right)\left(\sin a \cosh b + \frac{b}{a}\cos a \sinh b\right),\\
H &= \frac{1}{\Delta}\left(\frac{a^2}{\alpha^2}-\frac{b^2}{\alpha^2}\right)\left(\sin a + \frac{b}{a}\sinh b\right),
\end{aligned} \tag{3.7}$$

Δ being determined by (3.5) and $D = F$.

In view of the notation (2.3), the functions (3.7) are the functions of two dimensionless parameters α and λ only, thus they can be presented in tables or by nomograms. The tables of the values of the functions A, B, C, $D = F$,

G and H, whose application considerably facilitates the solution of numerical examples, are given on pp. 311—440 of the book. The values the function $E=E(\alpha, \lambda)$ can be determined from the relation

$$E=C+1, \tag{3.8}$$

which can be deduced from the formulae for the functions C and E in (3.7), with the aid of the relations (2.3). The tables can be also used for the investigation of forced vibrations of air-supported cylindrical structures; this will be shown in Chapter V.

4

Construction of Deformation Equations

The resulting system of homogeneous deformation equations for freely vibrating air-supported structures of cylindrical form can be constructed by the substitution of the realations of the form (3.6) into the equations of equilibrium and the equations of continuity valid for the boundary between two adjacent circular arcs.

For the boundary i of the arcs $i-1$, i and i, $i+1$, which form part of the vibrating air-supported structure, the condition of equilibrium for the amplitudes of dynamic forces is given by the equation (Fig. IV-6b)

$$S_{i,\,i-1} = S_{i,\,i+1}, \tag{4.1}$$

and the condition of continuity has the form (Fig. IV-6a)

$$\varphi_{i,\,i-1} = \varphi_{i,\,i+1}. \tag{4.2}$$

Equation (4.1) expresses the fact that the amplitude of the dynamic axial force on the left-hand side of the boundary i is equal to the amplitude of this force on the right-hand side of the boundary i. Equation (4.2) states that at

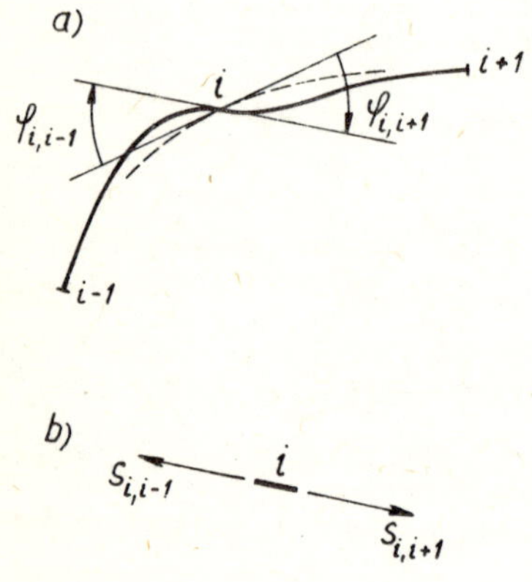

Fig. IV-6. a) Deformation of the cross-section of a vibrating cylindrical membrane at the boundary of adjacent circular arcs. b) Amplitudes of end axial forces affecting the boundary of two adjacent circular arcs of the membrane.

the point i the tangent to the arc $i, i-1$ is identical with the tangent to the arc $i, i+1$, which means that the vibrating membrane has no sudden change of direction at the point i.

For the amplitude of the angle of rotation $\varphi_{i-1,i}$ and the amplitude of the dynamic axial force $S_{i-1,i}$, pertaining to the arc $i-1, i$ at the point i, we have relations of the form of the second and the fourth relations (3.6), while for the amplitude of the angle of rotation $\varphi_{i,i+1}$ and the amplitude of the dynamic force $S_{i,i+1}$ the relations have the form of the first and the third relations (3.6).

Thus we have

$$\begin{aligned}
\varphi_{i,i-1} &= \frac{1}{R_{i-1,i}}(B_{i-1,i}w_{i-1} + A_{i-1,i}w_i + D_{i-1,i}v_{i-1} - C_{i-1,i}v_i), \\
\varphi_{i,i+1} &= \frac{1}{R_{i,i+1}}(-A_{i,i+1}w_i - B_{i,i+1}w_{i+1} - C_{i,i+1}v_i + D_{i,i+1}v_{i+1}), \\
S_{i,i-1} &= q_{i-1,i}(-F_{i-1,i}w_{i-1} - E_{i-1,i}w_i - H_{i-1,i}v_{i-1} + G_{i-1,i}v_i), \\
S_{i,i+1} &= q_{i,i+1}(-E_{i,i+1}w_i - F_{i,i+1}w_{i+1} - G_{i,i+1}v_i + H_{i,i+1}v_{i+1}),
\end{aligned} \tag{4.3}$$

where the subscript $i-1, i$ at the functions A to H indicates that these functions, as given by (3.7), are functions of the parameters $\alpha_{i-1,i}$ and $\lambda_{i-1,i}$, pertaining to the arc $i-1, i$; the index $i, i+1$ indicates that they are functions of the parameters $\alpha_{i,i+1}$ and $\lambda_{i,i+1}$, pertaining to the arc $i, i+1$.

Substituting the expressions (4.3) into the equations (4.2) and (4.1), we obtain two homogeneous deformation equations

$$\begin{aligned}
&B_{i-1,i}w_{i-1} + \left(A_{i-1,i} + \frac{R_{i-1,i}}{R_{i,i+1}}A_{i,i+1}\right)w_i + \frac{R_{i-1,i}}{R_{i,i+1}}B_{i,i+1}w_{i+1} + \\
&+ D_{i-1,i}v_{i-1} - \left(C_{i-1,i} - \frac{R_{i-1,i}}{R_{i,i+1}}C_{i,i+1}\right)v_i - \frac{R_{i-1,i}}{R_{i,i+1}}D_{i,i+1}v_{i+1} = 0, \\
&F_{i-1,i}w_{i-1} + \left(E_{i-1,i} - \frac{q_{i,i+1}}{q_{i-1,i}}E_{i,i+1}\right)w_i - \frac{q_{i,i+1}}{q_{i-1,i}}F_{i,i+1}w_{i+1} + \\
&+ H_{i-1,i}v_{i-1} - \left(G_{i-1,i} + \frac{q_{i,i+1}}{q_{i-1,i}}G_{i,i+1}\right)v_i + \frac{q_{i,i+1}}{q_{i-1,i}}H_{i,i+1}v_{i+1} = 0.
\end{aligned} \tag{4.4}$$

Equations (4.4) are valid for the boundary of any two adjacent arcs representing the cross-section of the investigated air-supported membrane structure. The overall number of deformation equations is twice the number of boundaries of arcs into which the cross-section of the air-supported structure has been divided.

Thus, for example, for the freely vibrating air-supported structure in Fig. IV-4a, whose cross-section has been replaced by three circular arcs, the resulting deformation equations can be written directly as indicated by (4.4). Putting

$$w_1=0, \quad v_1=0, \quad w_4=0, \quad v_4=0, \tag{4.5}$$

we have

$$\begin{gathered}
\left(A_{1,2}+\frac{R_{1,2}}{R_{2,3}}A_{2,3}\right)w_2+\\
+\frac{R_{1,2}}{R_{2,3}}B_{2,3}w_3-\left(C_{1,2}-\frac{R_{1,2}}{R_{2,3}}C_{2,3}\right)v_2-\frac{R_{1,2}}{R_{2,3}}D_{2,3}v_3=0,\\
B_{2,3}w_2+\left(A_{2,3}+\frac{R_{2,3}}{R_{3,4}}A_{3,4}\right)w_3+D_{2,3}v_2-\left(C_{2,3}-\right.\\
\left.-\frac{R_{2,3}}{R_{3,4}}C_{3,4}\right)v_3=0,\\
\left(E_{1,2}-\frac{q_{2,3}}{q_{1,2}}E_{2,3}\right)w_2-\\
-\frac{q_{2,3}}{q_{1,2}}F_{2,3}w_3-\left(G_{1,2}+\frac{q_{2,3}}{q_{1,2}}G_{2,3}\right)v_2+\frac{q_{2,3}}{q_{1,2}}H_{2,3}v_3=0,\\
F_{2,3}w_2+\left(E_{2,3}-\frac{q_{3,4}}{q_{2,3}}E_{3,4}\right)w_3+\\
+H_{2,3}v_2-\left(G_{2,3}+\frac{q_{3,4}}{q_{2,3}}G_{3,4}\right)v_3=0.
\end{gathered} \tag{4.6}$$

Thus the application of the semi-deformation method yields, for the air-supported structure from Fig. IV-4a, four resulting homogeneous equations instead of the twelve equations constructed in Section 2 by means of the general solution of the differential equation (1.10).

The characteristic numbers of the homogeneous system (4.6) can be determined from the equation

$$\begin{vmatrix} A_{1,2}+\frac{R_{1,2}}{R_{2,3}}A_{2,3}, & \frac{R_{1,2}}{R_{2,3}}B_{2,3}, & -\left(C_{1,2}-\frac{R_{1,2}}{R_{2,3}}C_{2,3}\right), & -\frac{R_{1,2}}{R_{2,3}}D_{2,3} \\ B_{2,3}, & A_{2,3}+\frac{R_{2,3}}{R_{3,4}}A_{3,4}, & D_{2,3}, & -\left(C_{2,3}-\frac{R_{2,3}}{R_{3,4}}C_{3,4}\right) \\ E_{1,2}-\frac{q_{2,3}}{q_{1,2}}E_{2,3}, & \frac{q_{2,3}}{q_{1,2}}F_{2,3}, & -\left(G_{1,2}+\frac{q_{2,3}}{q_{1,2}}G_{2,3}\right), & \frac{q_{2,3}}{q_{1,2}}H_{2,3} \\ F_{2,3}, & E_{2,3}-\frac{q_{3,4}}{q_{2,3}}E_{3,4}, & H_{2,3}, & -\left(G_{2,3}+\frac{q_{3,4}}{q_{2,3}}G_{3,4}\right) \end{vmatrix} = 0, \tag{4.7}$$

whose left-hand side is the determinant consisting of the coefficients of the unknown displacements

$$w_2, \quad w_3, \quad v_2, \quad v_3. \tag{4.8}$$

The ratios of the displacements (4.8), characterizing the natural mode of vibrations of the given air-supported structure, can be determined by the solution of any three equations of the system (4.6), if we substitute into them the respective characteristic number.

The example illustrates how the use of the semi-deformation method and the tables of functions (3.7) considerably facilitates the investigation of natural vibrations of air-supported structures. The calculation of characteristic numbers and the characteristic vectors from a major number of homogeneous equations is advatageous, if numerical methods are used [14].

The amplitudes of radial displacement (deflection) w, tangential displacement v, the angle of rotation of the arc cross-section φ and the dynamic axial force S at any point of the air-supported structure vibrating in a certain mode, can be determined from the equations (2.4), (2.5), (2.8) and (2.10); we have to substitute into them the respective characteristic number of the homogeneous system of deformation equations and the corresponding ratios of integration constants. These are determined by calculating the ratios of the displacements (4.8), with the aid of the relations (3.3).

5

Natural Vibrations of Cylindrical Membranes whose Cross-Section is Replaced by Circular Arcs and Straight Sectors

In a particular case, when the symmetric snow load $p_s < 0$ in the sector *2—3* (Fig. IV-7a) is of the same intensity as the internal overpressure $q_0 > 0$, the resulting load $q = q_0 + p_s$ in the sector *2—3* equals zero ($q = 0$) and the cross-section of the cylindrical membrane consists of two circular arcs *1—2* and *3—4* and a straight sector *2—3.* During the tests on a model of a semi-cylindrical membrane in a wind tunnel it was observed that on the windward side of the membrane, due to the deformation of its cross-section,

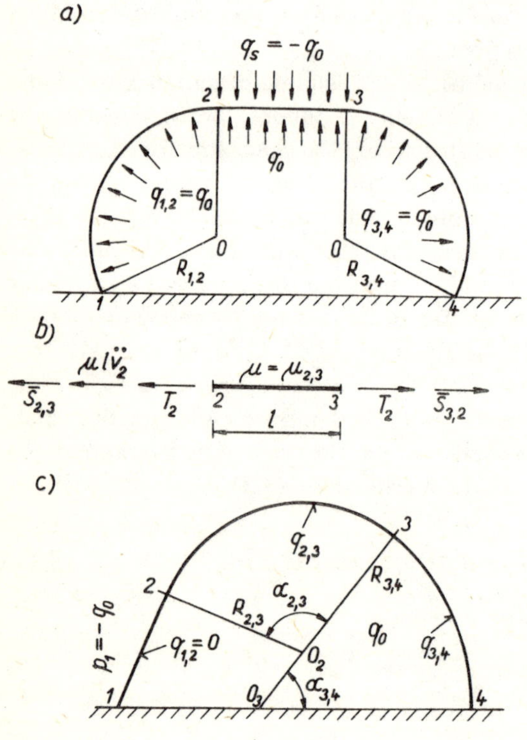

Fig. IV-7. a) Deformed cross-section of a cylindrical membrane affected by the internal overpressure and symmetric snow load. b) Effect of forces on a straight sector of the cross-section of a freely vibrating cylindrical membrane. c) Deformation of the cross-section of a cylindrical membrane under static wind load and internal overpressure.

a change in the wind pressure distribution occurs to the effect that, for some values of the ratio of the internal overpressure and wind velocity, the cross-section in the sector *1—2* becomes a straight sector (Fig. IV-7c). Then we have in the sector *1—2* that $q=q_0+p_1=0$, i.e. $p_1=-q_0$, p_1 being the dynamic wind pressure.

We shall present now a method of investigation of natural vibrations of cylindrical membranes under these particular ways of loading.

For the amplitude of the deflection w of the straight sector i—j of length l (Fig. IV-8) we get the same differential equation as for the free transverse vibrations of a string [14]:

$$\frac{d^2w}{dx^2}+\frac{\mu\omega^2}{T_2}w=0, \tag{5.1}$$

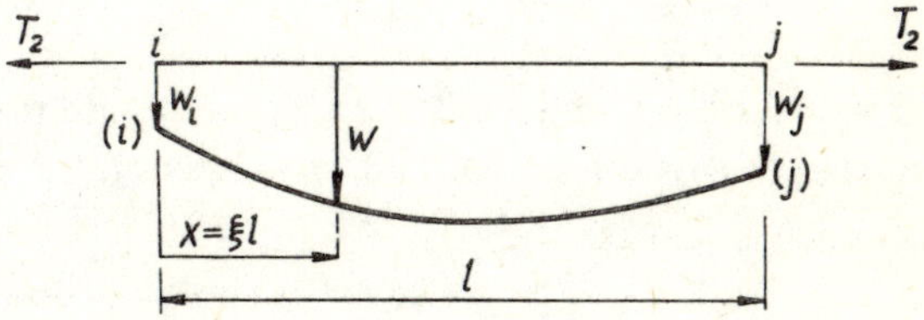

Fig. IV-8. Marking of the amplitude of deflection in a straight sector of the cross-section of a vibrating cylindrical membrane.

Here T_2 is the static axial (membrane) force in the straight sector ($T_2=q_{1,2}R_{1,2}=q_{3,4}R_{3,4}$).

If we put $x=\xi l$ and

$$\Lambda^2=\frac{\mu\omega^2 l^2}{T_2}, \tag{5.2}$$

we obtain from (5.1) the differential equation

$$\frac{d^2w}{d\xi^2}+\Lambda^2 w=0 \tag{5.3}$$

whose general solution is

$$w(\xi)=C_1\sin\Lambda\xi+C_2\cos\Lambda\xi. \tag{5.4}$$

For the angle of rotation $\varphi(\xi)$ we obtain

$$l\varphi(\xi)=\frac{dw}{d\xi}=C_1\Lambda\cos\Lambda\xi-C_2\Lambda\sin\Lambda\xi. \tag{5.5}$$

Let us express the integration constants C_1 and C_2 in terms of the end displacements w_i and w_j of the sector i—j. If we put $\xi=0$ and $\xi=1$ in equation (5.4), we obtain $[w(0)=w_i,\ w(1)=w_j]$

$$w_i=C_2, \quad w_j=C_1 \sin \Lambda + C_2 \cos \Lambda, \tag{5.6}$$

which implies

$$\begin{aligned} C_1 &= -w_i \cot \Lambda + w_j \frac{1}{\sin \Lambda}, \\ C_2 &= w_i . \end{aligned} \tag{5.7}$$

Substituting (5.7) into (5.5), we obtain

$$l\varphi(\xi) = -w_i(\sin \Lambda\xi + \cot \Lambda \cos \Lambda\xi)\Lambda + w_j \frac{\Lambda}{\sin \Lambda} \cos \Lambda\xi . \tag{5.8}$$

If we take $\xi=0$ and $\xi=1$ in (5.8), we obtain the following formulae for the angles of rotation of the end cross-sections of the straight sector i—j:

$$\begin{aligned} \varphi_{i,j} &= \frac{1}{l_{i,j}}(-\Phi w_i + \Psi w_j), \\ \varphi_{j,i} &= \frac{1}{l_{i,j}}(-\Psi w_i + \Phi w_j), \end{aligned} \tag{5.9}$$

where

$$\Phi = \Lambda \cot \Lambda, \quad \Psi = \frac{\Lambda}{\sin \Lambda}. \tag{5.10}$$

The dynamic axial force in the transversely vibrating straight sector equals zero; consequently

$$S_{i,j}=0, \quad S_{j,i}=0. \tag{5.11}$$

In the case when, during the vibrations of the cylindrical membrane, the straight sector moves also in the direction of its axis, as is the case, for example, for the structure in Fig. IV-7a, the straight sector *2*—*3* is affected by the end dynamic forces $\bar{S}_{2,3}$ and $\bar{S}_{3,2}$ and the force of inertia $\mu l \ddot{\bar{v}}_2$ (Fig. IV-7b), where $\ddot{\bar{v}}_2 = \ddot{\bar{v}}_3$ is the acceleration of the sector *2*—*3* in the direction of its axis. According to d'Alembert's principle, we have the equation of dynamic equilibrium

$$\bar{S}_{2,3}-\bar{S}_{3,2}+\mu l\ddot{\bar{v}}_2=0; \qquad (5.12)$$

substituting here the expressions

$$\bar{S}_{2,3}=S_{2,3}\sin\omega t, \quad \bar{S}_{3,2}=S_{3,2}\sin\omega t, \quad \bar{v}_2=v_2\sin\omega t,$$

we arrive at

$$S_{2,3}-S_{3,2}-\mu l\omega^2 v_2=0, \qquad (5.13)$$

where $S_{2,3}$, $S_{3,2}$ and v_2 are the amplitudes of the respective quantities.

The relations (5.9) and (5.11) or (5.13) are the principal equations of the semi-deformation method for the straight sector of the cross-section of the cylindrical membrane. They can be used together with the equations (3.6) for the construction of the deformation equations for cylindrical membranes whose cross-sections, after the deformation due to static load, are approximated by circular arcs and straight sectors.

Thus, for example, for a freely vibrating cylindrical air-supported structure, deformed by the internal overpressure q_0 and the symmetric snow load q_s we have, according to Fig. IV-7a, the following conditions of continuity at the points *2* and *3*:

$$\varphi_{2,1}=\varphi_{2,3}, \quad \varphi_{3,2}=\varphi_{3,4}. \qquad (5.14)$$

For the boundary *2* and *3* the equations of equilibrium are

$$S_{2,1}=S_{2,3}, \quad S_{3,2}=S_{3,4}, \qquad (5.15)$$

whence

$$S_{2,3}-S_{3,2}=S_{2,1}-S_{3,4}, \qquad (5.16)$$

and combining this with equation (5.13), we have

$$S_{2,1}-S_{3,4}-\mu l\omega^2 v_2=0. \qquad (5.17)$$

If we substitute in the equations (5.14) the first two formulae (3.6) for $\varphi_{2,1}$ and $\varphi_{3,4}$, and the expressions of the form (5.9) for $\varphi_{2,3}$, $\varphi_{3,2}$, and further substitute in (5.17) the last two expressions (3.6) for $S_{2,1}$ and $S_{3,4}$, we obtain three deformation equations

$$\left(\frac{1}{R_{1,2}}A_{2,1}+\frac{1}{l_{2,3}}\Phi_{2,3}\right)w_2-\frac{1}{l_{2,3}}\Psi_{2,3}w_3-\frac{1}{R_{1,2}}C_{2,1}v_2=0,$$

$$\frac{1}{l_{2,3}}\Psi_{2,3}w_2+\left(\frac{1}{R_{3,4}}A_{3,4}+\frac{1}{l_{2,3}}\Phi_{2,3}\right)w_3+\frac{1}{R_{3,4}}C_{3,4}v_2=0,$$

$$-q_{1,2}E_{1,2}w_2+q_{3,4}E_{3,4}w_3+(q_{1,2}G_{1,2}+q_{3,4}G_{3,4}-\mu_{2,3}l_{2,3}\omega^2)v_2=0, \tag{5.18}$$

as

$$w_1=w_4=v_1=v_4=0. \tag{5.19}$$

The value of $\mu_{2,3}$ includes also the mass of the membrane and the mass of the snow in the sector *2—3*.

The three homogeneous equations (5.18) are linear with respect to the three unknown displacements w_2, w_3 and v_2.

The natural circular frequency of the air-supported structure according to Fig. IV-7a can be determined from the equation

$$\begin{vmatrix} A_{2,1}+\dfrac{R_{1,2}}{l_{2,3}}\Phi_{2,3}, & -\dfrac{R_{1,2}}{l_{2,3}}\Psi_{2,3}, & -C_{2,1} \\ -\dfrac{R_{3,4}}{l_{2,3}}\Psi_{2,3}, & A_{3,4}+\dfrac{R_{3,4}}{l_{2,3}}\Phi_{2,3}, & -C_{3,4} \\ -E_{1,2}, & \dfrac{q_{3,4}}{q_{1,2}}E_{3,4}, & G_{1,2}+\dfrac{q_{3,4}}{q_{1,2}}G_{3,4}-\dfrac{\mu_{2,3}l_{2,3}}{q_{1,2}}\omega^2 \end{vmatrix}=0. \tag{5.20}$$

The mode of the natural vibrations of the investigated air-supported structure can be determined from any two equations of the system (5.18), if we substitute in them the respective characteristic number satisfying equation (5.20). The first natural frequency and the first natural mode correspond to the minimum positive characteristic number.

Let us note that the semi-deformation method and the tables on pp. 311—440 can be also used for the investigation of natural and forced vibrations of ropes in their planes, provided we accept that the form of the rope is a system of circular arcs of different diameters and straight sectors.

6

Numerical Example

In this section we present the solution of the system of homogeneous equations (4.6) which is valid for the freely vibrating air-supported cylindrical structure loaded according to Fig. IV-4a.

Using (1.11), we can express the parameters $\lambda_{1,2}$ and $\lambda_{3,4}$ by means of the parameter $\lambda_{2,3}$:

$$\lambda_{1,2}=\frac{q_{2,3}\mu_{1,2}R_{1,2}}{q_{1,2}\mu_{2,3}R_{2,3}}\lambda_{2,3},$$
$$\lambda_{3,4}=\frac{q_{2,3}\mu_{3,4}R_{3,4}}{q_{3,4}\mu_{2,3}R_{2,3}}\lambda_{2,3}, \tag{6.1}$$

while

$$\lambda_{2,3}=\frac{\mu_{2,3}\omega^2 R_{2,3}}{q_{2,3}}. \tag{6.2}$$

Let us determine now the deformation of the air-supported structure due to static load.

The cross-section of an air-supported cylindrical structure affected by the internal overpressure q_0 and the static radial wind load according to Fig. IV-4a is deformed so that the sectors *1—2*, *2—3* and *3—4* of lengths $s_{1,2}$, $s_{2,3}$ and $s_{3,4}$ respectively, form circular arcs of radii $R_{1,2}$, $R_{2,3}$ and $R_{3,4}$ respectively (Fig. IV-4b).

The individual sectors 1 m wide are affected by the resulting radial load given by

$$q_{1,2}=q_0+p_1,\quad q_{2,3}=q_0+p_2,\quad q_{3,4}=q_0+p_3, \tag{6.3}$$

where, according to (2.22),

$$p_1=-0.7\times 0.613v_0^2,\quad p_2=1.2\times 0.613v_0^2,\quad p_3=0.4\times 0.613v_0^2, \tag{6.4}$$

v_0 being the wind velocity in m/sec and q_0, p_i, $i=1, 2, 3$, are in N/m^2. The wind load has a negative sign, if it is pressure, and a positive sign, if it is suction.

The tensile force $T_{2,v}$ in the membrane strip 1 m wide, due to the load (6.4), is given approximately by the formula [11]

$$T_{2,v}=R\times 0.613v_0^2\left(-0.7\frac{\bar{\alpha}_{1,2}}{\pi}+1.2\frac{\bar{\alpha}_{2,3}}{\pi}+0.4\frac{\bar{\alpha}_{3,4}}{\pi}\right), \tag{6.5}$$

where $T_{2,v}$ is in N/m, v_0 in m/sec, R is the radius of the circular arc forming the cross-section of the non-deformed cylindrical membrane, and $\bar{\alpha}_{1,2}$, $\bar{\alpha}_{2,3}$, $\bar{\alpha}_{3,4}$ are the central angles of the sectors *1—2, 2—3* and *3—4* of the non-deformed membrane (Fig. IV-4a).

The resulting tensile force affecting the membrane in the cross-section, is

$$T_2=T_{2,q}+T_{2,v}, \tag{6.6}$$

where (Chapter II, Section 6)

$$T_{2,q}=q_0R. \tag{6.7}$$

The radii of the circular arcs forming the deformed cross-section of the air-supported structure, are

$$R_{1,2}=\frac{T_2}{q_{1,2}},\quad R_{2,3}=\frac{T_2}{q_{2,3}},\quad R_{3,4}=\frac{T_2}{q_{3,4}} \tag{6.8}$$

and the central angles of these arcs are given by

$$\alpha_{1,2}=\frac{s_{1,2}}{R_{1,2}},\quad \alpha_{2,3}=\frac{s_{2,3}}{R_{2,3}},\quad \alpha_{3,4}=\frac{s_{3,4}}{R_{3,4}}. \tag{6.9}$$

The following values referring to a membrane strip 1 m wide have been found $\mu_{1,2}=\mu_{2,3}=\mu_{3,4}=\mu$):

$$v_0=15 \text{ m/sec},\quad q_0=200 \text{ N/m}^2,\quad \mu=1 \text{ kg/m}^2=1 \text{ Nm}^{-3}\text{ sec}^2,$$

$$\begin{gathered} R=10 \text{ m},\quad \bar{\alpha}_{1,2}=\bar{\alpha}_{3,4}=0.66,\quad \alpha_{2,3}=1.43, \\ s_{1,2}=s_{3,4}=R\bar{\alpha}_{1,2}=10\times 0.66=6.6 \text{ m}, \\ s_{2,3}=R\bar{\alpha}_{2,3}=10\times 1.43=14.3 \text{ m}. \end{gathered} \tag{6.10}$$

The radial wind loads, according to (6.4), are

$$p_1 = -0.7 \times 0.613 \times 15^2 = -98.4 \ \text{N/m}^2,$$
$$p_2 = 1.2 \times 0.613 \times 15^2 = 168.75 \ \text{N/m}^2, \qquad (6.11)$$
$$p_3 = 0.4 \times 0.613 \times 15^2 = 56.25 \ \text{N/m}^2,$$

and the tensile forces $T_{2,v}$, $T_{2,q}$ and T_2, according to (6.5), (6.6) and (6.7), have the following magnitudes:

$$T_{2,v} = 10 \times 0.613 \times 15^2 \left(-0.7\frac{0.66}{3.14} + 1.2\frac{1.43}{3.14} + 0.4\frac{0.66}{3.14}\right) = 679.5 \ \text{N/m}$$
$$T_{2,q} = 200 \times 10 = 2000 \ \text{N/m}, \qquad (6.12)$$
$$T_2 = 2000 + 679.5 = 2679.5 \ \text{N/m}.$$

The sectors *1—2*, *2—3* and *3—4* are affected by the resulting radial loads given by (see (6.3))

$$q_{1,2} = 200 - 98.4 = 101.6 \ \text{N/m}^2,$$
$$q_{2,3} = 200 + 168.75 = 368.75 \ \text{N/m}^2, \qquad (6.13)$$
$$q_{3,4} = 200 + 56.25 = 256.25 \ \text{N/m}^2.$$

and, according to (6.8) and (6.9), we have

$$R_{1,2} = \frac{2679.5}{101.6} = 26.37 \ \text{m}, \qquad \alpha_{1,2} = \frac{6.6}{26.37} = 0.25,$$
$$R_{2,3} = \frac{2679.5}{368.75} = 7.27 \ \text{m}, \qquad \alpha_{2,3} = \frac{14.3}{7.27} = 1.97, \qquad (6.14)$$
$$R_{3,4} = \frac{2679.5}{256.25} = 10.46 \ \text{m}, \qquad \alpha_{3,4} = \frac{6.6}{10.46} = 0.63.$$

Substituting the known numerical values into the relations (6.1), we obtain

$$\lambda_{1,2} = 12.60134\lambda, \qquad \lambda_{3,4} = 1.98361\lambda \qquad (6.15)$$

and the ratios between the radii of circular arcs and the ratios between the radial loads appearing in the equations (4.6) have the following values:

$$\frac{R_{1,2}}{R_{2,3}} = \frac{26.37}{7.27} = 3.628, \qquad \frac{R_{2,3}}{R_{3,4}} = \frac{7.27}{10.46} = 0.695,$$
$$\frac{q_{2,3}}{q_{1,2}} = \frac{368.75}{101.6} = 3.628, \qquad \frac{q_{3,4}}{q_{2,3}} = \frac{256.25}{368.75} = 0.695. \qquad (6.16)$$

For this case we shall solve the system of homogeneous equations (4.6) by the method discussed by the author in [14]. We divide the homogeneous system (4.6) by an unknown quantity $w_2 \neq 0$, thus transforming it into a non-homogeneous system of equations. The left-hand side of the first transformed equation is a function $\Phi(\lambda)$ given by the equation

$$\Phi(\lambda)=A_{1,2}+\frac{R_{1,2}}{R_{2,3}}A_{2,3}+\frac{R_{1,2}}{R_{2,3}}B_{2,3}\frac{w_3}{w_2}-\left(C_{1,2}-\right.$$
$$\left.-\frac{R_{1,2}}{R_{2,3}}C_{2,3}\right)\frac{v_2}{w_2}-\frac{R_{1,2}}{R_{2,3}}D_{2,3}\frac{v_3}{w_2}. \tag{6.17}$$

The remaining transformed equations have the form

$$\left(A_{2,3}+\frac{R_{2,3}}{R_{3,4}}A_{3,4}\right)\frac{w_3}{w_2}+$$
$$+D_{2,3}\frac{v_2}{w_2}-\left(C_{2,3}-\frac{R_{2,3}}{R_{3,4}}C_{3,4}\right)\frac{v_3}{w_2}=-B_{2,3},$$

$$-\frac{q_{2,3}}{q_{1,2}}F_{2,3}\frac{w_3}{w_2}-\left(G_{1,2}+\frac{q_{2,3}}{q_{1,2}}G_{2,3}\right)\frac{v_2}{w_2}+\frac{q_{2,3}}{q_{1,2}}H_{2,3}\frac{v_3}{w_2}=$$
$$=-E_{1,2}+\frac{q_{2,3}}{q_{1,2}}E_{2,3},$$

$$\left(E_{2,3}-\frac{q_{3,4}}{q_{2,3}}E_{3,4}\right)\frac{w_3}{w_2}+H_{2,3}\frac{v_2}{w_2}-\left(G_{2,3}+\frac{q_{3,4}}{q_{2,3}}G_{3,4}\right)\frac{v_3}{w_2}=-F_{2,3}. \tag{6.18}$$

To obtain the required solution, we proceed as follows:

1. We select the value of λ_1, pertaining to the sector *2—3*, and, according to (6.15), calculate the values of $\lambda_{1,2}$ and $\lambda_{3,4}$, pertaining to the sectors *1—2* and *3—4* of the cross-section of the investigated membrane structure (Fig. IV-4b).

2. From the tables on pp. 311—440 we determine, by interpolation, for $\alpha_{2,3}$ and λ_1 the values of the functions $A_{2,3}$, $B_{2,3}$, $C_{2,3}$, $D_{2,3}$, $E_{2,3}$, $F_{2,3}$, $G_{2,3}$ and $H_{2,3}$; for $\alpha_{1,2}$ and $\lambda_{1,2}$ we determine the values of the functions $A_{1,2}$, $C_{1,2}$, $E_{1,2}$ and $G_{1,2}$, and for $\alpha_{3,4}$ and $\lambda_{3,4}$ we shall the values of the functions $A_{3,4}$, $C_{3,4}$, $E_{3,4}$ and $G_{3,4}$. Afterwards we calculate the coefficients in the equations (6.17) and (6.18).

3. Solving the three linear non-homogeneous equations (6.18), we determine the three ratios w_3/w_2, v_2/w_2 and v_3/w_2.

4. Substituting the values of w_3/w_2, v_2/w_2 and v_3/w_2 into equation (6.17), we calculate the value of the function $\Phi(\lambda)$ at the point λ_{I}. By the same method we shall determine the values of the function $\Phi(\lambda)$ at the selected points $\lambda_{II}, \lambda_{III}, \ldots, \lambda_{N}$, pertaining to the sector *2—3.* Interpolating between the values of the function $\Phi(\lambda)$ with opposite signs, pertaining to adjacent, sufficiently close values of the parameter λ, we obtain the characteristic numbers $\lambda_{(1)}, \lambda_{(2)}, \lambda_{(3)}$ of the homogeneous system (4.6). Also by interpolation we can determine the values of the ratios w_3/w_2, v_2/w_2 and v_3/w_2 pertaining to these characteristic numbers and characterizing the respective mode of the natural vibrations of the investigated membrane structure.

It is advisable to select the numbers $\lambda_{I}, \lambda_{II}, \ldots, \lambda_{N}$ as close as possible to the sought for characteristic numbers $\lambda_{(1)}, \lambda_{(2)}, \lambda_{(3)}, \ldots$. Therefore, we suggest to determine first, with the aid of equation (1.11), the approximate values of the characteristic numbers, using the natural circular frequencies of an arc of the tensile membrane loaded by the internal overpressure q only, where

$$q = \frac{T_2}{R}. \tag{6.19}$$

Here T_2 is the axial force according to (6.6), pertaining to the resulting radial load of the arc (6.13) as shown in Fig. IV-4a, and R is the radius of a circular arc of length

$$s = s_{1,2} + s_{2,3} + s_{3,4} = R\alpha, \tag{6.20}$$

while

$$\alpha = \bar{\alpha}_{1,2} + \bar{\alpha}_{2,3} + \bar{\alpha}_{3,4}. \tag{6.21}$$

For $R = 10$ m, $\mu = 1$ N m^{-3} sec^2,

$$q = \frac{2679.5}{10} = 268 \text{ N/m}^2,$$

$$\alpha = 0.66 + 1.43 + 0.66 = 2.75,$$

we obtain the following values of natural circular frequencies (see the Section 7 below):

$$\bar{\omega}_1 = 8.46 \text{ sec}^{-1}, \quad \bar{\omega}_2 = 14.98 \text{ sec}^{-1}, \quad \bar{\omega}_3 = 21.84 \text{ sec}^{-1}. \tag{6.22}$$

According to (1.11), the values (6.22) correspond approximately to the values of characteristic numbers ($R_{2,3} = 7.27$ m, $q_{2,3} = 368.75$ N/m^2)

$$\bar{\lambda}_{(1)}=\frac{1\times 8.46^2\times 7.27}{368.75}=1.4,$$

$$\bar{\lambda}_{(2)}=\frac{1\times 14.98^2\times 7.27}{368.75}=4.4, \tag{6.23}$$

$$\bar{\lambda}_{(3)}=\frac{1\times 21.84^2\times 7.27}{368.75}=9.4,$$

pertaining to the sector *2—3* of the investigated air-supported structure, loaded as shown on Fig. IV-4a.

In the process of solving the system of homogeneous equations (4.6) by the above mentioned method we use the values (see (6.23) and (6.15)):

$$\lambda_{\mathrm{I}}=\lambda_{(1)}=1.4, \quad \lambda_{1,2}=17.6545, \quad \lambda_{3,4}=2.7771; \tag{6.24}$$

then the solution of the system (6.18) with coefficients corresponding to the values in (6.24) and the angles (6.14) yields the equalities

$$\frac{w_3}{w_2}=-0.974, \quad \frac{v_2}{w_2}=0.125, \quad \frac{v_3}{w_2}=0.377. \tag{6.25}$$

According to (6.17), we have

$$\Phi(\lambda_{\mathrm{I}})=+0.1565. \tag{6.26}$$

If we select $\lambda_{\mathrm{II}}=1.6$, then according to (6.15), $\lambda_{1,2}=20.1765$, $\lambda_{3,4}=3.1738$ and from the equations (6.18) and (6.17) we obtain

$$\frac{w_3}{w_2}=-1.189, \quad \frac{v_2}{w_2}=0.128, \quad \frac{v_3}{w_2}=0.472, \tag{6.27}$$

$$\Phi(\lambda_{\mathrm{II}})=-2.5201. \tag{6.28}$$

To solve approximately the equation $\Phi(\lambda)=0$, we interpolate between the values (6.26) and (6.28), and we obtain the characteristic number $\lambda_{(1)}=1.4116$ corresponding to the first natural circular frequency ω_1 which, according to (2.14), is

$$\omega_1=\sqrt{\frac{q_{2,3}}{\mu R_{2,3}}\lambda_{(1)}}=\sqrt{1.4116\,\frac{368.75}{1\times 7.27}}=8.46\ \mathrm{sec}^{-1}. \tag{6.29}$$

The first natural frequency f_1 of the structure loaded as in Fig. IV-4a, is given by

$$f_1 = \frac{\omega_1}{2\pi} = 1.35 \text{ sec}^{-1}. \tag{6.30}$$

Interpolation between the values (6.25) and (6.27) yields, for $\Phi(\lambda)=0$, the ratios

$$\frac{w_3}{w_2} = -0.986, \quad \frac{v_2}{w_2} = 0.125, \quad \frac{v_3}{w_2} = 0.383. \tag{6.31}$$

These ratios characterize the first mode of natural vibrations of the investigated structure, as shown in Fig. IV-9. Further characteristic numbers and the corresponding modes of natural vibrations are determined analogously.

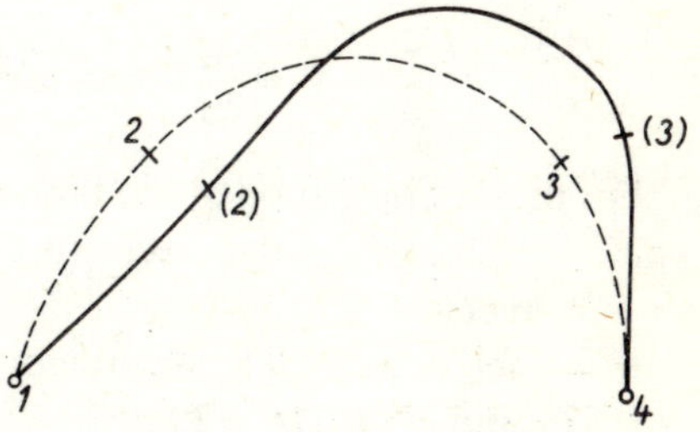

Fig. IV-9. The first mode of natural vibrations of a cylindrical membrane loaded according to Fig. IV-4.

For investigating the second natural frequency and the corresponding natural mode of the structure in Fig. IV-4a we use the value $\lambda_{III} = \bar{\lambda}_{(2)} = 4.4$ (see (6.23)), for which we obtain $\Phi(\lambda_{III}) = +7.0429$.

If we select $\lambda_{IV} = 5.0$, we obtain $w_3/w_2 = 0.685$, $v_2/w_2 = 0.163$, $v_3/w_2 = -0.448$ and $\Phi(\lambda_{IV}) = +1.6137$.

If we put $\lambda_V = 5.2$, we obtain $w_3/w_2 = 0.647$, $v_2/w_2 = 0.168$, $v_3/w_2 = -0.447$ and $\Phi(\lambda_V) = -0.3318$.

To solve approximately the equation $\Phi(\lambda)=0$, we interpolate between the above values. Thus we determine the second characteristic number $\lambda_{(2)} = 5.1659$ of the system of homogeneous deformation equations (4.6); we also get the ratios

$$w_3/w_2 = 0.653, \quad v_2/w_2 = 0.167, \quad v_3/w_2 = -0.448 \tag{6.32}$$

characterizing the second mode of natural vibrations, represented in Fig. IV-10.

The second natural circular frequency ω_2 and the second natural frequency f_2 are

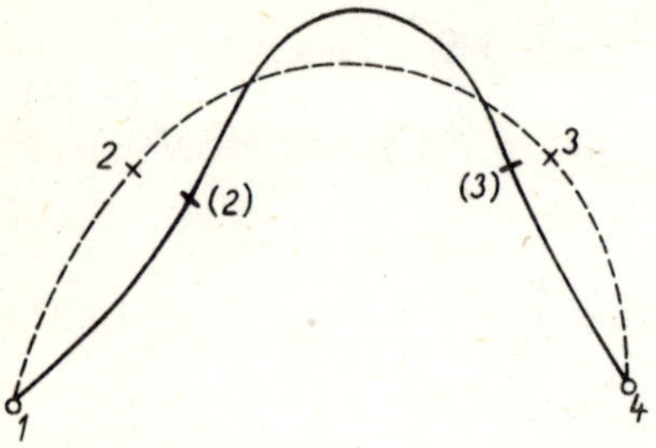

Fig. IV-10. The second mode of natural vibrations of a cylindrical membrane loaded according to Fig. IV-4.

$$\omega_2=\sqrt{5.1659\,\frac{368.75}{1\times 7.27}}=16.187\ \mathrm{sec}^{-1},$$
$$f_2=\frac{\omega_2}{2\pi}=2.58\ \mathrm{sec}^{-1}. \tag{6.33}$$

For the determination of the third natural frequency we shall use the initial value of $\lambda_{VI}=\bar{\lambda}_{(3)}=9.4$ (see (6.23)), which corresponds to $\Phi(\lambda_{VI})=$ $=+16.8666$.

If we select $\lambda_{VII}=9.8$, we obtain $w_3/w_2=0.282$, $v_2/w_2=0.273$, $v_3/w_2=$ $=0.534$ and $\Phi(\lambda_{VII})=+3.8657$.

If we put $\lambda_{VIII}=10.0$, we obtain $w_3/w_2=0.320$, $v_2/w_2=0.296$, $v_2/w_2=$ $=0.477$, and $\Phi(\lambda_{VIII})=-1.2721$.

Just as before, interpolation between the last values yields the third characteristic number $\lambda_{(3)}=9.9505$ and the ratios

$$w_3/w_2=0.311,\quad v_2/w_2=0.290,\quad v_3/w_2=0.491 \tag{6.34}$$

characterizing the third mode of natural vibrations of the investigated structure (Fig. IV-11).

The third natural circular frequency ω_3 and the third natural frequency f_3 are

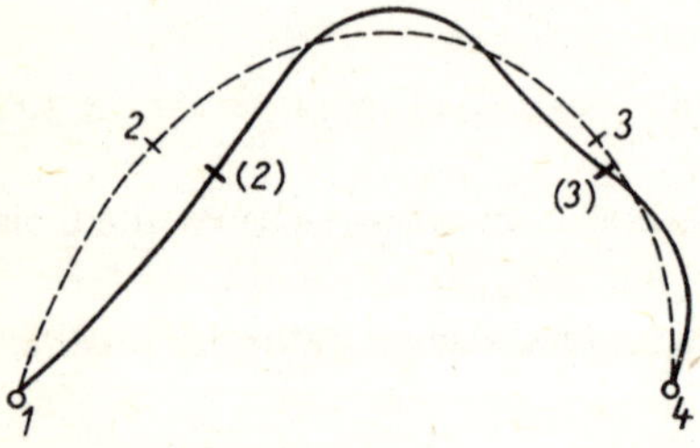

Fig. IV-11. The third mode of natural vibrations of a cylindrical membrane loaded according to Fig. IV-4.

$$\omega_3 = \sqrt{9.9505 \frac{368.75}{1 \times 7.27}} = 22.47 \text{ sec}^{-1},$$

$$f_3 = \frac{\omega_3}{2\pi} = 3.58 \text{ sec}^{-1}. \qquad (6.35)$$

A comparison of the approximate values of natural circular frequencies (6.22) of the investigated structure with their accurate values (6.29), (6.33) and (6.35) leads to the conclusion that the natural frequencies of an air-supported cylindrical structure are essentially influenced by the axial force T_2 in the membrane strip representing the cross-section of the structure, and not so much by its exact geometric form.

7

Roots of the Characteristic Equation (2.19)

For the natural vibrations of an air-supported structure as shown in Fig. IV-3, loaded by the internal overpressure only, the characteristic equation (2.19) is valid; its roots are identical with those of the equation $\Delta = 0$, where Δ is given by (3.5).

Table III contains the roots λ_{0j} of equation (2.19) for various values of the central angle α of the circular arc representing the cross-section of the structure.

The use of Table III facilitates the determination of the natural circular frequencies ω_j, $j = 1, 2, 3, \ldots$, and the natural frequencies f_j of the air-supported cylindrical structure.

If the jacket of the structure is affected by the internal overpressure q_0 only, the values of natural frequencies are theoretically accurate.

The natural circular frequency ω_j and the natural frequency f_j can be determined according to the relations (2.14) and (2.15), where λ_{0j} are the values obtained from Table III for the given values of the angle α.

If the structure is loaded simultaneously by the internal overpressure and outer static load, we can determine the approximate values of natural frequencies by the method described in Section 6 above. Using this method, we first calculate the axial force T_2 due to the resulting static load and then determine the equivalent internal overpressure q according to (6.19). After the substitution of q for q_0 in (2.14) we obtain approximate values of natural circular frequencies of the investigated structure.

In the case of an air-supported structure loaded by snow it is necessary, to replace in formula (2.14), apart from the equivalent internal overpressure also μ by an equivalent mass $\bar{\mu}$, comprizing, apart from the mass of the membrane, the mass of the snow on the jacket of the structure.

By way of example we shall determine the characteristic numbers, i.e. the roots of equation (2.19), for $\alpha = 2.75$.

From Table III we find the values of λ_{0j} for $\alpha = 2.70$ and $\alpha = 2.80$; by

Table III. The radices λ_{oj} of the characteristic equation (2.19)

j \ α	0.5	0.6	0.7	0.8	0.9	1.0	1.1	1.2	1.3	1.4
1	1 159.0036	105.7885	76.7092	57.8662	44.9644	35.7561	28.9554	23.8037	19.8083	16.6557
2	—	—	162.1746	123.5492	97.0834	78.1488	64.1429	53.4934	45.2098	38.6397
3			—	—	191.1219	154.1082	126.7553	105.9431	89.7314	76.8982
4					—	—	194.6481	163.1377	138.6182	119.1654
5							—	—	—	177.5974
6										—
7										
8										
9										
10										
11										
12										
13										
14										
15										

Table III. (Continued)

j \ α	1.5	1.6	1.7	1.8	1.9	2.0	2.1	2.2	2.3
1	14.1214	12.0618	10.3777	8.9608	7.7824	6.7873	5.9417	5.2178	4.5953
2	33.3435	29.0126	25.4257	22.4238	19.8861	17.7226	15.8645	14.2587	12.8572
3	66.5558	58.0973	51.0959	45.2372	40.2853	36.0616	32.4371	29.2988	26.5662
4	103.4710	90.6448	80.0049	71.0900	63.5466	57.1084	51.5687	46.7687	42.5830
5	154.2536	135.1689	119.3551	106.1460	94.9080	85.3511	77.1303	70.0093	63.8028
6	—	183.1520	161.9529	144.1671	129.1255	116.3160	105.2626	95.6789	87.3272
7		—	—	191.4071	171.4651	154.4146	139.7689	127.1130	116.0548
8				—	—	195.2526	176.8658	160.9348	147.0250
9						—	—	—	183.1755
10									—
11									
12									
13									
14									
15									

Table III. (Continued)

j \ α	2.4	2.5	2.6	2.7	2.8	2.9	3.0	3.1	3.2
1	4.0581	3.5908	3.1843	2.8289	2.5173	2.2427	1.9999	1.7859	1.5959
2	11.6320	10.5544	9.6022	8.7557	8.0011	7.3257	6.7194	6.1733	5.6801
3	24.1723	22.0634	20.2013	18.5445	17.0628	15.7378	14.5486	13.4744	12.5012
4	38.9107	35.6717	32.8005	30.2456	27.9581	25.9053	24.0575	22.3861	20.8671
5	58.3604	53.5607	49.3082	45.5222	42.1373	39.0995	36.3647	33.8887	31.6474
6	79.9980	73.5319	67.7981	62.6909	58.1226	54.0187	50.3202	46.9731	43.9365
7	106.3539	97.7458	90.1553	83.3959	77.3501	71.9201	67.0279	62.6042	58.5902
8	134.8071	124.0240	114.4794	106.0312	98.3864	91.5509	85.3888	79.8130	74.7531
9	167.9386	154.5726	142.7137	132.1475	122.6673	114.1271	106.5428	99.5572	93.2678
10	—	187.2117	172.8862	160.1468	148.7075	138.4588	129.2107	120.9227	113.2646
11		—	—	191.7057	178.0358	165.8095	154.7184	144.7220	135.7053
12				—	—	194.7991	181.8905	170.1807	159.5280
13						—	—	198.1293	185.8091
14								—	—
15									

linear interpolation between these two values we obtain for $\alpha=2.75$ the characteristic numbers

$$\begin{aligned} \lambda_{01} &= 2.673, & \lambda_{02} &= 8.378, \\ \lambda_{03} &= 17.804, & \lambda_{04} &= 29.102. \end{aligned} \tag{7.1}$$

If $q_0=200\ \text{N/m}^2$, $R=10\ \text{m}$, $\mu=1\ \text{N m}^{-3}\ \text{sec}^2$, we obtain, according to (2.14) and (2.15), the natural circular frequencies and natural frequencies:

$$\begin{aligned} \omega_1 &= 7.31\ \text{sec}^{-1} & (f_1 &= 1.16\ \text{sec}^{-1}), \\ \omega_2 &= 12.94\ \text{sec}^{-1} & (f_2 &= 2.06\ \text{sec}^{-1}), \\ \omega_3 &= 18.87\ \text{sec}^{-1} & (f_3 &= 3.00\ \text{sec}^{-1}), \\ \omega_4 &= 24.13\ \text{sec}^{-1} & (f_4 &= 3.84\ \text{sec}^{-1}). \end{aligned}$$

If $q_0=268\ \text{N/m}^2$, $R=10\ \text{m}$, $\mu=1\ \text{N m}^{-3}\ \text{sec}^2$, we obtain, by substituting these values and the values (7.1) into (2.14) and (2.15), the natural circular frequencies and natural frequencies

$$\begin{aligned} \omega_1 &= 8.46\ \text{sec}^{-1} & (f_1 &= 1.35\ \text{sec}^{-1}), \\ \omega_2 &= 14.98\ \text{sec}^{-1} & (f_2 &= 2.39\ \text{sec}^{-1}), \\ \omega_3 &= 21.84\ \text{sec}^{-1} & (f_3 &= 3.48\ \text{sec}^{-1}), \\ \omega_4 &= 27.93\ \text{sec}^{-1} & (f_4 &= 4.45\ \text{sec}^{-1}). \end{aligned}$$

8

Free Vibrations of Inflatable Plates

When inflatable plates are applied as construction elements, e.g. in gliders, it is necessary to investigate their vibrations [35], [36]. In this section we shall present some results of the lenear theory of inflatable plates, due to McComb [37] and experimentally verified by Stroud [55].

An inflatable plate of the type Airmat (developed by Goodyear Aircraft Corporation, USA) consists of two woven covers of an impermeable industrial fabric which are held a fixed distance apart by the combined action of cords and internal pressure (Fig. IV-12).

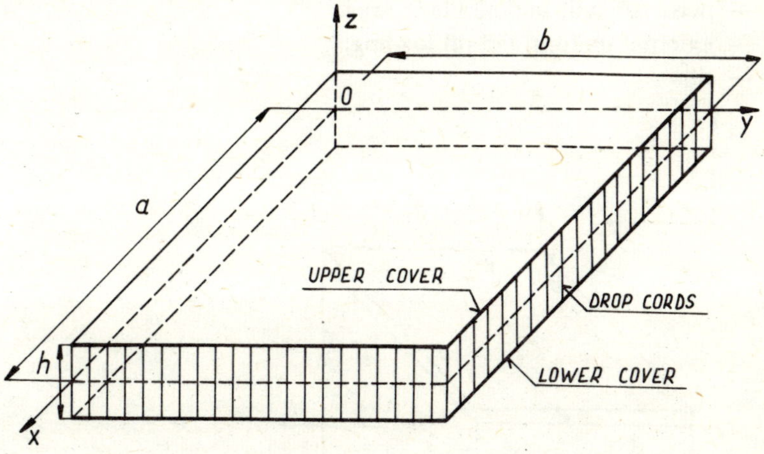

Fig. IV-12. Rectangular inflatable plate.

The principal differential equations of an inflatable plate have the form [36], [37]

$$q_0 h \frac{\partial}{\partial x}\left(\alpha + \frac{\partial w}{\partial x}\right) + q_0 h \frac{\partial}{\partial y}\left(\beta + \frac{\partial w}{\partial y}\right) + p_z = \mu \frac{\partial^2 w}{\partial t^2},$$

$$\frac{h^2}{2}\left(A_{11}\frac{\partial^2 \alpha}{\partial x^2} + A_{12}\frac{\partial^2 \beta}{\partial x\,\partial x}\right) + \frac{h^2}{2}A_{33}\left(\frac{\partial^2 \alpha}{\partial y^2} + \frac{\partial^2 \beta}{\partial x\,\partial y}\right) - q_0 h\left(\alpha + \frac{\partial w}{\partial x}\right) =$$
$$= \frac{1}{4}\mu h^2 \frac{\partial^2 \alpha}{\partial t^2}, \tag{8.1}$$

$$\frac{h^2}{2}\left(A_{21}\frac{\partial^2 \alpha}{\partial x\,\partial y} + A_{22}\frac{\partial^2 \beta}{\partial y^2}\right) + \frac{h^2}{2}A_{33}\left(\frac{\partial^2 \alpha}{\partial y\,\partial x} + \frac{\partial^2 \beta}{\partial x^2}\right) - q_0 h\left(\beta + \frac{\partial w}{\partial y}\right) =$$
$$= \frac{1}{4}\mu h^2 \frac{\partial^2 \beta}{\partial t^2},$$

where the lateral deflection w and the angles α and β between the drop cords and z-axis in xz-planes are variable quantities. α and β are the rotation angles of the drop cords during deformation (Fig. IV-13).

The symbols used in the equations (8.1) are:

q_0 — internal pressure,
h — plate depth,
μ — mass per unit middle plane area,
p_z — external uniform lateral loading,
t — time

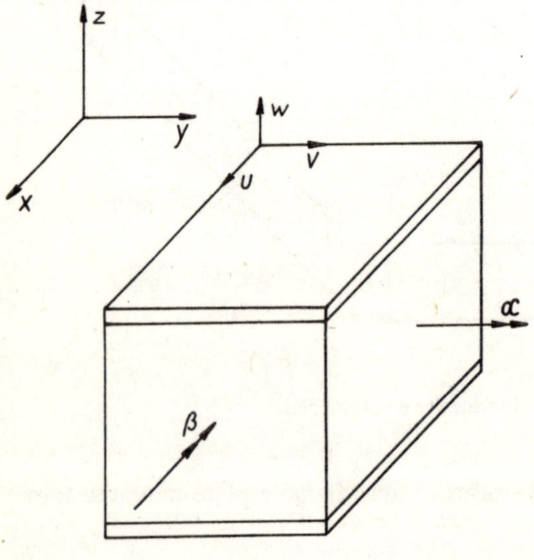

Fig. IV-13. Plate displacements and rotations applied to an element in their positive senses.

and

$$
\begin{aligned}
A_{11} &= \frac{E_w}{1-\nu_w \nu_f}, \qquad A_{22} = \frac{E_f}{1-\nu_w \nu_f}, \\
A_{12} &= \frac{\nu_f E_w}{1-\nu_w \nu_f}, \qquad A_{21} = \frac{\nu_w E_f}{1-\nu_w \nu_f}, \\
A_{33} &= G
\end{aligned} \tag{8.2}
$$

are the orthotropic elastic constants for the cover materials (see Chapter II, Subsection 8.2). The linear equations (8.1) are valid under the following assumptions:

1. The plate is of constant depth.
2. The material of the woven covers is elastic and orthotropic.
3. The internal pressure is constant during deformation.
4. The drop cords are inextensional and have the direction of the normal to the middle surface before deformation.

In the case of a simply supported rectangular plate with sides of lengths a and b, the boundary conditions along $x=0$ and $x=a$ can be written in the form (Fig. IV-12)

$$
\begin{aligned}
& w(0, y) = w(a, y) = 0, \\
& \beta(0, y) = \beta(a, y) = 0, \\
& \frac{\partial \alpha(0, y)}{\partial x} = \frac{\partial \alpha(a, y)}{\partial x} = 0
\end{aligned} \tag{8.3}
$$

and along $y=0$ and $y=b$, in the form

$$
\begin{aligned}
& w(x, 0) = w(x, b) = 0, \\
& \alpha(x, 0) = \alpha(x, b) = 0, \\
& \frac{\partial \beta(x, 0)}{\partial y} = \frac{\partial \beta(x, b)}{\partial y} = 0.
\end{aligned} \tag{8.4}
$$

During free vibrations of the inflatable plate we have $p_z = 0$ and the functions

$$
\begin{aligned}
w &= \sum_{m=1}^{\infty} \sum_{n=1}^{\infty} w_{mn} e^{i\omega t} \sin \frac{m\pi x}{a} \sin \frac{n\pi y}{b}, \\
\alpha &= \sum_{m=1}^{\infty} \sum_{n=1}^{\infty} \alpha_{mn} e^{i\omega t} \cos \frac{m\pi x}{a} \sin \frac{n\pi y}{b}, \\
\beta &= \sum_{m=1}^{\infty} \sum_{n=1}^{\infty} \beta_{mn} e^{i\omega t} \sin \frac{m\pi x}{a} \cos \frac{m\pi y}{b}
\end{aligned} \tag{8.5}
$$

satisfy the boundary conditions (8.3) and (8.4). Substituting the formulae (8.5) into (8.1) and applying the Galerkin method, we get a system of homogeneous algebraic equations which are linear with respect to the unknown quantities w_{mn}, α_{mn} and β_{mn}. Equating the determinant of these equations to zero, one obtains the frequency equation

$$\begin{aligned}&k^6-[\bar{q}_0(2\bar{R}+a_{mn})+b_{mn}+c_{mn}]k^4+\\&+[\bar{q}_0^2\bar{R}(\bar{R}+a_{mn})+\bar{q}_0(\bar{R}+a_{mn})\,(b_{mn}+c_{mn})+(b_{mn}c_{mn}-d_{mn}^2)]k^2-\\&-\bar{q}_0^2\bar{R}f_{mn}-\bar{q}_0a_{mn}(b_{mn}c_{mn}-d_{mn}^2)=0,\end{aligned} \tag{8.6}$$

where

$$k^2=\frac{\mu b^2\omega^2}{\pi A_{11}} \tag{8.7}$$

is the frequency parameter and

$$\begin{aligned}&f_{mn}=\lambda^2 mb_{mn}+n^2c_{mn}+2\lambda mnd_{mn},\\&\lambda=\frac{b}{a},\quad \bar{q}_0=\frac{q_0h}{A_{11}},\quad \bar{R}=\frac{4b^2}{\pi^2h^2},\\&a_{mn}=\lambda^2m^2+n^2,\\&b_{mn}=2(\lambda^2m^2+n^2A_{33}/A_{11}),\\&c_{mn}=2(n^2A_{22}/A_{11}+\lambda^2m^2A_{33}/A_{11}),\\&d_{mn}=\lambda^2m^2n^2(1+A_{22}/A_{11}-2A_{12}/A_{11}).\end{aligned} \tag{8.8}$$

The free frequencies ω_j, $j=1, 2, \ldots$, of the simply supported inflatable plate can be determined from the positive roots k_j^2, $j=1, 2, \ldots$, of equation (8.6). According to (8.7) we have

$$\omega_j=\frac{\pi^2A_{11}}{\mu b^2}k_j,\qquad j=1,2,\ldots. \tag{8.9}$$

The corresponding mode shapes are determined from the equations (8.5) by substituting $\omega=\omega_j$.

When the right-hand sides of the equations (8.1) are set equal to zero and one takes $\omega=0$ in (8.5), one obtains the basic equations of the calculus of static deflections. Assuming a uniform constant static load p_z, one gets the following formula for w_{mn} [37]:

$$\frac{w_{mn}}{h}=\frac{\bar{R}\bar{p}_z}{mn}\left\{\left(1-\frac{\bar{R}^2\bar{q}_0^2}{4\Lambda_{mn}}\right)a_{mn}-\frac{\bar{R}\bar{q}_0}{\Lambda_{mn}}\left[e_{mn}+(\lambda^2m^2-m^2)^2\frac{A_{33}}{A_{11}}\right]\right\}^{-1}, \tag{8.10}$$

where

$$\bar{p}_z=\frac{4p_z}{\pi^2 q_0},$$

$$\Lambda_{mn}=\frac{1}{4}\left[(\bar{R}\bar{q}_0+b_{mn})(\bar{R}\bar{q}_0+c_{mn})-d_{mn}^2\right], \tag{8.11}$$

$$e_{mn}=\lambda^2m^2n^2(1+A_{22}/A_{11}+2A_{12}/A_{11})$$

and the other symbols are as in (8.8).

Inflatable plates are a special case of the geometrically non-linear theory of inflatable shallow shells which has been developed by Nash and Ho [42], [43].

Chapter V

FORCED VIBRATIONS OF AIR-SUPPORTED STRUCTURES

Introduction

Among the dynamic factors which can affect air-supported membrane structures the following are of principal importance:

1. dynamic wind effects,
2. impact of pressure waves [25],
3. dynamic effects of a harmonic vibration exciter used for snow removal (see Chapter I, Section 1),
4. dynamic effects due to seismic factors.

The dynamic effects of wind manifest themselves chiefly by the fluttering of the membrane and its non-periodic motion. If the membrane jacket of the air-supported structure is not sufficiently pretensioned by tensile membrane forces due to internal overpressure, the afore-mentioned effects can bring about its failure. Since both the operational and the economic criteria make it desirable that the internal overpressure be as low as possible, it is necessary to determine the minimum overpressure which is not only suitable for calm weather, but also sufficient to make the structure resist major wind velocities. This can be achieved, however, only by taking into account not only static, but also dynamic wind effects. Considering the fact that the air-supported structures are usually not protected from wind by surrounding buildings or natural terrain obstacles and the dynamic forces of wind can affect them fairly often, it is necessary to analyse them quite thoroughly.

Air-supported structures can well resist impacts of pressure waves because of their suitable shape and smooth surface and also because major deflections of the membrane are not unfavourable, if there are not sharp objects in its proximity. But to ensure a sufficient rigidity of the air-supported structure during the impact of a pressure wave, a sufficiently high internal overpressure is of particular importance.

Let us consider the dynamic effects of the harmonic vibrations exciter. These vibrations are in resonance with the natural frequency of the snow-loaded air-supported structure, and their effects at the time of excitation depend on the amplitude of excited vibrations and on the quantity of snow and its adhesion to the jacket surface. When using this exciter, therefore, it is necessary to set an adequate amplitude of the motion of to shaft, follow closely the vibrations of the whole air-supported structure and switch off the motor at an adequate time.

The dynamic effects of seismic ground movements can manifest themselves only in a snow loaded structure in the case of a heavy earthquake. If the structure is not loaded by snow, these effects are negligible, as the weight of the membrane is very low.

In this chapter we shall deal with a method of analysis of forced vibrations of air-supported membrane structures of cylindrical shape, whose jacket is affected by a harmonic load or a dynamic load arbitrarily variable with time. The method can be used for the analysis of all afore-mentioned dynamic effects on cylindrical membranes, provided one knows the magnitude of excitation forces and their variability in time or the magnitude of the amplitude and the frequency of the harmonic motion of some place in the membrane.

The symbols employed and the assumptions of the analysis of forced vibrations of air-supported structures are the same as those used in Chapter IV. The meaning of new symbols is explained in the text.

1

Differential Equations of Forced Vibrations of a Cylindrical Membrane

If $\bar{p}_v^* = \bar{p}_v^*(\vartheta, t)$ is the component of continuous dynamic load in the direction of the tangent (displacement v) to a circular arc of the membrane of unit width, and $\bar{p}_w^* = \bar{p}_w^*(\vartheta, t)$ is the component of this load in the direction of the normal to this arc, the equations of the dynamic equilibrium for the deformed moving element of the arc have the form (Fig. V-1)

$$-\bar{S} - qR + \left(\bar{S} + \frac{\partial \bar{S}}{\partial \vartheta}\,\mathrm{d}\vartheta + qR\right)\cos \mathrm{d}\vartheta + \bar{p}_v^*\,\mathrm{d}s - \mu\,\mathrm{d}s\,\frac{\partial^2 \bar{v}}{\partial t^2} = 0,$$
$$(\bar{S} + qR)\cos\left(\frac{\pi}{2} - \mathrm{d}\vartheta\right) - q\,\mathrm{d}s + \bar{p}_w^*\,\mathrm{d}s - \mu\,\mathrm{d}s\,\frac{\partial^2 \bar{w}}{\partial t^2} = 0, \tag{1.1}$$

where the quantities with a bar depend both on the angle ϑ and the time t.

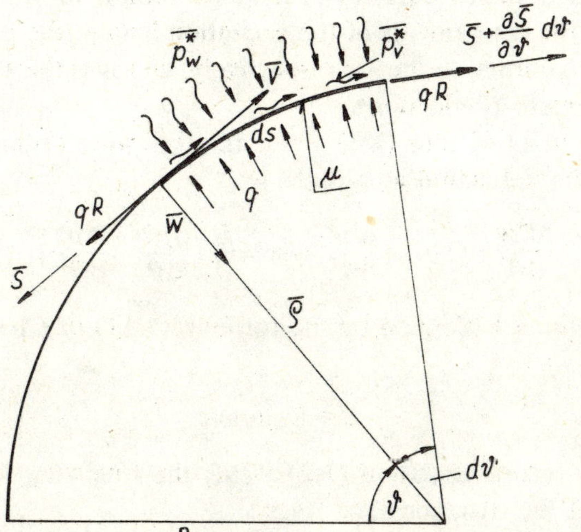

Fig. V-1. Static and dynamic loads and axial forces affecting an element of a membrane circular arc.

The positive orientation of the load components $\bar{p}_v^*$ and $\bar{p}_w^*$ is identical with the positive orientation of the displacements $\bar{v}$ and $\bar{w}$.

Due to the relations (1.3) in Chapter IV, the equations (1.1) assume the form

$$\frac{1}{R}\frac{\partial^2 \bar{S}}{\partial \vartheta^2}-\mu\frac{\partial^2 \bar{w}}{\partial t^2}=-\bar{p}_v^*,$$
$$\frac{1}{R}\bar{S}+\frac{q}{R}\left(\frac{\partial^2 \bar{w}}{\partial \vartheta^2}+\bar{w}\right)-\mu\frac{\partial^2 \bar{w}}{\partial t^2}=-\bar{p}_w^*. \tag{1.2}$$

The elimination of the dynamic axial force $\bar{S}$ from the system (1.2) yields the following linear partial differential equation for the radial displacement (deflection) $\bar{w}$:

$$\frac{\partial^4 \bar{w}}{\partial \vartheta^4}+\frac{\partial^2}{\partial \vartheta^2}\left(\bar{w}-\frac{\mu R}{q}\frac{\partial^2 \bar{w}}{\partial t^2}\right)+\frac{\mu R}{q}\frac{\partial^2 \bar{w}}{\partial t^2}=\frac{R}{q}\left(\frac{\partial \bar{p}_v^*}{\partial \vartheta}-\frac{\partial^2 \bar{p}_w^*}{\partial \vartheta^2}\right). \tag{1.3}$$

This is the basic differential equation for undamped forced vibrations of the circular arc of a membrane.

For stationary forced harmonic vibrations we have

$$\bar{p}_v^*=p_v^*\sin \omega t, \quad \bar{p}_w^*=p_w^*\sin \omega t, \quad \bar{w}=w\sin \omega t, \tag{1.4}$$

where ω is the circular frequency of the excitation load, p_v^* and p_w^* are the amplitudes of the components of the excitation load in the directions of the tangent and the normal to the arc respectively, and w is the amplitude of the radial displacement (deflection).

Substitution of (1.4) into (1.3) yields the following ordinary differential equation for the deflection amplitude w:

$$\frac{d^4 w}{d\vartheta^4}+(1+\lambda)\frac{d^2 w}{d\vartheta^2}-\lambda w=\frac{R}{q}\left(\frac{dp_v^*}{d\vartheta}-\frac{d^2 p_w^*}{d\vartheta^2}\right), \tag{1.5}$$

The parameter λ is defined by the formula (1.11) of Chapter IV.

In view of

$$\bar{S}=S\sin \omega t \tag{1.6}$$

and (1.4), the second equation (1.2) yields the following formula for the amplitude S of the dynamic axial force:

$$S=-\left[p_w^* R+q\frac{d^2 w}{d\vartheta^2}+(1+\lambda)qw\right]. \tag{1.7}$$

In a particular case, when the part $i-j$ of the cross-section of the cylindrical membrane deformed by static load is a straight sector of a length l (Fig. V-2), the differential equation for transverse forced vibrations assumes the form

$$\frac{\partial^2 \bar{w}}{\partial x^2}-\frac{\mu}{T_2}\frac{\partial^2 \bar{w}}{\partial t^2}=-\frac{1}{T_2}\bar{p}_w^*(x,t), \tag{1.8}$$

or, for $x=\xi l$,

$$\frac{\partial^2 \bar{w}}{\partial \xi^2}-\frac{\mu l^2}{T_2}\frac{\partial^2 \bar{w}}{\partial t^2}=-\frac{l^2}{T^2}\bar{p}_w^*(\xi,t)\,. \tag{1.9}$$

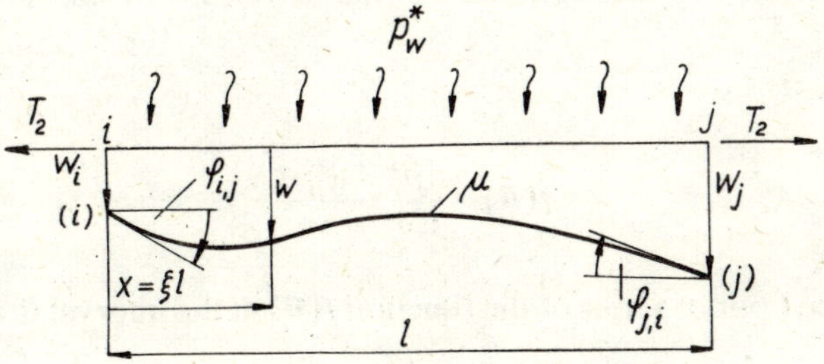

Fig. V-2. Positive sense of the amplitudes of end deformations in a straight sector of a vibrating membrane.

For harmonic vibrations we obtain, as a result of substituting (1.4) into (1.9), the ordinary differential equation

$$\frac{d^2 w}{d\xi^2}+\Lambda^2 w=-\frac{l^2}{T_2}p_w^*(\xi), \tag{1.10}$$

where the parameter Λ is defined by the formula (5.2) in Chapter IV.

2

General Solutions of Differential Equations for Forced Harmonic Vibrations

The non-homogeneous equation (1.5) with arbitrary right-hand side can be solved by the method of variation of constants. In this section we present a solution of (1.5) by means of the Fourier series, which suites our purposes better.

Let us set

$$f(\vartheta)=\frac{dp_v^*}{d\vartheta}-\frac{d^2p_w^*}{d\vartheta^2}, \tag{2.1}$$

and write the Fourier series of the function $f(\vartheta)$ in the interval $0\leqq\vartheta\leqq\alpha$:

$$f(\vartheta)=\frac{a_0}{2}+\sum_{j=1}^{\infty}a_j\cos\frac{2j}{\alpha}\vartheta+\sum_{j=1}^{\infty}b_j\sin\frac{2j\pi}{\alpha}\vartheta. \tag{2.2}$$

Here

$$\begin{aligned} a_j&=\frac{2}{\alpha}\int_0^{\alpha}f(\vartheta)\cos\frac{2j\pi}{\alpha}\vartheta\,d\vartheta,\\ b_j&=\frac{2}{\alpha}\int_0^{\alpha}f(\vartheta)\sin\frac{2j\pi}{\alpha}\vartheta\,d\vartheta. \end{aligned} \tag{2.3}$$

A particular solution of equation (1.5) with the right-hand side $f(\vartheta)R/q$, and $f(\vartheta)$ expanded as in (2.2), has the form

$$W=\frac{A_0}{2}+\sum_{j=1}^{\infty}A_j\cos\frac{2j\pi}{\alpha}\vartheta+\sum_{j=1}^{\infty}B_j\sin\frac{2j\pi}{\alpha}\vartheta. \tag{2.4}$$

We can determine the coefficients A_0, A_j and B_j, $j=1, 2, \ldots$ by substituting the expressions (2.2) and (2.4) into (1.5) and a comparing the corresponding coeficients in the series.

We obtain

$$A_0 = -\frac{R}{q\lambda} = -\frac{1}{\mu\omega^2},$$

$$A_j = \frac{R}{q} \frac{a_j}{\left(\frac{2j\pi}{\alpha}\right)^4 - (1+\lambda)\left(\frac{2j\pi}{\alpha}\right)^2 - \lambda}, \tag{2.5}$$

$$B_j = \frac{R}{q} \frac{b_j}{\left(\frac{2j\pi}{\alpha}\right)^4 - (1+\lambda)\left(\frac{2j\pi}{\alpha}\right)^2 - \lambda}.$$

The general solution of equation (1.5) equals the sum of the particular solution (2.4) and the general solution (2.4) given in Chapter IV of the homogeneous equation (1.10) in Chapter IV. Thus the general solution is

$$w = W(\vartheta) + C_1 \sin a\frac{\vartheta}{\alpha} + C_2 \cos a\frac{\vartheta}{\alpha} + C_3 \sinh b\frac{\vartheta}{\alpha} + C_4 \cosh b\frac{\vartheta}{\alpha}, \tag{2.6}$$

where C_i, $i=1, 2, 3, 4$, are integration constants, and a, b are the parameters given by the formulae (2.3) in Chapter IV.

The amplitude of displacement v in the direction of the tangent to the circular arc is

$$\begin{aligned} v = \int w \,\mathrm{d}\vartheta = V(\vartheta) - C_1\frac{\alpha}{a}\cos a\frac{\vartheta}{\alpha} + C_2\frac{\alpha}{a}\sin a\frac{\vartheta}{\alpha} + \\ + C_3\frac{\alpha}{b}\cosh b\frac{\vartheta}{\alpha} + C_4\frac{\alpha}{b}\sinh b\frac{\vartheta}{\alpha}, \end{aligned} \tag{2.7}$$

where

$$\begin{aligned} V(\vartheta) = \int W(\vartheta)\,\mathrm{d}\vartheta = \frac{A_0}{2}\vartheta + \frac{\alpha}{2\pi}\sum_{j=1}^{\infty} A_j \frac{1}{j}\sin\frac{2j\pi}{\alpha}\vartheta - \\ - \frac{\alpha}{2\pi}\sum_{j=1}^{\infty} B_j \frac{1}{j}\cos\frac{2j\pi}{\alpha}\vartheta. \end{aligned} \tag{2.8}$$

For the amplitude of the angle φ of rotation of the arc cross-section we obtain

$$R\varphi=\frac{\mathrm{d}w}{\mathrm{d}\vartheta}+v=W'+V+C_1\left(\frac{a}{\alpha}-\frac{\alpha}{a}\right)\cos a\frac{\vartheta}{\alpha}-$$
$$-C_2\left(\frac{a}{\alpha}-\frac{\alpha}{a}\right)\sin a\frac{\vartheta}{\alpha}+C_3\left(\frac{b}{\alpha}+\frac{\alpha}{b}\right)\cosh b\frac{\vartheta}{\alpha}+ \tag{2.9}$$
$$+C_4\left(\frac{b}{\alpha}+\frac{\alpha}{b}\right)\sinh\frac{\vartheta}{\alpha},$$

and for the amplitude of the force S we obtain, by substituting (2.6) into (1.7),

$$\frac{S}{q}=-\frac{p_w^*}{q}R-W''-(1+\lambda)W+$$
$$+C_1\left(-1-\lambda+\frac{a^2}{\alpha^2}\right)\sin a\frac{\vartheta}{\alpha}+C_2\left(-1-\lambda+\frac{a^2}{\alpha^2}\right)\cos a\frac{\vartheta}{\alpha}+ \tag{2.10}$$
$$+C_3\left(-1-\lambda-\frac{b^2}{\alpha^2}\right)\sinh b\frac{\vartheta}{\alpha}+C_4\left(-1-\lambda-\frac{b^2}{a^2}\right)\cosh b\frac{\vartheta}{\alpha},$$

where, according to (2.4),

$$W'=-\frac{2\pi}{\alpha}\sum_{j=1}^{\infty}A_j j\sin\frac{2j\pi}{\alpha}\vartheta+\frac{2\pi}{\alpha}\sum_{j=1}^{\infty}B_j j\cos\frac{2j\pi}{\alpha}\vartheta,$$
$$W''=-\left(\frac{2\pi}{\alpha}\right)^2\sum_{j=1}^{\infty}A_j j^2\cos\frac{2j\pi}{\alpha}\vartheta-\left(\frac{2\pi}{\alpha}\right)^2\sum_{j=1}^{\infty}B_j j^2\sin\frac{2j\pi}{\alpha}\vartheta. \tag{2.11}$$

The general solution of equation (1.10), whose right-hand side is an arbitrary function of the variable ξ, can be achieved by the method of variation of constants.

If we view C_1 and C_2 in the formula (5.4) of Chapter IV as functions of the variable ξ, we obtain

$$\sin\Lambda\xi\frac{\mathrm{d}C_1}{\mathrm{d}\xi}+\cos\Lambda\xi\frac{\mathrm{d}C_2}{\mathrm{d}\xi}=0,$$
$$\Lambda\cos\Lambda\xi\frac{\mathrm{d}C_1}{\mathrm{d}\xi}-\Lambda\sin\Lambda\xi\frac{\mathrm{d}C_2}{\mathrm{d}\xi}=-\frac{l^2}{T_2}p_w^*(\xi), \tag{2.12}$$

whence

$$\frac{\mathrm{d}C_1}{\mathrm{d}\xi}=-\frac{l^2}{\Lambda T_2}p_w^*\cos\Lambda\xi,$$
$$\frac{\mathrm{d}C_2}{\mathrm{d}\xi}=\frac{l^2}{\Lambda T_2}p_w^*\sin\Lambda\xi. \tag{2.13}$$

Thus, integrating, we obtain

$$
\begin{aligned}
C_1 &= -\frac{l^2}{\Lambda T_2} \int p_w^* \cos \Lambda\xi \, d\xi + D_1, \\
C_2 &= \frac{l^2}{\Lambda T_2} \int p_w^* \sin \Lambda\xi \, d\xi + D_2,
\end{aligned}
\qquad (2.14)
$$

where D_1 and D_2 are integration constants.

Substituting (2.14) into (5.4) of Chapter IV, we obtain a general solution of equation (1.10) in the form

$$w(\xi) = D_1 \sin \Lambda\xi + D_2 \cos \Lambda\xi + \frac{l^2}{\Lambda T_2} Q(\xi), \qquad (2.15)$$

where

$$Q(\xi) = \cos \Lambda\xi \int p_w^* \sin \Lambda\xi \, d\xi + \sin \Lambda\xi \int p_w^* \cos \Lambda\xi \, d\xi. \qquad (2.16)$$

For the angle of rotation of the cross-section we have the formula

$$l\varphi(\xi) = \frac{dw}{d\xi} = D_1\Lambda \cos \Lambda\xi - D_2\Lambda \sin \Lambda\xi + \frac{l^2}{\Lambda T_2} Q'(\xi). \qquad (2.17)$$

The amplitude of the dynamic axial force S in transverse vibrations of a straight membrane sector equals zero ($S=0$).

The above solutions can be used for the construction of the resulting system of algebraic equations by the method shown in Section 2 of Chapter IV. A considerable reduction of the number of resulting equations can be achieved by the application of the semi-deformation method, described in Sections 3 and 4 of this chapter.

3

Principal Equations of the Semi-Deformation Method for Forced Harmonic Vibrations of a Cylindrical Membrane

In the semi-deformation method we choose as unknown quantities the amplitudes (3.1) in Chapter IV of the displacements of the ends i and j of the circular arc i—j, or of the straight sector approximating the part of the cross-section of the cylindrical air-supported structure deformed by static load. Since our analysis of forced harmonic vibrations is based on the linear equations (1.5) and (1.10), we can use the law of superposition for the resulting quantities.

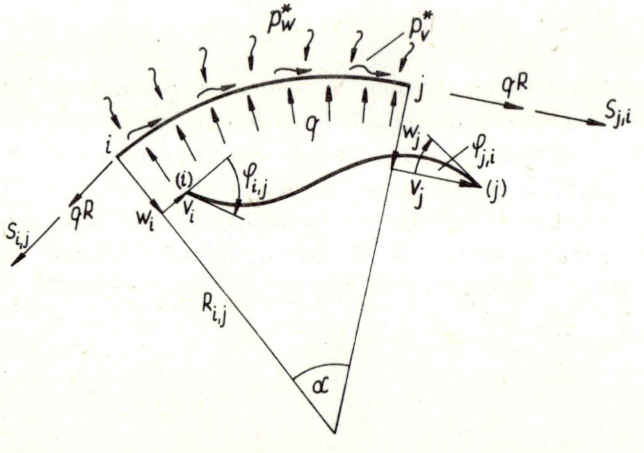

Fig. V-3. Static and dynamic loads and the positive sense of the amplitudes of end deformations in a circular arc of a membrane.

Thus, using the equalities (3.6) from Chapter IV, we find the following formulae for the amplitudes $\varphi_{i,j}$ and $\varphi_{j,i}$ of the rotation of the end cross-sections of the circular arc and the amplitudes $S_{i,j}$ and $S_{j,i}$ of the end dynamic axial forces of the circular arc in forced vibrations (Fig. V-3):

$$\begin{aligned}
\varphi_{i,j} &= \varphi^0_{i,j} + \frac{1}{R_{i,j}}(-Aw_i - Bw_j - Cv_i + Dv_j),\\
\varphi_{j,i} &= \varphi^0_{j,i} + \frac{1}{R_{i,j}}(Bw_i + Aw_j + Dv_i - Cv_j),\\
S_{i,j} &= S^0_{i,j} + q_{i,j}(-Ew_i - Fw_j - Gv_i + Hv_j),\\
S_{j,i} &= S^0_{j,i} + q_{i,j}(-Fw_i - Ew_j - Hv_i + Gv_j).
\end{aligned} \tag{3.1}$$

Here $\varphi^0_{i,j}$ and $\varphi^0_{j,i}$ are the amplitudes of the angles of rotation of the end cross-sections i and j due to harmonic loads of amplitudes p^*_w and p^*_v affecting the circular arc i—j whose ends are undisplaceable (i.e. hinged) and $S^0_{i,j}$, $S^0_{j,i}$ are the amplitudes of the end axial forces of this arc (Fig. V-4). The terms in brackets in the equations (3.1) represent the influence of the amplitudes of the end displacements on the magnitude of the quantities $\varphi_{i,j}$, $\varphi_{j,i}$, $S_{i,j}$ and $S_{j,i}$. The relations (3.1) are the principal equations of the semi-deformation method for the forced vibrations of a cylindrical membrane.

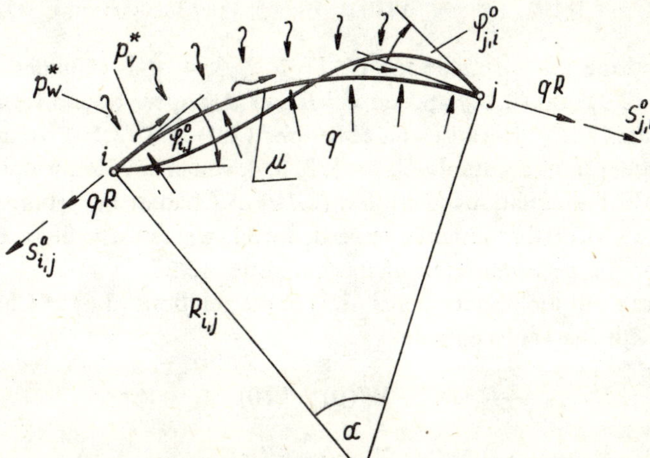

Fig. V-4. Static and dynamic forces and amplitudes of the angles of rotation of the end sections of a hinged circular arc of a membrane.

Let us determine now in an analogous way the quantities

$$\varphi^0_{i,j},\ \varphi^0_{j,i},\ S^0_{i,j},\ S^0_{j,i}. \tag{3.2}$$

If we set $\vartheta = 0$ and $\vartheta = \alpha$ in the equations (2.6) and (2.7), and put

$$w(0)=v(0)=w(\alpha)=v(\alpha)=0, \tag{3.3}$$

then, according to Fig. V-4, we obtain for a hinged circular arc

$$\begin{aligned}
&-W(0)=C_2+C_4,\\
&-V(0)=-C_1\frac{\alpha}{a}+C_3\frac{\alpha}{b},\\
&-W(\alpha)=C_1\sin a+C_2\cos a+C_3\sinh b+C_4\cosh b,\\
&-V(\alpha)=-C_1\frac{\alpha}{a}\cos a+C_2\frac{\alpha}{a}\sin a+C_3\frac{\alpha}{b}\cosh b+C_4\frac{\alpha}{b}\sinh b.
\end{aligned} \tag{3.4}$$

Equations (3.4) have the same form as the equations (3.2) of Chapter IV and, therefore, the integration constants C_i, $i=1, 2, 3, 4$, in (3.4) can be determined by means of the formulae (3.3) of Chapter IV, where we have to perform the following replacement of quantities:

$$w_i\to -W(0),\quad v_i\to -V(0),\quad w_j\to -W(\alpha),\quad v_j\to -V(\alpha). \tag{3.5}$$

If we substitute the formulae for C_i, $i=1, 2, 3, 4$, thus obtained, into the equations (2.9) and (2.10) and set $\vartheta=0$ and $\vartheta=\alpha$, we obtain formulae for the quantities (3.2). Because the equations (2.9) and (2.10) contain terms with the integration constants C_i, $i=1, 2, 3, 4$, which are identical with those appearing in the equations (2.8) and (2.10) of Chapter IV, from which the equalities (3.6) of that chapter were deduced, we can use these equalities directly for the determination of the quantities (3.2).

If we carry out the replacements (3.5) in the relations (3.6) of Chapter IV, together with the replacements

$$\begin{aligned}
&R_{i,j}\varphi_{i,j}\to R_{i,j}\varphi^0_{i,j}-W'(0)-V(0),\\
&R_{i,j}\varphi_{j,i}\to R_{i,j}\varphi^0_{j,i}-W'(\alpha)-V(\alpha),\\
&\frac{S_{i,j}}{q_{i,j}}\to\frac{S^0_{i,j}}{q_{i,j}}+\frac{p^*_w(0)}{q_{i,j}}R_{i,j}+W''(0)+(1+\lambda)W(0),\\
&\frac{S_{j,i}}{q_{i,j}}\to\frac{S^0_{j,i}}{q_{i,j}}+\frac{p^*_w(\alpha)}{q_{i,j}}R_{i,j}+W''(\alpha)+(1+\lambda)W(\alpha),
\end{aligned} \tag{3.6}$$

we get the following formulae for the quantities (3.2):

$$
\begin{aligned}
\varphi^0_{i,j} &= \frac{1}{R_{i,j}}[W'(0)+AW(0)+BW(\alpha)+EV(0)-DV(\alpha)],\\
\varphi^0_{j,i} &= \frac{1}{R_{i,j}}[W'(\alpha)-BW(0)-AW(\alpha)-DV(0)+EV(\alpha)],\\
S^0_{i,j} &= q_{i,j}[-p^*_w(0)R_{i,j}-W''(0)+(C-\lambda)W(0)+FW(\alpha)+GV(0)-HV(\alpha)],\\
S^0_{j,i} &= q_{i,j}[-p^*_w(\alpha)R_{i,j}-W''(\alpha)+FW(0)+(C-\lambda)W(\alpha)+HV(0)\\
&\qquad -GV(\alpha)].
\end{aligned}
\tag{3.7}
$$

Here A, B, C, ..., H are the functions defined by the equations (3.7) of Chapter IV; and according to (2.4), (2.8) and (2.11),

$$
\begin{aligned}
W(0) &= W(\alpha)=\frac{A_0}{2}+\sum_{j=1}^{\infty}A_j,\\
W'(0) &= W'(\alpha)=\frac{2\pi}{\alpha}\sum_{j=1}^{\infty}jB_j,\\
W''(0) &= W''(\alpha)=-\left(\frac{2\pi}{\alpha}\right)^2\sum_{j=1}^{\infty}j^2A_j,\\
V(0) &= -\frac{\alpha}{2\pi}\sum_{j=1}^{\infty}\frac{1}{j}B_j,\\
V(\alpha) &= \frac{A_0}{2}\alpha-\frac{\alpha}{2\pi}\sum_{j=1}^{\infty}\frac{1}{j}B_j.
\end{aligned}
\tag{3.8}
$$

Thus, for the calculation of the quantities (3.2) with the aid of the formulae (3.7) it is possible to use the tables of the functions A, B, ..., H given on pp. 311—440.

The principal equations of the semi-deformation method for a straight membrane sector i—j have the form

$$
\begin{aligned}
\varphi_{i,j} &= \varphi^0_{i,j}+\frac{1}{l_{i,j}}(-\Phi w_i+\Psi w_j),\\
\varphi_{j,i} &= \varphi^0_{j,i}+\frac{1}{l_{i,j}}(-\Psi w_i+\Phi w_j),\\
S_{i,j} &= 0,\quad S_{j,i}=0.
\end{aligned}
\tag{3.9}
$$

Here Φ and Ψ are the functions defined by the equations (5.10) of Chapter IV, and $\varphi^0_{i,j}$ and $\varphi^0_{j,i}$ are the amplitudes of the angles of rotation of the end cross-sections of the straight sector i—j hinged at the ends; they are affected by the stationary harmonic load of amplitude $p_w^* = p_w^*(\xi)$ and by a constant static axial force T_2.

The angles $\varphi_{i,j}^0$ and $\varphi_{j,i}^0$ can be determined from equation (2.17) by putting $\xi = 0$ and $\xi = 1$ there, and evaluating the integration constants with the aid of formula (2.15) and the boundary conditions

$$w(0) = 0, \quad w(1) = 0. \tag{3.10}$$

According to (2.15) and (3.10), we have

$$\begin{aligned} D_2 + \frac{l^2}{\Lambda T_2} Q(0) = 0, \\ D_1 \sin \Lambda + D_2 \cos \Lambda + \frac{l^2}{\Lambda T_2} Q(1) = 0, \end{aligned} \tag{3.11}$$

whence

$$\begin{aligned} D_2 &= -\frac{l^2}{\Lambda T_2} Q(0), \\ D_1 &= \frac{l^2}{T_2 \Lambda \sin \Lambda} [Q(0) \cos \Lambda - Q(1)]. \end{aligned} \tag{3.12}$$

Substituting (3.12) and $\xi = 0$, $\xi = 1$ into (2.17), we obtain

$$\begin{aligned} \varphi_{i,j}^0 &= \frac{l}{T_2 \sin \Lambda} [Q(0) \cos \Lambda - Q(1)] + \frac{l}{\Lambda T_2} Q'(0). \\ \varphi_{j,i}^0 &= \frac{l}{T_2 \sin \Lambda} [Q(0) - Q(1) \cos \Lambda] + \frac{l}{\Lambda T_2} Q'(1). \end{aligned} \tag{3.13}$$

4

Construction of Deformation Equations

The deformation equations for an air-supported structure affected by a continuous harmonic load are obtained by a substitution of formulae of the form (3.1) or (3.9) into equations of the form (4.1) and (4.2) of Chapter IV, expressing the continuity between the sectors $i—1$, i and i, $i+1$ within the range i (Figs. IV-6a, b).

For the circular sectors $i—1$, i and i, $i+1$, using the relations (3.1), we obtain the equations for the amplitudes of the angles of rotation of end cross-sections and the equations for the amplitudes of the end axial forces in the form (4.3) of Chapter IV, however the right-hand sides have to be supplemented by the terms $\varphi^0_{i,\,i-1}$, $\varphi^0_{i,\,i+1}$, $S^0_{i,\,i-1}$ and $S^0_{i,\,i+1}$. If we combine these relations with the equations (4.1) and (4.2) of Chapter IV, we obtain two non-homogeneous deformation equations for the boundary i of two adjoining sectors:

$$\begin{gathered}
B_{i-1,\,i}w_{i-1}+\left(A_{i-1,\,i}+\frac{R_{i,\,i-1}}{R_{i,\,i+1}}A_{i,\,i+1}\right)w_i+\frac{R_{i,\,i-1}}{R_{i,\,i+1}}B_{i,\,i+1}w_{i+1}+\\
+D_{i-1,\,i}v_{i-1}-\left(C_{i-1,\,i}-\frac{R_{i-1,\,i}}{R_{i,\,i+1}}C_{i,\,i+1}\right)v_i-\frac{R_{i-1,\,i}}{R_{i,\,i+1}}D_{i,\,i+1}v_{i+1}=\\
=R_{i-1,\,i}(\varphi^0_{i,\,i+1}-\varphi^0_{i,\,i-1}),\\
F_{i-1,\,i}w_{i-1}+\left(E_{i-1,\,i}-\frac{q_{i,\,i+1}}{q_{i-1,\,i}}E_{i,\,i+1}\right)w_i-\frac{q_{i,\,i+1}}{q_{i-1,i}}F_{i,\,i+1}w_{i+1}+\\
+H_{i-1,\,i}v_{i-1}-\left(G_{i-1,\,i}+\frac{q_{i,\,i+1}}{q_{i-1,\,i}}G_{i,\,i+1}\right)v_i+\frac{q_{i,\,i+1}}{q_{i-1,\,i}}H_{i,\,i+1}v_{i+1}=\\
=-\frac{1}{q_{i-1,\,i}}(S^0_{i,\,i+1}-S^0_{i,\,i-1}).
\end{gathered}\tag{4.1}$$

The values $\varphi^0_{i,\,i+1}$ and $S^0_{i,\,i+1}$ can be determined from the first and the third equations (3.7) and the values $\varphi^0_{i,\,i-1}$ and $S^0_{i,\,i-1}$ from the second and the fourth equations (3.7).

For example, for forced harmonic vibrations of the air-supported structure in Fig. IV-4a, whose cross-section under static load consists of three circular arcs, the deformation equations according to (4.1), with the boundary conditions (4.5) of Chapter IV, have the form

$$\begin{aligned}
&\left(A_{1,2}+\frac{R_{1,2}}{R_{2,3}}A_{2,3}\right)w_2+\frac{R_{1,2}}{R_{2,3}}B_{2,3}w_3-\left(C_{1,2}-\frac{R_{1,2}}{R_{2,3}}C_{2,3}\right)v_2-\\
&\qquad-\frac{R_{1,2}}{R_{2,3}}D_{2,3}v_3=\\
&\qquad=R_{1,2}(\varphi^0_{2,3}-\varphi^0_{2,1}),\\
&B_{2,3}w_2+\left(A_{2,3}+\frac{R_{2,3}}{R_{3,4}}A_{3,4}\right)w_3+D_{2,3}v_2-\left(C_{2,3}-\frac{R_{2,3}}{R_{3,4}}C_{3,4}\right)v_3=\\
&\qquad=R_{2,3}(\varphi^0_{3,4}-\varphi^0_{3,2}),\\
&\left(E_{1,2}-\frac{q_{2,3}}{q_{1,2}}E_{2,3}\right)w_2-\frac{q_{2,3}}{q_{1,2}}F_{2,3}w_3-\left(G_{1,2}+\frac{q_{2,3}}{q_{1,2}}G_{2,3}\right)v_2+\\
&\qquad+\frac{q_{2,3}}{q_{1,2}}H_{2,3}v_3=\\
&\qquad=-\frac{1}{q_{1,2}}(S^0_{2,3}-S^0_{2,1}),\\
&F_{2,3}w_2+\left(E_{2,3}-\frac{q_{3,4}}{q_{2,3}}E_{3,4}\right)w_3+H_{2,3}v_2-\left(G_{2,3}+\frac{q_{3,4}}{q_{2,3}}G_{3,4}\right)v_3=\\
&\qquad=-\frac{1}{q_{2,3}}(S^0_{3,4}-S^0_{3,2}).
\end{aligned}\tag{4.2}$$

The solution of the four deformation equations (4.2) yields the four unknown amplitudes of displacements w_2, w_3, v_2 and v_3. From the last two equations (3.1) we can determine the amplitudes of the end axial forces of every circular arc which we shall use in Section 7 below for the assessment of the dynamic stability of the investigated air-supported structure.

5

Orthogonality of Characteristic Functions

The amplitudes of displacements w and v, defined by the equations (2.4) and (2.5) of Chapter IV, which satisfy the homonegeous differential equation (1.10) of Chapter IV and the boundary conditions, shall be called hereafter the characteristic functions. For every characteristic number $\lambda_{(j)}$ there are the characteristic functions $w_{(j)}$ and $v_{(j)}$ which we shall mark with bracketed suffixes.

In this section we shall discuss the orthogonality of the characteristic functions $w_{(j)}$ and $v_{(j)}$ in a generalized sense [8] and we shall present a table of parameter values facilitating the determination of these functions. We shall use the characteristic functions for the investigation of forced vibrations of air-supported cylindrical structures in Section 6 below.

Due to the second equation (1.5) of Chapter IV and the constraint (2.6) in Chapter IV we obtain by integrating equation (1.10) of Chapter IV the following homogeneous differential equation for the displacement v

$$v^{\mathrm{IV}}+(1+\lambda)v''-\lambda v=0. \tag{5.1}$$

Let us note that this has the same form as equation (1.10) of Chapter IV.

Equation (5.1) is satisfied by the characteristic number $\lambda_{(i)}$ and the corresponding characteristic function $v_{(i)}$:

$$v_{(i)}^{\mathrm{IV}}+(1+\lambda_{(i)})v''_{(i)}-\lambda_{(i)}v_{(i)}=0\,. \tag{5.2}$$

If we multiply (5.2) by the characteristic function $v_{(k)}$, pertaining to another characteristic number $\lambda_{(k)}$, and integrate along the whole length of the arc, i.e. from 0 to α, we obtain

$$\int_0^\alpha v_{(i)}^{\mathrm{IV}}v_{(k)}\,\mathrm{d}\vartheta+(1+\lambda_{(i)})\int_0^\alpha v''_{(i)}v_{(k)}\,\mathrm{d}\vartheta-\lambda_{(i)}\int_0^\alpha v_{(i)}v_{(k)}\,\mathrm{d}\vartheta=0. \tag{5.3}$$

Integration by parts yields for the first two integrals in equation (5.3)

$$\int_0^\alpha v_{(i)}^{\mathrm{IV}} v_{(k)}\,\mathrm{d}\vartheta = [v_{(i)}''' v_{(k)}]_0^\alpha - [v_{(i)}'' v_{(k)}']_0^\alpha + \int_0^\alpha v_{(i)}'' v_{(i)}''\,\mathrm{d}\vartheta\,,$$
$$\int_0^\alpha v_{(i)}'' v_{(k)}\,\mathrm{d}\vartheta = [v_{(i)}' v_{(k)}]_0^\alpha - \int_0^\alpha v_{(i)}' v_{(k)}'\,\mathrm{d}\vartheta. \tag{5.4}$$

If the characteristic functions $v_{(i)}$ and $v_{(k)}$ satisfy the boundary conditions

$$v(0)=0, \qquad v(\alpha)=0,$$
$$w(0)=0, \qquad w(\alpha)=0, \tag{5.5}$$

the terms in square brackets in (5.4) vanish, as according to the second equation (1.5) in Chapter IV, $w=v'$ and, consequently,

$$v'(0)=w(0), \qquad v'(\alpha)=w(\alpha). \tag{5.6}$$

Substituting (5.4) into (5.3) and using (5.5) and (5.6), we get ($v_{(i)}'=w_{(i)}$, $v_{(k)}'=w_{(k)}$)

$$\int_0^\alpha v_{(i)}'' v_{(k)}''\,\mathrm{d}\vartheta - (1+\lambda_{(i)}) \int_0^\alpha w_{(i)} w_{(k)}\,\mathrm{d}\vartheta - \lambda_{(i)} \int_0^\alpha v_{(i)} v_{(k)}\,\mathrm{d}\vartheta = 0. \tag{5.7}$$

Interchanging the suffixes i and k in (5.7), we obtain

$$\int_0^\alpha v_{(k)}'' v_{(i)}''\,\mathrm{d}\vartheta - (1+\lambda_{(k)}) \int_0^\alpha w_{(k)} w_{(i)}\,\mathrm{d}\vartheta - \lambda_{(k)} \int_0^\alpha v_{(k)} v_{(i)}\,\mathrm{d}\vartheta = 0. \tag{5.8}$$

If we substract equation (5.8) from equation (5.7), we arrise at

$$(\lambda_{(k)} - \lambda_{(i)}) \left[\int_0^\alpha w_{(i)} w_{(k)}\,\mathrm{d}\vartheta + \int_0^\alpha v_{(i)} v_{(k)}\,\mathrm{d}\vartheta\right] = 0, \tag{5.9}$$

whence, for $\lambda_{(i)} \neq \lambda_{(k)}$, we obtain the condition of orthogonality of the characteristic functions:

$$\int_0^\alpha w_{(i)} w_{(k)}\,\mathrm{d}\vartheta + \int_0^\alpha v_{(i)} v_{(k)}\,\mathrm{d}\vartheta = 0, \quad i \neq k. \tag{5.10}$$

If the cross-section of the air-supported structure consists of several circular arcs and straight sectors, the integration must be carried out along the whole cross-section. This means that instead of (5.10) one has

$$\sum \int_0^\alpha w_{(i)} w_{(k)}\,\mathrm{d}\vartheta + \sum \int_0^\alpha v_{(i)} v_{(k)}\,\mathrm{d}\vartheta = 0, \qquad i \neq k, \tag{5.11}$$

where the summation Σ refers to all circular arcs and straight sectors of the cross-section.

The formulae for the characteristic functions $w_{(i)}$ and $v_{(i)}$ can be obtained from the formulae (2.4) and (2.5) of Chapter IV, if we put $C_5=0$ and $a=a_{(i)}$, $b=b_{(i)}$, where $a_{(i)}$, $b_{(i)}$ are the parameters given by the formulae (2.3) of Chapter IV, in which $\lambda=\lambda_{(i)}$. Thus

$$
\begin{aligned}
w_{(i)} &= C_1 \sin a_{(i)} \frac{\vartheta}{\alpha} + C_2 \cos a_{(i)} \frac{\vartheta}{\alpha} + C_3 \sinh b_{(i)} \frac{\vartheta}{\alpha} + C_4 \cosh b_{(i)} \frac{\vartheta}{\alpha}, \\
v_{(i)} &= -C_1 \frac{\alpha}{a_{(i)}} \cos a_{(i)} \frac{\vartheta}{\alpha} + C_2 \frac{\alpha}{a_{(i)}} \sin a_{(i)} \frac{\vartheta}{\alpha} + C_3 \frac{\alpha}{b_{(i)}} \cosh b_{(i)} \frac{\vartheta}{\alpha} + \\
&\quad + C_4 \frac{\alpha}{b_{(i)}} \sinh b_{(i)} \frac{\vartheta}{\alpha},
\end{aligned}
\tag{5.12}
$$

while the relations between the integration constants depend on the values of the characteristic number $\lambda_{(i)}$ (see Section 8 below).

The orthogonality of the characteristic functions will be used in Section 6 below for the calculation of forced vibrations of an air-supported structure.

6

Solution of Forced Vibrations by Modal Analysis

Using the second relation (1.5) of Chapter IV we obtain from equation (1.3) the following differential equation for the displacement $\bar{v}=\bar{v}(\vartheta, t)$:

$$\frac{\partial^4 \bar{v}}{\partial \vartheta^4}+\frac{\partial^2}{\partial \vartheta^2}\left(\bar{v}-\frac{\mu R}{q}\frac{\partial^2 \bar{v}}{\partial t^2}\right)+\frac{\mu R}{q}\frac{\partial^2 \bar{v}}{\partial t^2}=\frac{R}{q}\left(\bar{p}_v^*-\frac{\partial \bar{p}_w^*}{\partial \vartheta}\right). \tag{6.1}$$

Let us write

$$\Phi(\vartheta, t)=\bar{p}_v^*-\frac{\partial \bar{p}_w^*}{\partial \vartheta}, \tag{6.2}$$

assuming that the function Φ can be expressed in the form

$$\Phi(\vartheta, t)=\Psi(t)P(\vartheta), \tag{6.3}$$

where $\Psi(t)$ is a given function of time t only and $P(\vartheta)$ is a given function of the variable ϑ only.

Let us express the unknown displacement $\bar{v}$ by an infinite series

$$\bar{v}=\sum_{i=1}^{\infty} F_i(t)v_{(i)}(\vartheta), \tag{6.4}$$

where $v_{(i)}$ are the characteristic functions and $F_i(t)$ are functions of time t, which can be determined as follows:

We substitute (6.3) and (6.4) into (6.1), thus obtaining

$$\sum_{i=1}^{\infty} F_i(v_{(i)}^{\mathrm{IV}}+v_{(i)}'')+\frac{\mu R}{q}\sum_{i=1}^{\infty}\ddot{F}_i(v_{(i)}-v_{(i)}'')=\frac{R}{q}\Psi(t)P(\vartheta), \tag{6.5}$$

where the dots denote derivatives with respect to time.

From (5.2) we obtain

$$v_{(i)}^{\mathrm{IV}}+v_{(i)}''=\lambda_{(i)}(v_{(i)}-v_{(i)}''), \tag{6.6}$$

where, according to (1.11) from Chapter IV,

$$\lambda_{(i)} = \frac{\mu R}{q} \omega_{(i)}^2, \tag{6.7}$$

$\omega_{(i)}$ being the i-th circular eigenfrequency of the investigated air-supported structure.

Substituting (6.6) into (6.5) and aplying formula (6.7), we obtain

$$\sum_{i=1}^{\infty} (\omega_{(i)}^2 F_i + \ddot{F}_i)(v_{(i)} - v''_{(i)}) = \frac{1}{\mu} \Psi(t) P(\vartheta). \tag{6.8}$$

If we multiply equation (6.8) by the characteristic function $v_{(k)}$ and integrate along the whole cross-section, we obtain, due to (5.5) and the second relation (5.4), the equation

$$\sum_{i=1}^{\infty} (\ddot{F}_i + \omega_{(i)}^2 F_i) \int_0^{\alpha} (v_{(i)} v_{(k)} + w_{(i)} w_{(k)})\, \mathrm{d}\vartheta = \Psi(t) \int_0^{\alpha} \frac{1}{\mu} P(\vartheta) v_{(k)}\, \mathrm{d}\vartheta \tag{6.9}$$

as $v' = w$. Since, according to (5.10), the value of the integral on the left-hand side of equation (6.9) is different from zero only for $i = k$, we obtain from (6.9) for the function $F_i(t)$ the differential equation

$$\ddot{F}_i + \omega_{(i)}^2 F_i = K_i \Psi(t), \tag{6.10}$$

where

$$K_i = \frac{\displaystyle\int_0^{\alpha} \frac{1}{\mu} P(\vartheta) v_{(i)}\, \mathrm{d}\vartheta}{\displaystyle\int_0^{\alpha} v_{(i)}^2\, \mathrm{d}\vartheta + \int_0^{\alpha} w_{(i)}^2\, \mathrm{d}\vartheta}. \tag{6.11}$$

If the cross-section of the air-supported structure consists of several circular arcs and straight parts,

$$K_i = \frac{\displaystyle\sum \int_0^{\alpha} \frac{1}{\mu} P(\vartheta) v_{(i)}\, \mathrm{d}\vartheta}{\displaystyle\sum \int_0^{\alpha} v_{(i)}^2\, \mathrm{d}\vartheta + \sum \int_0^{\alpha} w_{(i)}^2\, \mathrm{d}\vartheta}, \tag{6.12}$$

where the summation Σ is over all circular arcs and straight parts of the cross-section.

Equation (6.10) can be solved by the method of variation of constants.

The general solution of the homogeneous equation

$$\ddot{F}_i + \omega^2_{(i)} F_i = 0 \tag{6.13}$$

has the form

$$F_i = C_{1,i} \sin \omega_{(i)} t + C_{2,i} \cos \omega_{(i)} t. \tag{6.14}$$

If we consider $C_{1,i}$ and $C_{2,i}$ as functions of time t, we obtain

$$\begin{aligned} \sin \omega_{(i)} t \frac{\mathrm{d}C_{1,i}}{\mathrm{d}t} + \cos \omega_{(i)} t \frac{\mathrm{d}C_{2,i}}{\mathrm{d}t} &= 0, \\ \omega_{(i)} \cos \omega_{(i)} t \frac{\mathrm{d}C_{1,i}}{\mathrm{d}t} - \omega_{(i)} \sin \omega_{(i)} t \frac{\mathrm{d}C_{2,i}}{\mathrm{d}t} &= K_i \Psi(t), \end{aligned} \tag{6.15}$$

whence

$$\frac{\mathrm{d}C_{1,i}}{\mathrm{d}t} = \frac{K_i}{\omega^2_{(i)}} \Psi(t) \cos \omega_{(i)} t, \quad \frac{\mathrm{d}C_{2,i}}{\mathrm{d}t} = -\frac{K_i}{\omega^2_{(i)}} \Psi(t) \sin \omega_{(i)} t. \tag{6.16}$$

Integration of (6.16) yields

$$\begin{aligned} C_{1,i} &= \frac{K_i}{\omega^2_{(i)}} \int \Psi(t) \cos \omega_{(i)} t \,\mathrm{d}t + D_i, \\ C_{2,i} &= -\frac{K_i}{\omega^2_{(i)}} \int \Psi(t) \sin \omega_{(i)} t \,\mathrm{d}t + E_i, \end{aligned} \tag{6.17}$$

where D_i and E_i are integration constants.

Substituting (6.17) into (6.14), we obtain the general solution of the non-homogeneous differential equation (6.10) in the form

$$F_i(t) = D_i \sin \omega_{(i)} t + E_i \cos \omega_{(i)} t + \frac{K_i}{\omega^2_{(i)}} M_{(i)}(t), \tag{6.18}$$

where

$$M_i(t) = \sin \omega_{(i)} t \int \Psi(t) \cos \omega_{(i)} t \,\mathrm{d}t - \cos \omega_{(i)} t \int \Psi(t) \sin \omega_{(i)} t \,\mathrm{d}t. \tag{6.19}$$

The integration constants D_i and E_i are determined from the initial conditions of the motion of the air-supported structure.

If the initial conditions at time $t = 0$ are

$$\bar{v}(\vartheta, 0) = f_1(\vartheta), \quad \frac{\partial \bar{v}(\vartheta, 0)}{\partial t} = f_2(\vartheta), \tag{6.20}$$

we obtain, after the substituting (6.20) and $t=0$ into (6.4),

$$f_1(\vartheta)=\sum_{i=1}^{\infty} F_i(0)v_{(i)}(\vartheta),$$
$$f_2(\vartheta)=\sum_{i=1}^{\infty} \dot{F}_i(0)v_{(i)}(\vartheta). \tag{6.21}$$

The values $F_i(0)$, $i=1, 2, \ldots$, can be determined as follows:

We multiply the first equation (6.21) by the function $v''_{(k)}$ and integrate along the whole length of the cross-section of the investigated air-supported structure. We have (see Section 5)

$$\int_0^\alpha f_1(\vartheta)v''_{(k)}\,\mathrm{d}\vartheta=\sum_{i=1}^{\infty} F_i(0)\int_0^\alpha v_{(i)}v''_{(k)}\,\mathrm{d}\vartheta=$$
$$=\sum_{i=1}^{\infty} F_i(0)\,[v_{(i)}v'_{(k)}]_0^\alpha-\sum_{i=1}^{\infty} F_i(0)\int_0^\alpha v'_{(i)}v'_{(k)}\,\mathrm{d}\vartheta=-\sum_{i=1}^{\infty} F_i(0)\int_0^\alpha w_{(i)}w_{(k)}\mathrm{d}\vartheta \tag{6.22}$$

and, multiplying the above equation by the characteristic function $v_{(k)}$ and then integrating, we obtain

$$\int_0^\alpha f_1(\vartheta)v_{(k)}\,\mathrm{d}\vartheta=\sum_{i=1}^{\infty} F_i(0)\int_0^\alpha v_{(i)}v_{(k)}\,\mathrm{d}\vartheta. \tag{6.23}$$

If we substract equation (6.22) from (6.23), we obtain

$$\int_0^\alpha f_1(\vartheta)\,(v_{(k)}-v''_{(k)})\,\mathrm{d}\vartheta=\sum_{i=1}^{\infty} F_i(0)\int_0^\alpha (v_{(i)}v_{(k)}+w_{(i)}w_{(k)})\,\mathrm{d}\vartheta. \tag{6.24}$$

Because of the condition of orthogonality (5.10), the right-hand side of (6.24) is different from zero only for $i=k$; consequently, for the values of $F_i(0)$, $i=1, 2, \ldots$, we obtain from (6.24) the formula

$$F_i(0)=\frac{\displaystyle\int_0^\alpha f_1(\vartheta)\,(v_{(i)}-v''_{(i)})\,\mathrm{d}\vartheta}{\displaystyle\int_0^\alpha v_{(i)}^2\,\mathrm{d}\vartheta+\int_0^\alpha w_{(i)}^2\,\mathrm{d}\vartheta}. \tag{6.25}$$

In the same way we obtain the formula

$$\dot{F}_i(0)=\frac{\int_0^\alpha f_2(\vartheta)\,(v_{(i)}-v''_{(i)})\,\mathrm{d}\vartheta}{\int_0^\alpha v_{(i)}^2\,\mathrm{d}\vartheta+\int_0^\alpha w_{(i)}^2\,\mathrm{d}\vartheta},\qquad i=1,2,\ldots, \tag{6.26}$$

for the values $\dot{F}_i(0)$ occuring in the second equation (6.21).

If the cross-section of the air-supported structure consists of several circular arcs and straight parts, the integrals in the numerator and the denominator of the formulae (6.25) and (6.26) will be preceded by the symbol Σ, similarly as in (6.12).

Differentiation of equation (6.18) with respect to t yields

$$\dot{F}_i(t)=D_i\,\omega_{(i)}\cos\omega_{(i)}t-E_i\omega_{(i)}\sin\omega_{(i)}t+\frac{K_i}{\omega_{(i)}^2}\dot{M}_i(t). \tag{6.27}$$

Putting $t=0$ in (6.18) and (6.27), we get

$$\begin{aligned}D_i&=\frac{1}{\omega_{(i)}}\dot{F}_i(0)-\frac{K_i}{\omega_{(i)}}\dot{M}_i(0),\\ E_i&=F_i(0)-\frac{K_i}{\omega_{(i)}^2}M_i(0).\end{aligned} \tag{6.28}$$

Substituting (6.28) into (6.18) and then (6.18) into (6.4), we obtain for the displacement $\bar{v}$ due to the excitation load (6.2) the formula

$$\begin{aligned}\bar{v}(\vartheta,t)=\sum_{i=1}^{\infty}v_{(i)}(\vartheta)\Big\{&\frac{1}{\omega_{(i)}}\left[\dot{F}_i(0)-\frac{K_i}{\omega_{(i)}^2}\dot{M}_i(0)\right]\sin\omega_{(i)}t+\\ &+\left[F_i(0)-\frac{K_i}{\omega_{(i)}^2}M_i(0)\right]\cos\omega_{(i)}t+\frac{K_i}{\omega_{(i)}^2}M_i(t)\Big\},\end{aligned} \tag{6.29}$$

where the values $F_i(0)$ and $\dot{F}_i(0)$, $i=1,2,\ldots$, can be determined from the formulae (6.25) and (6.26) respectively. If

$$\bar{v}(\vartheta,0)\equiv0,\quad \frac{\partial\bar{v}(\vartheta,0)}{\partial t}\equiv0 \tag{6.30}$$

i.e. if $f_1(\vartheta)\equiv0$ and $f_2(\vartheta)\equiv0$, then, according to (6.25) and (6.26),

$$F_i(0)=0,\quad \dot{F}_i(0)=0,\qquad i=1,2,\ldots. \tag{6.31}$$

To obtain a formula for the displacement $\bar{w}(\vartheta,t)=\partial\bar{v}/\partial\vartheta$, we have to replace $v_{(i)}(\vartheta)$ by the characteristic function $w_{(i)}(\vartheta)$ in the right-hand side of (6.29).

7

Criteria for Ultimate Dynamic Stresses of Cylindrical Membranes

The tests carried out by the author on a model of an air-supported cylindrical structure in a wind tunnel for various values of internal overpressure and wind velocity have shown that the aerodynamic instability of a cylindrical membrane under a wind pressure perpendicular to the longitudinal axis of the model can be overall or local. In the case of overall aerodynamic instability the whole jacket of the model flutters, thus producing speedily increasing stresses in the membrane; therefore, overall aerodynamic instability occurs simultaneously with the ultimate dynamic stresses of the membrane.

In this section we shall endeavour to determine the criteria for the overall aerodynamic instability of a cylindrical membrane.

Let us assume that the dynamic component of the wind load during membrane fluttering acts in the direction of the normal to the membrane, deformed by static load, and that this component can be expressed in the form

$$\bar{p}_1^*(\vartheta, t) = c\mu\bar{w} + k_1\mu\frac{\partial\bar{w}}{\partial t}, \tag{7.1}$$

where c and k_1 are parameters which must be determinated experimentally, and $\bar{w}$ is the radial displacement (deflection) of the vibrating cylindrical membrane ($\bar{w} = \bar{w}(\vartheta, t)$).

The damping of the air-supported structure is due primarily to the viscoelastic properties of the membrane and the air resistance, and it manifests itself as a radial load $\bar{p}_2^*$ satisfying

$$\bar{p}_2^* = -k_2\mu\frac{\partial\bar{w}}{\partial t}, \tag{7.2}$$

where $k_2 > 0$ is the damping coefficient.

The resulting dynamic radial load $\bar{p}_w^*$, applied to the cylindrical membrane during wind action, is

$$\bar{p}_w^* = \bar{p}_1^* + \bar{p}_2^* = c\mu\bar{w} - (k_2 - k_1)\mu\frac{\partial \bar{w}}{\partial t}. \tag{7.3}$$

For $k_2 - k_1 > 0$ the dynamic wind load is dampend and for $k_2 - k_1 < 0$ the amplitudes of the vibrating membrane are continously increasing, thus we have self-induced vibrations, i.e. aerodynamic instability.

If we substitute (7.3) into equation (1.3) and put $\bar{p}^*_v = 0$, we obtain

$$\begin{aligned}&\frac{\partial^4 \bar{w}}{\partial \vartheta^4} + \frac{\partial^2}{\partial \vartheta^2}\left(\bar{w} - \frac{\mu R}{q}\frac{\partial^2 \bar{w}}{\partial t^2}\right) + \frac{\mu R}{q}c\frac{\partial^2 \bar{w}}{\partial \vartheta^2} - \\ &-(k_2 - k_1)\frac{\mu R}{q}\frac{\partial^3 \bar{w}}{\partial \vartheta^2 \partial t} + \frac{\mu R}{q}\frac{\partial^2 \bar{w}}{\partial t^2} = 0.\end{aligned} \tag{7.4}$$

For harmonic self-induced vibrations we have

$$\bar{w} = w e^{i\omega t}, \tag{7.5}$$

where $i = \sqrt{-1}$, ω is the circular frequency and w is a quantity independent of the time t. If we substitute (7.5) into (7.4) and use the notation (1.11) of Chapter IV, we arrive at

$$\frac{d^4 w}{d\vartheta^4} + \left[1 + \lambda\left(1 - \frac{k_2 - k_1}{\omega}i + \frac{c}{\omega^2}\right)\right]\frac{d^2 w}{d\vartheta^2} - \lambda w = 0. \tag{7.6}$$

The general solution of this equation can be written in the form

$$w = \sum_{i=1}^{4} C_i e^{\beta_i \vartheta}, \tag{7.7}$$

where β_i, $i = 1, 2, 3, 4$, are the roots of the characteristic equation

$$\beta^4 + [1 + \lambda(1 - iK_1 + K_2)]\beta^2 - \lambda = 0, \tag{7.8}$$

pertaining to equation (7.6), in which the quantities

$$K_1 = (k_2 - k_1)/\omega, \tag{7.9}$$

and

$$K_2 = c/\omega^2 \tag{7.10}$$

are dimensionless parameters.

Let us note that for the natural vibrations of the cylindrical membrane the radial displacement (deflection) can be also expressed in the form (7.7), where β_i, $i = 1, 2, 3, 4$, are roots of the characteristic equation (2.1) of

Chapter IV, defined by the formulae (2.2) of Chapter IV. Thus (7.7) differs from the characteristic function $w_{(i)}$ in so far that the roots β_i in (7.7) comprize also the parameters K_1, K_2, and the deflection w is expressed in the complex form $w=w_1+iw_2$, where w_1 and w_2 are real quantities.

If we proceed in the same way as in Chapter IV, we obtain for the vibrating cylindrical membrane, replaced in a cross-section by three circular arcs, instead of equation (4.7) of Chapter IV, the equation

$$\Delta=0, \tag{7.11}$$

where Δ is the determinant of a system of four homogeneous equations, whose elements are complex-valued functions. If we develop the determinant Δ and require that the real and the imaginary parts of equation (7.11) are equal to zero, we obtain two equations

$$\Phi_1(\omega, \lambda)=0, \quad \Phi_2(\omega, \lambda)=0 \tag{7.12}$$

whose solution yields the values of λ_c and ω_c, pertaining to the critical wind welocity $v_{0,c}$, which produces self-induced vibrations. The dynamic effects of wind load manifests themselves together with the static effects, i.e. we have pressure and suction, and the latter exercizes considerable influence on the tensile force in the arc T_2.

Thus it is necessary to include into the radial static load q in equation (7.4), apart from the internal overpressure q_0, also the effect of static wind load.

If we replace, in the case of a semi-cylindrical membrane, the static wind effect by an equivalent internal overpressure $\bar{q}_0=0.613v_0^2$, then the resultant static load q, according to equation (6.98) of Chapter II, is

$$q=q_0+0.613v_0^2 \tag{7.13}$$

and the parameter λ in (7.8) is given by

$$\lambda=\frac{\mu\omega^2 R_0}{q_0+0.613v_0^2}, \tag{7.14}$$

where R_0 is the radius of the undeformed cylindrical membrane. In this case we obtain from equation (7.7) for the displacement v in the direction of the tangent to the arc the formula

$$v=\int w\,d\vartheta=\sum_{i=1}^{4}\frac{C_i}{\beta_i}e^{\beta_i\vartheta}. \tag{7.15}$$

Substitution of (7.7) and (7.15) into the boundary conditions (2.11) of Chapter IV yields four homogeneous equations

$$\begin{gathered} C_1+C_2+C_3+C_4=0, \\ \frac{1}{\beta_1}C_1+\frac{1}{\beta_2}C_2+\frac{1}{\beta_3}C_3+\frac{1}{\beta_4}C_4=0, \\ e^{\beta_1\alpha}C_1+e^{\beta_2\alpha}C_2+e^{\beta_3\alpha}C_3+e^{\beta_4\alpha}C_4=0, \\ \frac{e^{\beta_1\alpha}}{\beta_1}C_1+\frac{e^{\beta_2\alpha}}{\beta_2}C_2+\frac{e^{\beta_3\alpha}}{\beta_3}C_3+\frac{e^{\beta_4\alpha}}{\beta_4}C_4=0, \end{gathered} \tag{7.16}$$

where $\alpha=\pi$ is the central angle of the undeformed circular arc which represents the cross-section of the cylindrical membrane (Fig. IV-3).

The system of homogeneous equations (7.16) has a non-trivial solution only if

$$\begin{vmatrix} 1 & 1 & 1 & 1 \\ \frac{1}{\beta_1} & \frac{1}{\beta_2} & \frac{1}{\beta_3} & \frac{1}{\beta_4} \\ e^{\beta_1\alpha} & e^{\beta_2\alpha} & e^{\beta_3\alpha} & e^{\beta_4\alpha} \\ \frac{e^{\beta_1\alpha}}{\beta_1} & \frac{e^{\beta_2\alpha}}{\beta_2} & \frac{e^{\beta_3\alpha}}{\beta_3} & \frac{e^{\beta_4\alpha}}{\beta_4} \end{vmatrix}=0, \tag{7.17}$$

where, according to (7.8),

$$\begin{aligned} \beta_{1,2}&=\pm\frac{1}{\sqrt{2}}\sqrt{-1-\lambda(1-iK_1+K_2)+\sqrt{[1+\lambda(1-iK_1+K_2)]^2+4\lambda}}\,, \\ \beta_{3,4}&=\frac{i}{\sqrt{2}}\sqrt{1+\lambda(1-iK_1+K_2)+\sqrt{[1+\lambda(1-iK_1+K_2)]^2+4\lambda}}\,. \end{aligned} \tag{7.18}$$

Since some elements of the determinant in (7.17) are complex-valued, we obtain from (7.17) two equations of the form (7.12) from which it is possible to determine ω_c and λ_c. We can then calculate the critical wind velocity $v_{0,c}$ from equation (7.14).

If we ascertain on a model the values of ω_c and $v_{0,c}$ for various values of internal overpressure q_0, we can determine λ_c from formula (7.14) and the quantities K_1 and K_2 from two equations of the form (7.12). Moreover (7.9) and (7.10), determine the parameters k_2-k_1 and c.

Equation (7.14) implies

$$\frac{q_0}{q_v} = \left(\frac{\mu R_0}{q_0}\frac{\omega_c^2}{\lambda_c} - 1\right)^{-1}, \tag{7.19}$$

where $q_v = 0.613 v_{0c}^2$ is the dynamic wind pressure in N/m^2 corresponding to the critical wind velocity $v_{0,c}$ in m/sec.

On the model of an air-supported structure in the form of a semi-cylinder with quarter-spherical ends (Fig. I-7) the author determined experimentally that the self-induced vibrations of the whole PVC foil jacket occur when, approximately,

$$q_0/q_v \leqq 0.6 \text{ to } 0.5. \tag{7.20}$$

This was verified for

$$14 \text{ m/sec} \leqq v_0 \leqq 42 \text{ m/sec}, \tag{7.21}$$

also when the wind direction was perpendicular to the longitudinal axis of the model. The greatest vibrations of the membrane occured on the windward side of the model.

We propose to use the inequality

$$\mu_s S_{max} \leqq T_2 \tag{7.22}$$

as another criterion for the assessment of the aerodynamic instability of an air-supported cylindrical structure. Here S_{max} is the maximum amplitude of the axial force in the arc representing the cross-section of the jacket, T_2 is the static axial force due to overall static load and μ_s is the safety margin. The quantity S_{max} can be determined from equation (2.10) and the force T_2 from equation (6.6) of Chapter IV.

8

Calculation of the Integrals (8.1) and the Integration Constants

In calculating the forced vibrations of air-supported cylindrical structures by modal analysis it is necessary to determine the values of the following definite integrals (see Section 6):

$$\int_0^\alpha w_{(i)}^2 \, \mathrm{d}\vartheta =$$

$$= \int_0^\alpha \left[C_1 \sin a_{(i)} \frac{\vartheta}{\alpha} + C_2 \cos a_{(i)} \frac{\vartheta}{\alpha} + C_3 \sinh b_{(i)} \frac{\vartheta}{\alpha} + C_4 \cosh b_{(i)} \frac{\vartheta}{\alpha} \right]^2 \mathrm{d}\vartheta =$$

$$= U_1 C_1^2 + U_2 C_2^2 + U_3 C_3^2 + U_4 C_4^2 + U_5 C_1 C_2 + U_6 C_1 C_3 + U_7 C_1 C_4 + U_8 C_2 C_3 +$$

$$+ U_9 C_2 C_4 + U_{10} C_3 C_4,$$

$$\int_0^\alpha v_{(i)}^2 \, \mathrm{d}\vartheta = \int_0^\alpha \left[- C_1 \frac{\alpha}{a_{(i)}} \cos a_{(i)} \frac{\vartheta}{\alpha} + C_2 \frac{\alpha}{a_{(i)}} \sin a_{(i)} \frac{\vartheta}{\alpha} + C_3 \frac{\alpha}{b_{(i)}} \cosh b_{(i)} \frac{\vartheta}{\alpha} + \right.$$

$$\left. + C_4 \frac{\alpha}{b_{(i)}} \sinh b_{(i)} \frac{\vartheta}{\alpha} \right]^2 \mathrm{d}\vartheta = \tag{8.1}$$

$$= \frac{1}{\beta^2} U_2 C_1^2 + \frac{1}{\beta^2} U_1 C_2^2 + \frac{1}{\gamma^2} U_4 C_3^2 + \frac{1}{\gamma^2} U_3 C_4^2 -$$

$$- \frac{1}{\beta^2} U_5 C_1 C_2 - \frac{1}{\beta\gamma} U_9 C_1 C_3 - \frac{1}{\beta\gamma} U_8 C_1 C_4 + \frac{1}{\beta\gamma} U_7 C_2 C_3 +$$

$$+ \frac{1}{\beta\gamma} U_6 C_2 C_4 + \frac{1}{\gamma^2} U_{10} C_3 C_4,$$

where (see (2.3) in Chapter IV)

$$\beta=\frac{1}{\sqrt{2}}\sqrt{1+\lambda_{(i)}+\sqrt{1+6\lambda_{(i)}+\lambda_{(i)}^2}}\,,$$
$$\gamma=\frac{1}{\sqrt{2}}\sqrt{-1-\lambda_{(i)}+\sqrt{(1+6\lambda_{(i)}+\lambda_{(i)}^2)}}\,, \tag{8.2}$$

$$a=\alpha\beta, \quad b=\alpha\gamma$$

and

$$U_1=\frac{1}{2}\left(\alpha-\frac{1}{2\beta}\sin 2a\right),$$
$$U_2=\frac{1}{2}\left(\alpha+\frac{1}{2\beta}\sin 2a\right),$$
$$U_3=\frac{1}{2}\left(\frac{1}{2\gamma}\sinh 2b-\alpha\right),$$
$$U_4=\frac{1}{2}\left(\alpha+\frac{1}{2\gamma}\sinh 2b\right),$$
$$U_5=\frac{1}{2\beta}(1-\cos 2a), \tag{8.3}$$
$$U_6=\frac{2}{\beta^2+\gamma^2}(\gamma\cosh b\sin a-\beta\sinh b\cos \alpha),$$
$$U_7=\frac{2}{\beta^2+\gamma^2}[\gamma\sinh b\sin a+\beta(1-\cosh b\cos a)],$$
$$U_8=\frac{2}{\beta^2+\gamma^2}[\beta\sinh b\sin a-\gamma(1-\cosh b\cos a)],$$
$$U_9=\frac{2}{\beta^2+\gamma^2}(\gamma\sinh b\cos a+\beta\cosh b\sin a),$$
$$U_{10}=\frac{1}{2\gamma}(\cosh 2\mathrm{b}-1).$$

Though we have omitted the index i at the quantities a, b, β, γ, it is understood that they pertain to the characteristic number $\lambda_{(i)}$.

Using the equations (3.3) of Chapter IV, we can find relations between the integration constants C_r, $r=1, 2, 3, 4$, similarly as we found the relations

Table IV. The values of the parameters β and γ according to Eqs. (8.2)

λ	β	γ	λ	β	γ
0.2	1.161177	0.385138	8.4	3.197151	0.906518
0.4	1.281951	0.493354	8.6	3.228768	0.908265
0.6	1.383312	0.559958	8.8	3.260062	0.909946
0.8	1.472704	0.607337	9.0	3.291041	0.911566
1.0	1.553774	0.643594	9.2	3.321717	0.913127
1.2	1.628624	0.672620	9.4	3.352097	0.914634
1.4	1.698596	0.696584	9.6	3.382191	0.916089
1.6	1.764606	0.716823	9.8	3.412007	0.917494
1.8	1.827313	0.734215	10.0	3.441553	0.918852
2.0	1.887208	0.749368	11.0	3.585478	0.925016
2.2	1.944670	0.762720	12.0	3.723635	0.930301
2.4	2.000000	0.774597	13.0	3.856684	0.934884
2.6	2.053438	0.785245	14.0	3.985163	0.938897
2.8	2.105184	0.794857	15.0	4.109525	0.942441
3.0	2.155401	0.803587	16.0	4.230147	0.945594
3.2	2.204228	0.811556	17.0	4.347355	0.948417
3.4	2.251786	0.818865	18.0	4.461426	0.950961
3.6	2.298175	0.825597	19.0	4.572605	0.953264
3.8	2.343486	0.831820	20.0	4.681102	0.955360
4.0	2.387794	0.837593	21.0	4.787105	0.957275
4.2	2.431170	0.842964	22.0	4.890781	0.959032
4.4	2.473674	0.847977	23.0	4.992279	0.960650
4.6	2.515361	0.852665	24.0	5.091731	0.962144
4.8	2.556278	0.857063	25.0	5.189257	0.963529
5.0	2.596470	0.861195	26.0	5.284966	0.964816
5.2	2.635977	0.865087	27.0	5.378958	0.966015
5.4	2.674835	0.868760	28.0	5.471321	0.967134
5.6	2.713077	0.872232	29.0	5.562138	0.968183
5.8	2.750733	0.875519	30.0	5.651485	0.969166
6.0	2.787831	0.878636	31.0	5.739432	0.970090
6.2	2.824396	0.881597	32.0	5.826042	0.970960
6.4	2.860452	0.884413	33.0	5.911375	0.971781
6.6	2.896021	0.887095	34.0	5.995487	0.972557
6.8	2.931123	0.889653	35.0	6.078429	0.973291
7.0	2.965777	0.892094	36.0	6.160248	0.973987
7.2	3.000000	0.894427	37.0	6.240988	0.974647
7.4	3.033809	0.896660	38.0	6.320693	0.975275
7.6	3.067220	0.898798	39.0	6.399401	0.975872
7.8	3.100246	0.900847	40.0	6.477147	0.976441
8.0	3.132902	0.902814	41.0	6.553968	0.976984
8.2	3.165199	0.904703	42.0	6.629895	0.977503

Table IV. (Continued)

λ	β	γ	λ	β	γ
43.0	6.704959	0.977998	82.0	9.163867	0.988162
44.0	6.779189	0.978472	83.0	9.218283	0.988301
45.0	6.852612	0.978927	84.0	9.272379	0.988436
46.0	6.925255	0.979362	85.0	9.326160	0.988568
47.0	6.997140	0.979780	86.0	9.379633	0.988697
48.0	7.068292	0.980181	87.0	9.432803	0.988824
49.0	7.138733	0.980566	88.0	9.485674	0.988947
50.0	7.208484	0.980937	89.0	9.538252	0.989068
51.0	7.277564	0.981294	90.0	9.590542	0.989186
52.0	7.345993	0.981638	91.0	9.642547	0.989302
53.0	7.413789	0.981969	92.0	9.694274	0.989415
54.0	7.480969	0.982289	93.0	9.745725	0.989526
55.0	7.547549	0.982597	94.0	9.796907	0.989635
56.0	7.613546	0.982895	95.0	9.847821	0.989741
57.0	7.678974	0.983183	96.0	9.898474	0.989845
58.0	7.743849	0.983461	97.0	9.948869	0.989947
59.0	7.808183	0.983730	98.0	9.999010	0.990048
60.0	7.871991	0.983991	99.0	10.048900	0.990146
61.0	7.935284	0.984243	100.0	10.098543	0.990242
62.0	7.998076	0.984488	105.0	10.343185	0.990696
63.0	8.060377	0.984725	110.0	10.582169	0.991110
64.0	8.122200	0.984955	115.0	10.815870	0.991488
65.0	8.183555	0.985178	120.0	11.044625	0.991836
66.0	8.244453	0.985395	125.0	11.268734	0.992156
67.0	8.304903	0.985605	130.0	11.488471	0.992452
68.0	8.364916	0.985809	135.0	11.704081	0.992727
69.0	8.424501	0.986008	140.0	11.915788	0.992982
70.0	8.483666	0.986201	145.0	12.123798	0.993220
71.0	8.542421	0.986389	150.0	12.328298	0.993442
72.0	8.600775	0.986572	155.0	12.529459	0.993650
73.0	8.658734	0.986750	160.0	12.727440	0.993846
74.0	8.716308	0.986923	165.0	12.922387	0.994029
75.0	8.773503	0.987092	170.0	13.114436	0.994202
76.0	8.830327	0.987257	175.0	13.303712	0.994366
77.0	8.886788	0.987417	180.0	13.490332	0.994520
78.0	8.942891	0.987573	185.0	13.674405	0.994666
79.0	8.998645	0.987726	190.0	13.856033	0.994805
80.0	9.054054	0.987875	195.0	14.035309	0.994937
81.0	9.109126	0.988020	200.0	14.212324	0.995062

between the displacements w_i, w_j, v_i and v_j, which occur in these equations. The latter have to be determined in the investigation of the natural vibrations of an air-supported structure by the semi-deformation method, by solving the homogeneous deformation equations.

To facilitate the calculation of the intergration constants, which are the multiples of one selected integration constant of arbitrary value, and to facilitate the calculation of the integrals (8.1), we have collected in Table IV the numerical values of the parameters occuring in the respective equations.

To calculate by means of Table IV the values of the functions (3.4) of Chapter IV we use the equalities

$$\frac{\alpha}{a}=\frac{1}{\beta}, \quad \frac{\alpha}{b}=\frac{1}{\gamma}, \quad \frac{a}{b}=\frac{\beta}{\gamma}, \quad \frac{b}{a}=\frac{\gamma}{\beta},$$

$$\frac{b}{a}-\frac{a}{b}=\frac{\gamma^2-\beta^2}{\beta\gamma}. \tag{8.4}$$

TABLES OF THE VALUES OF THE FUNCTIONS A TO H ($D = F$, $E = C + 1$)

VALUES OF THE FUNCTIONS *A* TO *H*

$\alpha = 0.1$

λ	*A*	*B*	*C*	*D*	*G*	*H*
0.2	39.96030	19.99210	598.52366	599.52425	11978.4677	11978.4777
0.4	39.98265	20.00547	598.88085	599.88122	11983.6061	11983.6261
0.6	39.97835	20.00518	598.83464	599.83541	11980.6767	11980.7068
0.8	39.97726	20.00677	598.83929	599.84037	11978.7640	11978.8040
1.0	39.96768	20.00416	598.71786	599.71866	11974.3299	11974.3798
1.2	39.97208	20.00780	598.79790	599.79906	11973.9265	11973.9865
1.4	39.96711	20.00785	598.74890	599.75009	11970.9400	11971.0100
1.6	39.96626	20.00934	598.75532	599.75666	11969.0632	11969.1431
1.8	39.96319	20.00951	598.72597	599.72764	11966.4715	11966.5615
2.0	39.95939	20.00980	598.69101	599.69282	11963.7666	11963.8666
2.2	39.95469	20.00955	598.64120	599.64297	11960.7647	11960.8748
2.4	39.95442	20.01137	598.65662	599.65872	11959.0681	11959.1880
2.6	39.95121	20.01161	598.62692	599.62918	11956.4691	11956.5991
2.8	39.95030	20.01321	598.63364	599.63618	11954.5981	11954.7380
3.0	39.94593	20.01297	598.58741	599.59006	11951.6681	11951.8181
3.2	39.94349	20.01366	598.56997	599.57266	11949.3145	11949.4744
3.4	39.93957	20.01376	598.53170	599.53450	11946.5434	11946.7135
3.6	39.93856	20.01544	598.53827	599.54131	11944.6692	11944.8492
3.8	39.93482	20.01552	598.50147	599.50469	11941.9281	11942.1180
4.0	39.93358	20.01689	598.50255	599.50601	11939.9444	11940.1443
4.2	39.93005	20.01701	598.46850	599.47206	11937.2585	11937.4684
4.4	39.92725	20.01769	598.44731	599.45096	11934.8289	11935.0489
4.6	39.92512	20.01870	598.43597	599.43985	11932.5966	11932.8266
4.8	39.92190	20.01914	598.40802	599.41197	11930.0318	11930.2719
5.0	39.92038	20.02039	598.40500	599.40927	11927.9664	11928.2164
5.2	39.91666	20.02049	598.36890	599.37321	11925.2391	11925.4991
5.4	39.91480	20.02159	598.36107	599.36565	11923.0770	11923.3469
5.6	39.91131	20.02178	598.32814	599.33281	11920.4135	11920.6935
5.8	39.90929	20.02284	598.31824	599.32317	11918.2098	11918.4997
6.0	39.90609	20.02317	598.28961	599.29461	11915.6324	11915.9325
6.2	39.90412	20.02429	598.28100	599.28614	11913.4544	11913.7645
6.4	39,90101	20.02471	598.25396	599.25930	11910.9083	11911.2282
6.6	39.89881	20.02566	598.24136	599.24695	11908.6509	11908.9809
6.8	39.89598	20.02623	598.21865	599.22429	11906.1915	11906.5315
7.0	39.89302	20.02668	598.19346	599.19935	11903.6827	11904.0327
7.2	39.89068	20.02764	598.17955	599.18561	11901.3985	11901.7585
7.4	39.88806	20,02824	598.15923	599.16548	11898.9872	11899.3573
7.6	39.88472	20.02855	598.12897	599.13528	11896.3768	11896.7569
7.8	39.88254	20.02953	598.11687	599.12340	11894.1292	11894.5193
8.0	39.87954	20.02997	598.09104	599.09772	11891.6077	11892.0077
8.2	39.87673	20.03065	598.06972	599.07656	11889.1756	11889.5856

(continued 1) $\alpha = 0.1$

λ	A	B	C	D	G	H
8.4	39.87457	20.03154	598.05685	599.06388	11886.9130	11887.3331
8.6	39.87212	20.03239	598.04063	599.04793	11884.5827	11885.0127
8.8	39.86902	20.03279	598.01374	599.02109	11882.0401	11882.4801
9.0	39.86671	20.03366	597.99914	599.00669	11879.7427	11880.1927
9.2	39.86387	20.03426	597.97666	598.98439	11877.2876	11877.7476
9.4	39.86098	20.03483	597.95336	598.96118	11874.8163	11875.2864
9.6	39.85839	20.03555	597.93454	598.94260	11872.4347	11872.9147
9.8	39.85564	20.03613	597.91278	598.92090	11869.9942	11870.4842
10.0	39.85331	20.03702	597.89819	598.90660	11867.6971	11868.1970
11.0	39.83983	20.04030	597.79577	598.80500	11855.6223	11856.1723
12.0	39.82622	20.04354	597.69162	598.70160	11843.5127	11844.1128
13.0	39.81282	20.04692	597.59091	598.60177	11831.4720	11832.1220
14.0	39.79953	20.05038	597.49211	598.50383	11819.4689	11820.1689
15.0	39.78631	20.05386	597.39409	598.40667	11807.4816	11808.2315
16.0	39.77275	20.05712	597.29065	598.30400	11795.3864	11796.1864
17.0	39.75926	20.06045	597.18846	598.20270	11783.3162	11784.1662
18.0	39.74588	20.06381	597.08779	598.10289	11771.2768	11772.1767
19.0	39.73257	20.06730	596.98916	598.00502	11759.2766	11760.2266
20.0	39.71908	20.07065	596.88732	597.90404	11747.2130	11748.2130
21.0	39.70573	20.07407	596.78741	597.80498	11735.1882	11736.2383
22.0	39.69221	20.07739	596.68498	597.70338	11723.1135	11724.2135
23.0	39.67877	20.08077	596.58393	597.60317	11711.0659	11712.2159
24.0	39.66542	20.08420	596.48415	597.50427	11699.0435	11700.2435
25.0	39.65189	20.08753	596.38168	597.40260	11686.9676	11688.2177
26.0	39.63843	20.09091	596.28034	597.30210	11674.9146	11676.2146
27.0	39.62506	20.09437	596.18076	597.20338	11662.8957	11664.2457
28.0	39.61164	20.09779	596.08021	597.10366	11650.8581	11652.2581
29.0	39.59808	20.10110	595.97726	597.00156	11638.7732	11640.2232
30.0	39.58470	20.10455	595.87744	596.90257	11626.7497	11628.2496
31.0	39.57126	20.10796	595.77658	596.80257	11614.7061	11616.2561
32.0	39.55778	20.11137	595.67544	596.70225	11602.6565	11604.2566
33.0	39.54433	20.11479	595.57464	596.60229	11590.6138	11592.2638
34.0	39.53092	20.11823	595.47447	596.50297	11578.5836	11580.2836
35.0	39.51738	20.12161	595.37229	596.40163	11566.5140	11568.2641
36.0	39.50394	20.12505	595.27178	596.30197	11554.4770	11556.2770
37.0	39.49041	20.12841	595.16961	596.20063	11542.4075	11544.2575
38.0	39.47702	20.13191	595.07023	596.10210	11530.3924	11532.2924
39.0	39.46350	20.13531	594.96843	596.00113	11518.3300	11520.2801
40.0	39.44997	20.13870	594.86656	595.90014	11506.2669	11508.2670
41.0	39.43650	20.14214	594.76575	595.80018	11494.2238	11496.2738
42.0	39.42301	20.14558	594.66479	595.70005	11482.1779	11484.2779

(continued 2) $\alpha = 0.1$

λ	A	B	C	D	G	H
43.0	39.40953	20.14902	594.56385	595.59994	11470.1322	11472.2823
44.0	39.39601	20.15243	594.46219	595.49914	11458.0732	11460.2732
45.0	39.38248	20.15585	594.36074	595.39852	11446.0176	11448.2676
46.0	39.36898	20.15930	594.25970	595.29834	11433.9702	11436.2702
47.0	39.35550	20.16277	594.15905	595.19853	11421.9302	11424.2802
48.0	39.34203	20.16623	594.05840	595.09874	11409.8904	11412.2904
49.0	39.32845	20.16964	593.95619	594.99739	11397.8205	11400.2705
50.0	39.31495	20.17310	593.85527	594.89730	11385.7750	11388.2749
51.0	39.30143	20.17655	593.75417	594.79704	11373.7263	11376.2763
52.0	39.28787	20.17998	593.65234	594.69607	11361.6639	11364.2638
53.0	39.27435	20.18345	593.55128	594.59584	11349.6158	11352.2657
54.0	39.26080	20.18689	593.44972	594.49514	11337.5581	11340.2582
55.0	39.24730	20.19038	593.34913	594.39542	11325.5193	11328.2694
56.0	39.23370	20.19380	593.24676	594.29388	11313.4461	11316.2461
57.0	39.22017	20.19729	593.14587	594.19382	11301.4012	11304.2512
58.0	39.20661	20.20073	593.04420	594.09301	11289.3418	11292.2418
59.0	39.19307	20.20422	592.94319	593.99285	11277.2944	11280.2445
60.0	39.17949	20.20767	592.84133	593.89184	11265.2312	11268.2312
61.0	39.16595	20.21116	592.74023	593.79158	11253.1825	11256.2325
62.0	39.15244	20.21466	592.63967	593.69190	11241.1441	11244.2441
63.0	39.13885	20.21812	592.53786	593.59095	11229.0820	11232.2318
64.0	39.12525	20.22160	592.43613	593.49005	11217.0209	11220.2209
65.0	39.11168	20.22508	592.33475	593.38952	11204.9667	11208.2167
66.0	39.09807	20.22853	592.23266	593.28829	11192.8990	11196.1990
67.0	39.08449	20.23202	592.13128	593.18777	11180.8447	11184.1947
68.0	39.07093	20.23553	592.03025	593.08759	11168.7971	11172.1971
69.0	39.05733	20.23900	591.92840	592.98659	11156.7339	11160.1839
70.0	39.04377	20.24252	591.82744	592.88648	11144.6876	11148.1877
71.0	39.03011	20.24598	591.72506	592.78495	11132.6144	11136.1644
72.0	39.01649	20.24946	591.62313	592.68389	11120.5500	11124.1499
73.0	39.00297	20.25301	591.52287	592.58448	11108.5166	11112.1667
74.0	38.98937	20.25652	591.42140	592.48386	11096.4604	11100.1605
75.0	38.97573	20.26000	591.31939	592.38270	11084.3943	11088.1443
76.0	38.96205	20.26347	591.21675	592.28091	11072.3161	11076.1161
77.0	38.94849	20.26701	591.11603	592.18105	11060.2741	11064.1242
78.0	38.93485	20.27050	591.01407	592.07995	11048.2087	11052.1088
79.0	38.92124	20.27403	590.91269	591.97941	11036.1542	11040.1042
80.0	38.90754	20.27750	590.80996	591.87755	11024.0746	11028.0746
81.0	38.89395	20.28104	590.70896	591.77740	11012,0271	11016.0771
82.0	38.88029	20.28456	590.60691	591.67620	10999.9601	11004.0601
83.0	38.86667	20.28808	590.50543	591.57559	10987.9040	10992.0540

(continued 3) $\alpha = 0.1$

λ	A	B	C	D	G	H
84.0	38.85300	20.29159	590.40343	591.47444	10975.8375	10980.0376
85.0	38.83938	20.29513	590.30206	591.37393	10963.7831	10968.0331
86.0	38.82569	20.29865	590.19982	591.27255	10951.7126	10956.0127
87.0	38.81203	20.30217	590.09790	591.17149	10939.6480	10943.9980
88.0	38.79837	20.30571	589.99615	591.07060	10927.5865	10931.9866
89.0	38.78472	20.30924	589.89442	590.96974	10915.5254	10919.9754
90.0	38.77104	20.31277	589.79250	590.86867	10903.4606	10907.9605
91.0	38.75741	20.31633	589.69126	590.76829	10891.4084	10895.9584
92.0	38.74373	20.31987	589.58936	590.66725	10879.3440	10883.9440
93.0	38.73002	20.32338	589.48683	590.56558	10867.2679	10871.9180
94.0	38.71633	20.32694	589.38500	590.46461	10855.2048	10859.9048
95.0	38.70264	20.33048	589.28297	590.36343	10843.1379	10847.8879
96.0	38.68898	20.33404	589.18147	590.26280	10831.0807	10835.8808
97.0	38.67526	20.33758	589.07913	590.16133	10819.0083	10823.8583
98.0	38.66156	20.34114	588.97718	590.06023	10806.9427	10811.8427
99.0	38.64786	20.34470	588.87531	589.95921	10794.8787	10799.8288
100.0	38.63415	20.34824	588.77317	589.85793	10782.8096	10787.8097
105.0	38.56560	20.36609	588.26350	589.35259	10722.4833	10727.7335
110.0	38.49688	20.38394	587.75221	588.84562	10662.1268	10667.6269
115.0	38.42811	20.40189	587.24143	588.33917	10601.7789	10607.5291
120.0	38.35925	20.41988	586.73018	587.83226	10541.4220	10547.4221
125.0	38.29027	20.43791	586.21811	587.32454	10481.0497	10487.2998
130.0	38.22126	20.45605	585.70678	586.81756	10420.6897	10427.1898
135.0	38.15211	20.47422	585.19441	586.30955	10360.3104	10367.0606
140.0	38.08285	20.49243	584.68130	585.80080	10299.9172	10306.9174
145.0	38.01350	20.51072	584.16802	585.29191	10239.5204	10246.7705
150.0	37.94404	20.52905	583.65406	584.78233	10179.1106	10186.6106
155.0	37.87451	20.54745	583.14017	584.27283	10118.7014	10126.4514
160.0	37.80484	20.56589	582.62532	583.76237	10058.2742	10066.2743
165.0	37.73512	20.58441	582.11061	583.25208	9997.84899	10006.0992
170.0	37.66530	20.60299	581.59566	582.74155	9937.41839	9945.91857
175.0	37.59537	20.62163	581.08007	582.23038	9876.97570	9885.72595
180.0	37.52535	20.64033	580.56419	581.71892	9816.52691	9825.52716
185.0	37.45524	20.65910	580.04816	581.20734	9756.07489	9765.32507
190.0	37.38501	20.67792	579.53144	580.69506	9695.60974	9705.10992
195.0	37.31469	20.69679	579.01435	580.18243	9635.13732	9644.88757
200.0	37.24429	20.71575	578.49714	579.66968	9574,66156	9584.66186

VALUES OF THE FUNCTIONS *A* TO *H*

$\alpha = 0.2$

λ	*A*	*B*	*C*	*D*	*G*	*H*
0.2	19.96764	10.00782	148.87686	149.87751	1492.75793	1492.77793
0.4	19.96255	10.00933	148.85860	149.85991	1491.56465	1491.60466
0.6	19.95677	10.01049	148.83510	149.83709	1490.31902	1490.37902
0.8	19.95179	10.01203	148.81750	149.82018	1489.13238	1489.21239
1.0	19.94651	10.01341	148.79759	149.80093	1487.92254	1488.02255
1.2	19.94118	10.01477	148.77732	149.78134	1486.70922	1486.82921
1.4	19.93577	10.01611	148.75661	149.76130	1485.49144	1485.63143
1.6	19.93068	10.01761	148.73820	149.74356	1484.29674	1484.45673
1.8	19.92501	10.01880	148.71547	149.72148	1483.05879	1483.23880
2.0	19.91966	10.02016	148.69511	149.70179	1481.84445	1482.04445
2.2	19.91441	10.02158	148.67557	149.68292	1480.63842	1480.85841
2.4	19.90907	10.02293	148.65520	149.66323	1479.42411	1479.66410
2.6	19.90364	10.02427	148.63434	149.64305	1478.20484	1478.46484
2.8	19.89821	10.02559	148.61336	149.62272	1476.98441	1477.26441
3.0	19.89288	10.02698	148.59327	149.60330	1475.77280	1476.07280
3.2	19.88750	10.02833	148.57275	149.58346	1474.55702	1474.87702
3.4	19.88215	10.02969	148.55238	149.56376	1473.34268	1473.68268
3.6	19.87678	10.03106	148.53196	149.54400	1472.12776	1472.48777
3.8	19.87138	10.03241	148.51130	149.52401	1470.91051	1471.29051
4.0	19.86604	10.03379	148.49111	149.50449	1469.69798	1470.09799
4.2	19.86066	10.03514	148.47059	149.48465	1468.48219	1468.90219
4.4	19.85525	10.03649	148.44988	149.46460	1467.26441	1467.70442
4.6	19.84988	10.03787	148.42954	149.44493	1466.05027	1466.51027
4.8	19.84452	10.03924	148.40913	149.42520	1464.83560	1465.31561
5.0	19.83911	10.04059	148.38845	149.40518	1463.61811	1464.11813
5.2	19.83373	10.04195	148.36795	149.38536	1462.40251	1462.92252
5.4	19.82836	10.04332	148.34755	149.36564	1461.18787	1461.72788
5.6	19.82295	10.04468	148.32689	149.34564	1459.97060	1460.53060
5.8	19.81755	10.04604	148.30632	149.32574	1458.75428	1459.33429
6.0	19.81218	10.04742	148.28596	149.30606	1457.54005	1458.14006
6.2	19.80679	10.04879	148.26543	149.28620	1456.32415	1456.94416
6.4	19.80143	10.05017	148.24516	149.26661	1455.11074	1455.75075
6.6	19.79602	10.05153	148.22448	149.24660	1453.89332	1454.55332
6.8	19.79064	10.05291	148.20407	149.22686	1452.67858	1453.35858
7.0	19.78522	10.05427	148.18340	149.20686	1451.46125	1452.16126
7.2	19.77984	10.05564	148.16294	149.18708	1450.24600	1450.96601
7.4	19.77445	10.05703	148.14253	149.16733	1449.03119	1449.77120
7.6	19.76904	10.05839	148.12189	149.14737	1447.81420	1448.57421
7.8	19.76364	10.05976	148.10131	149.12746	1446.59777	1447.37779
8.0	19.75826	10.06115	148.08097	149.10780	1445.38378	1446.18379
8.2	19.75284	10.06251	148.06030	149.08780	1444.16645	1444.98647

(continued 1) $\alpha = 0.2$

λ	A	B	C	D	G	H
8.4	19.74744	10.06389	148.03971	149.06789	1442.94993	1443.78995
8.6	19.74206	10.06528	148.01937	149.04823	1441.73587	1442.59587
8.8	19.73664	10.06665	147.99875	149.02828	1440.51904	1441.39906
9.0	19.73124	10.06803	147.97824	149.00845	1439.30333	1440.20335
9.2	19.72584	10.06941	147.95771	148.98859	1438.08736	1439.00737
9.4	19.72042	10.07079	147.93712	148.96867	1436.87084	1437.81086
9.6	19.71500	10.07216	147.91650	148.94873	1435.65406	1436.61407
9.8	19.70961	10.07355	147.89603	148.92894	1434.43877	1435.41880
10.0	19.70419	10.07493	147.87543	148.90901	1433.22214	1434.22216
11.0	19.67711	10.08184	147.77262	148.80959	1427.14089	1428.24092
12.0	19.65001	10.08878	147.66979	148.71014	1421.05947	1422.25950
13.0	19.62289	10.09574	147.56696	148.61071	1414.97814	1416.27818
14.0	19.59572	10.10270	147.46386	148.51100	1408.89408	1410.29412
15.0	19.56855	10.10971	147.36101	148.41155	1402.81253	1404.31258
16.0	19.54133	10.11671	147.25785	148.31180	1396.72789	1398.32794
17.0	19.51408	10.12373	147.15463	148.21199	1390.64283	1392.34289
18.0	19.48680	10.13077	147.05145	148.11223	1384.55800	1386.35809
19.0	19.45950	10.13784	146.94822	148.01242	1378.47278	1380.37286
20.0	19.43216	10.14491	146.84485	147.91247	1372.38612	1374.38621
21.0	19.40479	10.15201	146.74143	147.81249	1366.29909	1368.39920
22.0	19.37740	10.15912	146.63799	147.71248	1360.21180	1362.41192
23.0	19.34998	10.16625	146.53447	147.61240	1354.12371	1356.42384
24.0	19.32252	10.17340	146.43086	147.51223	1348.03479	1350.43492
25.0	19.29504	10.18057	146.32724	147.41206	1341.94580	1344.44594
26.0	19.26753	10.18776	146.22354	147.31182	1335.85606	1338.45621
27.0	19.23998	10.19496	146.11980	147.21153	1329.76579	1332.46595
28.0	19.21241	10.20219	146.01596	147.11115	1323.67466	1326.47484
29.0	19.18480	10.20943	145.91205	147.01071	1317.58283	1320.48303
30.0	19.15717	10.21669	145.80813	146.91026	1311.49091	1314.49111
31.0	19.12950	10.22397	145.70408	146.80970	1305.39783	1308.49805
32.0	19.10180	10.23126	145.60001	146.70911	1299.30444	1302.50466
33.0	19.07407	10.23858	145.49585	146.60844	1293.21024	1296.51049
34.0	19.04632	10.24592	145.39169	146.50778	1287.11602	1290.51628
35.0	19.01852	10.25327	145.28736	146.40694	1281.02013	1284.52042
36.0	18.99071	10.26064	145.18306	146.30614	1274.92455	1278.52485
37.0	18.96285	10.26803	145.07864	146.20523	1268.82782	1272.52814
38.0	18.93497	10.27544	144.97422	146.10432	1262.73104	1266.53137
39.0	18.90706	10.28287	144.86970	146.00332	1256.63343	1260.53376
40.0	18.87911	10.29032	144.76507	145.90221	1250.53469	1254.53505
41.0	18.85114	10.29779	144.66046	145.80113	1244.43603	1248.53642
42.0	18.82313	10.30528	144.55574	145.69994	1238.33647	1242.53688

(continued 2) $\alpha = 0.2$

λ	A	B	C	D	G	H
43.0	18.79509	10.31279	144.45101	145.59875	1232.23668	1236.53712
44.0	18.76702	10.32031	144.34612	145.49741	1226.13550	1230.53594
45.0	18.73892	10.32785	144.24125	145.39608	1220.03438	1224.53485
46.0	18.71078	10.33542	144.13627	145.29465	1213.93226	1218.53276
47.0	18.68262	10.34301	144.03126	145.19321	1207.82986	1212.53038
48.0	18.65442	10.35061	143.92619	145.09170	1201.72682	1206.52736
49.0	18.62619	10.35823	143.82100	144.99008	1195.62271	1200.52326
50.0	18.59793	10.36588	143.71578	144.88843	1189.51830	1194.51889
51.0	18.56964	10.37354	143.61052	144.78675	1183.41342	1188.51402
52.0	18.54131	10.38122	143.50516	144.68497	1177.30767	1182.50830
53.0	18.51295	10.38893	143.39976	144.58317	1171.20149	1176.50216
54.0	18.48456	10.39665	143.29430	144.48129	1165.09465	1170.49533
55.0	18.45614	10.40440	143.18874	144.37934	1158.98700	1164.48770
56.0	18.42768	10.41216	143.08314	144.27734	1152.87889	1158.47963
57.0	18.39919	10.41995	142.97744	144.17525	1146.76990	1152.47066
58.0	18.37067	10.42775	142.87173	144.07316	1140.66069	1146.46147
59.0	18.34212	10.43558	142.76593	143.97098	1134.55066	1140.45148
60.0	18.31353	10.44343	142.66007	143.86874	1128.44001	1134.44087
61.0	18.28491	10.45129	142.55414	143.76644	1122.32874	1128.42961
62.0	18.25626	10.45918	142.44819	143.66413	1116.21720	1122.41811
63.0	18.22757	10.46709	142.34211	143.56170	1110.10456	1116.40549
64.0	18.19884	10.47502	142.23596	143.45919	1103.99118	1110.39216
65.0	18.17009	10.48297	142.12977	143.35665	1097.87737	1104.37837
66.0	18.14130	10.49094	142.02351	143.25406	1091.76298	1098.36402
67.0	18.11248	10.49893	141.91720	143.15140	1085.64795	1092.34902
68.0	18.08363	10.50695	141.81081	143.04869	1079.53231	1086.33342
69.0	18.05474	10.51498	141.70437	142.94592	1073.41604	1080.31718
70.0	18.02582	10.52304	141.59785	142.84309	1067.29913	1074.30030
71.0	17.99686	10.53112	141.49123	142.74015	1061.18128	1068.28248
72.0	17.96787	10.53922	141.38458	142.63719	1055.06305	1062.26430
73.0	17.93885	10.54734	141.27790	142.53421	1048.94448	1056.24576
74.0	17.90979	10.55548	141.17112	142.43114	1042.82506	1050.22637
75.0	17.88069	10.56365	141.06427	142.32799	1036.70491	1044.20625
76.0	17.85156	10.57183	140.95731	142.22475	1030.58384	1038.18523
77.0	17.82240	10.58004	140.85035	142.12150	1024.46257	1032.16400
78.0	17.79320	10.58827	140.74332	142.01820	1018.34063	1026.14209
79.0	17.76397	10.59652	140.63620	141.91481	1012.21794	1020.11944
80.0	17.73470	10.60479	140.52898	141.81132	1006.09432	1014.09587
81.0	17.70540	10.61309	140.42176	141.70785	999.97059	1008.07217
82.0	17.67606	10.62141	140.31441	141.60424	993.84570	1002.04732
83.0	17.64669	10.62975	140.20704	141.50063	987.72056	996.02223

(continued 3) $\alpha = 0.2$

λ	A	B	C	D	G	H
84.0	17.61728	10.63811	140.09959	141.39694	981.59464	989.99635
85.0	17.58784	10.74650	139.99209	141.29320	975.46822	983.96998
86.0	17.55836	10.65491	139.88446	141.18935	969.34073	977.94252
87.0	17.52884	10.66334	139.77680	141.08546	963.21281	971.91465
88.0	17.49929	10.67179	139.66909	140.98153	957.08435	965.88623
89.0	17.46971	10.68027	139.56131	140.87754	950.95526	959.85718
90.0	17.44009	10.68877	139.45344	140.77347	944.82539	953.82736
91.0	17.41043	10.69729	139.34554	140.66937	938.69509	947.79711
92.0	17.38074	10.70584	139.23754	140.56517	932.56393	941.76599
93.0	17.35101	10.71441	139.12947	140.46091	926.43207	935.73418
94.0	17.32124	10.72300	139.02132	140.35658	920.29953	929.70169
95.0	17.29144	10.73161	138.91312	140.25221	914.1664	923.66864
96.0	17.26160	10.74025	138.80488	140.14779	908.03282	917.63507
97.0	17.23172	10.74891	138.69651	140.04327	901.89824	911.60055
98.0	17.20181	10.75760	138.58810	139.93870	895.76314	905.56549
99.0	17.17186	10.76631	138.47964	139.83408	889.62747	899.52988
100.0	17.14187	10.77505	138.37110	139.72939	883.49111	893.49357
105.0	16.99138	10.81908	137.82735	139.20500	852.79917	863.30191
110.0	16.83995	10.86371	137.28184	138.67901	822.09041	833.09342
115.0	16.68757	10.90897	136.73459	138.15143	791.36499	802.86830
120.0	16.53421	10.95486	136.18557	137.62224	760.62257	772.62618
125.0	16.37987	11.00140	135.63478	137.09144	729.86313	742.36708
130.0	16.22454	11.04859	135.08221	136.55903	699.08660	712.09091
135.0	16.06820	11.09645	134.52782	136.02496	668.29273	681.79739
140.0	15.91082	11.14500	133.97160	135.48923	637.48138	651.48642
145.0	15.75240	11.19424	133.41354	134.95184	606.65246	621.15790
150.0	15.59292	11.24419	132.85363	134.41278	575.80586	590.81171
155.0	15.43236	11.29486	132.29188	133.87204	544.94154	560.44782
160.0	15.27070	11.34627	131.72822	133.32960	514.05920	530.06593
165.0	15.10794	11.39844	131.16269	132.78545	483.15881	499.66601
170.0	14.94404	11.45137	130.59525	132.23960	452.24026	469.24795
175.0	14.77900	11.50509	130.02588	131.69200	421.30332	438.81152
180.0	14.61279	11.55961	129.45458	131.14266	390.34794	408.35666
185.0	14.44540	11.61494	128.88133	130.59157	359.37393	377.88320
190.0	14.27680	11.67111	128.30610	130.03872	328.38117	347.39101
195.0	1410697	11.72812	127.72889	129.48408	297.36953	316.87995
200.0	13.93590	11.78601	127.14968	128.92765	266.33884	286.34988

$\alpha = 0.3$

λ	*A*	*B*	*C*	*D*	*G*	*H*
0.2	13.28518	6.67872	65.54542	66.54693	439.62018	439.65018
0.4	13.27713	6.68080	65.52463	66.52764	438.79888	438.85888
0.6	13.26907	6.68286	65.50371	66.50822	437.97676	438.06676
0.8	13.26098	6.68491	65.48268	66.48869	437.15389	437.27389
1.0	13.25288	6.68697	65.46164	66.46916	436.33098	436.48098
1.2	13.24482	6.68905	65.44083	66.44986	435.50959	435.68959
1.4	13.23670	6.69110	65.41967	66.43021	434.68595	434.89595
1.6	13.22859	6.69316	65.39861	66.41066	433.86290	434.10291
1.8	13.22051	6.69525	65.37773	66.39129	433.04107	433.31107
2.0	13.21239	6.69731	65.35663	66.37170	432.21774	432.51775
2.2	13.20429	6.69939	65.33567	66.35225	431.39536	431.72537
2.4	13.19617	6.70147	65.31462	66.33271	430.57242	430.93243
2.6	13.18804	6.70354	65.29349	66.31310	429.74893	430.13894
2.8	13.17991	6.70562	65.27243	66.29356	428.92593	429.34594
3.0	13.17179	6.70770	65.25140	66.27404	428.10310	428.55312
3.2	13.16366	6.70979	65.23035	66.25451	427.28015	427.76017
3.4	13.15553	6.71188	65.20933	66.23500	426.45738	426.96740
3.6	13.14739	6.71397	65.18822	66.21541	425.63406	426.17408
3.8	13.13925	6.71606	65.16715	66.19586	424.81094	425.38096
4.0	13.13110	6.71816	65.14610	66.17633	423.98798	424.58801
4.2	13.12295	6.72026	65.12500	66.15675	423.16475	423.79478
4.4	13.11480	6.72236	65.10390	66.13717	422.34146	423.00149
4.6	13.10664	6.72446	65.08279	66.11758	421.51813	422.20817
4.8	13.09847	6.72656	65.06169	66.09801	420.69486	421.41490
5.0	13.09031	6.72867	65.04056	66.07841	419.87145	420.62149
5.2	13.08214	6.73078	65.01946	66.05883	419.04816	419.82820
5.4	13.07396	6.73289	64.99833	66.03923	418.22473	419.03478
5.6	13.06579	6.73501	64.97721	66.01963	417.40134	418.24139
5.8	13.05761	6.73713	64.95610	66.00005	416.57800	417.44806
6.0	13.04942	6.73925	64.93496	65.98045	415.75454	416.65460
6.2	13.04123	6.74137	64.91379	65.96081	414.93081	415.86087
6.4	13.03303	6.74349	64.89264	65.94119	414.10724	415.06731
6.6	13.02483	6.74562	64.87149	65.92157	413.28364	414.27372
6.8	13.01663	6.74775	64.85032	65.90194	412.45999	413.48006
7.0	13.00843	6.74988	64.82915	65.88230	411.63628	412.68636
7.2	13.00021	6.75201	64.80796	65.86264	410.81245	411.89254
7.4	12.99200	6.75415	64.78680	65.84302	409.98884	411.09893
7.6	12.98378	6.75628	64.76562	65.82338	409.16505	410.30515
7.8	12.97556	6.75843	64.74444	65.80373	408.34128	409.51138
8.0	12.96734	6.76057	64.72323	65.78407	407.51737	408.71748
8.2	12.95911	6.76272	64.70204	65.76441	406.69353	407.92365

(continued 1) $\alpha = 0.3$

λ	A	B	C	D	G	H
8.4	12.95087	6.76486	64.68082	65.74473	405.86952	407.12964
8.6	12.94263	6.76702	64.65961	65.72506	405.04555	406.33568
8.8	12.93439	6.76917	64.63838	65.70538	404.22153	405.54166
9.0	12.92614	6.77133	64.61716	65.68570	403.39751	404.74765
9.2	12.91789	6.77348	64.59593	65.66602	402.57346	403.95360
9.4	12.90963	6.77564	64.57467	65.64631	401.74921	403.15936
9.6	12.90138	6.77781	64.55344	65.62662	400.92510	402.36526
9.8	12.89312	6.77997	64.53219	65.60692	400.10096	401.57113
10.0	12.88485	6.78214	64.51092	65.58720	399.27664	400.77681
11.0	12.84345	6.79302	64.40457	65.48861	395.15505	396.80526
12.0	12.80195	6.80396	64.29805	65.38987	391.03243	392.83268
13.0	12.76035	6.81496	64.19138	65.29101	386.90892	388.85921
14.0	12.71864	6.82603	64.08458	65.19204	382.78454	384.88488
15.0	12.67683	6.83716	63.97762	65.09294	378.65922	380.90961
16.0	12.63492	6.84836	63.87052	64.99372	374.53302	376.93347
17.0	12.59289	6.85962	63.76325	64.89436	370.40576	372.95627
18.0	12.55076	6.87095	63.65584	64.79489	366.27771	368.97828
19.0	12.50853	6.88234	63.54830	64.69531	362.14875	364.99939
20.0	12.46618	6.89380	63.44057	64.59557	358.01874	361.01944
21.0	12.42373	6.90533	63.33271	64.49572	353.88784	357.03862
22.0	12.38117	6.91693	63.22470	64.39574	349.75602	353.05688
23.0	12.33849	6.92859	63.11653	64.29564	345.62327	349.07422
24.0	12.29571	6.94032	63.00821	64.19541	341.48957	345.09060
25.0	12.25282	6.95212	62.89974	64.09506	337.35494	341.10606
26.0	12.20981	6.96400	62.79111	63.99457	333.21932	337.12053
27.0	12.16669	6.97594	62.68233	63.89397	329.08281	333.13412
28.0	12.12346	6.98795	62.57339	63.79322	324.94527	329.14668
29.0	12.08011	7.00004	62.46429	63.69236	320.80682	325.15834
30.0	12.03665	7.01220	62.35505	63.59137	316.66745	321.16908
31.0	11.99308	7.02443	62.24563	63.49024	312.52703	317.17877
32.0	11.94938	7.03673	62.13607	63.38898	308.38568	313.18754
33.0	11.90557	7.04911	62.02634	63.28759	304.24335	309.19533
34.0	11.86165	7.06156	61.91646	63.18608	300.10007	305.20218
35.0	11.81760	7.07409	61.80641	63.08443	295.95578	301.20802
36.0	11.77343	7.08670	61.69621	62.98265	291.81054	297.21291
37.0	11.72915	7.09938	61.58584	62.88074	287.66429	293.21681
38.0	11.68474	7.11214	61.47532	62.77871	283.51709	289.21975
39.0	11.64021	7.12497	61.36464	62.67653	279.36888	285.22169
40.0	11.59556	7.13789	61.25378	62.57423	275.21967	281.22263
41.0	11.55079	7.15088	61.14277	62.47179	271.06948	277.22261
42.0	11.50589	7.16396	61.03160	62.36922	266.91830	273.22158

(continued 2) $\alpha = 0.3$

λ	A	B	C	D	G	H
43.0	11.46087	7.17711	60.92027	62.26652	262.76612	269.21957
44.0	11.41572	7.19035	60.80876	62.16368	258.61291	265.21652
45.0	11.37044	7.20367	60.69709	62.06071	254.45872	261.21251
46.0	11.32504	7.21707	60.58526	61.95760	250.30348	257.20746
47.0	11.27951	7.23055	60.47326	61.85436	246.14727	253.20143
48.0	11.23385	7.24412	60.36109	61.75098	241.99002	249.19438
49.0	11.18806	7.25777	60.24875	61.64746	237.83175	245.18629
50.0	11.14214	7.27151	60.13625	61.54382	233.67245	241.17720
51.0	11.09609	7.28533	60.02358	61.44003	229.51216	237.16711
52.0	11.04991	7.29924	59.91074	61.33611	225.35080	233.15596
53.0	11.00359	7.31324	59.79773	61.23205	221.18842	229.14379
54.0	10.95714	7.32733	59.68455	61.12786	217.02501	225.13060
55.0	10.91055	7.34151	59.57120	61.02352	212.86055	221.11636
56.0	10.86383	7.35577	59.45767	60.91904	208.69504	217.10108
57.0	10.81697	7.37013	59.34397	60.81443	204.52849	213.08477
58.0	10.76997	7.38458	59.23011	60.70968	200.36089	209.06740
59.0	10.72283	7.39912	59.11606	60.60479	196.19224	205.04899
60.0	10.67555	7.41375	59.00184	60.49976	192.02252	201.02953
61.0	10.62813	7.42848	58.88745	60.39459	187.85175	197.00901
62.0	10.58058	7.44331	58.77289	60.28929	183.67993	192.98745
63.0	10.53287	7.45822	58.65814	60.18383	179.50701	188.96479
64.0	10.48502	7.47324	58.54322	60.07824	175.33303	184.94108
65.0	10.43703	7.48835	58.42812	59.97250	171.15797	180.91630
66.0	10.38889	7.50356	58.31284	59.86662	166.98184	176.89045
67.0	10.34061	7.51887	58.19738	59.76061	162.80463	172.86352
68.0	10.29218	7.53428	58.08175	59.65444	158.62633	168.83551
69.0	10.24359	7.54979	57.96593	59.54814	154.44694	164.80642
70.0	10.19486	7.56541	57.84993	59.44169	150.26646	160.77624
71.0	10.14598	7.58112	57.73375	59.33509	146.08488	156.74497
72.0	10.09694	7.59694	57.61739	59.22836	141.90219	152.71259
73.0	10.04776	7.61286	57.50084	59.12148	137.71841	148.67914
74.0	9.99841	7.62889	57.38412	59.01445	133.53353	144.64457
75.0	9.94891	7.64503	57.26720	58.90727	129.34752	140.60889
76.0	9.89926	7.66127	57.15010	58.79995	125.16040	136.57211
77.0	9.84944	7.67762	57.03282	58.69249	120.97216	132.53422
78.0	9.79947	7.69408	56.91535	58.58488	116.78280	128.49520
79.0	9.74934	7.71066	56.79769	58.47712	112.59231	124.45507
80.0	9.69904	7.72734	56.67984	58.36921	108.40069	120.41381
81.0	9.64859	7.74413	56.56181	58.26115	104.20794	116.37143
82.0	9.59796	7.76104	56.44358	58.15295	100.01405	112.32790
83.0	9.54718	7.77806	56.32517	58.04459	95.81903	108.28326

(continued 3) $\alpha = 0{,}3$

λ	A	B	C	D	G	H
84.0	9.49623	7.79520	56.20656	57.93609	91.62285	104.23747
85.0	9.44511	7.81246	56.08776	57.82744	87.42552	100.19053
86.0	9.39382	7.82983	55.96877	57.71864	83.22705	96.14245
87.0	9.34236	7.84732	55.84959	57.60969	79.02741	92.09321
88.0	9.29073	7.86494	55.73021	57.50058	74.82661	88.04282
89.0	9.23893	7.88267	55.61064	57.39133	70.62465	83.99128
90.0	9.18695	7.90053	55.49087	57.28192	66.42151	79.93856
91.0	9.13480	7.91851	55.37091	57.17237	62.21721	75.88469
92.0	9.08247	7.93661	55.25075	57.06266	58.01172	71.82964
93.0	9.02996	7.95484	55.13039	56.95279	53.80506	67.77342
94.0	8.97728	7.97320	55.00983	56.84277	49.59720	63.71601
95.0	8.92441	7.99168	54.88907	56.73260	45.38817	59.65743
96.0	8.87136	8.01029	54.76811	56.62228	41.17793	55.59766
97.0	8.81813	8.02904	54.64695	56.51180	36.96650	51.53669
98.0	8.76471	8.04792	54.52559	56.40117	32.75387	47.47454
99.0	8.71111	8.06693	54.40403	56.29038	28.54003	43.41119
100.0	8.65731	8.08607	54.28226	56.17943	24.32498	39.34663
105.0	8.38548	8.18386	53.67034	55.62236	3.23142	19.00563
110.0	8.10871	8.28521	53.05321	55.06132	–17.89303	–1.36607
115.0	7.82681	8.39032	52.43075	54.49628	–39.04893	–21.76902
120.0	7.53955	8.49936	51.80285	53.92718	–60.23683	–42.20378
125.0	7.24670	8.61254	51.16939	53.35396	–81.45733	–62.67091
130.0	6.94803	8.73010	50.53025	52.77657	–102.71100	–83.17099
135.0	6.64326	8.85225	49.88531	52.19497	–123.99844	–103.70461
140.0	6.33212	8.97925	49.23443	51.60909	–145.32030	–124.27240
145.0	6.01432	9.11137	48.57748	51.01889	–166.67719	–144.87496
150.0	5.68953	9.24890	47.91432	50.42431	–188.06980	–165.51298
155.0	5.35742	9.39214	47.24480	49.82530	–209.49876	–186.18705
160.0	5.01762	9.54143	46.56877	49.22182	–230.96481	–206.89792
165.0	4.66975	9.69714	45.88606	48.61381	–252.46863	–227.64624
170.0	4.31338	9.85963	45.19650	48.00122	–274.01095	–248.43271
175.0	3.94808	10.02933	44.49993	47.38401	–295.59255	–269.25812
180.0	3.57335	10.20669	43.79617	46.76213	–317.21417	–290.12317
185.0	3.18867	10.39219	43.08501	46.13554	–338.87661	–311.02863
190.0	2.79349	10.58636	42.36626	45.50420	–360.58071	–331.97534
195.0	2.38720	10.78978	41.63971	44.86809	–382.32728	–352.96408
200.0	1.96912	11.00305	40.90514	44.22715	–404.11721	–373.99568

$\alpha = 0.4$

λ	A	B	C	D	G	H
0.2	9.93572	5.01619	36.37815	37.38083	183.86944	183.90944
0.4	9.92487	5.01898	36.35641	37.36176	183.23943	183.31943
0.6	9.91401	5.02178	36.33465	37.34269	182.60934	182.72934
0.8	9.90315	5.02459	36.31291	37.32363	181.97934	182.13934
1.0	9.89227	5.02740	36.29112	37.30453	181.34911	181.54912
1.2	9.88138	5.03021	36.26934	37.28543	180.71891	180.95892
1.4	9.87048	5.03303	36.24752	37.26631	180.08855	180.36856
1.6	9.85958	5.03586	36.22573	37.24721	179.45831	179.77833
1.8	9.84867	5.03870	36.20392	37.22810	178.82799	179.18801
2.0	9.83775	5.04154	36.18209	37.20897	178.19758	178.59761
2.2	9.82681	5.04438	36.16024	37.18983	177.56709	178.00712
2.4	9.81587	5.04723	36.13839	37.17069	176.93662	177.41666
2.6	9.80492	5.05009	36.11652	37.15153	176.30604	176.82609
2.8	9.79396	5.05295	36.09465	37.13237	175.67543	176.23549
3.0	9.78299	5.05582	36.07277	37.11321	175.04481	175.64488
3.2	9.77200	5.05869	36.05087	37.09403	174.41409	175.05416
3.4	9.76102	5.06157	36.02897	37.07485	173.78338	174.46347
3.6	9.75001	5.06445	36.00704	37.05565	173.15254	173.87263
3.8	9.73900	5.06735	35.98511	37.03645	172.52170	173.28181
4.0	9.72799	5.07024	35.96318	37.01725	171.89088	172.69099
4.2	9.71696	5.07315	35.94122	36.99803	171.25989	172.10002
4.4	9.70592	5.07606	35.91926	36.97880	170.62891	171.50905
4.6	9.69487	5.07897	35.89729	36.95957	169.99789	170.91804
4.8	9.68381	5.08189	35.87530	36.94033	169.36679	170.32696
5.0	9.67274	5.08482	35.85331	36.92108	168.73565	169.73583
5.2	9.66167	5.08775	35.83130	36.90183	168.10446	169.14466
5.4	9.65058	5.09069	35.80928	36.88256	167.47323	168.55344
5.6	9.63948	5.09363	35.78725	36.86328	166.84192	167.96215
5.8	9.62837	5.09658	35.76521	36.84400	166.21059	167.37083
6.0	9.61726	5.09954	35.74316	36.82472	165.57923	166.77950
6.2	9.60613	5.10251	35.72109	36.80541	164.94777	166.18805
6.4	9.59499	5.10547	35.69902	36.78611	164.31631	165.59661
6.6	9.58385	5.10845	35.67694	36.76680	163.68479	165.00511
6.8	9.57269	5.11143	35.65483	36.74747	163.05318	164.41352
7.0	9.56152	5.11442	35.63272	36.72814	162.42155	163.82191
7.2	9.55035	5.11741	35.61061	36.70881	161.78989	163.23027
7.4	9.53916	5.12041	35.58847	36.68946	161.15815	162.63855
7.6	9.52796	5.12342	35.56633	36.67011	160.52637	162.04679
7.8	9.51676	5.12643	35.54418	36.65074	159.89453	161.45498
8.0	9.50554	5.12945	35.52201	36.63137	159.26266	160.86313
8.2	9.49431	5.13247	35.49984	36.61199	158.63073	160.27122

(continued 1) $\alpha = 0.4$

λ	A	B	C	D	G	H
8.4	9.48307	5.13550	35.47764	36.59260	157.99874	159.67926
8.6	9.47183	5.13854	35.45545	36.57321	157.36671	159.08726
8.8	9.46057	5.14159	35.43323	36.55380	156.73463	158.49520
9.0	9.44930	5.14464	35.41101	36.53439	156.10251	157.90311
9.2	9.43802	5.14769	35.38878	36.51497	155.47033	157.31095
9.4	9.42673	5.15075	35.36653	36.49554	154.83808	156.71873
9.6	9.41544	5.15382	35.34427	36.47610	154.20580	156.12648
9.8	9.40413	5.15690	35.32201	36.45666	153.57348	155.53419
10.0	9.39281	5.15998	35.29972	36.43721	152.94109	154.94183
11.0	9.33606	5.17549	35.18815	36.33981	149.77841	151.97931
12.0	9.27904	5.19115	35.07628	36.24221	146.61447	149.01554
13.0	9.22177	5.20699	34.96412	36.14441	143.44924	146.05051
14.0	9.16422	5.22299	34.85167	36.04639	140.28273	143.08421
15.0	9.10641	5.23916	34.73892	35.94816	137.11492	140.11663
16.0	9.04832	5.25550	34.62587	35.84972	133.94582	137.14777
17.0	8.98995	5.27202	34.51251	35.75107	130.77540	134.17760
18.0	8.931331	5.28872	34.39886	35.65220	127.60367	131.20615
19.0	8.87238	5.30560	34.28491	35.55312	124.43062	128.23341
20.0	8.81317	5.32266	34.17064	35.45383	121.25624	125.25933
21.0	8.75367	5.33990	34.05607	35.35432	118.08051	122.28394
22.0	8.69387	5.35734	33.94118	35.25459	114.90345	119.30723
23.0	8.63378	5.37496	33.82599	35.15465	111.72504	116.32919
24.0	8.57339	5.39278	33.71047	35.05449	108.54526	113.34980
25.0	8.51270	5.41080	33.59464	34.95410	105.36411	110.36906
26.0	8.45170	5.42902	33.47850	34.85350	102.18160	107.38697
27.0	8.39039	5.44744	33.36203	34.75268	98.99769	104.40350
28.0	8.32877	5.46607	33.24523	34.65164	95.81239	101.41867
29.0	8.26682	5.48490	33.12811	34.55037	92.62569	98.43246
30.0	8.20456	5.50395	33.01066	34.44888	89.43759	95.44486
31.0	8.14197	5.52322	32.89288	34.34716	86.24806	92.45587
32.0	8.07905	5.54270	32.77477	34.24522	83.05711	89.46546
33.0	8.01580	5.56241	32.65632	34.14305	79.86472	86.47365
34.0	7.95221	5.58234	32.53754	34.04066	76.67090	83.48042
35.0	7.88828	5.60251	32.41841	33.93804	73.47562	80.48575
36.0	7.82400	5.62290	32.29895	33.83519	70.27889	77.48966
37.0	7.75937	5.64354	32.17913	33.73211	67.08069	74.49211
38.0	7.69438	5.66441	32.05897	33.62879	63.88101	71.49312
39.0	7.62904	5.68553	31.93846	33.52525	60.67985	68.49266
40.0	7.56333	5.70690	31.81760	33.42147	57.47719	65.49073
41.0	7.49725	5.72853	31.69638	33.31746	54.27303	62.48732
42.0	7.43080	5.75040	31.57481	33.21322	51.06736	59.38242

(continued 2) $\alpha = 0.4$

λ	A	B	C	D	G	H
43.0	7.36397	5.77255	31.45287	33.10874	47.86017	56.47603
44.0	7.29675	5.79495	31.33057	33.00403	44.65144	53.46813
45.0	7.22915	5.81763	31.20790	32.89907	41.44118	50.45872
46.0	7.16115	5.84058	31.08487	32.79388	38.22937	47.44779
47.0	7.09275	5.86381	30.96146	32.68846	35.01600	44.43532
48.0	7.02395	5.88732	30.83768	32.58279	31.80106	41.42131
49.0	6.95474	5.91112	30.71353	32.47688	28.58454	38.40575
50.0	6.88511	5.93522	30.58899	32.37073	25.36644	35.38863
51.0	6.81506	5.95961	30.46407	32.26434	22.14674	32.36994
52.0	6.74459	5.98431	30.33876	32.15771	18.92543	29.34967
53.0	6.67368	6.00931	30.21307	32.05083	15.70251	26.32782
54.0	6.60233	6.03463	30.08698	31.94371	12.47796	23.30436
55.0	6.53053	6.06027	29.96050	31.83634	9.25177	20.27930
56.0	6.45828	6.08623	29.83362	31.72872	6.02394	17.25262
57.0	6.38558	6.11252	29.70634	31.62086	2.79445	14.22432
58.0	6.31241	6.13915	29.57865	31.51276	–0.43671	11.19437
59.0	6.23876	6.16612	29.45055	31.40440	–3.66955	8.16278
60.0	6 16464	6.19344	29.32205	31.29579	–6.90408	5.12953
61.0	6.09004	6.22112	29.19313	31.18694	–10.14030	2.09462
62.0	6.01494	6.24915	29.06379	31.07783	–13.37824	–0.94197
63.0	5.93934	6.27755	28.93403	30.96847	–16.61790	–3.98025
64.0	5.86323	6.30633	28.80385	30.85886	–19.85929	–7.02024
65.0	5.78661	6.33548	28.67323	30.74900	–23.10243	–10.06193
66.0	5.70946	6.36503	28.54219	30.63888	–26.34731	–13.10533
67.0	5.63179	6.39497	28.41071	30.52851	–29.59397	–16.15047
68.0	5.55357	6.42531	28.27879	30.41788	–32.84240	–19.19735
69.0	5.47481	6.45606	28.14642	30.30700	–36.09262	–22.24599
70.0	5.39549	6.48723	28.01361	30.19586	–39.34465	–25.29638
71.0	5.31561	6.51882	27.88035	30.08446	–42.59848	–28.34855
72.0	5.23515	6.55084	27.74664	29.97281	–45.85414	–31.40251
73.0	5.15411	6.58331	27.61246	29.86089	–49.11163	–34.45826
74.0	5.07248	6.61622	27.47782	29.74872	–52.37097	–37.51582
75.0	4.99025	6.64960	27.34272	29.63629	–55.63218	–40.57520
76.0	4.90741	6.68344	27.20714	29.52360	–58.89525	–43.63641
77.0	4.82394	6.71775	27.07109	29.41065	–62.16021	–46.69947
78.0	4.73984	6.75256	26.93456	29.29744	–65.42707	–49.76438
79.0	4.65511	6.78785	26.79755	29.18396	–68.69584	–52.83115
80.0	4.56972	6.82365	26.66004	29.07023	–71.96653	–55.89981
81.0	4.48366	6.85997	26.52205	28.95623	–75.23916	–58.97035
82.0	4.39693	6.89681	26.38356	28.84197	–78.51374	–62.04281
83.0	4.30952	6.93418	26.24456	28.72744	–81.79027	–65.11717

(continued 3) $\alpha = 0.4$

λ	A	B	C	D	G	H
84.0	4.22141	6.97210	26.10506	28.61266	–85.06879	–68.19346
85.0	4.13259	7.01058	25.96504	28.49760	–88.34929	–71.27169
86.0	4.04304	7.04962	25.82451	28.38229	–91.63180	–74.35188
87.0	3.95276	7.08925	25.68346	28.26671	–94.91632	–77.43403
88.0	3.86173	7.12946	25.54188	28.15087	–98.20287	–80.51816
89.0	3.76995	7.17028	25.39976	28.03476	–101.49147	–83.60428
90.0	3.67738	7.21172	25.25712	27.91838	–104.78212	–86.69241
91.0	3.58403	7.25379	25.11392	27.80174	–108.07485	–89.78255
92.0	3.48987	7.29650	24.97018	27.68484	–111.36966	–92.87473
93.0	3.39490	7.33986	24.82589	27.56767	–114.66658	–95.96895
94.0	3.29909	7.38390	24.68104	27.45024	–117.96561	–99.06523
95.0	3.20244	7.42862	24.53562	27.33254	–121.26677	–102.16359
96.0	3.10492	7.47404	24.38963	27.21458	–124.57008	–105.26402
97.0	3.00652	7.52017	24.24307	27.09635	–127.87556	–108.36657
98.0	2.90722	7.56704	24.09592	26.97786	–131.18320	–111.47122
99.0	2.80701	7.61465	23.94818	26.85911	–134.49304	–114.57800
100.0	2.70586	7.66303	23.79985	26.74009	–137.80508	–117.68692
105.0	2.18549	7.91696	23.04902	26.14107	–154.39899	–133.26418
110.0	1.63885	8.19275	22.28230	25.53557	–171.05054	–148.89725
115.0	1.06307	8.49310	21.49871	24.92372	–187.76188	–164.58805
120.0	0.45485	8.82114	20.69721	24.30574	–204.53529	–180.33862
125.0	–0.18968	9.18056	19.87661	23.68191	–221.37319	–196.15107
130.0	–0.87506	9.57573	19.03557	23.05264	–238.27811	–212.02761
135.0	–1.60665	10.01180	18.17256	22.41846	–255.25281	–227.97059
140.0	–2.39086	10.49498	17.28583	21.78007	–272.30017	–243.98240
145.0	–3.23536	11.03272	16.37333	21.13842	–289.42328	–260.06556
150.0	–4.14949	11.63413	15.43269	20.49471	–306.62545	–276.22269
155.0	–5.14468	12.31039	14.46106	19.85055	–323.91019	–292.45639
160.0	–6.23509	13.07540	13.45506	19.20804	–341.28144	–308.76956
165.0	–7.43850	13.94669	12.41053	18.56993	–358.74329	–325.16491
170.0	–8.77755	14.94659	11.32234	17.93990	–376.30031	–341.64531
175.0	–10.28151	16.10408	10.18401	17.32288	–393.95761	–358.21363
180.0	–11.98896	17.45740	8.98720	16.72555	–411.72078	–374.87250
185.0	–13.95174	19.05805	7.72091	16.15718	–429.59624	–391.62439
190.0	–16.24132	20.97712	6.37027	15.63077	–447.59145	–408.47131
195.0	–18.95896	23.31552	4.91448	15.16513	–465.71532	–425.41439
200.0	–22.25323	26.22135	3.32346	14.78825	–483.97883	–442.45327

$\alpha = 0.5$

λ	A	B	C	D	G	H
0.2	7.91944	4.02037	22.87711	23.88130	93.08182	93.13182
0.4	7.90575	4.02395	22.85438	23.86276	92.56428	92.66428
0.6	7.89205	4.02755	22.83164	23.84422	92.04671	92.19672
0.8	7.87832	4.03115	22.80887	23.82567	91.52904	91.72906
1.0	7.86457	4.03477	22.78608	23.80709	91.01131	91.26133
1.2	7.85081	4.03840	22.76328	23.78852	90.49355	90.79359
1.4	7.83703	4.04204	22.74046	23.76993	89.97570	90.32575
1.6	7.82322	4.04569	22.71762	23.75133	89.45780	89.85786
1.8	7.80940	4.04935	22.69476	23.73272	88.93984	89.38991
2.0	7.79556	4.05303	22.67188	23.71409	88.42181	88.92190
2.2	7.78170	4.05671	22.64899	23.69546	87.90372	88.45383
2.4	7.76782	4.06041	22.62607	23.67681	87.38556	87.98569
2.6	7.75392	4.06412	22.60313	23.65814	86.86733	87.51748
2.8	7.74001	4.06784	22.58018	23.63948	86.34906	87.04923
3.0	7.72607	4.07158	22.55721	23.62079	85.83071	86.58091
3.2	7.71211	4.07532	22.53421	23.60210	85.31230	86.11253
3.4	7.69814	4.07908	22.51120	23.58340	84.79384	85.64410
3.6	7.68414	4.08285	22.48817	23.56468	84.27530	85.17559
3.8	7.67013	4.08663	22.46512	23.54595	83.75670	84.70702
4.0	7.65609	4.09043	22.44205	23.52721	83.23804	84.23840
4.2	7.64204	4.09424	22.41896	23.50846	82.71932	83.76971
4.4	7.62796	4.09806	22.39585	23.48970	82.20052	83.30095
4.6	7.61387	4.10189	22.37273	23.47093	81.68167	82.83215
4.8	7.59975	4.10573	22.34958	23.45214	81.16276	82.36327
5.0	7.58562	4.10959	22.32641	23.43334	80.64377	81.89433
5.2	7.57146	4.11346	22.30323	23.41454	80.12472	81.42533
5.4	7.55728	4.11734	22.28002	23.39572	79.60561	80.95627
5.6	7.54309	4.12124	22.25680	23.37688	79.08643	80.48714
5.8	7.52887	4.12514	22.23355	23.35804	78.56720	80.01796
6.0	7.51464	4.12906	22.21029	23.33919	78.04789	79.54870
6.2	7.50038	4.13300	22.18701	23.32032	77.52851	79.07939
6.4	7.48610	4.13694	22.16370	23.30144	77.00908	78.61001
6.6	7.47180	4.14090	22.14038	23.28255	76.48958	78.14057
6.8	7.45748	4.14487	22.11703	23.26365	75.97001	77.67106
7.0	7.44314	4.14886	22.09367	23.24473	75.45038	77.20150
7.2	7.42878	4.15286	22.07029	23.22581	74.93068	76.73187
7.4	7.41440	4.15687	22.04689	23.20687	74.41092	76.26217
7.6	7.40000	4.16089	22.02346	23.18792	73.89109	75.79241
7.8	7.38557	4.16493	22.00002	23.16896	73.37119	75.32259
8.0	7.37113	4.16898	21.97656	23.14999	72.85123	74.85270
8.2	7.35666	4.17305	21.95307	23.13101	72.33121	74.38275

(continued 1) $\alpha = 0.5$

λ	A	B	C	D	G	H
8.4	7.34217	4.17712	21.92957	23.11201	71.81111	73.91274
8.6	7.32766	4.18122	21.90604	23.09300	71.29095	73.44265
8.8	7.31313	4.18532	21.88250	23.07399	70.77072	72.97251
9.0	7.29858	4.18944	21.85894	23.05495	70.25043	72.50230
9.2	7.28401	4.19357	21.83535	23.03591	69.73007	72.03203
9.4	7.26941	4.19772	21.81174	23.01686	69.20964	71.56169
9.6	7.25479	4.20188	21.78812	22.99779	68.68915	71.09128
9.8	7.24015	4.20605	21.76447	22.97871	68.16859	70.62082
10.0	7.22549	4.21024	21.74080	22.95962	67.64796	70.15028
11.0	7.15186	4.23140	21.62216	22.86399	65.04380	67.79664
12.0	7.07766	4.25291	21.50299	22.76806	62.43795	65.44134
13.0	7.00289	4.27479	21.38330	22.67184	59.83038	63.08439
14.0	6.92754	4.29705	21.26308	22.57531	57.22108	60.72576
15.0	6.85160	4.31970	21.14231	22.47848	54.61003	58.36544
16.0	6.77505	4.34273	21.02100	22.38134	51.99723	56.00343
17.0	6.69789	4.36618	20.89914	22.28390	49.38264	53.63969
18.0	6.62011	4.39004	20.77672	22.18616	46.76626	51.27422
19.0	6.54169	4.41432	20.65374	22.08810	44.14807	48.90700
20.0	6.46262	4.43903	20.53018	21.98974	41.52805	46.53801
21.0	6.38288	4.46420	20.40605	21.89107	38.90619	44.16725
22.0	6.30247	4.48982	20.28133	21.79209	36.28247	41.79469
23.0	6.22137	4.51590	20.15602	21.69279	33.65687	39.42032
24.0	6.13956	4.54247	20.03011	21.59319	31.02936	37.04412
25.0	6.05704	4.56953	19.90360	21.49327	28.39995	34.66608
26.0	5.97379	4.59709	19.77647	21.39303	25.76860	32.28618
27.0	5.88979	4.62517	19.64872	21.29248	23.13530	29.90439
28.0	5.80503	4.65378	19.52034	21.19161	20.50003	27.52072
29.0	5.71949	4.68294	19.39132	21.09042	17.86277	25.13513
30.0	5.63315	4.71265	19.26165	20.98891	15.22350	22.74761
31.0	5.54600	4.74294	19.13133	20.88709	12.58220	20.35815
32.0	5.45802	4.77381	19.00034	20.78494	9.93885	17.96671
33.0	5.36919	4.80530	18.86869	20.68248	7.29343	15.57330
34.0	5.27950	4.83740	18.73635	20.57969	4.64593	13.17788
35.0	5.18892	4.87014	18.60332	20.47658	1.99631	10.78044
36.0	5.09744	4.90354	18.46960	20.37315	–0.65543	8.38097
37.0	5.00502	4.93761	18.33516	20.26940	–3.30933	5.97943
38.0	4.91167	4.97237	18.20001	20.16532	–5.96540	3.57582
39.0	4.81734	5.00784	18.06412	20.06093	–8.62367	1.17011
40.0	4.72202	5.04405	17.92750	19.95621	–11.28415	–1.23772
41.0	4.62568	5.08101	17.79012	19.85116	–13.94688	–3.64769
42.0	4.52830	5.11874	17.65199	19.74580	–16.61187	–6.05981

(continued 2) $\alpha = 0.5$

λ	*A*	*B*	*C*	*D*	*G*	*H*
43.0	4.42986	5.15728	17.51308	19.64012	−19.27915	−8.47411
44.0	4.33033	5.19663	17.37339	19.53411	−21.94874	−10.89061
45.0	4.22968	5.23683	17.23290	19.42778	−24.62067	−13.30933
46.0	4.12788	5.27789	17.09161	19.32113	−27.29495	−15.73030
47.0	4.02491	5.31986	16.94949	19.21417	−29.97162	−18.15352
48.0	3.92073	5.36274	16.80655	19.10689	−32.65070	−20.57903
49.0	3.81531	5.40658	16.66275	18.99929	−35.33221	−23.00685
50.0	3.70863	5.45140	16.51810	18.89137	−38.01619	−25.43700
51.0	3.60065	5.49723	16.37258	18.78314	−40.70266	−27.86950
52.0	3.49133	5.54410	16.22617	18.67461	−43.39163	−30.30438
53.0	3.38064	5.59205	16.07886	18.56576	−46.08316	−32.74166
54.0	3.26854	5.64111	15.93063	18.45660	−48.77725	−35.18136
55.0	3.15499	5.69131	15.78147	18.34714	−51.47394	−37.62351
56.0	3.03995	5.74270	15.63136	18.23738	−54.17326	−40.06813
57.0	2.92338	5.79531	15.48029	18.12733	−56.87524	−42.51524
58.0	2.80524	5.84918	15.32824	18.01697	−59.57990	−44.96488
59.0	2.68547	5.90435	15.17519	17.90633	−62.28728	−47.41706
60.0	2.56404	5.96088	15.02112	17.79540	−64.99741	−49.87181
61.0	2.44089	6.01880	14.86602	17.68419	−67.71031	−52.32916
62.0	2.31597	6.07816	14.70986	17.57270	−70.42603	−54.78913
63.0	2.18923	6.13901	14.55263	17.46094	−73.14459	−57.25175
64.0	2.06061	6.20141	14.39431	17.34891	−75.86603	−59.71705
65.0	1.93006	6.26541	14.23487	17.23662	−78.59038	−62.18505
66.0	1.79750	6.33107	14.07429	17.12409	−81.31767	−64.65579
67.0	1.66289	6.39844	13.91256	17.01130	−84.04795	−67.12928
68.0	1.52614	6.46760	13.74964	16.89828	−86.78123	−69.60555
69.0	1.38720	6.53860	13.58551	16.78502	−89.51757	−72.08464
70.0	1.24598	6.61152	13.42015	16.67155	−92.25700	−74.56658
71.0	1.10242	6.68642	13.25353	16.55787	−94.99955	−77.05139
72.0	0.95643	6.76340	13.08562	16.44399	−97.74527	−79.53910
73.0	0.80793	6.84251	12.91640	16.32992	−100.49419	−82.02973
74.0	0.65682	6.92385	12.74584	16.21567	−103.24636	−84.52333
75.0	0.50303	7.00751	12.57390	16.10126	−106.00181	−87.01993
76.0	0.34644	7.09358	12.40055	15.98671	−108.76059	−89.51955
77.0	0.18696	7.18216	12.22577	15.87202	−111.52273	−92.02221
78.0	0.02447	7.27335	12.04951	15.75721	−114.28828	−94.52796
79.0	−0.14112	7.36725	11.87174	15.64230	−117.05730	−97.03683
80.0	−0.30995	7.46400	11.69242	15.52732	−119.82982	−99.54885
81.0	−0.48213	7.56369	11.51152	15.41227	−122.60588	−102.06403
82.0	−0.65781	7.66647	11.32898	15.29718	−125.38554	−104.58244
83.0	−0.83712	7.77246	11.14478	15.18208	−128.16884	−107.10408

(continued 3) $\alpha = 0.5$

λ	A	B	C	D	G	H
84.0	−1.02021	7.88182	10.95886	15.06699	−130.95583	−109.62900
85.0	−1.20723	7.99469	10.77117	14.95194	−133.74657	−112.15723
86.0	−1.39835	8.11123	10.58167	14.83695	−136.54111	−114.68881
87.0	−1.59375	8.23161	10.39030	14.72206	−139.33950	−117.22376
88.0	−1.79360	8.35602	10.19701	14.60729	−142.14179	−119.76211
89.0	−1.99812	8.48463	10.00174	14.49270	−144.94804	−122.30390
90.0	−2.20750	8.61767	9.80443	14.37831	−147.75831	−124.84918
91.0	−2.42197	8.75534	9.60502	14.26417	−150.57265	−127.39794
92.0	−2.64177	8.89787	9.40343	14.15032	−153.39113	−129.95026
93.0	−2.86714	9.04551	9.19960	14.03680	−156.21382	−132.50616
94.0	−3.09836	9.19853	8.99345	13.92369	−159.04077	−135.06567
95.0	−3.33571	9.35720	8.78490	13.81102	−161.87205	−137.62881
96.0	−3.57950	9.52182	8.57386	13.69886	−164.70772	−140.19562
97.0	−3.83005	9.69272	8.36024	13.58728	−167.54787	−142.76616
98.0	−4.08772	9.87024	8.14394	13.47634	−170.39256	−145.34042
99.0	−4.35288	10.05474	7.92487	13.36613	−173.24186	−147.91844
100.0	−4.62594	10.24663	7.70291	13.25672	−176.09585	−150.50027
105.0	−6.12604	11.33303	6.54536	12.72527	−190.43916	−163.46742
110.0	−7.90551	12.68501	5.29378	12.23158	−204.91329	−176.53353
115.0	−10.06926	14.40653	3.92094	11.79897	−219.53232	−189.70043
120.0	−12.78371	16.66296	2.38457	11.46540	−234.31472	−202.96720
125.0	−16.32979	19.73406	0.61450	11.29631	−249.28656	−216.32710
130.0	−21.22222	24.13328	−1.51621	11.41343	−264.48853	−229.76067
135.0	−28.51730	30.91550	−4.26576	12.06918	−279.99423	−243.21804
140.0	−40.78088	42.64499	−8.25774	13.88074	−295.96680	−256.56223
145.0	−66.28800	67.59506	−15.49693	18.84572	−312.90483	−269.32382
150.0	−154.83218	155.55727	−38.03016	39.00299	−333.73730	−278.60641
155.0	625.67054	−625.55454	149.57270	−151.08653	−303.95872	−338.95198
160.0	107.70509	−108.22774	23.07053	−27.19174	−350.38150	−323.58888
165.0	59.59872	−60.79238	10.14663	−17.00727	−369.58783	−335.98147
170.0	41.23186	−43.13202	4.34739	−14.09236	−387.34918	−350.41047
175.0	31.36083	−34.00655	0.52793	−13.31670	−404.93798	−365.66315
180.0	25.07263	−28.50706	−2.50783	−13.50087	−422.67543	−381.48617
185.0	20.62085	−24.89181	−5.19127	−14.23259	−440.70256	−397.81652
190.0	17.22723	−22.38793	−7.72098	−15.33531	−459.10976	−414.65383
195.0	14.49125	−20.60116	−10.20585	−16.72566	−477.97234	−432.02694
200.0	12.18456	−19.31045	−12.71509	−18.36419	−497.36419	−449.98344

VALUES OF THE FUNCTIONS *A* TO *H*

$a = 0.6$

λ	A	B	C	D	G	H
0.2	6.56970	3.35797	15.54250	16.54854	53.10973	53.16973
0.4	6.55308	3.36241	15.51856	16.53066	52.66469	52.78470
0.6	6.53643	3.36686	15.49459	16.51277	52.21956	52.39958
0.8	6.51975	3.37134	15.47059	16.49486	51.77437	52.01440
1.0	6.50303	3.37583	15.44656	16.47693	51.32907	51.62912
1.2	6.48627	3.38035	15.42250	16.45900	50.88372	51.24380
1.4	6.46949	3.38488	15.39841	16.44104	50.43828	50.85838
1.6	6.45267	3.38944	15.37429	16.42308	49.99277	50.47291
1.8	6.43581	3.39402	15.35014	16.40510	49.54718	50.08736
2.0	6.41892	3.39862	15.32596	16.38710	49.10150	49.70172
2.2	6.40200	3.40325	15.30175	16.36909	48.65575	49.31602
2.4	6.38504	3.40789	15.27751	16.35107	48.20992	48.93024
2.6	6.36805	3.41256	15.25324	16.33303	47.76401	48.54438
2.8	6.35102	3.41724	15.22894	16.31497	47.31801	48.15845
3.0	6.33396	3.42196	15.20460	16.29691	46.87194	47.77245
3.2	6.31686	3.42669	15.18024	16.27883	46.42579	47.38637
3.4	6.29972	3.43144	15.15584	16.26073	45.97956	47.00021
3.6	6.28255	3.43622	15.13142	16.24262	45.53325	46.61398
3.8	6.26534	3.44102	15.10696	16.22449	45.08686	46.22767
4.0	6.24810	3.44584	15.08247	16.20635	44.64038	45.84129
4.2	6.23082	3.45069	15.05795	16.18820	44.19383	45.45483
4.4	6.21351	3.45556	15.03340	16.17003	43.74719	45.06829
4.6	6.19615	3.46045	15.00882	16.15185	43.30047	44.68167
4.8	6.17877	3.46536	14.98420	16.13365	42.85367	44.29498
5.0	6.16134	3.47030	14.95955	16.11544	42.40679	43.90821
5.2	6.14388	3.47526	14.93487	16.09721	41.95982	43.52137
5.4	6.12637	3.48025	14.91016	16.07897	41.51278	43.13445
5.6	6.10884	3.48526	14.88542	16.06071	41.06565	42.74744
5.8	6.09126	3.49030	14.86064	16.04244	40.61844	42.36037
6.0	6.07365	3.49535	14.83584	16.02416	40.17114	41.97321
6.2	6.05599	3.50044	14.81099	16.00586	39.72376	41.58598
6.4	6.03830	3,50555	14.78612	15.98755	39.27630	41.19866
6.6	6.02057	3.51068	14.76122	15.96922	38.82875	40.81127
6.8	6.00280	3.51584	14.73628	15.95087	38.38112	40.42380
7.0	5.98500	3.52102	14.71130	15.93252	37.93340	40.03625
7.2	5.96715	3.52623	14.68630	15.91414	37.48560	39.64862
7.4	5.94926	3.53146	14.66126	15.89576	37.03772	39.26091
7.6	5.93134	3.53672	14.63619	15.87736	36.58975	38.87313
7.8	5.91337	3.54201	14.61108	15.85894	36.14170	38.48526
8.0	5.89537	3.54732	14.58595	15.84051	35.69356	38.09731
8.2	5.87732	3.55266	14.56077	15.82206	35.24533	37.70928

(continued 1) $\alpha = 0.6$

λ	A	B	C	D	G	H
8.4	5.85924	3.55802	14.53557	15.80360	34.79702	37.32117
8.6	5.84111	3.56341	14.51033	15.78513	34.34862	36.93298
8.8	5.82294	3.56883	14.48505	15.76664	33.90014	36.54472
9.0	5.80473	3.57427	14.45975	15.74813	33.45157	36.15637
9.2	5.78648	3.57974	14.43440	15.72962	33.00291	35.76793
9.4	5.76819	3.58524	14.40903	15.71108	32.55416	35.37942
9.6	5.74985	3.59077	14.38361	15.69253	32.10533	34.99082
9.8	5.73148	3.59632	14.35817	15.67397	31.65642	34.60215
10.0	5.71306	3.60190	14.33269	15.65539	31.20741	34.21339
11.0	5.62033	3.63023	14.20476	15.56228	28.96105	32.26836
12.0	5.52651	3.65929	14.07594	15.46881	26.71247	30.32126
13.0	5.43156	3.68910	13.94622	15.37497	24.46162	28.37204
14.0	5.33546	3.71968	13.81558	15.28075	22.20849	26.42070
15.0	5.23818	3.75107	13.68400	15.18618	19.95304	24.46720
16.0	5.13968	3.78330	13.55147	15.09123	17.69523	22.51151
17.0	5.03992	3.81639	13.41796	14.99592	15.43504	20.55361
18.0	4.93886	3.85038	13.28346	14.90024	13.17242	18.59347
19.0	4.83647	3.88530	13.14795	14.80420	10.90735	16.63106
20.0	4.73271	3.92119	13.01140	14.70780	8.63979	14.66636
21.0	4.62753	3.95809	12.87380	14.61104	6.36971	12.69932
22.0	4.52089	3.99603	12.73513	14.51391	4.09706	10.72993
23.0	4.41275	4.03505	12.59535	14.41644	1.82181	8.75815
24.0	4.30304	4.07520	12.45446	14.31861	−0.45607	6.78396
25.0	4.19173	4.11652	12.31242	14.22043	−2.73664	4.80731
26.0	4.07876	4.15906	12.16921	14.12190	−5.01992	2.82818
27.0	3.96406	4.20288	12.02481	14.02304	−7.30596	0.84653
28.0	3.84759	4.24801	11.87918	13.92384	−9.59480	−1.13766
29.0	3.72929	4.29453	11.73230	13.82431	−11.88649	−3.12443
30.0	3.60907	4.34248	11.58414	13.72445	−14.18107	−5.11382
31.0	3.48689	4.39193	11.43467	13.62428	−16.47857	−7.10585
32.0	3.36266	4.44294	11.28386	13.52380	−18.77906	−9.10057
33.0	3.23631	4.49559	11.13168	13.42302	−21.08258	−11.09801
34.0	3.10776	4.54994	10.97808	13.32195	−23.38917	−13.09821
35.0	2.97692	4.60608	10.82304	13.22060	−25.69889	−15.10120
36.0	2.84371	4.66408	10.66652	13.11899	−28.01179	−17.10702
37.0	2.70803	4.72403	10.50848	13.01711	−30.32791	−19.11570
38.0	2.56978	4.78602	10.34888	12.91500	−32.64732	−21.12730
39.0	2.42886	4.85016	10.18767	12.81265	−34.97007	−23.14183
40.0	2.28515	4.91653	10.02482	12.71010	−37.29622	−25.15935
41.0	2.13854	4.98526	9.86027	12.60735	−39.62581	−27.17990
42.0	1.98890	5.05646	9.69397	12.50444	−41.95893	−29.20351

(continued 2) $\alpha = 0.6$

λ	*A*	*B*	*C*	*D*	*G*	*H*
43.0	1.83610	5.13026	9.52587	12.40137	−44.29562	−31.23022
44.0	1.68000	5.20678	9.35593	12.29818	−46.63595	−33.26009
45.0	1.52044	5.28617	9.18407	12.19489	−48.97998	−35.29313
46.0	1.35727	5.36858	9.01025	12.09153	−51.32779	−37.32941
47.0	1.19031	5.45418	8.83439	11.98813	−53.67945	−39.36896
48.0	1.01937	5.54313	8.65643	11.88472	−56.03501	−41.41183
49.0	0.84428	5.63562	8.47629	11.78135	−58.39457	−43.45806
50.0	0.66481	5.73185	8.29391	11.67804	−60.75820	−45.50768
51.0	0.48074	5.83204	8.10919	11.57486	−63.12596	−47.56075
52.0	0.29183	5.93642	7.92206	11.47185	−65.49796	−49.61731
53.0	0.09782	6.04522	7.73243	11.36905	−67.87427	−51.67740
54.0	−0.10157	6.15873	7.54019	11.26654	−70.25498	−53.74107
55.0	−0.30664	6.27722	7.34524	11.16437	−72.64018	−55.80836
56.0	−0.51770	6.40102	7.14748	11.06261	−75.02997	−57.87930
57.0	−0.73512	6.53046	6.94677	10.96135	−77.42445	−59.95396
58.0	−0.95927	6.66590	6.74300	10.86067	−79.82371	−62.03235
59.0	−1.19056	6.80774	6.53603	10.76066	−82.22787	−64.11454
60.0	−1.42944	6.95643	6.32570	10.66142	−84.63704	−66.20055
61.0	−1.67639	7.11242	6.11187	10.56308	−87.05133	−68.29042
62.0	−1.93194	7.27624	5.89435	10.46576	−89.47086	−70.38420
63.0	−2.19668	7.44845	5.67296	10.36959	−91.89578	−72.48191
64.0	−2.47123	7.62968	5.44750	10.27474	−94.32621	−74.58360
65.0	−2.75630	7.82059	5.21774	10.18137	−96.76229	−76.68928
66.0	−3.05264	8.02195	4.98344	10.08967	−99.20417	−78.79899
67.0	−3.36110	8.23458	4.74433	9.99987	−101.65201	−80.91274
68.0	−3.68261	8.45940	4.50013	9.91219	−104.10598	−83.03055
69.0	−4.01821	8.69741	4.25052	9.82690	−106.56626	−85.15244
70.0	−4.36904	8.94976	3.99513	9.74430	−109.03303	−87.27841
71.0	−4.73637	9.21770	3.73358	9.66472	−111.50652	−89.40846
72.0	−5.12164	9.50265	3.46543	9.58854	−113.98692	−91.54256
73.0	−5.52646	9.80619	3.19020	9.51619	−116.47447	−93.68069
74.0	−5.95261	10.13010	2.90734	9.44814	−118.96944	−95.82283
75.0	−6.40215	10.47641	2.61623	9.38494	−121.47210	−97.96891
76.0	−6.87738	10.84739	2.31619	9.32722	−123.98276	−100.11886
77.0	−7.38092	11.24567	2.00643	9.27570	−126.50173	−102.27257
78.0	−7.91579	11.67423	1.68606	9.23119	−129.02939	−104.42993
79.0	−8.48544	12.13650	1.35406	9.19465	−131.56615	−106.59078
80.0	−9.09386	12.63645	1.00926	9.16718	−134.11247	−108.75493
81.0	−9.74569	13.17869	0.65028	9.15005	−136.66882	−110.92210
82.0	−10.44633	13.76861	0.27556	9.14478	−139.23582	−113.09202
83.0	−11.20213	14.41254	−0.11678	9.15315	−141.81410	−115.26428

(continued 3) $\alpha = 0.6$

λ	A	B	C	D	G	H
84.0	–12.02064	15.11799	–0.52891	9.17725	–144.40444	–117.43843
85.0	–12.91083	15.89391	–0.96345	9.21962	–147.00769	–119.61386
86.0	–13.88347	16.75106	–1.42352	9.28328	–149.62493	–121.78988
87.0	–14.95167	17.70249	–1.91291	9.37195	–152.25734	–123.96553
88.0	–16.13144	18.76421	–2.43624	9.49014	–154.90641	–126.13967
89.0	–17.44263	19.95605	–2.99925	9.64349	–157.57392	–128.31085
90.0	–18.91014	21.30285	–3.60909	9.83907	–160.26209	–130.47721
91.0	–20.56557	22.83621	–4.27486	10.08587	–162.97365	–132.63630
92.0	–22.44974	24.59690	–5.00827	10.39549	–165.71216	–134.78501
93.0	–24.61623	26.63847	–5.82472	10.78323	–168.48218	–136.91912
94.0	–27.13685	29.03271	–6.74480	11.26955	–171.28977	–139.03296
95.0	–30.11006	31.87803	–7.79673	11.88257	–174.14318	–141.11863
96.0	–33.67466	35.31322	–9.02027	12.66194	–177.05396	–143.16499
97.0	–38.03280	39.54036	–10.47332	13.66542	–180.03892	–145.15574
98.0	–43.49071	44.86567	–12.24356	14.98057	–183.12329	–147.06599
99.0	–50.53571	51.77643	–14.47046	16.74674	–186.34721	–148.85607
100.0	–59.99317	61.09797	–17.39023	19.19999	–189.77845	–150.45868
105.0	–474.32075	474.71913	–137.04378	136.42914	–237.13587	–128.58551
110.0	88.59467	–88.95152	22.82142	–26.02807	–204.78487	–186.99613
115.0	41.35683	–42.52442	8.03050	–14.01963	–222.70368	–195.79056
120.0	26.88809	–28.92990	2.51518	–11.50463	–238.31132	–207.64375
125.0	19.60202	–22.59114	–1.06011	–11.18050	–253.75844	–220.51830
130.0	15.02265	–19.04386	–3.99436	–11.78839	–269.47983	–234.11716
135.0	11.73066	–16.88315	–6.71631	–12.94899	–285.67590	–248.40957
140.0	9.12985	–15.53073	–9.42470	–14.52500	–302.50017	–263.45292
145.0	6.92099	–14.70990	–12.24064	–16.47276	–320.10912	–279.35652
150.0	4.93178	–14.27726	–15.25880	–18.79663	–338.68558	–296.27737
155.0	3.04969	–14.15773	–18.57034	–21.53409	–358.45792	–314.42849
160.0	1.19062	–14.31686	–22.27741	–24.75335	–379.72340	–334.09742
165.0	–0.71834	–14.74886	–26.50642	–28.55832	–402.88040	–355.67611
170.0	–2.75087	–15.47311	–31.42485	–33.10092	–428.47919	–379.71037
175.0	–4.99284	–16.53666	–37.26631	–38.60370	–457.30417	–406.98144
180.0	–7.55681	–18.02359	–44.37231	–45.39997	–490.51439	–438.64602
185.0	–10.60539	–20.07507	–53.26717	–54.00794	–529.89964	–476.49216
190.0	–14.39458	–22.93056	–64.80344	–65.27541	–578.37833	–523.43690
195.0	–19.36416	–27.01686	–80.46998	–80.68753	–641.05412	–584.58284
200.0	–26.35209	–33.16149	–103.12611	–103.10064	–727.73516	–669.73728

$\alpha = 0.7$

λ	*A*	*B*	*C*	*D*	*G*	*H*
0.2	5.60075	2.88616	11.11921	12.12746	32.87480	32.94481
0.4	5.58109	2.89152	11.09383	12.11036	32.47937	32.61939
0.6	5.56138	2.89692	11.06841	12.09324	32.08386	32.29390
0.8	5.54162	2.90235	11.04294	12.07610	31.68824	31.96832
1.0	5.52179	2.90781	11.01742	12.05895	31.29253	31.64265
1.2	5.50192	2.91331	10.99185	12.04178	30.89671	31.31689
1.4	5.48198	2.91885	10.96624	12.02460	30.50080	30.99104
1.6	5.46200	2.92442	10.94059	12.00740	30.10480	30.66511
1.8	5.44195	2.93002	10.91489	11.99019	29.70870	30.33909
2.0	5.42185	2.93567	10.88914	11.97296	29.31250	30.01298
2.2	5.40169	2.94135	10.86335	11.95571	28.91620	29.68678
2.4	5.38147	2.94706	10.83750	11.93845	28.51980	29.36050
2.6	5.36119	2.95281	10.81161	11.92117	28.12330	29.03412
2.8	5.34085	2.95860	10.78568	11.90387	27.72670	28.70766
3.0	5.32046	2.96443	10.75969	11.88656	27.33001	28.38111
3.2	5.30000	2.97030	10.73366	11.86924	26.93321	28.05447
3.4	5.27948	2.97620	10.70758	11.85189	26.53631	27.72773
3.6	5.25891	2.98215	10.68145	11.83453	26.13931	27.40091
3.8	5.23827	2.98813	10.65527	11.81716	25.74221	27.07399
4.0	5.21757	2.99415	10.62904	11.79977	25.34500	26.74699
4.2	5.19681	3.00022	10.60276	11.78236	24.94770	26.41990
4.4	5.17599	3.00632	10.57644	11.76494	24.55029	26.09271
4.6	5.15510	3.01246	10.55006	11.74751	24.15278	25.76543
4.8	5.13415	3.01865	10.52363	11.73005	23.75517	25.43806
5.0	5.11314	3.02488	10.49716	11.71258	23.35745	25.11059
5.2	5.09206	3.03114	10.47063	11.69510	22.95963	24.78304
5.4	5.07092	3.03745	10.44405	11.67760	22.56170	24.45539
5.6	5.04971	3.04381	10.41742	11.66008	22.16367	24.12765
5.8	5.02844	3.05020	10.39074	11.64255	21.76553	23.79981
6.0	5.00710	3.05664	10.36401	11.62501	21.36729	23.47188
6.2	4.98570	3.06313	10.33723	11.60744	20.96894	23.14386
6.4	4.96423	3.06965	10.31039	11.58987	20.57049	22.81574
6.6	4.94269	3.07623	10.28350	11.57227	20.17193	22.48753
6.8	4.92108	3.08284	10.25656	11.55466	19.77326	22.15922
7.0	4.89941	3.08951	10.22957	11.53704	19.37448	21.83081
7.2	4.87766	3.09622	10.20252	11.51940	18.97560	21.50232
7.4	4.85585	3.10297	10.17542	11.50174	18.57660	21.17372
7.6	4.83397	3.10977	10.14827	11.48407	18.17750	20.84503
7.8	4.81202	3.11662	10.12106	11.46639	17.77829	20.51624
8.0	4.78999	3.12352	10.09380	11.44869	17.37897	20.18736
8.2	4.76789	3.13046	10.06648	11.43097	16.97954	19.85837

(continued 1) $\alpha = 0.7$

λ	A	B	C	D	G	H
8.4	4.74572	3.13746	10.03911	11.41324	16.58000	19.52930
8.6	4.72348	3.14450	10.01168	11.39549	16.18034	19.20012
8.8	4.70116	3.15159	9.98420	11.37773	15.78058	18.87084
9.0	4.67877	3.15873	9.95666	11.35995	15.38070	18.54147
9.2	4.65631	3.16592	9.92907	11.34216	14.98072	18.21200
9.4	4.63377	3.17317	9.90142	11.32435	14.58062	17.88242
9.6	4.61116	3.18046	9.87371	11.30653	14.18040	17.55275
9.8	4.58847	3.18781	9.84595	11.28870	13.78007	17.22298
10.0	4.56570	3.19521	9.81813	11.27084	13.37963	16.89311
11.0	4.45069	3.23301	9.67815	11.18137	11.37570	15.24224
12.0	4.33368	3.27221	9.53670	11.09153	9.36885	13.58882
13.0	4.21456	3.31287	9.39372	11.00134	7.35902	11.93281
14.0	4.09327	3.35507	9.24918	10.91081	5.34615	10.27416
15.0	3.96970	3.39889	9.10303	10.81994	3.33019	8.61284
16.0	3.84375	3.44442	8.95521	10.72876	1.31106	6.94879
17.0	3.71533	3.49174	8.80569	10.63727	–0.71130	5.28197
18.0	3.58431	3.54097	8.65439	10.54549	–2.73696	3.61234
19.0	3.45058	3.59221	8.50128	10.45343	–4.76598	1.93985
20.0	3.31401	3.64556	8.34628	10.36112	–6.79845	0.26446
21.0	3.17447	3.70116	8.18933	10.26858	–8.83444	–1.41390
22.0	3.03180	3.75913	8.03037	10.17583	–10.87402	–3.09525
23.0	2.88586	3.81962	7.86933	10.08290	–12.91728	–4.77966
24.0	2.73646	3.88278	7.70612	9.98982	–14.96431	–6.46718
25.0	2.58344	3.94877	7.54068	9.89662	–17.01519	–8.15785
26.0	2.42660	4.01778	7.37291	9.80335	–19.07001	–9.85172
27.0	2.26572	4.08999	7.20272	9.71003	–21.12887	–11.54886
28.0	2.10059	4.16562	7.03002	9.61672	–23.19187	–13.24930
29.0	1.93095	4.24490	6.85471	9.52347	–25.25910	–14.95310
30.0	1.75654	4.32806	6.67666	9.43033	–27.33069	–16.66032
31.0	1.57707	4.41538	6.49577	9.33737	–29.40673	–18.37100
32.0	1.39223	4.50716	6.31190	9.24467	–31.48736	–20.08520
33.0	1.20168	4.60372	6.12492	9.15229	–33.57268	–21.80298
34.0	1.00504	4.70540	5.93468	9.06034	–35.66284	–23.52437
35.0	0.80191	4.81259	5.74101	8.96890	–37.75796	–25.24944
36.0	0.59184	4.92573	5.54374	8.87809	–39.85819	–26.97823
37.0	0.37434	5.04527	5.34268	8.78803	–41.96369	–28.71078
38.0	0.14887	5.17174	5.13762	8.69887	–44.07460	–30.44715
39.0	–0.08517	5.30571	4.92833	8.61075	–46.19110	–32.18738
40.0	–0.32844	5.44782	4.71457	8.52386	–48.31337	–33.93150
41.0	–0.58168	5.59877	4.49605	8.43840	–50.44159	–35.67954
42.0	–0.84569	5.75936	4.27248	8.35458	–52.57597	–37.43155

(continued 2) $\alpha = 0.7$

λ	A	B	C	D	G	H
43.0	−1.12141	5.93047	4.04352	8.27267	−54.71673	−39.18754
44.0	−1.40985	6.11311	3.80880	8.19296	−56.86409	−40.94752
45.0	−1.71216	6.30838	3.56790	8.11577	−59.01832	−42.71150
46.0	−2.02964	5.51757	3.32035	8.04148	−61.17967	−44.47948
47.0	−2.36377	6.74212	3.06562	7.97053	−63.34845	−46.25143
48.0	−2.71623	6.98367	2.80314	7.90341	−65.52498	−48.02732
49.0	−3.08893	7.24410	2.53222	7.84070	−67.70960	−49.80709
50.0	−3.48408	7.52560	2.25209	7.78306	−69.90270	−51.59066
51.0	−3.90421	7.83065	1.96187	7.73128	−72.10471	−53.37790
52.0	−4.35229	8.16219	1.66054	7.68625	−74.31612	−55.16867
53.0	−4.83177	8.52362	1.34693	7.64906	−76.53746	−56.96276
54.0	−5.34669	8.91896	1.01964	7.62095	−78.76933	−58.75992
55.0	−5.90185	9.35295	0.67705	7.60345	−81.01244	−60.55981
56.0	−6.50298	9.83129	0.31722	7.59836	−83.26759	−62.36202
57.0	−7.15695	10.36081	−0.06215	7.60786	−85.53571	−64.16598
58.0	−7.87211	10.94979	−0.46388	7.63462	−87.81790	−65.97100
59.0	−8.65869	11.60843	−0.89138	7.68191	−90.11547	−67.77621
60.0	−9.52934	12.34934	−1.34887	7.75382	−92.43000	−69.58045
61.0	−10.49997	13.18835	−1.84164	7.85547	−94.76342	−71.38225
62.0	−11.59076	14.14562	−2.37640	7.99340	−97.11815	−73.17965
63.0	−12.82783	15.24719	−2.96177	8.17610	−99.49724	−74.97012
64.0	−14.24546	16.52729	−3.60909	8.41471	−101.90465	−76.75020
65.0	−15.88967	18.03188	−4.33356	8.72425	−104.34558	−78.51520
66.0	−17.82363	19.82407	−5.15598	9.12536	−106.82711	−80.25860
67.0	−20.13655	21.99300	−6.10572	9.64720	−109.35916	−81.97108
68.0	−22.95861	24.66879	−7.22562	10.33239	−111.95603	−83.63892
69.0	−26.48753	28.04908	−8.58066	11.24573	−114.63934	−85.24112
70.0	−31.03892	32.44940	−10.27453	12.49068	−117.44343	−86.74402
71.0	−37.14955	38.40646	−12.48355	14.24331	−120.42647	−88.09013
72.0	−45.81231	46.91306	−15.53372	16.82941	−123.69578	−89.17281
73.0	−59.09132	60.03324	−20.10273	20.92640	−127.47414	−89.77003
74.0	−82.10035	82.88067	−27.86926	28.21271	−132.31293	−89.33119
75.0	−131.93430	132.55018	−44.44487	44.29963	−140.05452	−86.01476
76.0	−322.36923	322.81771	−107.17676	106.53405	−162.95855	−67.56198
77.0	787.49248	−787.21445	256.91078	−258.06002	−45.77072	−189.22797
78.0	180.09407	−179.98964	57.30206	−58.96720	−113.62903	−125.87565
79.0	102.43726	−102.50970	31.57715	−33.76789	−124.42267	−119.61684
80.0	71.87428	−72.12697	21.30636	−24.03276	−130.15677	−118.44746
81.0	55.49196	−55.92840	15.68575	−18.95823	−134.37819	−118.82167
82.0	45.25137	−45.87520	12.07637	−15.90571	−137.95370	−119.87389
83.0	38.22347	−39.03845	9.51658	−13.91397	−141.19946	−121.28908

(continued 3) $\alpha = 0.7$

λ	A	B	C	D	G	H
84.0	33.08576	−34.09580	7.57209	−12.54913	−144.25860	−122.92540
85.0	29.15358	−30.36273	6.01795	−11.58670	−147.20552	−124.70973
86.0	26.03698	−27.44945	4.72595	−10.89892	−150.08294	−126.60076
87.0	23.49747	−25.11764	3.61761	−10.40780	−152.91725	−128.57352
88.0	21.38105	−23.21346	2.64213	−10.06306	−155.72581	−130.61219
89.0	19.58376	−21.63316	1.76516	−9.83088	−158.52061	−132.70637
90.0	18.03292	−20.30421	0.96258	−9.68771	−161.31033	−134.84912
91.0	16.67613	−19.17445	0.21692	−9.61670	−164.10148	−137.03567
92.0	15.47469	−18.20537	−0.48482	−9.60549	−166.89920	−139.26283
93.0	14.39934	−17.36796	−1.15252	−9.64486	−169.70762	−141.52847
94.0	13.42757	−16.63994	−1.79390	−9.72781	−172.53022	−143.83124
95.0	12.54177	−16.00394	−2.41507	−9.84898	−175.36999	−146.17035
96.0	11.72792	−15.44623	−3.02100	−10.00422	−178.22963	−148.54551
97.0	10.97474	−14.95582	−3.61576	−10.19030	−181.11162	−150.95682
98.0	10.27305	−14.52381	−4.20278	−10.40470	−184.01826	−153.40460
99.0	9.61524	−14.14294	−4.78498	−10.64547	−186.95177	−155.88951
100.0	8.99502	−13.80725	−5.36487	−10.91110	−189.91436	−158.41240
105.0	6.29625	−12.65878	−8.30100	−12.58192	−205.24063	−171.64064
110.0	3.98730	−12.15715	−11.44888	−14.80470	−221.64577	−186.05939
115.0	1.82228	−12.13724	−14.99599	−17.63201	−239.50858	−202.00318
120.0	−0.36913	−12.54781	−19.15559	−21.20438	−259.36660	−219.98565
125.0	−2.74354	−13.41525	−24.22098	−25.77235	−282.03487	−240.80785
130.0	−5.49019	−14.84705	−30.64642	−31.76334	−308.82769	−265.77523
135.0	−8.89360	−17.07360	−39.20584	−39.93352	−342.02138	−297.15830
140.0	−13.46209	−20.56570	−51.36008	−51.73140	−385.94051	−339.27755
145.0	−20.26846	−26.36917	−70.26133	−70.30029	−449.93995	−401.48497
150.0	−32.12117	−37.27233	−104.22823	−103.95224	−558.72972	−508.48838
155.0	−59.61924	−63.85862	−184.73495	−184.15627	−805.82766	−753.80395
160.0	−210.45098	−213.80396	−631.75853	−630.88535	−2142.03323	−2088.22978
165.0	171.86032	169.37866	505.42047	506.58336	1229.17747	1284.75911
170.0	64.44408	62.82743	187.37665	188.82734	276.22349	333.58280
175.0	40.37730	39.62720	117.00582	118.74499	59.22517	118.36239
180.0	29.48721	29.61240	85.80069	87.83143	−41.38913	19.52710
185.0	23.10360	24.11975	68.00702	70.33469	−102.14452	−39.44743
190.0	18.78715	20.71680	56.38558	59.01783	−144.54922	−80.06869
195.0	15.58285	18.45561	48.10781	51.05465	−177.00510	−110.73775
200.0	13.03854	16.89152	41.84009	45.11401	−203.47171	−135.41334

$\alpha = 0.8$

λ	*A*	*B*	*C*	*D*	*G*	*H*
0.2	4.86969	2.53353	8.24755	9.25836	21.57637	21.65638
0.4	4.84687	2.53991	8.22050	9.24216	21.21627	21.37630
0.6	4.82396	2.54635	8.19338	9.22595	20.85605	21.09614
0.8	4.80097	2.55283	8.16619	9.20972	20.49572	20.81587
1.0	4.77789	2.55938	8.13893	9.19348	20.13527	20.53550
1.2	4.75473	2.56598	8.11161	9.17722	19.77469	20.25503
1.4	4.73147	2.57263	8.08421	9.16095	19.41400	19.97447
1.6	4.70813	2.57934	8.05675	9.14466	19.05319	19.69380
1.8	4.68470	2.58611	8.02922	9.12835	18.69226	19.41303
2.0	4.66118	2.59294	8.00162	9.11204	18.33120	19.13216
2.2	4.63756	2.59983	7.97395	9.09570	17.97003	18.85119
2.4	4.61386	2.60677	7.94621	9.07935	17.60873	18.57011
2.6	4.59006	2.61378	7.91840	9.06299	17.24730	18.28894
2.8	4.56617	2.62085	7.89051	9.04661	16.88576	18.00766
3.0	4.54218	2.62798	7.86256	9.03022	16.52409	17.72627
3.2	4.51809	2.63517	7.83453	9.01382	16.16229	17.44479
3.4	4.49391	2.64243	7.80643	8.99740	15.80037	17.16320
3.6	4.46964	2.64975	7.77825	8.98096	15.43832	16.88150
3.8	4.44526	2.65713	7.75000	8.96451	15.07614	16.59970
4.0	4.42079	2.66458	7.72167	8.94805	14.71384	16.31780
4.2	4.39621	2.67210	7.69327	8.93157	14.35141	16.03579
4.4	4.37153	2.67968	7.66480	8.91508	13.98885	15.75367
4.6	4.34675	2.68734	7.63625	8.89858	13.62616	15.47144
4.8	4.32187	2.69506	7.60762	8.88206	13.26333	15.18911
5.0	4.29688	2.70285	7.57891	8.86553	12.90038	14.90667
5.2	4.27179	2.71071	7.55012	8.84899	12.53730	14.62413
5.4	4.24659	2.71864	7.52126	8.83243	12.17408	14.34147
5.6	4.22129	2.72665	7.49232	8.81587	11.81073	14.05871
5.8	4.19587	2.73473	7.46330	8.79928	11.44725	13.77584
6.0	4.17035	2.74288	7.43419	8.78269	11.08363	13.49286
6.2	4.14471	2.75111	7.40501	8.76608	10.71988	13.20977
6.4	4.11897	2.75942	7.37575	8.74946	10.35599	12.92656
6.6	4.09311	2.76780	7.34640	8.73283	9.99196	12.64325
6.8	4.06713	2.77626	7.31697	8.71619	9.62780	12.35983
7.0	4.04104	2.78480	7.28746	8.69954	9.26350	12.07629
7.2	4.01484	2.79342	7.25786	8.68287	8.89906	11.79264
7.4	3.98851	2.80212	7.22818	8.66620	8.53448	11.50888
7.6	3.96207	2.81090	7.19841	8.64951	8.16976	11.22501
7.8	3.93551	2.81977	7.16856	8.63281	7.80490	10.94102
8.0	3.90883	2.82872	7.13862	8.61610	7.43990	10.65692
8.2	3.88202	2.83776	7.10860	8.59938	7.07475	10.37271

(continued 1) $\alpha = 0.8$

λ	A	B	C	D	G	H
8.4	3.85509	2.84688	7.07848	8.58265	6.70946	10.08838
8.6	3.82803	2.85609	7.04828	8.56591	6.34403	9.80393
8.8	3.80085	2.86539	7.01799	8.54916	5.97845	9.51937
9.0	3.77354	2.87479	6.98761	8.53240	5.61273	9.23470
9.2	3.74610	2.88427	6.95713	8.51563	5.24686	8.94990
9.4	3.71853	2.89384	6.92657	8.49885	4.88084	8.66499
9.6	3.69083	2.90351	6.89591	8.48207	4.51468	8.37996
9.8	3.66299	2.91327	6.86516	8.46527	4.14836	8.09482
10.0	3.63502	2.92313	6.83432	8.44847	3.78190	7.80955
11.0	3.49308	2.97393	6.67867	8.36433	1.94730	6.38144
12.0	3.34752	3.02734	6.52057	8.28003	0.10882	4.95030
13.0	3.19814	3.08355	6.35989	8.19561	–1.73364	3.51607
14.0	3.04469	3.14278	6.19654	8.11111	–3.58021	2.07870
15.0	2.88693	3.20525	6.03038	8.02660	–5.43101	0.63812
16.0	2.72458	3.27120	5.86128	7.94211	–7.28617	–0.80573
17.0	2.55733	3.34092	5.68910	7.85774	–9.14583	–2.25290
18.0	2.38486	3.41470	5.51368	7.77355	–11.01013	–3.70345
19.0	2.20680	3.49289	5.33486	7.68963	–12.87924	–5.15745
20.0	2.02275	3.57584	5.15244	7.60609	–14.75330	–6.61495
21.0	1.83225	3.66397	4.96623	7.52303	–16.63250	–8.07601
22.0	1.63483	3.75774	4.77601	7.44059	–18.51702	–9.54068
23.0	1.42993	3.85767	4.58154	7.35891	–20.40705	–11.00902
24.0	1.21693	3.96432	4.38254	7.27817	–22.30280	–12.48107
25.0	0.99517	4.07835	4.17873	7.19856	–24.20450	–13.95687
26.0	0.76386	4.20047	3.96977	7.12030	–26.11239	–15.43647
27.0	0.52214	4.33152	3.75530	7.04364	–28.02672	–16.91990
28.0	0.26903	4.47242	3.53491	6.96889	–29.94777	–18.40717
29.0	0.00342	4.62426	3.30813	6.89638	–31.87586	–19.89829
30.0	–0.27595	4.78824	3.07445	6.82651	–33.81131	–21.39326
31.0	–0.57056	4.96578	2.83328	6.75974	–35.75448	–22.89206
32.0	–0.88207	5.15850	2.58394	6.69661	–37.70579	–24.39464
33.0	–1.21243	5.36830	2.32565	6.63777	–39.66567	–25.90092
34.0	–1.56391	5.59739	2.05751	6.58395	–41.63462	–27.41080
35.0	–1.93917	5.84837	1.77848	6.53605	–43.61320	–28.92412
36.0	–2.34135	6.12432	1.48734	6.49514	–45.60204	–30.44067
37.0	–2.77420	6.42891	1.18263	6.46250	–47.60187	–31.96018
38.0	–3.24220	6.76658	0.86262	6.43967	–49.61353	–33.48226
39.0	–3.75078	7.14268	0.52524	6.42856	–51.63798	–35.00643
40.0	–4.30659	7.56378	0.16796	6.43150	–53.67638	–36.53205
41.0	–4.91785	8.03803	–0.21233	6.45139	–55.73009	–38.05827
42.0	–5.59483	8.57561	–0.61951	6.49190	–57.80079	–39.58399

(continued 2) $a = 0.8$

λ	A	B	C	D	G	H
43.0	−6.35055	9.18947	−1.05846	6.55770	−59.89052	−41.10775
44.0	−7.20180	9.89631	−1.53547	6.65483	−62.00187	−42.62757
45.0	−8.17057	10.71800	−2.05873	6.79124	−64.13811	−44.14082
46.0	−9.28622	11.68385	−2.63916	6.97760	−66.30358	−45.64386
47.0	−10.58896	12.83391	−3.29167	7.22854	−68.50405	−47.13160
48.0	−12.13522	14.22456	−4.03716	7.56468	−70.74754	−48.59679
49.0	−14.00693	15.93756	−4.90591	8.01599	−73.04555	−50.02872
50.0	−16.32761	18.09635	−5.94353	8.62777	−75.41521	−51.41110
51.0	−19.29275	20.89628	−7.22210	9.47178	−77.88343	−52.71788
52.0	−23.23113	24.66599	−8.86279	10.66881	−80.49521	−53.90500
53.0	−28.74178	30.00437	−11.08623	12.43913	−83.33217	−54.89181
54.0	−37.04365	38.13023	−14.34124	15.23118	−86.55996	−55.51371
55.0	−51.05751	51.96418	−19.70302	20.11972	−90.57477	−55.37560
56.0	−79.96878	80.69146	−30.55251	30.48525	−96.62150	−53.23375
57.0	−175.44447	175.97890	−65.90635	65.34396	−111.69988	−42.08965
58.0	1140.29614	−1139.95439	418.14377	−419.21296	64.37913	−222.13367
59.0	137.57271	−137.42827	48.86017	−50.44835	−73.63164	−88.11991
60.0	73.98449	−74.04222	25.22699	−27.34694	−84.43757	−81.34459
61.0	50.87091	−51.13589	16.48508	−19.15014	−89.77776	−80.07013
62.0	38.86580	−39.34335	11.82570	−15.04987	−93.63116	−80.31924
63.0	31.47764	−32.17333	8.85932	−12.65726	−96.87803	−81.21346
64.0	26.44719	−27.36687	6.75409	−11.14118	−99.82534	−82.44770
65.0	22.78163	−23.93145	5.14428	−10.13667	−102.60858	−83.88848
66.0	19.97627	−21.36267	3.84376	−9.45844	−105.29718	−85.46854
67.0	17.74728	−19.37706	2.74771	−9.00252	−107.93072	−87.15063
68.0	15.92280	−17.80311	1.79245	−8.70619	−110.53369	−88.91271
69.0	14.39262	−16.53099	0.93698	−8.52947	−113.12235	−90.74122
70.0	13.08277	−15.48715	0.15363	−8.44576	−115.70810	−92.62760
71.0	11.94174	−14.62051	−0.57709	−8.43676	−118.29944	−94.56643
72.0	10.93250	−13.89454	−1.26934	−8.48956	−120.90301	−96.55440
73.0	10.02773	−13.28243	−1.93375	−8.59488	−123.52424	−98.58961
74.0	9.20680	−12.76410	−2.57856	−8.74586	−126.16786	−100.67125
75.0	8.45379	−12.32425	−3.21031	−8.93787	−128.83805	−102.79930
76.0	7.75623	−11.95106	−3.83433	−9.16697	−131.53874	−104.97439
77.0	7.10413	−11.63524	−4.45510	−9.43068	−134.27368	−107.19765
78.0	6.48939	−11.36948	−5.07647	−9.72716	−137.04660	−109.47073
79.0	5.90534	−11.14795	−5.70187	−10.05522	−139.86127	−111.79570
80.0	5.34639	−10.96598	−6.33438	−10.41420	−142.72156	−114.17506
81.0	4.80780	−10.81983	−6.97688	−10.80385	−145.63152	−116.61168
82.0	4.28549	−10.70653	−7.63215	−11.22432	−148.59548	−119.10896
83.0	3.77586	−10.62369	−8.30286	−11.67612	−151.61799	−121.67063

(continued 3) $\alpha = 0.8$

λ	A	B	C	D	G	H
84.0	3.27572	–10.56947	–8.99169	–12.16006	–154.70405	–124.30100
85.0	2.78220	–10.54246	–9.70139	–12.67731	–157.85902	–127.00486
86.0	2.29263	–10.54162	–10.43476	–13.22931	–161.08883	–129.78764
87.0	1.80452	–10.56626	–11.19479	–13.81787	–164.39990	–132.65533
88.0	1.31546	–10.61601	–11.98461	–14.44513	–167.79941	–135.61471
89.0	0.82314	–10.69078	–12.80763	–15.11357	–171.29526	–138.67339
90.0	0.32524	–10.79078	–13.66754	–15.82614	–174.89630	–141.83991
91.0	–0.18058	–10.91647	–14.56839	–16.58617	–178.61239	–145.12388
92.0	–0.69672	–11.06864	–15.51466	–17.39757	–182.45466	–148.53621
93.0	–1.22573	–11.24834	–16.51135	–18.26477	–186.43563	–152.08923
94.0	–1.77028	–11.45696	–17.56406	–19.19292	–190.56948	–155.79694
95.0	–2.33329	–11.69626	–18.67911	–20.18789	–194.87237	–159.67533
96.0	–2.91793	–11.96836	–19.86371	–21.25652	–199.36278	–163.74274
97.0	–3.52774	–12.27588	–21.12609	–22.40668	–204.06202	–168.02037
98.0	–4.16669	–12.62191	–22.47570	–23.64753	–208.99457	–172.53258
99.0	–4.83927	–13.01021	–23.92354	–24.98977	–214.18901	–177.30784
100.0	–5.55066	–13.44524	–25.48240	–26.44594	–219.67872	–182.37944
105.0	–9.94120	–16.55259	–35.58050	–36.06676	–253.22690	–213.85032
110.0	–16.93328	–22.38425	–52.65930	–52.71321	–305.18429	–263.74686
115.0	–31.20930	–35.58077	–88.98552	–88.63663	–407.79334	–364.30583
120.0	–83.63733	–86.98053	–225.70462	–224.97148	–774.44352	–728.91266
125.0	223.24795	220.90421	582.49035	583.59746	1343.89458	1391.46513
130.0	52.63346	51.27825	134.91109	136.38851	159.70699	209.31606
135.0	30.57195	30.20963	78.01137	79.86111	2.95758	54.60602
140.0	21.51471	22.16348	55.33151	57.56074	–63.79011	–10.09947
145.0	16.34758	18.03907	42.91764	45.53855	–103.52592	–47.78840
150.0	12.84500	15.62464	34.93237	37.96228	–131.59565	–73.80464
155.0	10.19196	14.12012	29.24917	32.71105	–153.58541	–93.73218
160.0	8.01545	13.16947	24.90624	28.82939	–172.01750	–110.09095
165.0	6.11669	12.59394	21.40333	25.82459	–188.19898	–124.18517
170.0	4.37578	12.29805	18.45269	23.41809	–202.87626	–136.75777
175.0	2.71137	12.23117	15.87496	21.44212	–216.50797	–148.26307
180.0	1.06083	12.37033	13.55011	19.79156	–229.39300	–158.99430
185.0	–0.63128	12.71279	11.39172	18.39986	–241.73619	–169.14881
190.0	–2.42146	13.27400	9.33224	17.22658	–253.68471	–178.86358
195.0	–4.37547	14.08967	7.31303	16.25135	–265.34962	–188.23523
200.0	–6.57877	15.22260	5.27637	15.47232	–276.81981	–197.33159

$\alpha = 0.9$

λ	*A*	*B*	*C*	*D*	*G*	*H*
0.2	4.29714	2.26042	6.27795	7.29168	14.79211	14.88213
0.4	4.27100	2.26792	6.24898	7.27652	14.45781	14.63787
0.6	4.24474	2.27551	6.21992	7.26135	14.12335	14.39351
0.8	4.21835	2.28318	6.19076	7.24617	13.78876	14.14903
1.0	4.19182	2.29094	6.16150	7.23098	13.45402	13.90445
1.2	4.16517	2.29878	6.13214	7.21578	13.11913	13.65975
1.4	4.13839	2.30670	6.10268	7.20056	12.78409	13.41494
1.6	4.11147	2.31472	6.07313	7.18534	12.44891	13.17002
1.8	4.08441	2.32282	6.04347	7.17011	12.11358	12.92499
2.0	4.05722	2.33101	6.01371	7.15487	11.77809	12.67984
2.2	4.02989	2.33930	5.98385	7.13962	11.44246	12.43458
2.4	4.00242	2.34767	5.95388	7.12436	11.10667	12.18920
2.6	3.97480	2.35615	5.92381	7.10909	10.77072	11.94371
2.8	3.94704	2.36471	5.89363	7.09381	10.43463	11.69811
3.0	3.91914	2.37338	5.86335	7.07853	10.09837	11.45239
3.2	3.89108	2.38214	5.83296	7.06324	9.76196	11.20655
3.4	3.86288	2.39101	5.80246	7.04794	9.42539	10.96060
3.6	3.83453	2.39998	5.77185	7.03263	9.08866	10.71453
3.8	3.80602	2.40905	5.74113	7.01732	8.75178	10.46834
4.0	3.77736	2.41823	5.71029	7.00200	8.41473	10.22204
4.2	3.74854	2.42751	5.67935	6.98668	8.07752	9.97561
4.4	3.71956	2.43691	5.64829	6.97135	7.74014	9.72907
4.6	3.69041	2.44641	5.61711	6.95602	7.40260	9.48240
4.8	3.66111	2.45603	5.58582	6.94068	7.06489	9.23562
5.0	3.63164	2.46576	5.55441	6.92534	6.72702	8.98871
5.2	3.60200	2.47561	5.52288	6.90999	6.38898	8.74168
5.4	3.57219	2.48557	5.49123	6.89464	6.05077	8.49453
5.6	3.54221	2.49566	5.45946	6.87929	5.71239	8.24726
5.8	3.51205	2.50587	5.42757	6.86394	5.37384	7.99987
6.0	3.48172	2.51620	5.39555	6.84858	5.03511	7.75235
6.2	3.45121	2.52666	5.36341	6.83323	4.69621	7.50471
6.4	3.42051	2.53725	5.33114	6.81787	4.35713	7.25694
6.6	3.38963	2.54796	5.29875	6.80251	4.01788	7.00905
6.8	3.35857	2.55881	5.26622	6.78716	3.67845	6.76104
7.0	3.32731	2.56980	5.23357	6.77181	3.33884	6.51289
7.2	3.29586	2.58092	5.20078	6.75645	2.99905	6.26462
7.4	3.26422	2.59218	5.16786	6.74110	2.65907	6.01623
7.6	3.23238	2.60359	5.13480	6.72576	2.31891	5.76770
7.8	3.20034	2.61514	5.10161	6.71042	1.97857	5.51905
8.0	3.16809	2.62683	5.06828	6.69508	1.63804	5.27027
8.2	3.13564	2.63868	5.03482	6.67975	1.29732	5.02136

(continued 1) $\alpha = 0.9$

λ	A	B	C	D	G	H
8.4	3.10298	2.65068	5.00121	6.66442	0.95642	4.77232
8.6	3.07011	2.66283	4.96746	6.64910	0.61532	4.52315
8.8	3.03702	2.67514	4.93356	6.63379	0.27403	4.27385
9.0	3.00372	2.68762	4.89952	6.61848	–0.06745	4.02441
9.2	2.97019	2.70025	4.86533	6.60319	–0.40913	3.77485
9.4	2.93643	2.71305	4.83099	6.58790	–0.75101	3.52515
9.6	2.90245	2.72603	4.79651	6.57263	–1.09308	2.27532
9.8	2.86824	2.73917	4.76187	6.55737	–1.43536	3.02536
10.0	2.83379	2.75249	4.72707	6.54212	–1.77783	2.77526
11.0	2.65787	2.82185	4.55071	6.46610	–3.49329	1.52276
12.0	2.47547	2.89613	4.37021	6.39055	–5.21404	0.26683
13.0	2.28606	2.97582	4.18527	6.31562	–6.94032	–0.99257
14.0	2.08901	3.06149	3.99562	6.24149	–8.67237	–2.25550
15.0	1.88364	3.15377	3.80090	6.16836	–10.41045	–3.52201
16.0	1.66915	3.25339	3.60075	6.09647	–12.15485	–4.79213
17.0	1.44466	3.36118	3.39474	6.02609	–13.90590	–6.06590
18.0	1.20913	3.47810	3.18239	5.95755	–15.66392	–7.34334
19.0	0.96140	3.60527	2.96316	5.89124	–17.42931	–8.62444
20.0	0.70009	3.74396	2.73643	5.82760	–19.20248	–9.90920
21.0	0.42363	3.89570	2.50147	5.76718	–20.98388	–11.19756
22.0	0.13015	4.06227	2.25747	5.71061	–22.77405	–12.48944
23.0	–0.18252	4.24575	2.00345	5.65868	–24.57357	–13.78474
24.0	–0.51701	4.44868	1.73826	5.61231	–26.38311	–15.08326
25.0	–0.87641	4.67407	1.46056	5.57265	–28.20342	–16.38478
26.0	–1.26449	4.92559	1.16873	5.54107	–30.03539	–17.68894
27.0	–1.68585	5.20773	0.86081	5.51929	–31.88006	–18.99530
28.0	–2.14616	5.52605	0.53438	5.50947	–33.73865	–20.30324
29.0	–2.65248	5.88750	0.18647	5.51432	–35.61264	–21.61192
30.0	–3.21376	6.30092	–0.18667	5.53728	–37.50383	–22.92023
31.0	–3.84153	6.77768	–0.58983	5.58285	–39.41447	–24.22665
32.0	–4.55079	7.33268	–1.02924	5.65693	–41.34739	–25.52911
33.0	–5.36161	7.98581	–1.51316	5.76744	–43.30629	–26.82473
34.0	–6.30131	8.76425	–2.05283	5.92524	–45.29608	–28.10947
35.0	–7.40815	9.70611	–2.66392	6.14562	–47.32349	–29.37752
36.0	–8.73747	10.86653	–3.36902	6.45076	–49.39809	–30.62029
37.0	–10.37232	12.32838	–4.20198	6.87406	–51.53403	–31.82467
38.0	–12.44343	14.22220	–5.21600	7.46824	–53.75334	–32.96977
39.0	–15.16913	16.76610	–6.49977	8.32151	–56.09247	–34.02033
40.0	–18.94363	20.35406	–8.21331	9.59332	–58.61680	–34.91224
41.0	–24.55880	25.77771	–10.67782	11.60432	–61.45709	–35.51609
42.0	–33.87504	34.89719	–14.64765	15.10822	–64.91972	–35.52700

(continued 2) $\alpha = 0.9$

λ	*A*	*B*	*C*	*D*	*G*	*H*
43.0	–52.54175	53.36161	–22.41298	22.39455	–69.93363	–34.01758
44.0	–109.65203	110.26378	–45.76652	45.25529	–81.27950	–26.20885
45.0	2987.77835	–2987.38087	1209.51762	–1210.53621	425.63600	–536.69620
46.0	106.00067	–105.82395	41.21691	–42.75824	–49.82558	–64.84245
47.0	54.73119	–54.78211	20.21167	–22.29204	–60.26497	–58.04966
48.0	37.12888	–37.41473	12.84574	–15.48239	–65.19281	–56.80927
49.0	28.16613	–28.69464	8.97413	–12.18537	–68.72223	–57.01062
50.0	22.69363	–23.47301	6.50918	–10.31445	–71.70057	–57.80908
51.0	18.97420	–20.01319	4.74625	–9.16621	–74.41463	–58.92074
52.0	16.25810	–17.56599	3.38086	–8.43753	–76.98915	–60.22407
53.0	14.16856	–15.75527	2.25971	–7.97657	–79.48771	–61.65894
54.0	12.49546	–14.37159	1.29695	–7.69909	–81.94667	–63.19278
55.0	11.11223	–13.28912	0.44057	–7.55482	–84.38884	–64.80695
56.0	9.93803	–12.42781	–0.34301	–7.51210	–86.82972	–66.49048
57.0	8.91868	–11.73440	–1.07662	–7.55022	–89.28063	–68.23706
58.0	8.01645	–11.17213	–1.77651	–7.65526	–91.75045	–70.04333
59.0	7.20418	–10.71493	–2.45478	–7.81770	–94.24659	–71.90796
60.0	6.46171	–10.34385	–3.12076	–8.03104	–96.77564	–73.83112
61.0	5.77370	–10.04488	–3.78200	–8.29093	–99.34377	–75.81415
62.0	5.12817	–9.80753	–4.44479	–8.59457	–101.95699	–77.85932
63.0	4.51555	–9.62391	–5.11466	–8.94038	–104.62140	–79.96993
64.0	3.92801	–9.48802	–5.79662	–9.32779	–107.34337	–82.15005
65.0	3.35895	–9.39539	–6.49539	–9.75701	–110.12967	–84.40463
66.0	2.80272	–9.34272	–7.21558	–10.22905	–112.98767	–86.73959
67.0	2.25430	–9.32768	–7.96187	–10.74558	–115.92549	–89.16182
68.0	1.70914	–9.34876	–8.73908	–11.30899	–118.95218	–91.67936
69.0	1.16295	–9.40517	–9.55236	–11.92237	–122.07790	–94.30158
70.0	0.61162	–9.49677	–10.40732	–12.58959	–125.31421	–97.03930
71.0	0.05104	–9.62408	–11.31014	–13.31537	–128.67434	–99.90518
72.0	–0.52302	–9.78821	–12.26777	–14.10541	–132.17349	–102.91391
73.0	–1.11499	–9.99097	–13.28812	–14.96656	–135.82934	–106.08272
74.0	–1.72966	–10.23486	–14.38031	–15.90701	–139.66258	–109.43194
75.0	–2.37235	–10.52321	–15.55497	–16.93656	–143.69756	–112.98560
76.0	–3.04905	–10.86029	–16.82459	–18.06703	–147.96322	–116.77233
77.0	–3.76670	–11.25150	–18.20407	–19.31269	–152.49416	–120.82650
78.0	–4.53340	–11.70360	–19.71132	–20.69090	–157.33226	–125.18976
79.0	–5.35885	–12.22512	–21.36810	–22.22295	–162.52860	–129.91299
80.0	–6.25480	–12.82673	–23.20122	–23.93521	–168.14618	–135.05903
81.0	–7.23575	–13.52199	–25.24404	–25.86065	–174.26359	–140.70630
82.0	–8.31985	–14.32820	–27.53871	–28.04110	–180.98033	–146.95419
83.0	–9.53025	–15.26775	–30.13929	–30.53028	–188.42402	–153.93015

(continued 3) $\alpha = 0.9$

λ	A	B	C	D	G	H
84.0	–10.89703	–16.37003	–33.11630	–33.39845	–196.76136	–161.80079
85.0	–12.46007	–17.67429	–36.56354	–36.73915	–206.21415	–170.78780
86.0	–14.27351	–19.23409	–40.60863	–40.67976	–217.08426	–181.19297
87.0	–16.41282	–21.12440	–45.42970	–45.39821	–229.79319	–193.43770
88.0	–18.98658	–23.45329	–51.28332	–51.15088	–244.94814	–208.12911
89.0	–22.15693	–26.38249	–58.55291	–58.32099	–263.45657	–226.17459
90.0	–26.17748	–30.16517	–67.83843	–67.50834	–286.73864	–248.99423
91.0	–31.46836	–35.22109	–80.13408	–79.70698	–317.14777	–278.94140
92.0	–38.78020	–42.30055	–97.21654	–96.69344	–358.88978	–320.22185
93.0	–49.59871	–52.88891	–122.60258	–121.98437	–420.28989	–381.16074
94.0	–67.33659	–70.39856	–164.37084	–163.65827	–520.47262	–480.88255
95.0	–101.95807	–104.79345	–246.10954	–245.30325	–715.27862	–675.22787
96.0	–200.21151	–202.82166	–478.48650	–477.58702	–1266.69328	–1226.18209
97.0	–2647.72715	–2650.11315	–6272.30886	–6271.31658	–14983.8036	–14942.8320
98.0	245.56199	243.39930	577.23680	578.32153	1229.92499	1271.35670
99.0	119.31844	117.37844	278.59589	279.77287	521.64156	563.53341
100.0	79.54677	77.82909	184.66080	185.92991	297.96127	340.31321
105.0	30.84479	30.24057	70.45446	72.18546	21.06843	65.72150
110.0	19.25961	19.79041	44.13597	46.33849	–47.75693	–0.79844
115.0	13.68978	15.40149	32.10218	34.79589	–82.74686	–33.47453
120.0	10.16569	13.12938	24.98035	28.19536	–106.07941	–54.48049
125.0	7.55055	11.86577	20.10716	23.88540	–124.04794	–70.10479
130.0	5.38628	11.18698	16.43173	20.82943	–139.12627	–82.81526
135.0	3.44214	10.90627	13.45074	18.54249	–152.48802	–93.77791
140.0	1.57674	10.94130	10.88704	16.77205	–164.76531	–103.61456
145.0	–0.31696	11.26785	8.56801	15.38008	–176.33306	–112.68572
150.0	–2.34099	11.90542	6.37055	14.29423	–187.43320	–121.21209
155.0	–4.61657	12.92021	4.19038	13.48852	–198.23627	–129.33150
160.0	–7.31632	14.44621	1.91775	12.98127	–208.87729	–137.12549
165.0	–10.72454	16.73928	–0.59386	12.85165	–219.48308	–144.62658
170.0	–15.37761	20.31381	–3.58998	13.29316	–230.20697	–151.80466
175.0	–22.46150	26.33792	–7.58353	14.77643	–241.30451	–158.50436
180.0	–35.27987	38.09999	–13.96974	18.66049	–253.39392	–164.19310
185.0	–67.93264	69.68608	–28.65329	30.81727	–268.97423	–166.44946
190.0	–389.68387	390.34704	–164.15018	163.73143	–335.08710	–118.30599
195.0	112.63249	–113.09653	44.62629	–47.71539	–257.44232	–214.12800
200.0	49.38036	–51.02277	17.04865	–22.92932	–278.66861	–211.36616

$\alpha = 1.0$

λ	A	B	C	D	G	H
0.2	3.83544	2.04304	4.86828	5.88530	10.48354	10.58357
0.4	3.80581	2.05180	4.83715	5.87132	10.16834	10.36846
0.6	3.77600	2.06067	4.80589	5.85733	9.85297	10.15323
0.8	3.74600	2.06967	4.77449	5.84335	9.53742	9.93789
1.0	3.71582	2.07880	4.74295	5.82936	9.22169	9.72242
1.2	3.68544	2.08805	4.71127	5.81538	8.90578	9.50684
1.4	3.65487	2.09743	4.67945	5.80139	8.58969	9.29113
1.6	3.62411	2.10694	4.64749	5.78741	8.27341	9.07531
1.8	3.59314	2.11658	4.61538	5.77343	7.95694	8.85936
2.0	3.56197	2.12636	4.58313	5.75945	7.64029	8.64329
2.2	3.53060	2.13628	4.55073	5.74548	7.32344	8.42710
2.4	3.49902	2.14634	4.51817	5.73151	7.00641	8.21078
2.6	3.46722	2.15655	4.48547	5.71755	6.68918	7.99434
2.8	3.43521	2.16690	4.45261	5.70359	6.37175	7.77778
3.0	3.40298	2.17740	4.41959	5.68964	6.05413	7.56109
3.2	3.37053	2.18806	4.38642	5.67570	5.73631	7.34428
3.4	3.33785	2.19887	4.35309	5.66176	5.41830	7.12734
3.6	3.30494	2.20984	4.31959	5.64784	5.10007	6.91027
3.8	3.27179	2.22098	4.28594	5.63392	4.78165	6.69308
4.0	3.23841	2.23228	4.25211	5.62002	4.46302	6.47575
4.2	3.20479	2.24375	4.21812	5.60613	4.14418	6.25830
4.4	3.17092	2.25540	4.18395	5.59226	3.82513	6.04073
4.6	3.13680	2.26722	4.14962	5.57840	3.50586	5.82302
4.8	3.10243	2.27922	4.11510	5.56456	3.18639	5.60518
5.0	3.06779	2.29141	4.08041	5.55073	2.86670	5.38722
5.2	3.03290	2.30378	4.04554	5.53692	2.54679	5.16912
5.4	2.99774	2.31635	4.01049	5.52314	2.22666	4.95089
5.6	2.96231	2.32911	3.97525	5.50937	1.90630	4.73253
5.8	2.92659	2.34207	3.93983	5.49563	1.58573	4.51404
6.0	2.89060	2.35524	3.90421	5.48191	1.26492	4.29541
6.2	2.85432	2.36862	3.86840	5.46821	0.94389	4.07665
6.4	2.81775	2.38221	3.83240	5.45455	0.62262	3.85776
6.6	2.78088	2.39602	3.79619	5.44091	0.30112	3.63874
6.8	2.74371	2.41005	3.75979	5.42730	–0.02062	3.41958
7.0	2.70623	2.42432	3.72318	5.41372	–0.34259	3.20029
7.2	2.66843	2.43881	3.68636	5.40018	–0.66481	2.98086
7.4	2.63031	2.45355	3.64933	5.38667	–0.98727	2.76129
7.6	2.59187	2.46853	3.61208	5.37320	–1.30997	2.54159
7.8	2.55310	2.48376	3.57462	5.35977	–1.63293	2.32176
8.0	2.51398	2.49924	3.53694	5.34638	–1.95614	2.10179
8.2	2.47452	2.51499	3.49903	5.33304	–2.27960	1.88168

(continued 1) $\alpha = 1.0$

λ	A	B	C	D	G	H
8.4	2.43471	2.53101	3.46090	5.31974	−2.60332	1.66143
8.6	2.39453	2.54730	3.42253	5.30648	−2.92730	1.44105
8.8	2.35399	2.56387	3.38393	5.29328	−3.25155	1.22053
9.0	2.31308	2.58073	3.34509	5.28013	−3.57606	0.99987
9.2	2.27178	2.59788	3.30600	5.26704	−3.90083	0.77908
9.4	2.23010	2.61534	3.26667	5.25400	−4.22589	0.55814
9.6	2.18802	2.63310	3.22709	5.24102	−4.55121	0.33707
9.8	2.14553	2.65119	3.18725	5.22810	−4.87682	0.11586
10.0	2.10262	2.66959	3.14715	5.21525	−5.20271	−0.10549
11.0	1.88156	2.76679	2.94259	5.15209	−6.83649	−1.21432
12.0	1.64869	2.87343	2.73079	5.09105	−8.47786	−2.32661
13.0	1.40257	2.99084	2.51104	5.03263	−10.12732	−3.44232
14.0	1.14155	3.12059	2.28251	4.97744	−11.78542	−4.56140
15.0	0.86361	3.26456	2.04422	4.92619	−13.45281	−5.68374
16.0	0.56639	3.42502	1.79503	4.87978	−15.13022	−6.80916
17.0	0.24699	3.60471	1.53355	4.83931	−16.81850	−7.93743
18.0	−0.09810	3.80702	1.25812	4.80615	−18.51863	−9.06819
19.0	−0.47324	4.03616	0.96669	4.78204	−20.23178	−10.20095
20.0	−0.88388	4.29741	0.65674	4.76915	−21.95936	−11.33504
21.0	−1.33690	4.59751	0.32509	4.77031	−23.70307	−12.46954
22.0	−1.84120	4.94517	−0.03233	4.78919	−25.46504	−13.60315
23.0	−2.40842	5.35183	−0.42080	4.83068	−27.24793	−14.73408
24.0	−3.05415	5.83288	−0.84736	4.90136	−29.05524	−15.85982
25.0	−3.79974	6.40946	−1.32158	5.01033	−30.89158	−16.97677
26.0	−4.67526	7.11139	−1.85691	5.17052	−32.76331	−18.07969
27.0	−5.72447	7.98218	−2.47282	5.40085	−34.67947	−19.16073
28.0	−7.01365	9.08783	−3.19869	5.73012	−36.65350	−20.20772
29.0	−8.64827	10.53352	−4.08119	6.20433	−38.70647	−21.20099
30.0	−10.80703	12.49761	−5.19921	6.90169	−40.87371	−22.10673
31.0	−13.81889	15.30873	−6.69776	7.96646	−43.21965	−22.86213
32.0	−18.36326	19.64590	−8.87608	9.69704	−45.87664	−23.33660
33.0	−26.10645	27.17503	−12.46758	12.82597	−49.16811	−23.20867
34.0	−42.51673	43.36391	−19.87832	19.75831	−54.15338	−21.42108
35.0	−102.11267	102.73065	−46.29453	45.67926	−67.51429	−11.29432
36.0	319.37432	−318.99390	138.92583	−140.05441	12.21498	−94.29670
37.0	64.64377	−64.50984	26.64878	−28.30998	−38.95098	−46.44555
38.0	36.47799	−36.60014	14.03655	−16.25109	−46.28903	−42.46707
39.0	25.54789	−25.93644	8.99799	−11.78812	−50.31433	−41.84943
40.0	19.67443	−20.34046	6.17444	−9.56413	−53.38578	−42.23739
41.0	15.96258	−16.91807	4.29156	−8.30666	−56.06564	−43.07278
42.0	13.37070	−14.62857	2.89025	−7.55871	−58.55769	−44.15636

(continued 2) $\alpha = 1.0$

λ	*A*	*B*	*C*	*D*	*G*	*H*
43.0	11.43171	–13.00599	1.76411	–7.11620	–60.95472	–45.40035
44.0	9.90477	–11.81068	0.80613	–6.87475	–63.30606	–46.76107
45.0	8.65286	–10 90698	–0.04511	–6.77583	–65.64124	–48.21528
46.0	7.59210	–10.21253	–0.82783	–6.78451	–67.97990	–49.75045
47.0	6.66808	–9.67462	–1.56744	–6.87910	–70.33640	–51.36019
48.0	5.84375	–9.25815	–2.28191	–7.04580	–72.72224	–53.04206
49.0	5.09282	–8.93903	–2.98471	–7.27592	–75.14745	–54.79633
50.0	4.39591	–8.70041	–3.68651	–7.56421	–77.62141	–56.62536
51.0	3.73820	–8.53035	–4.39634	–7.90791	–80.15345	–58.53331
52.0	3.10795	–8.42044	–5.12221	–8.30618	–82.75329	–60.52600
53.0	2.49547	–8.36485	–5.87172	–8.75977	–85.43143	–62.61093
54.0	1.89246	–8.35979	–6.65237	–9.27088	–88.19955	–64.79741
55.0	1.29148	–8.40309	–7.47199	–9.84310	–91.07087	–67.09683
56.0	0.68556	–8.49400	–8.33899	–10.48145	–94.06067	–69.52297
57.0	0.06786	–8.63308	–9.26275	–11.19254	–97.18682	–72.09248
58.0	–0.56863	–8.82220	–10.25398	–11.98483	–100.47052	–74.82556
59.0	–1.23141	–9.06459	–11.32521	–12.86895	–103.93720	–77.74681
60.0	–1.92881	–9.36499	–12.49135	–13.85825	–107.61773	–80.88642
61.0	–2.67042	–9.72996	–13.77050	–14.96950	–111.55005	–84.28174
62.0	–3.46757	–10.16828	–15.18498	–16.22386	–115.78136	–87.97946
63.0	–4.33411	–10.69158	–16.76273	–17.64834	–120.37125	–92.03877
64.0	–5.28730	–11.31523	–18.53945	–19.27776	–125.39605	–96.53560
65.0	–6.34924	–12.05968	–20.56145	–21.15775	–130.95521	–101.56908
66.0	–7.54895	–12.95249	–22.89016	–23.34907	–137.18084	–107.27106
67.0	–8.92551	–14.03148	–25.60889	–25.93449	–144.25241	–113.82076
68.0	–10.53305	–15.34967	–28.83363	–29.02949	–152.42003	–121.46807
69.0	–12.44909	–16.98357	–32.73095	–32.80020	–162.04310	–130.57219
70.0	–14.78889	–19.04756	–37.54902	–37.49441	–173.65736	–141.66869
71.0	–17.73191	–21.72027	–43.67479	–43.49871	–188.09829	–155.59288
72.0	–21.57402	–25.29687	–51.74673	–51.45124	–206.74467	–173.72343
73.0	–26.84065	–30.30212	–62.89917	–62.48606	–232.04639	–198.51007
74.0	–34.56344	–37.76707	–79.35991	–78.83068	–268.81522	–234.76448
75.0	–47.08405	–50.03279	–106.18639	–105.54230	–327.97489	–293.41024
76.0	–71.10213	–73.79846	–157.84928	–157.09137	–440.78150	–405.70340
77.0	–136.63986	–139.08574	–299.19529	–298.32438	–747.30212	–711.71088
78.0	–1079.64772	–1081.84470	–2335.85632	–2334.87304	–5147.65625	–5111.55230
79.0	193.20900	191.25985	413.69411	414.78935	790.08619	826.70302
80.0	90.56777	88.86572	192.20250	193.40943	310.45195	347.58137
81.0	59.81585	58.36057	125.98970	127.30822	166.21589	203.85793
82.0	44.94975	43.74129	94.08989	95.52009	96.09516	134.24985
83.0	36.14484	35.18359	75.28241	76.82451	54.25366	92.92115

(continued 3) $\alpha = 1.0$

λ	A	B	C	D	G	H
84.0	30.29290	29.57958	62.85385	64.50825	26.19081	65.37129
85.0	26.10036	25.63607	54.01047	55.77770	5.87230	45.56604
86.0	22.93265	22.71878	47.38186	49.26261	–9.661780	30.54555
87.0	20.44190	20.48017	42.21690	44.21199	–22.03373	18.68758
88.0	18.42139	18.71387	38.06941	40.17983	–32.20715	9.02862
89.0	16.74059	17.28965	34.65767	36.88454	–40.79040	0.96035
90.0	15.31294	16.12129	31.79501	34.13960	–48.18660	–5.92027
91.0	14.07876	15.14945	29.35283	31.81658	–54.67335	–11.89076
92.0	12.99558	14.33198	27.23966	29.82412	–60.44813	–17.14857
93.0	12.03229	13.63814	25.38861	28.09553	–65.65526	–21.83792
94.0	11.16560	13.04497	23.74967	26.58093	–70.40265	–26.06665
95.0	10.37768	12.53503	22.28465	25.24233	–74.77260	–29.91699
96.0	9.65466	12.09482	20.96392	24.05024	–78.82901	–33.45275
97.0	8.98557	11.71376	19.76410	22.98147	–82.62223	–36.72422
98.0	8.36157	11.38344	18.66648	22.01751	–86.19247	–39.77151
99.0	7.77548	11.09713	17.65593	21.14341	–89.57227	–42.62706
100.0	7.22140	10.84936	16.72006	20.34701	–92.78820	–45.31737
105.0	4.78204	10.05777	12.85459	17.23223	–107.06459	–56.94110
110.0	2.63660	9.81199	9.83827	15.08205	–119.40866	–66.57964
115.0	0.55384	9.98895	7.26378	13.53846	–130.65824	–75.04827
120.0	–1.64352	10.57917	4.89111	12.43805	–141.26647	–82.76522
125.0	–4.14984	11.66806	2.53874	11.72701	–151.51670	–89.95552
130.0	–7.25414	13.47556	0.01588	11.44692	–161.62310	–96.72722
135.0	–11.49949	16.49756	–2.96565	11.78405	–171.80412	–103.08215
140.0	–18.15822	21.97166	–7.01647	13.27395	–182.40252	–108.83964
145.0	–31.25653	33.89643	–13.99873	17.71911	–194.31954	–113.22744
150.0	–74.41023	75.86380	–34.71734	35.87282	–212.53341	–111.37885
155.0	350.50054	–350.26856	158.28342	–159.76924	–133.21402	–207.22896
160.0	54.38374	–55.43159	21.96492	–26.21808	–204.39147	–152.85499
165.0	28.98505	–31.39628	9.20308	–16.40436	–219.29899	–155.14208
170.0	18.92833	–22.81623	3.35072	–13.74514	–231.26027	–160.90544
175.0	13.14205	–18.65671	–0.67959	–13.23264	–242.66651	–167.92611
180.0	9.08228	–16.42159	–4.09032	–13.76772	–254.15496	–175.79029
185.0	5.82700	–15.25350	–7.35769	–15.01400	–266.06234	–184.46361
190.0	2.93605	–14.80342	–10.76042	–16.88981	–278.68404	–194.07688
195.0	0.14029	–14.93623	–14.53573	–19.44778	–292.37560	–204.90133
200.0	–2.77830	–15.64092	–18.95977	–22.85943	–307.64497	–217.39714

VALUES OF THE FUNCTIONS *A* TO *H*

$\alpha = 1.1$

λ	*A*	*B*	*C*	*D*	*G*	*H*
0.2	3.45424	1.86626	3.82444	4.84513	7.62254	7.73259
0.4	3.42093	1.87642	3.79090	4.83246	7.32160	7.54178
0.6	3.38735	1.88674	3.75717	4.81982	7.02043	7.35085
0.8	3.35351	1.89725	3.72325	4.80719	6.71904	7.15980
1.0	3.31941	1.90793	3.68913	4.79458	6.41743	6.96862
1.2	3.28503	1.91879	3.65483	4.78199	6.11559	6.77732
1.4	3.25038	1.92985	3.62032	4.76942	5.81353	6.58589
1.6	3.21544	1.94109	3.58561	4.75687	5.51123	6.39434
1.8	3.18022	1.95253	3.55069	4.74435	5.20870	6.20266
2.0	3.14470	1.96418	3.51557	4.73185	4.90593	6.01086
2.2	3.10888	1.97603	3.48024	4.71938	4.60292	5.81893
2.4	3.07276	1.98809	3.44469	4.70694	4.29967	5.62687
2.6	3.03633	2.00037	3.40892	4.69453	3.99617	5.43468
2.8	2.99959	2.01287	3.37293	4.68215	3.69242	5.24237
3.0	2.96252	2.02560	3.33672	4.66981	3.38843	5.04993
3.2	2.92511	2.03856	3.30028	4.65751	3.08418	4.85736
3.4	2.88738	2.05177	3.26360	4.64524	2.77967	4.66466
3.6	2.84929	2.06521	3.22669	4.63302	2.47490	4.47183
3.8	2.81086	2.07891	3.18953	4.62084	2.16986	4.27887
4.0	2.77207	2.09287	3.15213	4.60871	1.86456	4.08578
4.2	2.73291	2.10710	3.11448	4.59662	1.55899	3.89257
4.4	2.69338	2.12159	3.07658	4.58459	1.25314	3.69922
4.6	2.65347	2.13637	3.03841	4.57261	0.94702	3.50574
4.8	2.61316	2.15143	2.99999	4.56069	0.64061	3.31214
5.0	2.57246	2.16679	2.96129	4.54883	0.33392	3.11840
5.2	2.53135	2.18245	2.92232	4.53703	0.02694	2.92453
5.4	2.48982	2.19842	2.88308	4.52529	–0.28034	2.73053
5.6	2.44786	2.21471	2.84354	4.51363	–0.58791	2.53640
5.8	2.40547	2.23133	2.80373	4.50204	–0.9578	2.34214
6.0	2.36263	2.24828	2.76361	4.49052	–1.20396	2.14775
6.2	2.31933	2.26559	2.72320	4.47908	–1.51244	1.95323
6.4	2.27556	2.28325	2.68248	4.46773	–1.82125	1.75859
6.6	2.23131	2.30128	2.64144	4.45647	–2.13037	1.56381
6.8	2.18657	2.31968	2.60009	4.44529	–2.43981	1.36890
7.0	2.14133	2.33848	2.55841	4.43422	–2.74958	1.17386
7.2	2.09557	2.35767	2.51640	4.42324	–3.05969	0.97870
7.4	2.04928	2.37728	2.47405	4.41237	–3.37013	0.78341
7.6	2.00245	2.39731	2.43135	4.40161	–3.68092	0.58799
7.8	1.95506	2.41778	2.38830	4.39097	–3.99205	0.39244
8.0	1.90710	2.43870	2.34489	4.38045	–4.30354	0.19677
8.2	1.85855	2.46008	2.30111	4.37005	–4.61539	0.00097

(continued 1) $\alpha = 1.1$

λ	A	B	C	D	G	H
8.4	1.80940	2.48194	2.25695	4.35979	–4.92760	–0.19495
8.6	1.75964	2.50430	2.21241	4.34966	–5.24018	–0.39100
8.8	1.70924	2.52716	2.16746	4.33968	–5.55314	–0.58716
9.0	1.65818	2.55056	2.12212	4.32985	–5.86649	–0.78345
9.2	1.60646	2.57449	2.07636	4.32018	–6.18022	–0.97985
9.4	1.55404	2.59898	2.03017	4.31068	–6.49435	–1.17638
9.6	1.50092	2.62405	1.98356	4.30135	–6.80889	–1.37302
9.8	1.44707	2.64972	1.93649	4.29220	–7.12383	–1.56977
10.0	1.39247	2.67601	1.88897	4.28324	–7.43919	–1.76664
11.0	1.10738	2.81746	1.64412	4.24162	–9.02254	–2.75259
12.0	0.79977	2.97788	1.38592	4.20626	–10.61757	–3.74097
13.0	0.46568	3.16103	1.11238	4.17877	–12.22552	–4.73130
14.0	0.10016	3.37166	0.82099	4.16125	–13.84790	–5.72293
15.0	–0.30316	3.61590	0.50857	4.15642	–15.48654	–6.71491
16.0	–0.75260	3.90184	0.17103	4.16789	–17.14375	–7.70590
17.0	–1.25920	4.24025	–0.19706	4.20056	–18.82248	–8.69400
18.0	–1.83802	4.64590	–0.60302	4.26117	–20.52663	–9.67646
19.0	–2.51010	5.13952	–1.05700	4.35927	–22.26146	–10.64925
20.0	–3.30585	5.75116	–1.57362	450877	–24.03438	–11.60633
21.0	–4.27084	6.52605	–2.17466	4.73073	–25.85624	–12.53832
22.0	–5.47682	7.53552	–2.89426	5.05848	–27.74375	–13.43011
23.0	–7.04356	8.89890	–3.78929	5.54798	–29.72441	–14.25599
24.0	–9.18741	10.83203	–4.96216	6.30069	–31.84729	–14.96883
25.0	–12.34329	13.76930	–6.61845	7.52111	–34.21017	–15.47299
26.0	–17.53458	18.73351	–9.24141	9.69128	–37.04194	–15.54195
27.0	–27.87857	28.84127	–14.30209	14.28095	–41.03755	–14.48337
28.0	–59.55495	60.27155	–29.43373	28.92191	–49.79905	–8.69811
29.0	1292.67447	–1292.21466	608.27637	–609.30016	249.39771	–310.91354
30.0	57.87135	–57.67993	25.55940	–27.11829	–27.15282	–37.42779
31.0	30.19998	–30.28956	12.29052	–14.40965	–35.08850	–32.60697
32.0	20.56760	–20.95195	7.52095	–10.22774	–39.03553	–31.82944
33.0	15.59099	–16.28516	4.93626	–8.26070	–41.97468	–32.11957
34.0	12.49630	–13.51680	3.22677	–7.20178	–44.52600	–32.86312
35.0	10.34436	–11.70937	1.94819	–6.60998	–46.90159	–33.85457
36.0	8.72863	–10.45822	0.90727	–6.29585	–49.19549	–35.00742
37.0	7.44404	–9.56046	0.00564	–6.16533	–51.45740	–36.28060
38.0	6.37556	–8.90357	–0.81289	–6.16725	–53.71764	–37.65376
39.0	5.45326	–8.42051	–1.58352	–6.27220	–55.99721	–39.11743
40.0	4.63175	–8.06930	–2.33022	–6.46293	–58.31256	–40.66862
41.0	3.87985	–7.82270	–3.07074	–6.72963	–60.67805	–42.30869
42.0	3.17496	–7.66283	–3.81928	–7.06742	–63.10751	–44.04245

(continued 2) $\alpha = 1.1$

λ	*A*	*B*	*C*	*D*	*G*	*H*
43.0	2.49982	−7.57797	−4.58825	−7.47499	−65.61521	−45.87774
44.0	1.84043	−7.56081	−5.38930	−7.95399	−68.21679	−47.82543
45.0	1.18476	−7.60738	−6.23427	−8.50868	−70.93004	−49.89976
46.0	0.52174	−7.71647	−7.13584	−9.14596	−73.77586	−52.11891
47.0	−0.15946	−7.88937	−8.10838	−9.87568	−76.77944	−54.50592
48.0	−0.87021	−8.12987	−9.16876	−10.71115	−79.97168	−57.09001
49.0	−1.62313	−8.44450	−10.33749	−11.66999	−83.39129	−59.90853
50.0	−2.43296	−8.84306	−11.64021	−12.77553	−87.08754	−63.00967
51.0	−3.31775	−9.33954	−13.10978	−14.05869	−91.12449	−66.45657
52.0	−4.30048	−9.95354	−14.78943	−15.56112	−95.58705	−70.33338
53.0	−5.41149	−10.71257	−16.73746	−17.33977	−100.59040	−74.75465
54.0	−6.69231	−11.65575	−19.03481	−19.47445	−106.29494	−79.88026
55.0	−8.20195	−12.84005	−21.79750	−22.08022	−112.93113	−85.94019
56.0	−10.02774	−14.35103	−25.19821	−25.32891	−120.84244	−93.27752
57.0	−12.30532	−16.32278	−29.50605	−29.48889	−130.56453	−102.42757
58.0	−15.25802	−18.97728	−35.16502	−35.00355	−142.98169	−114.27432
59.0	−19.28228	−22.70980	−42.96420	−42.66139	−159.66584	−130.38944
60.0	−25.15664	−28.29779	−54.45413	−54.01244	−183.70633	−153.86202
61.0	−34.64965	−37.50887	−73.15823	−72.57968	−222.12381	−191.71250
62.0	−52.82984	−55.41067	−109.17572	−108.46191	−295.04230	−264.06472
63.0	−102.49964	−104.80488	−207.94366	−207.09582	−493.00247	−461.45915
64.0	−843.69980	−845.73150	−1684.70230	−1683.72132	−3436.83874	−3404.73005
65.0	144.70035	142.94084	285.05387	286.16747	487.07785	519.75171
66.0	68.31982	66.83176	133.05708	134.30302	183.04901	216.28795
67.0	45.28489	44.06815	87.36011	88.73843	90.84020	124.64429
68.0	34.10109	33.15610	65.27875	66.78977	45.68964	80.05906
69.0	27.44751	26.77527	52.22532	53.86961	18.52912	53.46422
70.0	23.00294	22.60499	43.57482	45.35322	0.14278	35.64401
71.0	19.80010	19.67851	37.40018	39.31379	−13.30919	22.75873
72.0	17.36402	17.52138	32.75535	34.80551	−23.71177	12.92355
73.0	15.43414	15.87355	29.12148	31.30980	−32.09865	5.10489
74.0	13.85548	14.58060	26.19021	28.51855	−39.08413	−1.31142
75.0	12.53010	13.54511	23.76673	26.23721	−45.05616	−6.71322
76.0	11.39297	12.70260	21.72189	24.33689	−50.27192	−11.35754
77.0	10.39921	12.00876	19.96668	22.72887	−54.90857	−15.42141
78.0	9.51679	11.43214	18.43772	21.35005	−59.09218	−19.03075
79.0	8.72220	10.94987	17.08864	20.15438	−62.91491	−22.27759
80.0	7.99779	10.54492	15.88473	19.10744	−66.44571	−25.23073
81.0	7.32998	10.20440	14.79945	18.18303	−69.73721	−27.94260
82.0	6.70813	9.91840	13.81215	17.36087	−72.83028	−30.45394
83.0	6.12375	9.67922	12.90652	16.62502	−75.75715	−32.79676

(continued 3) $\alpha = 1.1$

λ	A	B	C	D	G	H
84.0	5.56996	9.48080	12.06944	15.96278	–78.54363	–34.99669
85.0	5.04109	9.31837	11.29029	15.36396	–81.21058	–37.07435
86.0	4.53236	9.18813	10.56029	14.82027	–83.77512	–39.04664
87.0	4.03972	9.08708	9.87216	14.32493	–86.25138	–40.92744
88.0	3.55964	9,01286	9.21973	13.87238	–88.65120	–42.72829
89.0	3.08902	8.96362	8.59778	13.45798	–90.98454	–44.45886
90.0	2.62506	8.93798	8.00178	13.07792	–93.25989	–46.12729
91.0	2.16522	8.93491	7.42780	12.72902	–95.48452	–47.74048
92.0	1.70711	8.95374	6.87237	12.40864	–97.66473	–49.30431
93.0	1.24844	8.99410	6.33239	12.11462	–99.80599	–50.82381
94.0	0.78702	9.05590	5.80503	11.84519	–101.91313	–52.30326
95.0	0.32065	9.13933	5.28772	11.59894	–103.99041	–53.74638
96.0	–0.15291	9.24485	4.77803	11.37478	–106.04168	–55.15631
97.0	–0.63594	9.37321	4.27366	11.17191	–108.07038	–56.53581
98.0	–1.13086	9.52544	3.77240	10.98980	–110.07967	–57.88716
99.0	–1.64027	9.70291	3.27203	10.82822	–112.07247	–59.21234
100.0	–2.16697	9.90732	2.77036	10.68717	–114.05150	–60.51296
105.0	–5.18381	11.42614	0.15767	10.31214	–123.82903	–66.68387
110.0	–9.35651	14.22696	–2.93110	10.66439	–133.65346	–72.30771
115.0	–16.27946	19.84139	–7.32984	12.45258	–144.03018	–77.12866
120.0	–32.18370	34.45413	–16.14809	18.69397	–156.56290	–79.72704
125.0	–135.33859	136.29604	–68.03864	67.96455	–190.38338	–61.12423
130.0	73.55428	–73.96758	34.01465	–36.82426	–148.32362	–118.63264
135.0	28.90532	–30.78570	10.87909	–16.61664	–168.18608	–114.60327
140.0	17.07588	–20.56503	3.83250	–12.78096	–180.32182	–118.86572
145.0	11.03761	–16.33530	–0.50652	–12.05170	–191.42125	–124.96082
150.0	6.94868	–14.33427	–4.09362	–12.63212	–202.56884	–132.12135
155.0	3.63721	–13.50622	–7.59434	–14.08848	–214.26712	–140.30735
160.0	0.56795	–13.49588	–11.40805	–16.38070	–226.98313	–149.76905
165.0	–2.62187	–14.23904	–15.93806	–19.70141	–241.36364	–161.04921
170.0	–6.31699	–15.88903	–21.77852	–24.53053	–258.50879	–175.19161
175.0	–11.12763	–18.91858	–30.04422	–31.91500	–280.59291	–194.33709
180.0	–18.37689	–24.56279	–43.31011	–44.38632	–312.72330	–223.57152
185.0	–32.02256	–36.71930	–69.43132	–69.76995	–370.55755	–278.53781
190.0	–72.89958	–76.17924	–149.96434	–149.60072	–537.03191	–442.16155
195.0	1031.62820	1029.72764	2055.85945	2056.90686	3860.36183	3958.07402
200.0	69.43851	68.90761	136.22371	137.95053	22.76863	123.32031

$\alpha = 1.2$

λ	A	B	C	D	G	H
0.2	3.13333	1.72001	3.02965	4.05439	5.65261	5.77268
0.4	3.09609	1.73173	2.99343	4.04320	5.36225	5.60254
0.6	3.05849	1.74370	2.95695	4.03205	5.07162	5.43228
0.8	3.02053	1.75591	2.92021	4.02095	4.78072	5.26190
1.0	2.98221	1.76837	2.88321	4.00990	4.48953	5.09139
1.2	2.94350	1.78110	2.84593	3.99890	4.19806	4.92077
1.4	2.90440	1.79409	2.80838	3.98795	3.90630	4.75002
1.6	2.86490	1.80736	2.77055	3.97706	3.61425	4.57914
1.8	2.82500	1.82092	2.73243	3.96624	3.32190	4.40815
2.0	2.78468	1.83478	2.69402	3.95547	3.02925	4.23703
2.2	2.74393	1.84893	2.65531	3.94478	2.73629	4.06579
2.4	2.70275	1.86340	2.61629	3.93415	2.44303	3.89443
2.6	2.66111	1.87819	2.57697	3.92361	2.14944	3.72295
2.8	2.61902	1.89331	2.53732	3.91314	1.85554	3.55134
3.0	2.57645	1.90877	2.49735	3.90275	1.56131	3.37962
3.2	2.53340	1.92459	2.45705	3.89246	1.26675	3.20777
3.4	2.48986	1.94076	2.41641	3.88225	0.97185	3.03580
3.6	2.44580	1.95732	2.37543	3.87215	0.67661	2.86372
3.8	2.40122	1.97426	2.33409	3.86214	0.38102	2.69151
4.0	2.35611	1.99160	2.29239	3.85225	0.08508	2.51919
4.2	2.31044	2.00936	2.25031	3.84246	–0.21123	2.34675
4.4	2.26421	2.02754	2.20786	3.83280	–0.50790	2.17419
4.6	2.21739	2.04616	2.16502	3.82326	–0.80494	2.00152
4.8	2.16997	2.06524	2.12178	3.81386	–1.10236	1.82874
5.0	2.12194	2.08479	2.07813	3.80459	–1.40017	1.65584
5.2	2.07327	2.10483	2.03407	3.79547	–1.69837	1.48283
5.4	2.02395	2.12538	1.98958	3.78650	–1.99698	1.30971
5.6	1.97396	2.14644	1.94465	3.77769	–2.29599	1.13648
5.8	1.92327	2.16805	1.89926	3.76905	–2.59542	0.96315
6.0	1.87187	2.19022	1.85342	3.76059	–2.89527	0.78971
6.2	1.81973	2,21297	1.80710	3.75231	–3.19556	0.61617
6.4	1.76684	2.23632	1.76030	3.74423	–3.49629	0.44254
6.6	1.71316	2.26029	1.71299	3.73635	–3.79747	0.26880
6.8	1.65867	2.28491	1.66517	3.72869	–4.09911	0.09497
7.0	1.60334	2.31021	1.61682	3.72126	–4.40122	–0.07895
7.2	1.54716	2.33620	1.56793	3.71406	–4.70381	–0.25296
7.4	1.49008	2.36292	1.51848	3.70712	–5.00689	–0.42705
7.6	1.43208	2.39039	1.46845	3.70044	–5.31047	–0.60122
7.8	1.37313	2.41864	1.41783	3.69403	–5.61456	–0.77547
8.0	1.31319	2.44771	1.36659	3.68792	–5.91918	–0.94979
8.2	1.25223	2.47763	1.31472	3.68211	–6.22433	–1.12418

(continued 1) $\alpha = 1.2$

λ	A	B	C	D	G	H
8.4	1.19021	2.50842	1.26221	3.67663	−6.53003	−1.29862
8.6	1.12709	2.54014	1.20902	3.67148	−6.83630	−1.47313
8.8	1.06283	2.57281	1.15514	3.66669	−7.14314	−1.64768
9.0	0.99739	2.60648	1.10054	3.66227	−7.45057	−1.82227
9.2	0.93073	2.64119	1.04520	3.65825	−7.75861	−1.99690
9.4	0.86279	2.67698	0.98909	3.65465	−8.06727	−2.17156
9.6	0.79352	2.71391	0.93219	3.65149	−8.37656	−2.34626
9.8	0.72288	2.75202	0.87447	3.64879	−8.68651	−2.52094
10.0	0.65081	2.79137	0.81590	3.64657	−8.99714	−2.69563
11.0	0.26671	3.00881	0.50910	3.64381	−10.56105	−3.56874
12.0	−0.16385	3.26741	0.17543	3.65809	−12.14509	−4.43997
13.0	−0.65309	3.57904	−0.19148	3.6909	−13.75290	−5.30694
14.0	−1.21818	3.96042	−0.60050	3.76292	−15.38944	−6.16610
15.0	−1.88396	4.43599	−1.06436	3.87352	−17.06164	−7.01204
16.0	−2.68797	5.04278	−1.60221	4.04508	−18.77971	−7.83625
17.0	−3.68968	5.83970	−2.24418	4.30674	−20.55943	−8.62485
18.0	−4.98939	6.92645	−3.04091	4.70800	−22.42691	−9.35379
19.0	−6.77076	8.48604	−4.08482	5.34005	−24.42927	−9.97829
20.0	−9.41048	10.89437	−5.56503	6.39058	−26.66218	−10.40530
21.0	−13.82283	15.06488	−7.93991	8.31639	−29.35619	−10.40718
22.0	−22.95305	23.94185	−12.68480	12.59105	−33.25122	−9.24730
23.0	−54.47240	55.19547	−28.64855	28.06140	−42.77932	−2.49735
24.0	227.14299	−226.69937	112.16697	−113.27295	26.17573	−74.27774
25.0	39.11731	−38.96824	17.81992	−19.472[illegible]9	−22.63564	−28.34375
26.0	21.82196	−21.98416	8.94731	−11.17806	−28.67042	−25.24387
27.0	15.19391	−15.68593	5.40330	−8.24635	−32.06046	−24.85258
28.0	11.60558	−12.44809	3.36726	−6.86097	−34.72145	−25.26154
29.0	9.29559	−10.51174	1.95558	−6.14292	−37.09892	−26.03370
30.0	7.63788	−9.25373	0.85285	−5.78213	−39.35445	−27.01747
31.0	6.35314	−8.39813	−0.08311	−5.64275	−41.56350	−28.14909
32.0	5.29712	−7.80475	−0.92727	−5.65732	−43.76868	−29.39987
33.0	4.38701	−7.39558	−1.72434	−5.79005	−45.99845	−30.75794
34.0	3.57101	−7.12464	−2.50421	−6.02185	−48.27507	−32.22105
35.0	2.81413	−6.96403	−3.28918	−6.34353	−50.61859	−33.79347
36.0	2.09090	−6.89693	−4.09789	−6.75255	−53.04925	−35.48485
37.0	1.38116	−6.91398	−4.94773	−7.25152	−55.58930	−37.31003
38.0	0.66751	−7.01133	−5.85661	−7.84772	−58.26476	−39.28970
39.0	−0.06654	−7.18981	−6.84449	−8.55325	−61.10737	−41.45166
40.0	−0.83811	−7.45471	−7.93515	−9.38596	−64.15720	−43.83303
41.0	−1.66657	−7.81627	−9.15832	−10.37103	−67.46635	−46.48361
42.0	−2.57544	−8.29087	−10.55283	−11.54368	−71.10439	−49.47117

(continued 2) $\alpha = 1.2$

λ	*A*	*B*	*C*	*D*	*G*	*H*
43.0	−3.59512	−8.90322	−12.17119	−12.95359	−75.16686	−52.88983
44.0	−4.76707	−9.69017	−14.08723	−14.67227	−79.78912	−56.87378
45.0	−6.15075	−10.70743	−16.40872	−16.80562	−85.16987	−61.62079
46.0	−7.83605	−12.04178	−19.30033	−19.51671	−91.61354	−67.43449
47.0	−9.96703	−13.83462	−23.02751	−23.06971	−99.61172	−74.80581
48.0	−12.79143	−16.33149	−28.04841	−27.92161	−110.01364	−84.58341
49.0	−16.77636	−19.99758	−35.22834	−34.93679	−124.42402	−98.37154
50.0	−22.92247	−25.83185	−46.42261	−45.96970	−146.28438	−119.61129
51.0	−33.83374	−36.43686	−66.46134	−65.84973	−184.55534	−157.26289
52.0	−59.08392	−61.38502	−113.10539	−112.33708	−272.18743	−244.27656
53.0	−185.75453	−187.75668	−347.91272	−346.98910	−708.91204	−680.38338
54.0	187.51660	185.81140	344.79975	345.87782	575.15177	604.29790
55.0	64.80247	63.39321	117.33052	118.56269	152.04032	181.80379
56.0	29.86982	38.75643	71.27110	72.65750	65.49102	95.87198
57.0	29.04751	28.23078	51.39045	52.93165	27.51043	58.50927
58.0	22.94002	22.42161	40.25803	41.95504	5.76071	37.37804
59.0	18.97724	18.75962	33.10619	34.96044	−8.60309	23.63353
60.0	16.16897	16.25542	28.09837	30.11172	−18.98798	13.86897
61.0	14.05245	14.44708	24.37662	26.55132	−26.98543	6.49312
62.0	12.38236	13.09009	21.48630	23.82506	−33.43858	0.66303
63.0	11.01641	12.04298	19.16400	21.66995	−38.83586	−4.10949
64.0	9.86629	11.21832	17.24649	19.92321	−43.47981	−8.12675
65.0	8.87425	10.55926	15.62717	18.47871	−47.56786	−11.58593
66.0	8.00079	10.02725	14.23346	17.26437	−51.23437	−14.62112
67.0	7.21798	9.59536	13.01416	16.22950	−54.57395	−17.32669
68.0	6.50538	9.24423	11.93209	15.33750	−57.65523	−19.77093
69.0	5.84765	8.95968	10.95955	14.56128	−60.52917	−22.00451
70.0	5.23297	8.73115	10.07545	13.88040	−63.23447	−24.06577
71.0	4.65200	8.55069	9.26340	13.27921	−65.80107	−25.98428
72.0	4.09721	8.41227	8.51041	12.74551	−68.25250	−27.78314
73.0	3.56237	8.31134	7.80605	12.26974	−70.60756	−29.48069
74.0	3.04221	8.24448	7.14175	11.84433	−72.88149	−31.09164
75.0	2.53212	8.20919	6.51039	11.46324	−75.08676	−32.62792
76.0	2.02803	8.20372	5.90593	11.12169	−77.23377	−34.09924
77.0	1.52617	8.22696	5.32317	10.81584	−79.33125	−35.51359
78.0	1.02301	8.27838	4.75752	10.54273	−81.38663	−36.87759
79.0	0.51510	8.35796	4.20488	10.30004	−83.40633	−38.19669
80.0	−0.00102	8.46623	3.66148	10.08610	−85.39597	−39.47542
81.0	−0.52889	8.60420	3.12375	9.89976	−87.36054	−40.71746
82.0	−1.07227	8.77344	2.58827	9.74042	−89.30455	−41.92587
83.0	−1.63524	8.97612	2.05163	9.60794	−91.23213	−43.10303

(continued 3) $\alpha = 1.2$

λ	A	B	C	D	G	H
84.0	–2.22231	9.21510	1.51032	9.50272	–93.14720	–44.25080
85.0	–2.83860	9.49402	0.96067	9.42572	–95.05349	–45.37048
86.0	–3.49003	9.81748	0.39869	9.37850	–96.95474	–46.46285
87.0	–4.18352	10.19127	–0.18005	9.36335	–98.85471	–47.52815
88.0	–4.92740	10.62264	–0.78063	9.38340	–100.75736	–48.56599
89.0	–5.73172	11.12073	–1.40900	9.44284	–102.66701	–49.57532
90.0	–6.60895	11.69713	–2.07231	9.54724	–104.58848	–50.55427
91.0	–7.57471	12.36668	–2.77933	9.70392	–106.52737	–51.49992
92.0	–8.64897	13.14864	–3.54108	9.92256	–108.49039	–52.40807
93.0	–9.85774	14.06835	–4.37175	10.21610	–110.48586	–53.27269
94.0	–11.23562	15.15979	–5.29001	10.60208	–112.52440	–54.08529
95.0	–12.82975	16.46951	–6.32116	11.10472	–114.62009	–54.83378
96.0	–14.70615	18.06299	–7.50044	11.75822	–116.79220	–55.50077
97.0	–16.96034	20.03522	–8.87862	12.61240	–119.06824	–56.06061
98.0	–19.73590	22.52931	–10.53190	13.74253	–121.48902	–56.47415
99.0	–23.25940	25.77131	–12.58024	15.26768	–124.11852	–56.67908
100.0	–27.91039	30.14032	–15.22444	17.38778	–127.06336	–56.57039
105.0	–197.51674	198.31386	–105.76095	105.26090	–182.65296	–15.24172
110.0	45.59384	–46.30898	21.63853	–24.95001	–123.23192	–89.19968
115.0	19.93562	–22.30377	6.98548	–13.37041	–139.01504	–88.44117
120.0	11.60241	–15.84107	1.33811	–11.20120	–150.31122	–92.92260
125.0	6.83029	–13.26434	–2.64484	–11.30069	–161.13082	–99.00470
130.0	3.20976	–12.32857	–6.34557	–12.59266	–172.40756	–106.32271
135.0	–0.12527	–12.44287	–10.40472	–14.94820	–184.82051	–115.14923
140.0	–3.73103	–13.55833	–15.45384	–18.67923	–199.33135	–126.27713
145.0	–8.28743	–16.05610	–22.55847	–24.69347	–217.81664	–141.49959
150.0	–15.21260	–21.18757	–34.24494	–35.42917	–244.91601	–165.40970
155.0	–29.16606	–33.50939	–59.11022	–59.42893	–296.58124	–213.93052
160.0	–84.08074	–86.88497	–160.21401	–159.71569	–490.06179	–404.29206
165.0	151.70522	150.39919	279.27857	280.57277	321.80519	410.68297
170.0	42.04421	42.23797	76.33484	78.42625	–61.08338	30.90347
175.0	24.02184	25.75593	43.84064	46.75129	–126.96786	–31.86002
180.0	16.03807	19.39284	30.06860	33.84166	–158.00573	–59.75371
185.0	11.14706	16.24790	22.13127	26.83385	–178.14822	–76.71608
190.0	7.55309	14.58072	16.72580	22.45449	–193.51818	–88.85432
195.0	4.56073	13.76900	12.60743	19.49775	–206.40469	–98.43675
200.0	1.81644	13.56192	9.18681	17.42903	–217.86746	–106.49391

$\alpha = 1.3$

λ	*A*	*B*	*C*	*D*	*G*	*H*
0.2	2.85866	1.59733	2.41023	3.43942	4.25477	4.38487
0.4	2.81724	1.61081	2.37105	3.42987	3.97217	4.23261
0.6	2.77533	1.62462	2.33153	3.42039	3.68924	4.08023
0.8	2.73293	1.63878	2.29165	3.41101	3.40596	3.92775
1.0	2.69002	1.65329	2.25141	3.40172	3.12233	3.77515
1.2	2.64659	1.66818	2.21081	3.39253	2.83834	3.62244
1.4	2.60262	1.68344	2.16982	3.38345	2.55398	3.46962
1.6	2.55809	1.69910	2.12844	3.37448	2.26925	3.31670
1.8	2.51299	1.71517	2.08666	3.36564	1.98413	3.16367
2.0	2.46731	1.73166	2.04448	3.35691	1.69863	3.01053
2.2	2.42102	1.74860	2.00188	3.34832	1.41274	2.85729
2.4	2.37411	1.76599	1.95884	3.33987	1.12644	2.70394
2.6	2.32655	1.78385	1.91536	3.33157	0.83973	2.55050
2.8	2.27834	1.80221	1.87143	3.32342	0.55260	2.39695
3.0	2.22944	1.82107	1.82704	3.31544	0.26503	2.24331
3.2	2.17983	1.84047	1.78216	3.30763	–0.02297	2.08958
3.4	2.12949	1.86042	1.73679	3.30000	–0.31142	1.93575
3.6	2.07840	1.88094	1.69091	3.29257	–0.60032	1.78184
3.8	2.02653	1.90205	1.64451	3.28534	–0.88970	1.62784
4.0	1.97386	1.92379	1.59758	3.27832	–1.17956	1.47376
4.2	1.92035	1.94617	1.55009	3.27153	–1.46990	1.31960
4.4	1.86598	1.96922	1.50202	3.26498	–1.76075	1.16536
4.6	1.81072	1.99298	1.45338	3.25869	–2.05212	1.01105
4.8	1.75453	2.01746	1.40412	3.25266	–2.34401	0.85668
5.0	1.69739	2.04271	1.35424	3.24691	–2.63644	0.70225
5.2	1.63924	2.06875	1.30371	3.24146	–2.92942	0.54776
5.4	1.58007	2.09562	1.25251	3.23632	–3.22297	0.39322
5.6	1.51982	2.12335	1.20061	3.23152	–3.51711	0.23864
5.8	1.45845	2.15199	1.14801	3.22706	–3.81185	0.08403
6.0	1.39593	2.18158	1.09466	3.22298	–4.10720	–0.07062
6.2	1.33220	2.21216	1.04054	3.21929	–4.40318	–0.22529
6.4	1.26721	2.24378	0.98563	3.21601	–4.69982	–0.37997
6.6	1.20091	2.27649	0.92990	3.21318	–4.99712	–0.53465
6.8	1.13325	2.31033	0.87330	3.21080	–5.29511	–0.68932
7.0	1.06416	2.34538	0.81582	3.20893	–5.59382	–0.84398
7.2	0.99358	2.38168	0.75741	3.20757	–5.89326	–0.99860
7.4	0.92145	2.41930	0.69804	3.20677	–6.19345	–1.15318
7.6	0.84770	2.45830	0.63767	3.20655	–6.49442	–1.30770
7.8	0.77224	2.49877	0.57625	3.20696	–6.79621	–1.46215
8.0	0.69499	2.54077	0.51374	3.20802	–7.09882	–1.61652
8.2	0.61588	2.58440	0.45010	3.20979	–7.40230	–1.77077

(continued 1) $\alpha = 1.3$

λ	A	B	C	D	G	H
8.4	0.53481	2.62973	0.38526	3.21231	−7.70667	−1.92490
8.6	0.45167	2.67687	0.31919	3.21563	−8.01197	−2.07888
8.8	0.36637	2.72591	0.25181	3.21979	−8.31824	−2.23269
9.0	0.27879	2.77697	0.18307	3.22485	−8.62550	−2.38630
9.2	0.18880	2.83017	0.11290	3.23087	−8.93380	−2.53969
9.4	0.09628	2.88562	0.04122	3.23792	−9.24318	−2.69283
9.6	0.00108	2.94348	−0.03203	3.24607	−9.55369	−2.84569
9.8	−0.09695	3.00388	−0.10695	3.25539	−9.86537	−2.99823
10.0	−0.19797	3.06699	−0.18362	3.26596	−10.17828	−3.15041
11.0	−0.75484	3.42982	−0.59695	3.34103	−11.76336	−3.90431
12.0	−1.42127	3.89431	−1.07283	3.46487	−13.38976	−4.64087
13.0	−2.24388	4.50638	−1.63624	3.66126	−15.07138	−5.34830
14.0	−3.30122	5.34379	−2.32902	3.97068	−16.83109	−6.00608
15.0	−4.73692	6.54929	−3.22739	4.46780	−18.70997	−6.57583
16.0	−6.84522	8.41611	−4.48694	5.30649	−20.79107	−6.97755
17.0	−10.33942	11.65641	−6.48415	6.86119	−23.27389	−7.01519
18.0	−17.51314	18.56250	−10.42807	10.33869	−26.79676	−6.05443
19.0	−42.06740	42.83387	−23.53717	22.95483	−35.16390	−0.29561
20.0	181.92445	−181.45786	94.27506	−95.37996	25.41907	−63.53843
21.0	30.98280	−30.83513	14.5636	−16.22495	−18.07127	−22.76562
22.0	17.30140	−17.49407	7.14798	−9.40157	−23.50031	−20.11905
23.0	12.02436	−12.58161	4.14419	−7.03297	−26.63049	−19.84479
24.0	9.13166	−10.08106	2.37967	−5.95165	−29.13700	−20.27780
25.0	7.23599	−8.60914	1.12116	−5.43133	−31.41035	−21.03968
26.0	5.84394	−7.67724	0.10581	−5.21755	−33.59411	−22.00150
27.0	4.73445	−7.07021	−0.78642	−5.20054	−35.75774	−23.11166
28.0	3.79220	−6.67998	−1.62068	−5.32778	−37.94322	−24.35016
29.0	2.94954	−6.44795	−2.43810	−5.57388	−40.18180	−25.71317
30.0	2.16236	−6.34141	−3.26884	−5.92854	−42.50140	−27.20733
31.0	1.39865	−6.34299	−4.13866	−6.39149	−44.93107	−28.84795
32.0	0.63225	−6.44558	−5.07289	−6.97054	−47.50427	−30.65927
33.0	−0.16109	−6.65021	−6.09966	−7.68157	−50.26253	−32.67636
34.0	−1.00689	−6.96565	−7.25323	−8.55002	−53.26015	−34.94888
35.0	−1.93521	−7.40958	−8.57841	−9.61417	−56.57116	−37.54740
36.0	−2.98497	−8.01153	−10.13726	−10.93112	−60.30057	−40.57433
37.0	−4.21068	−8.81875	−12.02027	−12.58754	−64.60360	−44.18286
38.0	−5.69433	−9.90751	−14.36668	−14.71965	−69.72078	−48.61194
39.0	−7.56829	−11.40558	−17.40413	−17.55264	−76.04692	−54.25507
40.0	−10.06436	−13.54101	−21.53385	−21.48578	−84.27924	−61.80843
41.0	−13.63247	−16.76062	−27.53715	−27.29873	−95.77679	−72.63021
42.0	−19.28097	−22.07014	−37.16775	−36.74379	−113.59148	−89.77156

(continued 2) $\alpha = 1.3$

λ	A	B	C	D	G	H
43.0	–29.84068	–32.29814	–55.35196	–54.74611	–146.29801	–121.80656
44.0	–57.49607	–59.62713	–103.29852	–102.51335	–230.81448	–205.65275
45.0	–346.42483	–348.23308	–606.05464	–605.09184	–1107.03417	–1081.20293
46.0	95.63654	94.14911	163.64679	164.78643	231.75481	258.25529
47.0	43.51989	42.35268	73.11559	74.43198	73.10806	100.27788
48.0	28.61085	27.76463	47.35324	48.84706	27.20694	55.04660
49.0	21.45549	20.93227	35.08905	36.76165	4.80326	33.31363
50.0	17.19588	16.99892	27.86740	29.72081	–8.82157	20.36075
51.0	14.32976	14.46354	23.07401	25.11092	–18.21792	11.63795
52.0	12.23952	12.70972	19.63447	21.85821	–25.25566	5.27572
53.0	10.62432	11.43785	17.02586	19.44046	–30.84464	0.36459
54.0	9.31992	10.48498	14.96303	17.57323	–35.48081	–3.59101
55.0	8.22888	9.75501	13.27717	16.08843	–39.45760	–6.88410
56.0	7.28959	9.18773	11.86186	14.88044	–42.95978	–9.69901
57.0	6.46103	8.74367	10.64661	13.87961	–46.10956	–12.15748
58.0	5.71470	8.39597	9.58277	13.03822	–48.99105	–14.34312
59.0	5.03003	8.12588	8.63564	12.32256	–51.66411	–16.31520
60.0	4.39166	7.92006	7.77966	11.70822	–54.17253	–18.11691
61.0	3.78775	7.76891	6.99554	11.17715	–56.54916	–19.78039
62.0	3.20885	7.66551	6.26834	10.71582	–58.81919	–21.33003
63.0	2.64718	7.60495	5.58619	10.31396	–61.00230	–22.78464
64.0	2.09609	7.58384	4.93946	9.96379	–63.11419	–24.15888
65.0	1.54969	7.60009	4.32011	9.65938	–65.16760	–25.46429
66.0	1.00253	7.65261	3.72128	9.39630	–67.17304	–26.71003
67.0	0.44940	7.74133	3.13689	9.17136	–69.13939	–27.90337
68.0	–0.11492	7.86704	2.56141	8.98241	–71.07427	–29.05003
69.0	–0.69587	8.03150	1.98962	8.82824	–72.98441	–30.15448
70.0	–1.29931	8.23744	1.41641	8.70856	–74.87588	–31.22010
71.0	–1.93177	8.48874	0.83652	8.62397	–76.75435	–32.24929
72.0	–2.60078	8.79063	0.24441	8.57604	–78.62528	–33.24349
73.0	–3.31519	9.15001	–0.36608	8.56749	–80.49416	–34.20324
74.0	–4.08577	9.57588	–1.00199	8.60233	–82.36671	–35.12805
75.0	–4.92581	10.08003	–1.67168	8.68626	–84.24920	–36.01631
76.0	–5.85220	10.67797	–2.38542	8.82720	–86.14877	–36.86498
77.0	–6.88680	11.39036	–3.15620	9.03599	–88.07397	–37.66921
78.0	–8.05868	12.24517	–4.00095	9.32768	–90.03542	–38.42171
79.0	–9.40748	13.28102	–4.94249	9.72334	–92.04694	–39.11170
80.0	–10.98884	14.55266	–6.01259	10.25310	–94.12730	–39.72322
81.0	–12.88369	16.14013	–7.25711	10.96136	–96.30309	–40.23225
82.0	–15.21453	18.16514	–8.74522	11.91589	–98.61394	–40.60164
83.0	–18.17642	20.82200	–10.58676	13.22520	–101.12220	–40.77128

(continued 3) $\alpha = 1.3$

λ	A	B	C	D	G	H
84.0	−22.10028	24.44091	−12.96772	15.07403	−103.93293	−40.63837
85.0	−27.59654	29.63159	−16.23081	17.80386	−107.23901	−40.01209
86.0	−35.92980	37.65796	−21.08555	22.12305	−111.43902	−38.49597
87.0	−50.21700	51.63630	−29.28021	29.77868	−117.51399	−35.11098
88.0	−80.78076	81.88858	−46.60364	46.55848	−128.71075	−26.61230
89.0	−194.49738	195.29042	−110.55759	109.96302	−166.05933	8.02810
90.0	532.80397	−532.32968	297.09275	−298.24367	61.04333	−221.79491
91.0	114.33769	−114.18678	62.24319	−63.95859	−72.14515	−91.34103
92.0	64.38266	−64.56048	34.03566	−36.32489	−89.46322	−76.77388
93.0	44.84401	−45.35663	22.88071	−25.75438	−97.22526	−71.78135
94.0	34.34874	−35.20298	16.79284	−20.26290	−102.15299	−69.64401
95.0	27.74620	−28.94970	12.88331	−16.96309	−105.86713	−68.74357
96.0	23.17136	−24.73258	10.10579	−14.81009	−108.95527	−68.49498
97.0	19.78512	−21.71341	7.98921	−13.33439	−111.68251	−68.63592
98.0	17.15393	−19.45964	6.28991	−12.29401	−114.18647	−69.03167
99.0	15.03109	−17.72557	4.86891	−11.55180	−116.54615	−69.60639
100.0	13.26575	−16.36149	3.64097	−11.02445	−118.81052	−70.31453
105.0	7.27273	−12.60914	−1.04889	−10.24707	−129.56230	−75.14068
110.0	3.23962	−11.37716	−4.97229	−11.21513	−140.46493	−81.52603
115.0	−0.36486	−11.51340	−9.21260	−13.50729	−152.51834	−89.62671
120.0	−4.38615	−12.94780	−14.70359	−17.54390	−167.04884	−100.48111
125.0	−9.97653	−16.43076	−23.21929	−24.87362	−187.02094	−116.92736
130.0	−20.39033	−25.00808	−40.31937	−40.93914	−222.07444	−148.53963
135.0	−56.19020	−59.11869	−101.84034	−101.50751	−334.75175	−257.82135
140.0	161.01043	159.70754	278.43119	279.68158	324.07964	404.38642
145.0	35.62034	35.94410	60.41527	62.58463	−61.91504	21.76927
150.0	19.50828	21.51806	33.27310	36.39568	−114.55506	−27.47345
155.0	12.48495	16.30111	22.07385	26.21836	−139.31263	−48.79447
160.0	8.08465	13.90048	15.56911	20.84551	−155.80720	−61.78997
165.0	4.70390	12.80994	11.01605	17.58935	−168.79628	−71.18624
170.0	1.70668	12.53545	7.38766	15.50331	−180.02594	−78.68380
175.0	−1.27535	12.93367	4.17341	14.20437	−190.31253	−85.02677
180.0	−4.57019	14.06511	1.02966	13.56906	−200.12633	−90.56061
185.0	−8.62765	16.22396	−2.39036	13.67235	−209.82419	−95.40315
190.0	−14.31703	20.18051	−6.63779	14.89244	−219.82055	−99.44201
195.0	−23.91429	28.14213	−13.04320	18.44014	−230.94024	−102.06148
200.0	−46.57632	49.21391	−26.82771	29.44606	−246.19623	−100.40669

$\alpha = 1.4$

λ	*A*	*B*	*C*	*D*	*G*	*H*
0.2	2.62020	1.49325	1.91782	2.95190	3.23735	3.37751
0.4	2.57430	1.50871	1.87540	2.94414	2.96029	3.24093
0.6	2.52775	1.52463	1.83252	2.93652	2.68281	3.10427
0.8	2.48053	1.54101	1.78916	2.92907	2.40489	2.96751
1.0	2.43263	1.55789	1.74532	2.92178	2.12652	2.83067
1.2	2.38400	1.57529	1.70097	2.91466	1.84769	2.69375
1.4	2.33464	1.59322	1.65610	2.90773	1.56839	2.55674
1.6	2.28452	1.61171	1.61070	2.90099	1.28862	2.41966
1.8	2.23360	1.63079	1.56474	2.89446	1.00835	2.28251
2.0	2.18185	1.65047	1.51822	2.88815	0.72757	2.14528
2.2	2.12926	1.67079	1.47110	2.88208	0.44628	2.00799
2.4	2.07578	1.69178	1.42339	2.87625	0.16446	1.87064
2.6	2.02139	1.71346	1.37504	2.87068	–0.11791	1.73324
2.8	1.96604	1.73587	1.32604	2.86539	–0.40084	1.59578
3.0	1.90970	1.75904	1.27638	2.86040	–0.68435	1.45829
3.2	1.85234	1.78301	1.22602	2.85572	–0.96844	1.32075
3.4	1.79389	1.80781	1.17493	2.85137	–1.25315	1.18319
3.6	1.73433	1.83350	1.12310	2.84738	–1.53849	1.04561
3.8	1.67361	1.86010	1.07050	2.84376	–1.82447	0.90801
4.0	1.61167	1.88767	1.01709	2.84054	–2.11112	0.77041
4.2	1.54847	1.91626	0.96284	2.83774	–2.39845	0.63282
4.4	1.48393	1.94592	0.90773	2.83539	–2.68650	0.49524
4.6	1.41802	1.97670	0.85170	2.83352	–2.97528	0.35770
4.8	1.35065	2.00867	0.79474	2.83215	–3.26481	0.22020
5.0	1.28176	2.04189	0.73679	2.83133	–3.55513	0.08275
5.2	1.21128	2.07643	0.67782	2.83109	–3.84626	–0.05462
5.4	1.13914	2.11237	0.61777	2.83146	–4.13823	–0.19190
5.6	1.06523	2.14977	0.55660	2.83249	–4.43107	–0.32908
5.8	0.98948	2.18873	0.49426	2.83422	–4.72481	–0.46612
6.0	0.91179	2.22935	0.43069	2.83670	–5.01949	–0.60302
6.2	0.83206	2.27171	0.36583	2.83998	–5.31515	–0.73974
6.4	0.75017	2.31592	0.29962	2.84412	–5.61183	–0.87626
6.6	0.66600	2.36211	0.23199	2.84917	–5.90957	–1.01255
6.8	0.57942	2.41040	0.16286	2.85521	–6.20841	–1.14858
7.0	0.49029	2.46091	0.09215	2.86230	–6.50841	–1.28432
7.2	0.39846	2.51381	0.01978	2.87053	–6.80962	–1.41973
7.4	0.30376	2.56924	–0.05436	2.87997	–7.11209	–1.55476
7.6	0.20601	2.62738	–0.13035	2.89072	–7.41589	–1.68937
7.8	0.10502	2.68843	–0.20832	2.90289	–7.72108	–1.82352
8.0	0.00057	2.75259	–0.28838	2.91657	–8.02773	–1.95713
8.2	–0.10758	2.82009	–0.37067	2.93190	–8.33594	–2.09017

(continued 1) $\alpha = 1.4$

λ	A	B	C	D	G	H
8.4	–0.21967	2.89118	–0.45534	2.94901	–8.64577	–2.22254
8.6	–0.33600	2.96613	–0.54253	2.96804	–8.95733	–2.35419
8.8	–0.45687	3.04526	–0.63243	2.98917	–9.27071	–2.48503
9.0	–0.58263	3.12888	–0.72522	3.01256	–9.58603	–2.61496
9.2	–0.71365	3.12738	–0.82113	3.03842	–9.90341	–2.74388
9.4	–0.85036	3.31117	–0.92038	3.06698	–10.22300	–2.87168
9.6	–0.99322	3.41071	–1.02324	3.09847	–10.54493	–2.99822
9.8	–1.14276	3.51652	–1.13000	3.13319	–10.86937	–3.12336
10.0	–1.29956	3.62916	–1.24098	3.17145	–11.19653	–3.24694
11.0	–2.21872	4.32094	–1.87388	3.42982	–12.88114	–3.83383
12.0	–3.46717	5.33012	–2.69444	3.85631	–14.67968	–4.33894
13.0	–5.31069	6.92117	–3.84955	4.59564	–16.67413	–4.68383
14.0	–8.41551	9.75878	–5.70649	6.01257	–19.06787	–4.66928
15.0	–15.06264	16.12218	–9.52080	9.35970	–22.55308	–3.60795
16.0	–41.59770	42.35489	–24.28377	23.62486	–32.07686	3.44159
17.0	81.61319	–81.17948	43.30234	–44.49379	3.67253	–34.83916
18.0	22.15111	–22.06496	10.40458	–12.16817	–15.86617	–17.89693
19.0	13.04474	–13.33381	5.18741	–7.56863	–20.20538	–16.22896
20.0	9.20427	–9.90054	2.84895	–5.90037	–22.99012	–16.20196
21.0	6.98462	–8.12539	1.38147	–5.16444	–25.32939	–16.72124
22.0	5.46256	–7.09171	0.27360	–4.86025	–27.51040	–17.51743
23.0	4.29332	–6.46299	–0.66881	–4.80725	–29.64734	–18.49863
24.0	3.31630	–6.08913	–1.53985	–4.92863	–31.80142	–19.63206
25.0	2.44371	–5.89595	–2.39523	–5.19102	–34.01551	–20.91156
26.0	1.62024	–5.84598	–3.27516	–5.58358	–36.32779	–22.34726
27.0	0.80535	–5.92273	–4.21507	–6.11037	–38.77936	–23.96300
28.0	–0.03596	–6.12420	–5.25229	–6.78837	–41.42045	–25.79779
29.0	–0.93961	–6.46120	–6.43214	–7.64906	–44.31802	–27.91098
30.0	–1.94862	–6.95905	–7.81566	–8.74375	–47.56709	–30.39224
31.0	–3.12146	–7.66338	–9.49196	–10.15447	–51.31030	–33.38031
32.0	–4.54632	–8.65278	–11.60078	–12.01568	–55.77451	–37.09913
33.0	–6.36938	–10.06608	–14.37847	–14.55965	–61.34606	–41.93280
34.0	–8.85887	–12.16574	–18.26285	–18.22103	–68.74302	–48.59763
35.0	–12.57398	–15.50634	–24.17075	–23.91409	–79.47135	–58.59815
36.0	–18.91702	–21.48639	–34.40559	–33.94015	–97.31689	–75.71905
37.0	–32.70463	–34.91937	–56.88458	–56.21464	–135.29512	–112.97482
38.0	–88.82436	–90.69010	–148.96383	–148.09218	–287.70274	–264.66133
39.0	165.98179	164.46177	270.13922	271.21113	400.30873	424.07069
40.0	45.60491	44.42949	72.41387	73.68579	74.21026	98.69281
41.0	27.06893	26.23894	42.12245	43.59525	23.38539	48.58927
42.0	19.42933	18.94744	29.74669	31.42227	2.01604	27.94255

(continued 2) $\alpha = 1.4$

λ	A	B	C	D	G	H
43.0	15.18534	15.05599	22.95574	24.83702	–10.17004	16.48100
44.0	12.43360	12.66301	18.62130	20.71221	–18.31545	9.06256
45.0	10.46736	11.06349	15.58233	17.88781	–24.32848	3.77956
46.0	8.96344	9.93611	13.30852	15.83458	–29.08045	–0.23872
47.0	7.75283	9.11375	11.52299	14.27674	–33.02598	–3.44623
48.0	6.73823	8.50118	10.06688	13.05664	–36.42560	–6.10283
49.0	5.85942	8.04041	8.84235	12.07774	–39.43947	–8.36791
50.0	5.07685	7.69435	7.78570	11.27779	–42.17156	–10.34459
51.0	4.36322	7.43846	6.85349	10.61495	–44.69244	–12.10248
52.0	3.69878	7.25611	6.01490	10.06026	–47.05180	–13.69022
53.0	3.06862	7.13595	5.24730	9.59316	–49.28576	–15.14266
54.0	2.46096	7.07031	4.53348	9.19888	–51.42133	–16.48540
55.0	1.86600	7.05423	3.85990	8.86671	–53.47926	–17.73753
56.0	1.27524	7.08485	3.21554	8.58899	–55.47592	–18.91342
57.0	0.68082	7.16110	2.59105	8.36036	–57.42456	–20.02398
58.0	0.07511	7.28349	1.97819	8.17738	–59.33622	–21.07739
59.0	–0.54974	7.45410	1.36927	8.03825	–61.22042	–22.07972
60.0	–1.20219	7.67669	0.75682	7.94270	–63.08567	–23.03521
61.0	–1.89184	7.95694	0.13310	7.89202	–64.93994	–23.94648
62.0	–2.63000	8.30286	–0.51030	7.88922	–66.79102	–24.81459
63.0	–3.43048	8.72549	–1.18301	7.93936	–68.64704	–25.63899
64.0	–4.31073	9.23989	–1.89661	8.05009	–70.51691	–26.41722
65.0	–5.29350	9.86673	–2.66563	8.23252	–72.41102	–27.14449
66.0	–6.40936	10.63476	–3.50912	8.50269	–74.34219	–27.81291
67.0	–7.70073	11.58479	–4.45299	8.88387	–76.32717	–28.41013
68.0	–9.22865	12.77639	–5.53410	9.41053	–78.38898	–28.91702
69.0	–11.08461	14.29976	–6.80724	10.13528	–80.56116	–29.30362
70.0	–13.41271	16.29777	–8.35822	11.14196	–82.89550	–29.52143
71.0	–16.45430	19.01064	–10.33032	12.57199	–85.47770	–29.48782
72.0	–20.64732	22.87527	–12.98303	14.68309	–88.46206	–29.05134
73.0	–26.87858	28.77743	–16.84115	17.99839	–92.15975	–27.90359
74.0	–37.25829	38.82635	–23.15275	23.76435	–97.30945	–25.30855
75.0	–58.34380	59.57842	–35.79502	35.85656	–106.20421	–18.97582
76.0	–125.91267	126.81025	–75.92001	75.42551	–131.34986	3.59785
77.0	1121.50273	–1120.94676	661.36619	–662.42432	303.10289	–433.43951
78.0	105.45174	–105.24294	60.49678	–62.12772	–53.56162	–79.37472
79.0	55.72840	–55.87332	30.91306	–33.12768	–72.49005	–63.06402
80.0	37.90277	–38.40903	20.18308	–22.99397	–80.27893	–57.91372
81.0	28.64515	–29.52146	14.51383	–17.93541	–85.08359	–55.77148
82.0	22.91450	–24.17076	10.92445	–14.97306	–88.66732	–54.87718
83.0	18.97362	–20.62095	8.38725	–13.08130	–91.63880	–54.62557

(continued 3) $\alpha = 1.4$

λ	A	B	C	D	G	H
84.0	16.06325	–18.11416	6.45264	–11.81279	–94.26499	–54.75338
85.0	13,79856	–16.26701	4.89233	–10.94164	–96.68215	–55.12835
86.0	11.96344	–14.86501	3.57770	–10.34192	–98.96773	–55.67743
87.0	10.42690	–13.77895	2.43036	–9.93818	–101.16963	–56.35758
88.0	9.10472	–12.92660	1.39949	–9.68285	–103.31962	–57.14240
89.0	7.94009	–12.25335	0.45032	–9.54481	–105.44019	–58.01543
90.0	6.89303	–11.72169	–0.44210	–9.50322	–107.54827	–58.96653
91.0	5.93439	–11.30531	–1.29650	–9.54400	–109.65738	–59.98981
92.0	5.04214	–10.98537	–2.12759	–9.65771	–11.77902	–61.08252
93.0	4.19911	–10.74837	–2.94747	–9.83832	–113.92357	–62.24429
94.0	3.39141	–10.58465	–3.76659	–10.08236	–116.10090	–63.47678
95.0	2.60741	–10.48749	–4.59443	–10.38845	–118.32090	–64.78344
96.0	1.83698	–10.45249	–5.44001	–10.75703	–120.59392	–66.16957
97.0	1.07091	–10.47719	–6.31236	–11.19023	–122.93121	–67.64233
98.0	0.30047	–10.56087	–7.22088	–11.69185	–125.34533	–69.21097
99.0	–0.48301	–10.70443	–8.17577	–12.26750	–127.85067	–70.88718
100.0	–1.28853	–10.91042	–9.18847	–12.92482	–130.46412	–72.68558
105.0	–6.04351	–13.11107	–15.68808	–17.90921	–146.04133	–84.34200
110.0	–14.05163	–19.02780	–27.80451	–28.78336	–171.03709	–105.57790
115.0	–37.94864	–41.06894	–66.27933	–66.15430	–239.62512	–170.48740
120.0	213.76514	212.40072	350.23480	351.40559	442.95735	515.73983
125.0	31.23266	31.62113	49.79181	52.00777	–58.18072	18.24687
130.0	16.21862	18.44191	25.96989	29.28096	–102.58239	–22.47924
135.0	9.82628	14.05735	16.47329	20.98371	–123.27791	–39.43644
140.0	5.71380	12.24182	10.89266	16.77603	–137.41695	–49.73291
145.0	2.37661	11.66109	6.83360	14.36542	–148.89907	–57.20731
150.0	–0.82619	11.95456	3.38550	13.00894	–159.13947	–63.17452
155.0	–4.37684	13.15706	0.02427	12.49196	–168.84056	–68.15174
160.0	–8.93960	15.69898	–3.77071	12.94901	–178.51717	–72.26171
165.0	–15.99570	20.92319	–8.95883	15.11102	–188.86376	–75.18713
170.0	–30.72309	33.91583	–18.68772	21.97397	–201.84414	–75.21458
175.0	–98.10853	99.59480	–59.92049	60.38728	–233.65701	–56.33318
180.0	107.81128	–108.06219	62.41770	–64.82133	–167.59128	–135.41495
185.0	34.72885	–36.80584	17.71158	–23.13273	–201.58794	–114.67773
190.0	19.41299	–23.47049	7.50019	–16.19424	–215.07248	–114.87492
195.0	12.01686	–18.29199	1.90404	–14.26301	–226.00136	–118.27591
200.0	7.09837	–15.94290	–2.39466	–14.21074	–236.46031	–123.10812

$\alpha = 1.5$

λ	A	B	C	D	G	H
0.2	2.41059	1.40415	1.51964	2.55902	2.48056	2.63079
0.4	2.35987	1.42184	1.47368	2.55325	2.20724	2.50816
0.6	2.30829	1.44015	1.42710	2.54770	1.93338	2.38547
0.8	2.22582	1.45910	1.37988	2.54241	1.65897	2.26274
1.0	2.20242	1.47873	1.33201	2.53738	1.38399	2.13996
1.2	2.14806	1.49907	1.28346	2.53263	1.10842	2.01715
1.4	2.09268	1.52016	1.23421	2.52818	0.83224	1.89431
1.6	2.03625	1.54204	1.18422	2.52405	0.55544	1.77145
1.8	1.97873	1.56475	1.13347	2.52026	0.27800	1.64857
2.0	1.92006	1.58832	1.08193	2.51683	–0.00011	1.52569
2.2	1.86020	1.61282	1.02957	2.51379	–0.27890	1.40282
2.4	1.79908	1.63829	0.97635	2.51117	–0.55840	1.27996
2.6	1.73665	1.66478	0.92224	2.50899	–0.83863	1.15714
2.8	1.67285	1.69234	0.86720	2.50728	–1.11963	1.03435
3.0	1.60761	1.72105	0.81118	2.50607	–1.40141	0.91163
3.2	1.54085	1.75097	0.75415	2.50541	–1.68401	0.78898
3.4	1.47250	1.78217	0.69606	2.50532	–1.96746	0.66641
3.6	1.40248	1.81473	0.63686	2.50585	–2.25179	0.54396
3.8	1.33070	1.84873	0.57648	2.50705	–2.53703	0.42164
4.0	1.25706	1.88426	0.51489	2.50896	–2.82323	0.29947
4.2	1.18145	1.92141	0.45200	2.51163	–3.11042	0.17749
4.4	1.10378	1.96031	0.38776	2.51513	–3.39865	0.05570
4.6	1.02390	2.00105	0.32210	2.51951	–3.68796	–0.06584
4.8	0.94170	2.04376	0.25493	2.52484	–3.97841	–0.18712
5.0	0.85703	2.08858	0.18617	2.53120	–4.27004	–0.30809
5.2	0.76973	2.13566	0.11573	2.53867	–4.56292	–0.42871
5.4	0.67964	2.18516	0.04350	2.54733	–4.85711	–0.54894
5.6	0.58657	2.23725	–0.03061	2.55729	–5.15268	–0.66872
5.8	0.49032	2.29214	–0.10674	2.56864	–5.44969	–0.78801
6.0	0.39067	2.35003	–0.18501	2.58151	–5.74824	–0.90675
6.2	0.28736	2.41116	–0.26556	2.59604	–6.04841	–1.02485
6.4	0.18014	2.47579	–0.34855	2.61235	–6.35029	–1.14225
6.6	0.06871	2.54422	–0.43416	2.63062	–6.65400	–1.25887
6.8	–0.04727	2.61675	–0.52258	2.65102	–6.95965	–1.37460
7.0	–0.16816	2.69376	–0.61403	2.67376	–7.26736	–1.48935
7.2	–0.29437	2.77563	–0.70873	2.69904	–7.57729	–1.60299
7.4	–0.42636	2.86281	–0.80696	2.72713	–7.88958	–1.71538
7.6	–0.56463	2.95581	–0.90901	2.75831	–8.20443	–1.82638
7.8	–0.70977	3.05519	–1.01522	2.79289	–8.52202	–1.93581
8.0	–0.86242	3.16159	–1.12596	2.83123	–8.84257	–2.04348
8.2	–1.02332	3.27574	–1.24166	2.87374	–9.16635	–2.14915

(continued 1) $\alpha = 1.5$

λ	A	B	C	D	G	H
8.4	−1.19332	3.39847	−1.36280	2.92089	−9.49364	−2.25258
8.6	−1.37337	3.53073	−1.48994	2.97322	−9.82476	−2.35346
8.8	−1.56458	3.67360	−1.62372	3.03134	−10.16008	−2.45146
9.0	−1.76823	3.82837	−1.76488	3.09597	−10.50004	−2.54617
9.2	−1.98581	3.99650	−1.91426	3.16794	−10.84513	−2.63713
9.4	−2.21906	4.17970	−2.07286	3.24823	−11.19593	−2.72379
9.6	−2.47001	4.38003	−2.24186	3.33799	−11.55312	−2.80550
9.8	−2.74108	4.59987	−2.42264	3.43859	−11.91750	−2.88151
10.0	−3.03517	4.84210	−2.61685	3.55165	−12.29003	−2.95088
11.0	−5.00377	6.54155	−3.87703	4.39059	−14.32525	−3.14932
12.0	−8.65639	9.90690	−6.10677	6.17082	−16.92756	−2.82484
13.0	−18.48892	19.43151	−11.87171	11.45400	−21.56231	−0.51830
14.0	−192.50037	193.11131	−111.29284	110.35629	−79.63508	55.16834
15.0	27.13722	−26.88541	13.80179	−15.30010	−9.66993	−17.24996
16.0	13.19462	−13.33401	5.65101	−7.76119	−15.73015	−13.72116
17.0	8.72016	−9.28857	2.88185	−5.66293	−18.75013	−13.32481
18.0	6.37895	−7.42138	1.30705	−4.82934	−21.12378	−13.68459
19.0	4.84072	−6.41140	0.16328	−4.51145	−23.29362	−14.38044
20.0	3.67511	−5.84022	−0.80087	−4.47657	−25.41230	−15.28889
21.0	2.69638	−5.53794	−1.69918	−4.63561	−27.55698	−16.37137
22.0	1.80619	−5.42762	−2.59836	−4.95528	−29.78228	−17.62545
23.0	0.94143	−5.47581	−3.54911	−5.43120	−32.14009	−19.07160
24.0	0.05216	−5.67491	−4.60093	−6.08002	−34.69139	−20.75206
25.0	−0.91124	−6.03793	−5.81315	−6.94020	−37.51852	−22.73725
26.0	−2.00932	−6.60105	−7.26824	−8.08028	−40.74338	−25.14110
27.0	−3.32940	−7.43460	−9.09418	−9.61851	−44.56098	−28.15308
28.0	−5.01543	−8.67029	−11.50968	−11.76662	−49.30943	−32.10728
29.0	−7.33668	−10.56827	−14.93074	−14.93538	−55.63574	−37.64774
30.0	−10.87799	−13.70639	−20.26717	−20.03058	−64.95868	−46.19090
31.0	−17.21546	−19.65520	−29.97748	−29.50753	−81.11585	−61.57254
32.0	−32.60732	−34.66846	−53.83221	−53.13419	−119.36355	−99.04744
33.0	−136.06481	−137.75362	−215.21246	−214.28947	−372.37103	−351.28360
34.0	75.13121	73.81172	114.76442	115.91118	141.91771	163.77613
35.0	30.97170	30.02139	45.98101	47.35203	33.50755	56.13757
36.0	19.87788	9.29930	28.83350	30.43087	5.73218	29.13537
37.0	14.71421	14.51244	20.94860	22.77589	−7.57728	16.60155
38.0	11.65613	11.83872	16.35513	18.41739	−15.74463	9.21320
39.0	9.58293	10.15993	13.30417	15.60799	−21.50022	4.24086
40.0	8.04655	9.03063	11.09732	13.65083	−25.93393	0.59562
41.0	6.83223	8.23883	9.40054	12.21356	−29.56661	−2.24234
42.0	5.82368	7.67125	8.03362	11.11784	−32.67874	−4.55240

(continued 2) $\alpha = 1.5$

λ	A	B	C	D	G	H
43.0	4.95194	7.26228	6.89052	10.25964	−35.43500	−6.49796
44.0	4.17308	6.97179	5.90439	9.57447	−37.93864	−8.18085
45.0	3.45727	6.77431	5.03073	9.02051	−40.25789	−9.66764
46.0	2.78303	6.65346	4.23840	8.56977	−42.43989	−11.00354
47.0	2.13394	6.59882	3.50461	8.20314	−44.51851	−12.22010
48.0	1.49661	6.60425	2.81184	7.90760	−46.51896	−13.33981
49.0	0.85933	6.66682	2.14598	7.67445	−48.46075	−14.37880
50.0	0.21111	6.78636	1.49494	7.49832	−50.35957	−15.34862
51.0	−0.45916	6.96529	0.84777	7.37660	−52.22860	−16.25726
52.0	−1.16350	7.20876	0.19376	7.30924	−54.07953	−17.10985
53.0	−1.91579	7.52508	−0.47828	7.29878	−55.92340	−17.90889
54.0	−2.73288	7.92657	−1.18094	7.35070	−57.77133	−18.65433
55.0	−3.63627	8.43096	−1.92931	7.47420	−59.63543	−19.34325
56.0	−4.65455	9.06366	−2.74258	7.68353	−61.52996	−19.96916
57.0	−5.82736	9.86161	−3.64661	8.00029	−63.47293	−20.52068
58.0	−7.21205	10.87978	−4.67803	8.45742	−65.48874	−20.97914
59.0	−8.89550	12.20300	−5.89164	9.10649	−67.61291	−21.31412
60.0	−11.01689	13.96857	−7.37453	10.03168	−69.90074	−21.47485
61.0	−13.81480	16.41339	−9.27532	11.37897	−72.44555	−21.37214
62.0	−17.73655	19.98322	−11.87185	13.42375	−75.42062	−20.83653
63.0	−23.73199	25.62643	−15.75327	16.75288	−79.19203	−19.50556
64.0	−34.23746	35.77796	−22.42957	22.87415	−84.69629	−16.44616
65.0	−57.95958	59.14308	−37.29530	37.17995	−95.27638	−8.31866
66.0	−166.49319	167.31525	−104.70508	104.02281	−138.41164	32.35290
67.0	210.28807	−209.83324	128.54166	−129.80001	4.70961	−113.24648
68.0	65.68266	−65.60224	38.77864	−40.62438	−52.28361	−58.74922
69.0	39.09266	−39.39528	22.12367	−24.57039	−63.99187	−49.55831
70.0	27.77973	−28.47555	14.92819	−17.99187	−69.84748	−46.24519
71.0	21.43145	−22.53223	10.80275	−14.50189	−73.81000	−44.85429
72.0	17.30862	−18.82788	8.04954	−12.40541	−76.93254	−44.33679
73.0	14.37113	−16.32430	6.02331	−11.06015	−79.61627	−44.29621
74.0	12.13695	−14.54155	4.42443	−10.16978	−82.04837	−44.55054
75.0	10.35172	−13.22761	3.09420	−9.57927	−84.32891	−45.00549
76.0	8.86811	−12.23777	1.94011	−9.20023	−86.51714	−45.60825
77.0	7.59451	−11.48340	0.90400	−8.97916	−88.65131	−46.32787
78.0	6.47059	−10.90756	−0.05311	−8.88238	−90.75810	−47.14603
79.0	5.45460	−10.47237	−0.95896	−8.88829	−92.85760	−48.05225
80.0	4.51629	−10.15213	−1.83441	−8.98311	−94.96612	−49.04138
81.0	3.63275	−9.92921	−2.69618	−9.15844	−97.09793	−50.11213
82.0	2.78586	−9.79173	−3.55847	−9.40988	−99.26650	−51.26646
83.0	1.96053	−9.73201	−4.43414	−9.73619	−101.48534	−52.50912

(continued 3) $\alpha = 1.5$

λ	A	B	C	D	G	H
84.0	1.14350	−9.74571	−5.33564	−10.13892	−103.76880	−53.84772
85.0	0.32248	−9.83135	−6.27572	−10.62230	−106.13286	−55.29295
86.0	−0.51465	−9.99006	−7.26820	−11.19332	−108.59595	−56.85903
87.0	−1.38054	−10.22562	−8.32871	−11.86212	−111.18002	−58.56450
88.0	−2.28913	−10.54474	−9.47571	−12.64269	−113.91192	−60.43343
89.0	−3.25658	−10.95759	−10.73172	−13.55383	−116.82523	−62.49710
90.0	−4.30243	−11.47872	−12.12506	−14.62077	−119.96297	−64.79662
91.0	−5.45129	−12.12853	−13.69236	−15.87753	−123.38137	−67.38660
92.0	−6.73528	−12.93556	−15.48233	−17.37060	−127.15591	−70.34114
93.0	−8.19779	−13.94013	−17.56167	−19.16478	−131.39036	−73.76285
94.0	−9.89968	−15.20043	−20.02464	−21.35268	−136.23197	−77.79793
95.0	−11.92965	−16.80284	−23.00942	−24.07103	−141.89697	−82.66173
96.0	−14.42306	−18.88067	−26.72754	−27.53010	−148.71688	−88.68497
97.0	−17.59794	−21.65014	−31.52048	−32.07023	−157.22733	−96.40260
98.0	−21.83057	−25.48589	−37.97824	−38.28041	−168.35415	−106.73979
99.0	−27.83193	−31.09742	−47.21639	−47.27530	−183.84842	−121.44709
100.0	−37.12799	−40.00935	−61.63015	−61.44929	−207.47040	−144.28422
105.0	95.55289	94.54739	147.57310	148.92605	116.02077	183.11520
110.0	22.63916	23.52776	33.98803	36.52618	−67.27974	3.73218
115.0	11.87260	14.79437	18.04258	21.85452	−97.23892	−22.25313
120.0	6.65723	11.89244	10.94430	16.20698	−113.22460	−34.15319
125.0	2.90837	10.94105	6.37601	13.39424	−125.09484	−41.74591
130.0	−0.52949	11.13134	2.68682	11.98302	−135.33914	−47.38412
135.0	−4.36212	12.44366	−0.90825	11.61931	−144.98699	−51.82665
140.0	−9.57302	15.52462	−5.18567	12.55069	−154.77529	−55.16705
145.0	−18.82169	22.84869	−11.90343	16.24453	−165.94535	−56.62034
150.0	−46.80228	48.99134	−30.34904	31.80198	−184.41983	−50.55652
155.0	280.93517	−280.58725	174.68391	−176.12449	−62.49650	−184.89951
160.0	36.70210	−38.28020	20.33522	−24.80306	−166.41856	−93.60774
165.0	18.33320	−22.01134	7.80369	−15.57257	−181.37327	−91.71412
170.0	10.59752	−16.66357	1.81641	−13.33912	−192.43044	−94.43016
175.0	5.61215	−14.51931	−2.65706	−13.33214	−202.87313	−98.88141
180.0	1.49866	−13.96754	−6.92307	−14.66570	−213.72043	−104.70967
185.0	−2.59614	−14.63659	−11.74490	−17.33358	−225.80920	−112.26835
190.0	−7.43702	−16.74649	−18.06785	−21.93881	−240.46413	−122.66709
195.0	−14.36870	−21.35895	−27.87253	−30.28420	−260.71529	−138.82425
200.0	−27.42058	−32.33759	−47.41589	−48.52252	−296.35087	−170.46257

$\alpha = 1.6$

λ	A	B	C	D	G	H
0.2	2.22427	1.32731	1.19280	2.23794	1.90693	2.06725
0.4	2.16835	1.34752	1.14297	2.23435	1.63587	1.95715
0.6	2.11131	1.36856	1.09232	2.23110	1.36413	1.84706
0.8	2.05307	1.39048	1.04084	2.22824	1.09169	1.73699
1.0	1.99360	1.41332	0.98848	2.22577	0.81851	1.62694
1.2	1.93282	1.43714	0.93521	2.22374	0.54458	1.51693
1.4	1.87067	1.46200	0.88098	2.22218	0.26987	1.40697
1.6	1.80708	1.48796	0.82575	2.22112	–0.00567	1.29709
1.8	1.74198	1.51510	0.76947	2.22059	–0.28205	1.18729
2.0	1.67528	1.54349	0.71209	2.22065	–0.55931	1.07759
2.2	1.60689	1.57321	0.65356	2.22132	–0.83750	0.96803
2.4	1.53673	1.60434	0.59382	2.22267	–1.11664	0.85862
2.6	1.46468	1.63698	0.53280	2.22474	–1.39679	0.74938
2.8	1.39064	1.67123	0.47043	2.22759	–1.67799	0.64035
3.0	1.31448	1.70722	0.40664	2.23128	–1.96029	0.53155
3.2	1.23608	1.74505	0.34135	2.23589	–2.24373	0.42303
3.4	1.15529	1.78487	0.27446	2.24149	–2.52839	0.31481
3.6	1.07195	1.82683	0.20589	2.24815	–2.81432	0.20695
3.8	0.98590	1.87108	0.13553	2.25598	–3.10159	0.09949
4.0	0.89694	1.91780	0.06326	2.26507	–3.39027	–0.00751
4.2	0.80486	1.96720	–0.01104	2.27553	–3.68045	–0.11400
4.4	0.70944	2.01949	–0.08751	2.28750	–3.97221	–0.21992
4.6	0.61043	2.07491	–0.16631	2.30110	–4.26565	–0.32518
4.8	0.50754	2.13373	–0.24760	2.31649	–4.56088	–0.42970
5.0	0.40047	2.19625	–0.33157	2.33384	–4.85801	–0.53340
5.2	0.28887	2.26281	–0.41844	2.35334	–5.15717	–0.63616
5.4	0.17235	2.33379	–0.50843	2.37521	–5.45852	–0.73787
5.6	0.05047	2.40960	–0.60180	2.39970	–5.76221	–0.83839
5.8	–0.07725	2.49072	–0.69886	2.42707	–6.06842	–0.93756
6.0	–0.21137	2.57770	–0.79992	2.45765	–6.37736	–1.03522
6.2	–0.35253	2.67115	–0.90538	2.49178	–6.68926	–1.13116
6.4	–0.50142	2.77177	–1.01565	2.52987	–7.00438	–1.22516
6.6	–0.65889	2.88038	–1.13123	2.57240	–7.32301	–1.31694
6.8	–0.82587	2.99790	–1.25267	2.61989	–7.64550	–1.40620
7.0	–1.00346	3.12541	–1.38064	2.67299	–7.97225	–1.49258
7.2	–1.19293	3.26417	–1.51588	2.73242	–8.30370	–1.57566
7.4	–1.39577	3.41566	–1.65928	2.79904	–8.64038	–1.65495
7.6	–1.61374	3.58161	–1.81188	2.87387	–8.98293	–1.72985
7.8	–1.84894	3.76409	–1.97492	2.95812	–9.33207	–1.79968
8.0	–2.10384	3.96559	–2.14987	3.05323	–9.68870	–1.86360
8.2	–2.38147	4.18908	–2.33851	3.16095	–10.05386	–1.92058

(continued 1) $\alpha = 1.6$

λ	A	B	C	D	G	H
8.4	−2.68549	4.43823	−2.54301	3.28342	−10.42885	−1.96938
8.6	−3.02041	4.71751	−2.76602	3.42328	−10.81525	−2.00848
8.8	−3.39186	5.03254	−3.01087	3.58380	−11.21501	−2.03595
9.0	−3.80693	5.39038	−3.28174	3.76913	−11.63061	−2.04938
9.2	−4.27467	5.80006	−3.58395	3.98457	−12.06522	−2.04565
9.4	−4.80689	6.27336	−3.92447	4.23704	−12.52291	−2.02073
9.6	−5.41920	6.82588	−4.31248	4.53569	−13.00913	−1.96924
9.8	−6.13276	7.47874	−4.76043	4.89294	−13.53121	−1.88391
10.0	−6.97693	8.26128	−5.28561	5.32602	−14.09932	−1.75461
11.0	−15.73677	16.69792	−10.58721	10.14468	−18.52333	0.44185
12.0	−313.02344	313.63319	−185.55064	184.58314	−122.72474	102.35304
13.0	21.08359	−20.85852	10.71986	−12.26194	−8.66921	−14.06668
14.0	10.61072	−10.81006	4.36502	−6.54090	−13.78084	−11.40754
15.0	7.03893	−7.71067	2.04611	−4.92735	−16.55178	−11.19575
16.0	5.08939	−6.29230	0.65439	−4.32863	−18.81720	−11.62013
17.0	3.75056	−5.55783	−0.41141	−4.16491	−20.94031	−12.34944
18.0	2.68444	−5.18883	−1.35888	−4.25779	−23.05636	−13.29195
19.0	1.73813	−5.05981	−2.29029	−4.54586	−25.24391	−14.42996
20.0	0.82270	−5.12137	−3.27580	−5.01790	−27.57082	−15.78347
21.0	−0.12981	−5.36480	−4.38212	−5.69550	−30.11618	−17.40503
22.0	−1.18895	−5.81359	−5.69212	−6.63445	−32.99129	−19.39000
23.0	−2.44686	−6.53046	−7.32988	−7.94150	−36.37330	−21.90530
24.0	−4.05265	−7.64470	−9.50875	−9.81829	−40.57490	−25.25674
25.0	−6.29122	−9.42728	−12.64709	−12.67496	−46.21601	−30.05937
26.0	−9.81461	−12.52002	−17.71218	−17.47274	−54.73564	−37.74861
27.0	−16.55931	−18.85164	−27.58691	−27.08992	−70.42873	−52.616665
28.0	−36.09524	−37.98595	−56.53268	−55.78421	−114.55044	−95.91649
29.0	−3121.34875	−3122.84422	−4654.97952	−4653.98263	−6969.58334	−6950.12887
30.0	41.95507	40.85275	60.14031	61.38516	57.39343	77.66869
31.0	21.67796	20.97052	30.08556	31.58024	11.61722	32.71489
32.0	14.79655	14.48929	19.99920	21.74779	−4.39431	17.52880
33.0	11.22764	11.32928	14.85357	16.86225	−13.03918	9.71368
34.0	8.97348	9.49623	11.67249	13.94962	−18.75249	4.83580
35.0	7.36967	8.32932	9.46722	12.02342	−23.00700	1.42382
36.0	6.13094	7.54718	7.81421	10.66257	−26.43296	−1.15094
37.0	5.11364	7.01048	6.50122	9.65755	−29.34556	−3.20193
38.0	4.23677	6.64312	5.40959	8.89285	−31.92033	−4.90269
39.0	3.45034	6.40081	4.46723	8.30004	−34.26316	−6.35675
40.0	2.72095	6.25693	3.62723	7.83653	−36.44214	−7.62940
41.0	2.02448	6.19551	2.85710	7.47511	−38.50346	−8.76350
42.0	1.34199	6.20767	2.13286	7.19825	−40.48013	−9.78784

(continued 2) $\alpha = 1.6$

λ	A	B	C	D	G	H
43.0	0.65722	6.28973	1.43561	6.99515	−42.39692	−10.72200
44.0	−0.04519	6.44233	0.74921	6.86001	−44.27350	−11.57897
45.0	−0.78118	6.67030	0.05867	6.79123	−46.12654	−12.36681
46.0	−1.56882	6.98311	−0.65129	6.79129	−47.97134	−13.08933
47.0	−2.43017	7.39603	−1.39780	6.86719	−49.82322	−13.74628
48.0	−3.39399	7.93238	−2.20172	7.03165	−51.69910	−14.33289
49.0	−4.49994	8.62732	−3.09054	7.30550	−53.61955	−14.83855
50.0	−5.80574	9.53482	−4.10313	7.72201	−55.61202	−15.24413
51.0	−7.40026	10.74059	−5.29822	8.33515	−57.71646	−15.51680
52.0	−9.42938	12.38772	−6.77118	9.23615	−59.99620	−15.59940
53.0	−12.15216	14.73281	−8.69034	10.58968	−62.56150	−15.38715
54.0	−16.07926	18.82430	−11.38640	12.72310	−65.62730	−14.67008
55.0	−22.37941	24.20884	−15.61492	16.38890	−69.68282	−12.96358
56.0	−34.44923	35.90111	−23.57009	23.77833	−76.14618	−8.85389
57.0	−68.07403	69.14453	−45.44599	45.08266	−91.60136	4.23873
58.0	−794.39257	795.07598	−514.62766	513.68413	−395.90558	306.16737
59.0	86.21289	−85.92417	53.84991	−55.38513	−30.15195	−61.97926
60.0	41.36293	−41.47844	24.71545	−26.85674	−50.30925	−44.23658
61.0	27.21220	−27.74354	15.39985	−18.16467	−57.67617	−39.31058
62.0	20.16194	−21.12291	10.66339	−14.07248	−62.09505	−37.36382
63.0	15.86229	−17.26904	7.69615	−11.77383	−65.38191	−36.58564
64.0	12.90964	−14.78098	5.59065	−10.36516	−68.12444	−36.39426
65.0	10.71268	−13.07035	3.96380	−9.46780	−70.57188	−36.54701
66.0	8.97843	−11.84745	2.62498	−8.89609	−72.84945	−36.92614
67.0	7.54448	−10.95371	1.46771	−8.54924	−75.02835	−37.46900
68.0	6.31294	−10.29560	0.42691	−8.36875	−77.15340	−38.14062
69.0	5.22065	−9.81509	−0.54042	−8.31933	−79.25566	−38.92149
70.0	4.22431	−9.47494	−1.46481	−8.37950	−81.35873	−39.80164
71.0	3.29249	−9.25093	−2.36964	−8.53674	−83.48235	−40.77760
72.0	2.40100	−9.12762	−3.27427	−8.78478	−85.64458	−41.85093
73.0	1.53000	−9.09577	−4.19605	−9.12221	−87.86348	−43.02752
74.0	0.66200	−9.15107	−5.15190	−9.55187	−90.15839	−44.31766
75.0	−0.21956	−9.29340	−6.15954	−10.08072	−92.55133	−45.73638
76.0	−1.13164	−9.52672	−7.23887	−10.72024	−95.06853	−47.30452
77.0	−2.09287	−9.85936	−8.41344	−11.48735	−97.74253	−49.05030
78.0	−3.12513	−10.30493	−9.71256	−12.40600	−100.61500	−51.01195
79.0	−4.25561	−10.88386	−11.17413	−13.50973	−103.74097	−53.24167
80.0	−5.51980	−11.62611	−12.84905	−14.84588	−107.19529	−55.81202
81.0	−6.96631	−12.57561	−14.80828	−16.48237	−11.08314	−58.82622
82.0	−8.66477	−13.79810	−17.15462	−18.51950	−115.55776	−62.43587
83.0	−10.71988	−15.39492	−20.04380	−21.11081	−120.85187	−66.87230

(continued 3) $\alpha = 1.6$

λ	A	B	C	D	G	H
84.0	–13.29814	–17.52967	–23.72441	–24.50303	–127.33776	–72.50659
85.0	–16.68275	–20.48304	–28.62038	–29.11847	–135.65091	–79.97316
86.0	–21.39952	–24.77861	–35.51985	–35.74381	–146.97347	–90.45323
87.0	–28.54992	–31.51584	–46.07490	–46.02985	–163.78412	–106.42464
88.0	–40.89938	–43.45836	–64.43481	–64.12469	–192.30911	–134.11289
89.0	–67.93500	–70.09163	–104.83604	–104.26371	–253.91244	–194.88128
90.0	–177.96573	–179.72307	–269.78813	–268.95550	–502.42824	–442.56329
91.0	330.55964	329.20000	493.51689	494.60889	642.07695	702.77518
92.0	88.05464	87.09245	129.76175	131.11306	95.21997	156.75142
93.0	51.28111	50.71747	74.74515	76.35658	11.68253	74.04782
94.0	36.26997	36.10726	52.38801	54.26121	–22.84270	40.35757
95.0	28.02780	28.26971	40.19046	42.32793	–42.12055	21.91639
96.0	22.76038	23.41189	32.45915	34.86423	–54.69462	10.18125
97.0	19.06130	20.12871	27.08403	29.76093	–63.73103	1.98654
98.0	16.28859	17.77958	23.10233	26.05613	–70.67425	–4.11157
99.0	14.10728	16.03095	20.01192	23.24867	–76.27738	–8.86561
100.0	12.32525	14.69223	17.52517	21.05189	–80.97168	–12.70617
105.0	6.40773	11.21198	9.67519	14.79729	–97.50323	–24.87309
110.0	2.41415	10.24250	4.95967	12.06307	–109.16845	–31.91990
115.0	–1.29413	10.64939	1.13260	10.93380	–119.19822	–36.86075
120.0	–5.73217	12.55221	–2.85003	11.10957	–128.84496	–40.46007
125.0	–12.76062	17.42351	–8.37264	13.39238	–139.26280	–42.43780
130.0	–30.58284	33.25836	–20.90441	22.93873	–154.17145	–39.54352
135.0	–1062.56501	1063.28537	–698.68624	697.78298	–605.44736	399.78936
140.0	36.06865	–37.38574	21.12667	–25.09162	–139.99000	–77.79887
145.0	16.41044	–19.96844	7.28194	–14.61474	–155.92325	–74.45685
150.0	8.81773	–14.97851	1.19828	–12.44317	–166.99631	–76.79356
155.0	3.89428	–13.26457	–3.39172	–12.67755	–177.47630	–81.09503
160.0	–0.38749	–13.23463	–8.00263	–14.45810	–188.63416	–87.07471
165.0	–5.07661	–14.73694	–13.70271	–18.06541	–201.74206	–95.48803
170.0	–11.55437	–18.63331	–22.34839	–25.01507	–219.44637	–108.75778
175.0	–23.81466	–28.64958	–39.83013	–41.02169	–250.52770	–135.54920
180.0	–70.21087	–72.97927	–108.70382	–108.53622	–358.91583	–239.72317
185.0	128.07523	127.30599	190.22417	191.70728	85.59295	208.97153
190.0	34.54620	35.79788	50.47576	53.28934	–129.35369	–1.77898
195.0	18.80757	22.18773	27.71466	31.93011	–168.40360	–36.58536
200.0	11.46538	17.17935	17.66230	23.41541	–188.29136	–52.13979

$\alpha = 1.7$

λ	*A*	*B*	*C*	*D*	*G*	*H*
0.2	2.05699	1.26068	0.92094	1.97231	1.46490	1.63533
0.4	1.99544	1.28375	0.86689	1.97111	1.19487	1.53663
0.6	1.93241	1.30792	0.81179	1.97042	0.92397	1.43801
0.8	1.86782	1.33326	0.75558	1.97029	0.65218	1.33950
1.0	1.80157	1.35987	0.69820	1.97076	0.37945	1.24112
1.2	1.73357	1.38783	0.63960	1.97187	0.10574	1.14290
1.4	1.66371	1.41723	0.57970	1.97369	−0.16900	1.04486
1.6	1.59188	1.44818	0.51844	1.97627	−0.44481	0.94703
1.8	1.51796	1.48079	0.45573	1.97969	−0.72176	0.84945
2.0	1.44180	1.51519	0.39149	1.98400	−0.99989	0.75215
2.2	1.36326	1.55153	0.32562	1.98929	−1.27927	0.65518
2.4	1.28217	1.58994	0.25802	1.99565	−1.55998	0.55858
2.6	1.19835	1.63062	0.18857	2.00317	−1.84208	0.46241
2.8	1.11161	1.67373	0.11716	2.01197	−2.12565	0.36671
3.0	1.02172	1.71949	0.04363	2.02217	−2.41080	0.27157
3.2	0.92843	1.76813	−0.03215	2.03389	−2.69762	0.17705
3.4	0.83148	1.81992	−0.11037	2.04730	−2.98621	0.08324
3.6	0.73056	1.87514	−0.19121	2.06256	−3.27670	−0.00977
3.8	0.62533	1.93411	−0.27488	2.07986	−3.56923	−0.10188
4.0	0.51540	1.99721	−0.36161	2.09941	−3.86395	−0.19295
4.2	0.40036	2.06486	−0.45168	2.12148	−4.16102	−0.28286
4.4	0.27970	2.13752	−0.54538	2.14632	−4.46064	−0.37144
4.6	0.15287	2.21574	−0.64306	2.17426	−4.76302	−0.45851
4.8	0.01924	2.30012	−0.74510	2.20567	−5.06841	−0.54386
5.0	−0.12192	2.39140	−0.85196	2.24096	−5.37708	−0.62724
5.2	−0.27144	2.49037	−0.96413	2.28063	−5.68936	−0.70838
5.4	−0.43031	2.59801	−1.08223	2.32524	−6.00561	−0.78694
5.6	−0.59966	2.71541	−1.20694	2.37546	−6.32627	−0.86253
5.8	−0.78081	2.84391	−1.33908	2.43207	−6.65184	−0.93469
6.0	−0.97535	2.98505	−1.47959	2.49600	−6.98290	−1.00288
6.2	−1.18516	3.14068	−1.62963	2.56835	−7.32016	−1.06643
6.4	−1.41247	3.31303	−1.79056	2.65047	−7.66445	−1.12456
6.6	−1.66003	3.50482	−1.96403	2.74397	−8.01679	−1.17631
6.8	−1.93117	3.71935	−2.15207	2.85084	−8.37841	−1.22049
7.0	−2.23002	3.96072	−2.35718	2.97354	−8.75084	−1.25563
7.2	−2.56175	4.23409	−2.58249	3.11517	−9.13597	−1.27988
7.4	−2.93292	4.54598	−2.83196	3.27964	−9.53621	−1.29090
7.6	−3.35198	4.90481	−3.11070	3.47204	−9.95467	−1.28564
7.8	−3.83003	5.32165	−3.42541	3.69900	−10.39540	−1.26011
8.0	−4.38188	5.81128	−3.78504	3.96944	−10.86380	−1.20897
8.2	−5.02784	6.39397	−4.20184	4.29555	−11.36730	−1.12487

(continued 1)

α = 1.7

λ	A	B	C	D	G	H
8.4	–5.79644	7.09823	–4.69304	4.69452	–11.91630	–0.99749
8.6	–6.72917	7.96549	–5.28366	5.19131	–12.52589	–0.81178
8.8	–7.88864	9.05835	–6.01146	5.82364	–13.21888	–0.54504
9.0	–9.37417	10.47605	–6.93634	6.65133	–14.03140	–0.16122
9.2	–11.35310	12.38593	–8.15921	7.77524	–15.02431	0.40046
9.4	–14.13134	15.09383	–9.86452	9.37975	–16.30884	1.25114
9.6	–18.33396	19.22478	–12.42913	11.84167	–18.11224	2.61797
9.8	–25.46869	26.28647	–16.76208	16.06995	–20.98258	5.04893
10.0	–40.33379	41.07709	–25.75625	24.95740	–26.66400	10.28804
11.0	25.63328	–25.28587	13.84508	–15.21106	–4.09674	–14.53874
12.0	10.46874	–10.56255	4.51353	–6.51145	–11.05784	–9.92978
13.0	6.56296	–7.15398	1.94654	–4.65641	–13.97377	–9.48032
14.0	4.58751	–5.74604	0.51412	–4.03640	–16.24731	–9.81684
15.0	3.25796	–5.07414	–0.56631	–3.89719	–18.36196	–10.49629
16.0	2.19051	–4.78267	–1.53781	–4.03599	–20.48091	–11.41306
17.0	1.21744	–4.74547	–2.51867	–4.39390	–22.70002	–12.55612
18.0	0.23669	–4.92440	–3.59683	–4.97429	–25.11002	–13.96521
19.0	–0.84133	–5.33349	–4.86913	–5.82839	–27.83045	–15.73235
20.0	–2.12890	–6.04048	–6.47760	–7.07136	–31.05315	–18.03314
21.0	–3.81136	–7.20260	–8.67547	–8.93940	–35.13292	–21.21207
22.0	–6.26935	–9.18164	–11.99883	–11.95719	–40.83098	–26.02346
23.0	–10.49265	–12.95409	–17.85635	–17.52531	–50.15665	–34.47169
24.0	–20.22645	–22.25531	–31.59490	–30.98464	–70.74674	–54.18991
25.0	–74.31494	–75.92186	–108.70950	–107.82549	–181.88709	–164.46106
26.0	59.17333	57.98396	82.17130	83.32746	89.86628	108.16125
27.0	22.66769	21.89688	30.17544	31.60551	14.61325	33.77900
28.0	14.30953	13.96314	18.39664	20.10546	–3.17866	16.86166
29.0	10.46423	10.55281	13.06913	15.06455	–11.75066	9.16994
30.0	8.16478	8.70358	9.95568	12.24855	–17.15548	4.65309
31.0	6.57155	7.58078	7.85859	10.46301	–21.10304	1.60328
32.0	5.35399	6.85929	6.30769	9.24134	–24.26370	–0.64750
33.0	4.35425	6.38747	5.07985	8.36448	–26.95471	–2.41380
34.0	3.48608	6.08634	4.05461	7.71681	–29.34632	–3.86267
35.0	2.69696	5.91211	3.16028	7.23242	–31.53876	–5.09047
36.0	1.95149	5.84017	2.35038	6.87205	–33.59568	–6.15606
37.0	1.22318	5.85762	1.59225	6.61209	–35.56031	–7.09660
38.0	0.48978	5.95973	0.86074	6.43913	–37.46412	–7.93574
39.0	–0.26986	6.14849	0.13443	6.34720	–39.33184	–8.68785
40.0	–1.07839	6.43252	–0.60719	6.33674	–41.18483	–9.36028
41.0	–1.96287	6.82827	–1.38623	6.41476	–43.04363	–9.95407
42.0	–2.95862	7.36290	–2.22947	6.59633	–44.93063	–10.46357

(continued 2) $\alpha = 1.7$

λ	*A*	*B*	*C*	*D*	*G*	*H*
43.0	–4.11546	8.07976	–3.17274	6.90802	–46.87334	–10.87451
44.0	–5.50881	9.04898	–4.26855	7.39483	–48.90981	–11.15960
45.0	–7.26132	10.38886	–5.60075	8.13431	–51.09857	–11.26927
46.0	–9.58981	12.31245	–7.31600	9.26778	–53.53992	–11.11088
47.0	–12.92072	15.24294	–9.70122	11.07751	–56.42787	–10.49713
48.0	–18.22924	20.15260	–13.41088	14.21372	–60.20246	–8.99400
49.0	–28.34962	29.87291	–20.34476	20.57230	–66.13793	–5.33287
50.0	–56.46538	57.58481	–39.33600	38.98267	–80.15335	6.40002
51.0	–671.61557	672.32484	–451.60675	450.66334	–357.64549	281.59623
52.0	71.85575	–71.56553	46.31184	–47.85823	–25.37188	–52.99211
53.0	34.53504	–34.67538	21.14327	–23.30932	–43.54098	–37.16185
54.0	22.70095	–23.28617	13.04356	–15.84998	–50.28305	–32.78857
55.0	16.76651	–17.81395	8.88951	–12.36132	–54.39151	–31.08508
56.0	13.11590	–14.64623	6.25744	–10.42451	–57.49190	–30.43275
57.0	10.58120	–12.61883	4.36348	–9.26104	–60.11146	–30.31211
58.0	8.66963	–11.24325	2.87586	–8.54531	–62.47481	–30.50739
59.0	7.13629	–10.27956	1.62859	–8.11847	–64.69590	–30.91494
60.0	5.84470	–9.59711	0.52801	–7.89524	–66.84073	–31.48085
61.0	4.71173	–9.11970	–0.48427	–7.82726	–68.95224	–32.17656
62.0	3.68269	–8.80105	–1.44809	–7.88675	–71.06181	–32.98813
63.0	2.71892	–8.61273	–2.39334	–8.05862	–73.19513	–33.91116
64.0	1.79083	–8.53795	–3.34472	–8.33653	–75.37585	–34.94845
65.0	0.87386	–8.56816	–4.32491	–8.72095	–77.62806	–36.10925
66.0	–0.05442	–9.70153	–5.35684	–9.21857	–79.97856	–37.40946
67.0	–1.01656	–8.94233	–6.46595	–9.84259	–82.45928	–38.87281
68.0	–2.03738	–9.30141	–7.68265	–10.61404	–85.11041	–40.53321
69.0	–3.14666	–9.79758	–9.04559	–11.56423	–87.98481	–42.43861
70.0	–4.38271	–10.46047	–10.60663	–12.73920	–91.15495	–44.65756
71.0	–5.79807	–11.33557	–12.43874	–14.20721	–94.72418	–47.29026
72.0	–7.46908	–12.49346	–14.64973	–16.07218	–98.84629	–50.48790
73.0	–9.51354	–14.04715	–17.40743	–18.49875	–103.76188	–54.48894
74.0	–12.12567	–16.18679	–20.99014	–21.76248	–109.87087	–59.69145
75.0	–15.65297	–19.25642	–25.89718	–26.36038	–117.89179	–66.81241
76.0	–20.78797	–23.94557	–33.12521	–33.28707	–129.25983	–77.28566
77.0	–29.14285	–31.86374	–44.99600	–44.86254	–147.31950	–94.45449
78.0	–45.54330	–47.83424	–68.46146	–68.03711	–182.08643	–128.33348
79.0	–94.01820	–95.88381	–138.14142	–137.42914	–283.44009	–228.80110
80.0	–19734.5749	–19736.0184	–28435.9581	–28434.9599	–41055.6988	–41000.1745
81.0	99.62454	98.60370	141.17990	142.46447	117.08404	173.49300
82.0	50.43273	49.83501	70.46595	72.03737	14.24004	71.53457
83.0	33.89932	33.72758	46.80701	48.66734	–20.80671	37.37490

(continued 3) $\alpha = 1.7$

λ	A	B	C	D	G	H
84.0	25.49793	25.75678	34.86631	37.01881	–38.96871	20.10226
85.0	20.34185	21.03766	27.60381	30.05292	–50.38876	9.57468
86.0	16.80424	17.94519	22.67623	25.42762	–58.44255	2.41727
87.0	14.18792	15.78409	19.07982	22.14044	–64.57581	–2.81481
88.0	12.14398	14.20743	16.31261	19.69077	–69.51123	–6.84333
89.0	10.47812	13.02306	14.09549	17.80095	–73.65043	–10.06893
90.0	9.07326	12.11622	12.26064	16.30474	–77.23458	–12.73170
91.0	7.85434	11.41441	10.70091	15.09673	–80.41741	–14.98419
92.0	6.77074	10.86987	9.34445	14.10700	–83.30171	–16.92787
93.0	5.78682	10.45012	8.14108	13.28756	–85.95893	–18.63271
94.0	4.87639	10.13263	7.05450	12.60456	–88.44046	–20.14844
95.0	4.01953	9.90159	6.05753	12.03365	–90.78437	–21.51118
96.0	3.20039	9.74598	5.12920	11.55712	–93.01963	–22.74769
97.0	2.40592	9.65830	4.25279	11.16207	–95.16890	–23.87804
98.0	1.62479	9.63384	3.41452	10.83919	–97.25033	–24.91732
99.0	0.84676	9.67015	2.60256	10.58199	–99.27893	–25.87689
100.0	0.06204	9.76684	1.80639	10.38633	–101.26745	–26.76516
105.0	–4.33817	11.28690	–2.28723	10.32749	–110.98110	–30.27353
110.0	–11.23697	15.84395	–7.88058	12.54362	–121.42357	–31.93676
115.0	–30.36849	32.85884	–21.72891	23.33855	–137.37171	–27.62857
120.0	224.04958	–223.64112	150.93184	–152.32624	–26.14646	–150.42366
125.0	25.46285	–27.25712	14.70324	–19.27672	–125.57837	–62.81511
130.0	11.84601	–16.13665	4.46830	–12.64525	–139.27440	–61.55580
135.0	5.60557	–12.93491	–0.94473	–11.61908	–149.98192	–64.41660
140.0	0.94895	–12.29649	–5.65155	–12.71365	–160.74731	–69.26229
145.0	–3.81745	–13.44950	–11.15803	–15.75391	–173.10435	–76.48064
150.0	–10.34495	–17.17711	–19.51769	–22.20146	–189.79253	–88.40756
155.0	–23.53215	–27.98336	–37.65325	–38.70996	–220.70499	–114.75272
160.0	–96.33612	–98.60802	–141.94642	–141.51287	–376.07246	–265.64570
165.0	64.94288	64.79289	91.72934	93.61480	–43.46322	71.41241
170.0	24.06114	26.09369	33.52874	36.90846	–132.27710	–12.92302
175.0	13.29535	17.69194	18.88087	23.87989	–158.15830	–34.23963
180.0	7.45279	14.54678	11.46769	18.31511	–173.50675	–44.86605
185.0	3.09441	13.44460	6.40907	15.48852	–185.31027	–51.68416
190.0	–0.92997	13.61929	2.18987	14.14833	–195.65673	–56.60129
195.0	–5.37749	15.06370	–1.99887	13.99992	–205.51922	–60.23776
200.0	–11.30104	18.45951	–7.01996	15.37010	–215.69265	–62.57263

$\alpha = 1.8$

λ	A	B	C	D	G	H
0.2	1.90542	1.20267	0.69211	1.75021	1.11926	1.29984
0.4	1.83775	1.22897	0.63347	1.75164	0.84920	1.21158
0.6	1.76814	1.25674	0.57347	1.75380	0.57805	1.12352
0.8	1.69648	1.28609	0.51202	1.75675	0.30576	1.03571
1.0	1.62262	1.31715	0.44901	1.76058	0.03225	0.94819
1.2	1.54641	1.35006	0.38437	1.76535	−0.24253	0.86099
1.4	1.46768	1.38497	0.31798	1.77117	−0.51865	0.77417
1.6	1.38625	1.42206	0.24971	1.77814	−0.79621	0.68777
1.8	1.30190	1.46152	0.17944	1.78636	−1.07529	0.60187
2.0	1.21440	1.50357	0.10703	1.79597	−1.35599	0.51654
2.2	1.12351	1.54845	0.03229	1.80710	−1.63842	0.43185
2.4	1.02893	1.59643	−0.04494	1.81993	−1.92269	0.34790
2.6	0.93034	1.64781	−0.12487	1.83462	−2.20895	0.26478
2.8	0.82737	1.70295	−0.20774	1.85140	−2.49734	0.18262
3.0	0.71962	1.76223	−0.29380	1.87049	−2.78803	0.10155
3.2	0.60661	1.82611	−0.38335	1.89216	−3.08122	0.02173
3.4	0.48781	1.89510	−0.47673	1.91672	−3.37712	−0.05666
3.6	0.36261	1.96980	−0.57431	1.94453	−3.67598	−0.13341
3.8	0.23030	2.05089	−0.67655	1.97601	−3.97809	−0.20827
4.0	0.09006	2.13917	−0.78396	2.01163	−4.28376	−0.28097
4.2	−0.05906	2.23556	−0.89712	2.05197	−4.59338	−0.35117
4.4	−0.21819	2.34118	−1.01675	2.09768	−4.90740	−0.41847
4.6	−0.38865	2.45731	−1.14367	2.14957	−5.22632	−0.48239
4.8	−0.57202	2.58552	−1.27885	2.20856	−5.55078	−0.54237
5.0	−0.77020	2.72766	−1.42348	2.27581	−5.88150	−0.59772
5.2	−0.98547	2.88602	−1.57897	2.35270	−6.21938	−0.64761
5.4	−1.22066	3.06336	−1.74707	2.44092	−6.56550	−0.69099
5.6	−1.47923	3.26312	−1.92993	2.54259	−6.92121	−0.72661
5.8	−1.76554	3.48964	−2.13023	2.66034	−7.28817	−0.75284
6.0	−2.08510	3.74839	−2.35139	2.79754	−7.66850	−0.76763
6.2	−2.44504	4.04645	−2.59779	2.95854	−8.06492	−0.76834
6.4	−2.85469	4.39314	−2.87521	3.14903	−8.48100	−0.75146
6.6	−3.32655	4.80089	−3.19133	3.37667	−8.92149	−0.71232
6.8	−3.87773	5.28679	−3.55671	3.65195	−9.39290	−0.64448
7.0	−4.53232	5.87489	−3.98618	3.98965	−9.90439	−0.53888
7.2	−5.32537	6.60017	−4.50132	4.41127	−10.46925	−0.38230
7.4	−6.30984	7.51555	−5.13474	4.94935	−11.10757	−0.15475
7.6	−7.56984	8.70511	−5.93819	5.65560	−11.85122	0.17554
7.8	−9.24748	10.31087	−6.99913	6.61737	−12.75402	0.66231
8.0	−11.60314	12.59319	−8.47778	7.99482	−13.91459	1.40404
8.2	−15.17114	16.08631	−10.70287	10.11660	−15.53519	2.60290

(continued 1) $\alpha = 1.8$

λ	A	B	C	D	G	H
8.4	−21.24835	22.08703	−14.47229	13.78051	−18.10673	4.74968
8.6	−34.00949	34.77002	−22.35444	21.55486	−23.20939	9.42441
8.8	−78.43716	79.11780	−49.71914	48.80937	−40.29901	26.08282
9.0	326.50621	−325.90727	199.37749	−200.39994	112.69493	−127.34575
9.2	55.02896	−54.51362	32.32594	−33.46369	9.66522	−24.75427
9.4	30.57798	−30.14821	17.24911	−18.50487	0.13235	−15.66336
9.6	21.38271	−21.04058	11.55711	−12.93372	−3.62787	−12.34900
9.8	16.53884	−16.28651	8.54134	−10.04177	−5.74275	−10.68408
10.0	13.53398	−13.37372	6.65610	−8.28347	−7.16330	−9.71776
11.0	7.14428	−7.48236	2.52751	−4.84183	−11.02681	−8.19730
12.0	4.69760	−5.61039	0.79548	−3.90176	−13.45596	−8.25557
13.0	3.22278	−4.81043	−0.37694	−3.65899	−15.59333	−8.79496
14.0	2.09167	−4.48934	−1.38954	−3.76188	−17.70314	−9.61738
15.0	1.07088	−4.46761	−2.40636	−4.12042	−19.91447	−10.69557
16.0	0.02740	−4.69954	−3.54244	−4.73962	−22.34368	−12.07816
17.0	−1.15977	−5.21581	−4.92983	−5.69592	−25.15120	−13.89131
18.0	−2.66038	−6.13538	−6.78262	−7.17208	−28.61945	−16.39812
19.0	−4.80289	−7.75557	−9.54055	−9.58859	−33.34562	−20.18389
20.0	−8.42663	−10.89529	−14.34965	−14.07882	−40.90028	−26.81132
21.0	−16.69627	−18.70495	−25.55473	−24.97868	−57.25248	−42.24391
22.0	−62.54758	−64.10968	−88.42541	−87.55114	−144.62748	−128.70272
23.0	49.84990	48.72937	66.23048	67.40132	67.00899	83.84991
24.0	19.21967	18.54281	24.28006	25.75037	8.39038	26.15036
25.0	12.13804	11.91336	14.70189	16.47876	−5.72722	12.95746
26.0	8.83999	9.08224	10.32915	12.42375	−12.68596	6.93176
27.0	6.83546	7.56578	7.74127	10.16906	−17.18624	3.37572
28.0	5.41809	6.66456	5.96969	8.75081	−20.55782	0.96275
29.0	4.30848	6.10710	4.63312	7.79307	−23.32297	−0.82577
30.0	3.37182	5.76803	3.54950	7.12025	−25.72971	−2.23344
31.0	2.53287	5.58361	2,61935	6.64076	−27.91209	−3.38887
32.0	1.74374	5.52059	1.78188	6.30387	−29.95021	−4.36525
33.0	0.96964	5.56316	0.99574	6.08136	−31.89643	−5.20588
34.0	0.18117	5.70739	0.22925	5.95911	−33.78830	−5.93616
35.0	−0.65067	5.95939	−0.54549	5.93351	−35.65586	−6.56937
36.0	−1.55909	6.33599	−1.35702	6.01061	−37.52657	−7.10901
37.0	−2.58720	6.86820	−2.23931	6.20773	−39.42960	−7.54878
38.0	−3.79701	7.60882	−3.23843	6.55824	−41.40116	−7.86992
39.0	−5.28628	8.64834	−4.42450	7.12230	−43.49322	−8.03423
40.0	−7.22385	10.14972	−5.91641	8.01069	−45.79081	−7.96779
41.0	−9.93468	12.43299	−7.94076	9.44321	−48.45312	−7.52080
42.0	−14.14195	16.21708	−10.99955	11.91600	−51.82961	−6.35186

(continued 2) $\alpha = 1.8$

λ	A	B	C	D	G	H
43.0	−21.85897	23.51148	−16.48764	16.81865	−56.88755	−3.50098
44.0	−41.61382	42.84073	−30.30900	30.05025	−67.71512	5.11326
45.0	−227.13473	227.92964	−158.86912	158.01158	−157.96338	93.13545
46.0	71.52355	−71.17040	47.69596	−49.16601	−16.25681	−50.81645
47.0	31.43755	−31.53945	19.79157	−21.89268	−36.84686	−32.49767
48.0	20.11591	−20.68982	11.79065	−14.54650	−43.66890	−27.97992
49.0	14.62991	−15.69685	7.82149	−11.26138	−47.71064	−26.28329
50.0	11.29593	−12.88147	5.33289	−9.49239	−50.74008	−25.64849
51.0	8.98420	−11.11907	3.54104	−8.46292	−53.30225	−25.54046
52.0	7.23090	−9.95190	2.12280	−7.85824	−55.62371	−25.74430
53.0	5.80894	−9.16006	0.91851	−7.52866	−57.81848	−26.15986
54.0	4.59243	−8.62638	−0.16202	−7.39609	−59.95317	−26.73731
55.0	3.50418	−8.28439	−1.17594	−7.41829	−62.07252	−27.45258
56.0	2.49231	−8.09563	−2.16384	−7.57328	−64.21112	−28.29696
57.0	1.51840	−8.03878	−3.15826	−7.85226	−66.39975	−29.27262
58.0	0.55079	−8.10438	−4.18883	−8.25643	−68.66967	−30.39111
59.0	−0.43986	−8.29251	−5.28613	−8.79608	−71.05626	−31.67379
60.0	−1.48423	−8.61241	−6.48529	−9.49149	−73.60315	−33.15401
61.0	−2.61820	−9.08383	−7.83052	−10.37574	−76.36784	−34.88156
62.0	−3.88795	−9.74061	−9.38165	−11.50009	−79.43044	−36.93062
63.0	−5.35806	−10.63761	−11.22499	−12.94412	−82.90848	−39.41406
64.0	−7.12577	−11.86433	−13.49287	−14.83482	−86.98387	−42.51006
65.0	−9.34886	−13.57223	−16.40222	−17.38472	−91.95632	−46.51533
66.0	−12.30599	−16.03477	−20.33822	−20.97538	−98.35951	−51.96102
67.0	−16.54538	−19.79573	−26.06064	−26.36353	−107.24748	−59.89905
68.0	−23.32154	−26.10584	−35.31039	−35.28743	−121.03700	−72.74436
69.0	−36.28612	−38.61344	−53.15748	−52.81480	−146.76819	−97.53548
70.0	−72.48490	−74.36133	−103.27305	−102.61470	−217.31340	−167.14332
71.0	−906.69925	−908.12818	−1261.62617	−1260.65433	−1826.91346	−1775.80738
72.0	91.82138	90.83909	125.21299	126.49790	98.09703	150.13892
73.0	44.47212	43.93800	59.61104	61.21028	6.04455	59.02329
74.0	29.45741	29.37533	38.91462	40.83108	−23.64488	30.27284
75.0	21.96507	22.34124	28.66685	30.90505	−38.82237	16.03761
76.0	17.39539	18.23837	22.48074	25.04686	−48.35723	7.44946
77.0	14.26090	15.58173	18.29145	21.19340	−55.11638	1.64270
78.0	11.93439	13.74671	15.22882	18.47633	−60.30761	−2.58921
79.0	10.10500	12.42525	12.86212	16.46689	−64.52843	−5.84239
80.0	8.60065	11.44833	10.95342	14.92931	−68.10853	−8.44499
81.0	7.31787	10.71592	9.36020	13.72348	−71.24497	−10.59240
82.0	6.19029	10.16546	7.99160	12.76123	−74.06306	−12.40802
83.0	5.17285	9.75630	6.78656	11.98459	−76.64651	−13.97339

(continued 3) $\alpha = 1.8$

λ	A	B	C	D	G	H
84.0	4.23342	9.46137	5.70217	11.35425	–79.05368	–15.34433
85.0	3.34792	9.26253	4.70702	10.84298	–81.32677	–16.56011
86.0	2.49735	9.14781	3.77715	10.43174	–83.49737	–17.64882
87.0	1.66589	9.10979	2.89346	10.10737	–85.58998	–18.63078
88.0	0.83958	9.14466	2.04000	9.86112	–87.62428	–19.52062
89.0	0.00523	9.25173	1.20275	9.68773	–89.61675	–20.32860
90.0	–0.85041	9.43324	0.36859	9.58504	–91.58187	–21.06150
91.0	–1.74152	9.69453	–0.47549	9.55380	–93.53303	–21.72295
92.0	–2.68421	10.04448	–1.34335	9.59781	–95.48340	–22.31374
93.0	–3.69773	10.49640	–2.25060	9.72440	–97.44684	–22.83149
94.0	–4.80621	11.06940	–3.21596	9.94537	–99.43895	–23.27026
95.0	–6.04112	11.79073	–4.26309	10.27851	–101.47841	–23.61950
96.0	–7.44523	12.69954	–5.42342	10.75019	–103.58912	–23.86232
97.0	–9.07901	13.85315	–6.74068	11.39980	–105.80345	–23.97243
98.0	–11.03177	15.33811	–8.27884	12.28745	–108.16782	–23.90869
99.0	–13.44207	17.29053	–10.13653	13.50838	–110.75303	–23.60510
100.0	–16.53826	19.93653	–12.47564	15.22138	–113.67435	–22.95062
105.0	–107.64147	108.83807	–77.51993	77.20297	–165.90544	18.03968
110.0	31.57278	–32.64902	19.94080	–23.42045	–103.49531	–55.75173
115.0	2.34219	–15.97962	5.49069	–12.53511	–120.19703	–50.99200
120.0	5.43423	–12.22564	–0.46767	–10.96738	–131.19036	–53.09116
125.0	0.52652	–11.62315	–5.41024	–12.01875	–142.01784	–57.58911
130.0	–4.67182	–13.17659	–11.39840	–15.43135	–154.75329	–64.92302
135.0	–12.70851	–18.40541	–21.62648	–23.67386	–173.59532	–78.77936
140.0	–35.53159	–38.81561	–52.57899	–52.91873	–221.38952	–121.78344
145.0	200.63879	199.61427	277.03316	278.29380	232.82272	337.14398
150.0	28.47746	29.72833	38.09624	40.96970	–104.59878	4.44720
155.0	13.92199	17.62076	18.66526	23.27493	–136.34251	–22.48352
160.0	7.34923	13.85878	10.47261	17.07691	–152.35038	–33.49438
165.0	2.71537	12.68740	5.20157	14.26392	–164.13250	–39.94993
170.0	–1.57655	13.03346	0.80721	13.16305	–174.40931	–44.30641
175.0	–6.56741	14.98925	–3.77330	13.51633	–184.38619	–47.19756
180.0	–14.02001	19.88566	–9.93945	16.12184	–195.26233	–48.26155
185.0	–30.42482	33.97697	–22.38226	25.34169	–210.48219	–44.51196
190.0	–154.17646	155.50394	–111.04853	110.90842	–279.80533	13.51239
195.0	60.27312	–61.20312	40.51397	–43.79966	–178.49872	–99.15765
200.0	23.96380	–27.30324	13.89846	–20.54176	–203.83301	–85.48262

$\alpha = 1.9$

λ	A	B	C	D	G	H
0.2	1.76692	1.15204	0.49740	1.56276	0.84542	1.03620
0.4	1.69255	1.18202	0.43378	1.56708	0.57443	0.95760
0.6	1.61568	1.21394	0.36839	1.57241	0.30206	0.87938
0.8	1.53611	1.24798	0.30109	1.57888	0.02822	0.80159
1.0	1.45363	1.28433	0.23176	1.58658	–0.24718	0.72430
1.2	1.36798	1.32322	0.16023	1.59566	–0.52424	0.64759
1.4	1.27891	1.36491	0.08632	1.60627	–0.80310	0.57155
1.6	1.18611	1.40967	0.00985	1.61858	–1.08387	0.49626
1.8	1.08922	1.45783	–0.06943	1.63278	–1.36671	0.42185
2.0	0.98787	1.50976	–0.15175	1.64910	–1.65178	0.34844
2.2	0.88160	1.56589	–0.23740	1.66780	–1.93928	0.27618
2.4	0.76990	1.62671	–0.32671	1.68918	–2.22943	0.20526
2.6	0.65219	1.69277	–0.42006	1.71357	–2.52245	0.13587
2.8	0.52778	1.76474	–0.51788	1.74139	–2.81865	0.06825
3.0	0.39587	1.84339	–0.62069	1.77310	–3.11835	0.00270
3.2	0.25554	1.92962	–0.72906	1.80926	–3.42194	–0.06047
3.4	0.10568	2.02450	–0.84371	1.85052	–3.72985	–0.12085
3.6	–0.05500	2.12930	–0.96545	1.89768	–4.04263	–0.17795
3.8	–0.22809	2.24557	–1.09527	1.95169	–4.36091	–0.23121
4.0	–0.41547	2.37516	–1.23438	2.01369	–4.68545	–0.27992
4.2	–0.61946	2.52036	–1.38424	2.08510	–5.01718	–0.32321
4.4	–0.84293	2.68399	–1.54665	2.16768	–5.35725	–0.36000
4.6	–1.08944	2.86960	–1.72384	2.26361	–5.70707	–0.38894
4.8	–1.36354	3.08166	–1.91869	2.37571	–6.06845	–0.40830
5.0	–1.67105	3.32597	–2.13483	2.50756	–6.44370	–0.41584
5.2	–2.01959	3.61011	–2.37706	2.66289	–6.83582	–0.40864
5.4	–2.41936	3.94421	–2.65175	2.85101	–7.24882	–0.38277
5.6	–2.88426	4.34214	–2.96763	3.07759	–7.68819	–0.33284
5.8	–3.43380	4.82336	–3.33690	3.35576	–8.16157	–0.25129
6.0	–4.09626	5.41607	–3.77726	3.70313	–8.68005	–0.12714
6.2	–4.91416	6.16276	–4.31530	4.14623	–9.26029	0.05617
6.4	–5.95465	7.13049	–4.99304	4.72698	–9.92856	0.32480
6.6	–7.33029	8.43178	–5.88083	5.51566	–10.72890	0.72267
6.8	–9.24529	10.27074	–7.10636	6.63985	–11.74112	1.32948
7.0	–12.11268	13.06034	–8.92782	8.35765	–13.12667	2.30652
7.2	–16.91292	17.78097	–11.95809	11.28183	–15.26866	4.03679
7.4	–26.67862	27.46515	–18.09278	17.30789	–19.35203	7.70505
7.6	–57.75609	58.45908	–37.54944	36.65326	–31.76343	19.69782
7.8	609.02322	–608.40586	379.19942	–380.20968	228.48124	–240.96918
8.0	51.00370	–50.47416	30.36787	–31.49513	10.18689	–23.10105
8.2	27.19631	–26.75692	15.45313	–16.70046	0.61137	–13.95583

(continued 1) $\alpha = 1.9$

λ	A	B	C	D	G	H
8.4	18.76198	−18.41517	10.14669	−11.51733	−2.95946	−10.81956
8.6	14.42054	−14.16886	7.39767	−8.89499	−4.93295	−9.28515
8.8	11.75825	−11.60439	5.69728	−7.32485	−6.25224	−8.40967
9.0	9.94699	−9.89378	4.52782	−6.28942	−7.24099	−7.86973
9.2	8.62587	−8.67631	3.66367	−5.56326	−8.04029	−7.52449
9.4	7.61242	−7.76965	2.99072	−5.03248	−8.72158	−7.30284
9.6	6.80442	−7.07178	2.44500	−4.63336	−9.32489	−7.16503
9.8	6.14015	−6.52117	1.98787	−4.32751	−9.87448	−7.08715
10.0	5.58009	−6.07851	1.59456	−4.09044	−10.38594	−7.05400
11.0	3.64532	−4.79544	0.14606	−3.50905	−12.64969	−7.29575
12.0	2.35547	−4.29465	−0.94790	−3.46449	−14.75609	−7.93678
13.0	1.26802	−4.19251	−1.98465	−3.73758	−16.91843	−8.86748
14.0	0.18587	−4.39076	−3.12209	−4.30154	−19.28145	−10.11710
15.0	−1.04696	−4.91302	−4.52080	−5.23412	−22.02918	−11.81781
16.0	−2.64696	−5.90714	−6.44398	−6.75551	−25.49579	−14.27683
17.0	−5.06838	−7.78779	−9.48053	−9.42971	−30.45396	−18.25117
18.0	−9.67903	−11.89776	−15.43657	−15.04708	−39.29457	−26.12177
19.0	−23.95482	−25.69599	−34.22536	−33.51007	−65.16695	−51.03105
20.0	208.72600	207.45181	274.64612	275.68239	343.70175	358.79902
21.0	22.77629	21.96849	28.07246	29.43141	15.59918	31.66035
22.0	12.5509	12.19177	14.61925	16.30831	−3.19474	13.83666
23.0	8.64691	8.80436	9.62612	11.65823	−10.76023	7.25148
24.0	6.48148	7.15434	6.91247	9.30629	−15.29595	3.71005
25.0	5.01046	6.23255	5.13006	7.91060	−18.58876	1.42992
26.0	3.87800	5.69375	3.81006	7.00975	−21.25837	−0.20343
27.0	2.92405	5.39089	2.74417	6.40459	−23.57833	−1.45708
28.0	2.06239	5.25426	1.82312	5.99750	−25.69003	−2.46419
29.0	1.23828	5.25085	0.98096	5.73796	−27.67619	−3.29654
30.0	0.41021	5.36886	0.17180	5.60117	−29.59123	−3.99354
31.0	−0.46003	5.61213	−0.64216	5.57925	−31.47608	−4.57486
32.0	−1.41509	5.99994	−1.49844	5.67877	−33.36698	−5.04535
33.0	−2.51100	6.57185	−2.44202	5.92279	−35.30274	−5.39534
34.0	−3.83188	7.39976	−3.53659	6.35880	−37.33368	−5.59597
35.0	−5.51934	8.61597	−4.88589	7.07814	−39.53752	−5.58600
36.0	−7.84168	10.48135	−6.68335	8.26441	−42.05514	−5.23767
37.0	−11.38883	13.57968	−9.35119	10.33162	−45.18986	−4.25853
38.0	−17.77842	19.52331	−14.04357	14.42687	−49.76618	−1.83350
39.0	−33.66453	34.96150	−25.50283	25.28614	−59.16490	5.40973
40.0	−159.89617	160.73877	−115.58308	114.75750	−124.58236	68.65943
41.0	64.85540	−64.47807	44.40083	−45.85014	−11.83947	−46.27139
42.0	27.52089	−27.62428	17.64916	−19.74313	−32.14833	−28.17880

(continued 2) $\alpha = 1.9$

λ	*A*	*B*	*C*	*D*	*G*	*H*
43.0	17.41500	−18.01946	10.29073	−13.05682	−38.64874	−23.93174
44.0	12.55275	−13.68399	6.66002	−10.13292	−42.50431	−22.37629
45.0	9.58739	−11.27727	4.37063	−8.59324	−45.41787	−21.82063
46.0	7.51080	−9.79837	2.70204	−7.72691	−47.90581	−21.76131
47.0	5.91223	−8.84515	1.35887	−7.25012	−50.18223	−21.99978
48.0	4.59051	−8.22701	0.19449	−7.03042	−52.35606	−22.44611
49.0	3.43295	−7.84451	−0.87557	−7.00111	−54.49308	−23.05839
50.0	2.36862	−7.64361	−1.90719	−7.12902	−56.64026	−23.82016
51.0	1.34723	−7.59614	−2.94316	−7.40107	−58.83746	−24.73144
52.0	0.32806	−7.69099	−4.02182	−7.81866	−61.12452	−25.80577
53.0	−0.72703	−7.93063	−5.18316	−8.39636	−63.54715	−27.07074
54.0	−1.85949	−8.33111	−6.47485	−9.16373	−66.16366	−28.57176
55.0	−3.12077	−8.92521	−7.96012	−10.17070	−69.05477	−30.38007
56.0	−4.58227	−9.77026	−9.73040	−11.49869	−72.34017	−32.60820
57.0	−6.35299	−10.96440	−11.92872	−13.28296	−76.20951	−35.44030
58.0	−8.61544	−12.68151	−14.79750	−15.75979	−80.98647	−39.19565
59.0	−11.70794	−15.25299	−18.78899	−19.37648	−87.27699	−44.47665
60.0	−16.34571	−19.38831	−24.86210	−25.08784	−96.36780	−52.56710
61.0	−24.36218	−26.91601	−35.47799	−35.35155	−111.56370	−66.76934
62.0	−42.32495	−44.39950	−59.45713	−58.98505	−144.71163	−98.92812
63.0	−124.82443	−126.42543	−170.12565	−169.31180	−294.30436	−247.53430
64.0	155.41181	154.28204	206.44130	207.59551	210.57003	258.32581
65.0	49.51985	48.86220	64.35067	65.84611	18.76988	67.51219
66.0	29.70064	29.51907	37.87635	39.71613	−17.73048	32.00076
67.0	21.17245	21.47398	26.57048	28.75989	−33.85464	16.86952
68.0	16.32419	17.11893	20.21060	22.75719	−43.33297	8.38972
69.0	13.12636	14.42768	16.07165	18.98529	−49.82511	2.90341
70.0	10.80631	12.63102	13.11684	16.40991	−54.72295	−0.97949
71.0	9.00492	11.37360	10.86496	14.55257	−58.67201	−3.90252
72.0	7.53185	10.46926	9.06164	13.16193	−62.01304	−6.20423
73.0	6.27608	9.81176	7.55922	12.09381	−64.94329	−8.07935
74.0	5.16776	9.33674	6.26558	11.26005	−67.58527	−9.64753
75.0	4.16000	9.00368	5.11969	10.60427	−70.01955	−10.98592
76.0	3.21925	8.78669	4.07885	10.08934	−72.30175	−12.14613
77.0	2.32003	8.66944	3.11165	9.69049	−74.47203	−13.16346
78.0	1.44154	8.64238	2.19361	9.39144	−76.56073	−14.06231
79.0	0.56554	8.70119	1.30454	9.18210	−78.59188	−14.85936
80.0	−0.32532	8.84606	0.42651	9.05731	−80.58562	−15.56544
81.0	−1.24893	9.08158	−0.45760	9.01633	−82.55990	−16.18661
82.0	−2.22524	9.41715	−1.36552	9.06276	−84.53199	−16.72459
83.0	−3.27801	9.86811	−2.31716	9.20516	−86.51986	−17.17665

(continued 3) $\alpha = 1.9$

λ	A	B	C	D	G	H
84.0	–4.43735	10.45769	–3.33655	9.45830	–88.54385	–17.53489
85.0	–5.74352	11.22041	–4.45474	9.84557	–90.62891	–17.78464
86.0	–7.25303	12.20799	–5.71433	10.40312	–92.80804	–17.90160
87.0	–9.04938	13.49985	–7.17737	11.18749	–95.12808	–17.84629
88.0	–11.26291	15.22277	–8.93979	12.28984	–97.66033	–17.55384
89.0	–14.11071	17.59071	–11.16055	13.86495	–100.52176	–16.91291
90.0	–17.98533	20.99344	–14.12621	16.19560	–103.92209	–15.71885
91.0	–23.67949	26.22112	–18.41449	19.85609	–108.28281	–13.55474
92.0	–33.07453	35.15274	–25.39531	26.21318	–114.60193	–9.42686
93.0	–51.99386	53.60948	–39.30753	39.50275	–125.96533	–0.25334
94.0	–111.91942	113.07114	–83.06388	82.63462	–159.04942	30.63835
95.0	1119.22354	–1118.53910	812.79753	–813.85590	491.70372	–622.31345
96.0	95.66962	–95.45798	67.72515	–69.42007	–51.79074	–81.02784
97.0	50.18714	–50.45591	34.47499	–36.81678	–77.23565	–57.80566
98.0	33.90804	–34.66704	22.47551	–25.47743	–87.21881	–50.06318
99.0	25.41883	–26.68021	16.14160	–19.82005	–93.08378	–46.46102
100.0	20.12511	–21.90348	12.12874	–16.50345	–97.26566	–44.56857
105.0	8.20540	–12.89281	2.55207	–10.84515	–110.67824	–43.19267
110.0	2.43768	–10.98300	–2.84196	–10.64861	–121.44372	–46.18816
115.0	–2.71821	–11.75823	–8.43369	–13.04836	–133.21934	–51.93886
120.0	–10.03587	–15.94909	–17.32742	–19.66175	–149.72729	–63.08471
125.0	–30.03025	–33.35802	–43.47094	–43.91788	–189.60097	–97.88153
130.0	163.10973	162.16892	217.51361	218.81087	157.37425	254.06735
135.0	23.91786	25.39020	30.70099	33.76319	–99.03409	2.64884
140.0	11.28888	15.40915	14.49846	19.49878	–125.49930	–18.69877
145.0	5.25321	12.52702	7.33500	14.64487	–139.67106	–27.48006
150.0	0.63587	12.02696	2.37831	12.70484	–150.65242	–32.55215
155.0	–4.17354	13.26306	–2.24520	12.51201	–160.75076	–35.70430
160.0	–10.88225	17.13171	–8.03609	14.47434	–171.40176	–36.97319
165.0	–24.99588	28.75914	–19.14687	22.23306	–185.80652	–33.84666
170.0	–125.68007	127.08987	–93.67872	93.59137	–246.64831	15.95700
175.0	52.01956	–52.99369	35.85204	–39.15447	–157.91293	–83.87316
180.0	20.16672	–23.71238	11.70994	–18.48087	–181.90751	–71.31565
185.0	10.34993	–16.84975	3.60350	–14.35939	–194.30634	–71.05101
190.0	4.43961	–14.57424	–1.84918	–13.81018	–205.04297	–73.68805
195.0	–0.55640	–14.43347	–7.00779	–15.20176	–216.07151	–78.25509
200.0	–6.03943	–16.18606	–13.25692	–18.71449	–228.88904	–85.36640

$\alpha = 2.0$

λ	*A*	*B*	*C*	*D*	*G*	*H*
0.2	1.63934	1.10782	0.33008	1.40325	0.62590	0.82691
0.4	1.55764	1.14200	0.26104	1.41075	0.35317	0.75734
0.6	1.47270	1.17873	0.18971	1.41966	0.07870	0.68837
0.8	1.38421	1.21829	0.11591	1.43012	–0.19765	0.62009
1.0	1.29186	1.26099	0.03940	1.44234	–0.47602	0.55263
1.2	1.19524	1.30718	–0.04006	1.45653	–0.75659	0.48609
1.4	1.09394	1.35728	–0.12275	1.47295	–1.03953	0.42064
1.6	0.98745	1.41175	–0.20899	1.49187	–1.32508	0.35645
1.8	0.87519	1.47116	–0.29918	1.51366	–1.61348	0.29371
2.0	0.75648	1.53614	–0.39373	1.53869	–1.90501	0.23267
2.2	0.63054	1.60745	–0.49317	1.56744	–2.20002	0.17362
2.4	0.49642	1.68600	–0.59809	1.60048	–2.49890	0.11689
2.6	0.35303	1.77286	–0.70920	1.63846	–2.80211	0.06288
2.8	0.19903	1.86932	–0.82736	1.68220	–3.11021	0.01211
3.0	0.03283	1.97694	–0.95360	1.73267	–3.42386	–0.03485
3.2	–0.14755	2.09764	–1.08917	1.79108	–3.74388	–0.07723
3.4	–0.34451	2.23381	–1.23561	1.85890	–4.07126	–0.11411
3.6	–0.56109	2.38841	–1.39485	1.93802	–4.40724	–0.14434
3.8	–0.80109	2.56523	–1.56932	2.03079	–4.75337	–0.16643
4.0	–1.06945	2.76912	–1.76215	2.14030	–5.11165	–0.17847
4.2	–1.37258	3.00646	–1.97742	2.27054	–5.48470	–0.17793
4.4	–1.71907	3.28578	–2.22058	2.42689	–5.87598	–0.16145
4.6	–2.12062	3.61872	–2.49905	2.61671	–6.29021	–0.12441
4.8	–2.59363	4.02162	–2.82323	2.85029	–6.73399	–0.06029
5.0	–3.16184	4.51814	–3.20815	3.14260	–7.21686	0.04032
5.2	–3.86093	5.14389	–3.67641	3.51613	–7.75315	0.19163
5.4	–4.74714	5.95505	–4.26361	4.00639	–8.36541	0.41606
5.6	–5.91458	7.04563	–5.02931	4.67283	–9.09134	0.75118
5.8	–7.53361	8.58591	–6.08129	5.62312	–9.99917	1.26508
6.0	–9.94764	10.91921	–7.63681	7.07441	–11.22636	2.09507
6.2	–13.96794	14.85670	–10.20909	9.53977	–13.09758	3.56565
6.4	–22.08185	22.88561	–15.37153	14.59249	–16.60815	6.67197
6.6	–47.40678	48.12324	–31.42167	30.52995	–27.00678	16.66254
6.8	716.09649	–715.46974	451.50549	–452.51300	278.22470	–288.98100
7.0	44.39503	–43.86058	26.58427	–27.71086	9.19005	–20.36263
7.2	23.48660	–23.04713	13.32527	–14.57440	0.55372	–12.14703
7.4	16.18122	–15.83957	8.67000	–10.04535	–2.64147	–9.37728
7.6	12.43615	–12.19533	6.26584	–7.77126	–4.41398	–8.03517
7.8	10.14114	–10.00435	4.77788	–6.41746	–5.60830	–7.27653
8.0	8.57787	–8.54848	3.75165	–5.52975	–6.51197	–6.81411
8.2	7.43467	–7.51628	2.98996	–4.91119	–7.24982	–6.52344

(continued 1) $\alpha = 2.0$

λ	A	B	C	D	G	H
8.4	6.55444	–6.75085	2.39334	–4.46260	–7.88487	–6.34187
8.6	5.84931	–6.16459	1.90615	–4.12865	–8.45241	–6.23450
8.8	5.26630	–5.70477	1.49479	–3.87608	–8.97381	–6.18041
9.0	4.77153	–5.33779	1.13770	–3.68372	–9.46282	–6.16633
9.2	4.34225	–5.04126	0.82043	–3.53751	–9.92873	–6.18350
9.4	3.96265	–4.79968	0.53283	–3.42777	–10.37810	–6.22593
9.6	3.62134	–4.60205	0.26759	–3.34765	–10.81571	–6.28946
9.8	3.30986	–4.44035	0.01922	–3.29223	–11.24522	–6.37115
10.0	3.02181	–4.30864	–0.21649	–3.25790	–11.66947	–6.46891
11.0	1.78867	–3.97559	–1.30298	–3.33043	–13.78236	–7.16156
12.0	0.67331	–4.03261	–2.40428	–3.73791	–16.01905	–8.17968
13.0	–0.53581	–4.43928	–3.71005	–4.51047	–18.57119	–9.61393
14.0	–2.07269	–5.31640	–5.48402	–5.84007	–21.75415	–11.73416
15.0	–4.40466	–7.07216	–8.30720	–8.27014	–26.30198	–15.25052
16.0	–9.00237	–11.14258	–14.05720	–13.65615	–34.61899	–22.55315
17.0	–25.36079	–27.00023	–34.92145	–34.17111	–62.35316	–49.28070
18.0	74.39577	73.24652	93.33849	94.43384	101.43364	115.51177
19.0	17.30350	16.64645	20.15341	21.59785	6.50065	21.58908
20.0	10.08101	9.92912	11.01966	12.82468	–6.16741	9.94100
21.0	7.04839	7.42525	7.27272	9.45716	–11.91245	5.23074
22.0	5.25012	6.19062	5.12001	7.71068	–15.60305	2.59533
23.0	3.96603	5.51804	3.64034	6.67329	–18.41971	0.86090
24.0	2.92915	5.15648	2.49530	6.01807	–20.79380	–0.39565
25.0	2.01237	4.99948	1.52743	5.60249	–22.92340	–1.36157
26.0	1.14119	5.00047	0.64883	5.35898	–24.91575	–2.12931
27.0	0.26124	5.14461	–0.19936	5.25747	–26.83847	–2.74548
28.0	–0.67804	5.43981	–1.06589	5.29191	–28.74232	–3.22970
29.0	–1.73616	5.91761	–2.00195	5.47758	–30.67467	–3.58078
30.0	–2.99801	6.64364	–3.07494	5.85708	–32.69163	–3.77445
31.0	–4.60508	7.74539	–4.39212	6.51970	–34.87661	–3.75045
32.0	–6.82935	9.48422	–6.15525	7.65354	–37.38084	–3.37510
33.0	–10.28990	12.47078	–8.81921	9.70258	–40.53988	–2.32689
34.0	–16.79825	18.50950	–13.71014	13.98384	–45.32566	0.35433
35.0	–34.98351	36.22321	–27.13747	26.79864	–56.35247	9.27242
36.0	–642.85439	643.61479	–471.61430	470.65256	–382.47007	333.26737
37.0	42.74366	–42.47608	29.40166	–31.00421	–17.45893	–33.89007
38.0	20.96933	–21.21405	13.33843	–15.60739	–30.42865	–23.10030
39.0	13.76336	–14.54636	7.91673	–10.88614	–35.62959	–20.12392
40.0	10.01662	–11.37123	5.01393	–8.72739	–39.00190	–19.03320
41.0	7.61257	–9.58073	3.08045	–7.59276	–41.68150	–18.70679
42.0	5.85615	–8.49023	1.60523	–6.98479	–44.04582	–18.78494

(continued 2) $\alpha = 2.0$

λ	A	B	C	D	G	H
43.0	4.44857	–7.81393	0.36616	–6.69824	–46.26206	–19.12247
44.0	3.23638	–7.41495	–0.75365	–6.63782	–48.42184	–19.65587
45.0	2.12842	–7.22387	–1.82708	–6.75900	–50.58651	–20.36070
46.0	1.06196	–7.20725	–2.90834	–7.04569	–52.80662	–21.23611
47.0	–0.01389	–7.35477	–4.04624	–7.50197	–55.13298	–22.30020
48.0	–1.14868	–7.67507	–5.29365	–8.15079	–57.62591	–23.59143
49.0	–2.39993	–8.19722	–6.71739	–9.03817	–60.36656	–25.17573
50.0	–3.84546	–8.97851	–8.41300	–10.24464	–63.47513	–27.16232
51.0	–5.60456	–10.12296	–10.53158	–11.91012	–67.14534	–29.73676
52.0	–7.88204	–11.82370	–13.33669	–14.28965	–71.71822	–33.23383
53.0	–11.07501	–14.46867	–17.34325	–17.89139	–77.86290	–38.31772
54.0	–16.08469	–18.95163	–23.72359	–23.88224	–87.10563	–46.51068
55.0	–25.51424	–27.86958	–35.86844	–35.64840	–103.86395	–62.22695
56.0	–51.35049	–53.20402	–69.39766	–68.80581	–148.49275	–105.81852
57.0	–548.45235	–549.80922	–717.08773	–716.12754	–993.50476	–949.79560
58.0	68.58363	67.72263	87.18010	88.50836	53.69266	98.43675
59.0	32.92817	32.56622	40.85303	42.55204	–7.61518	38.16606
60.0	21.69846	21.84271	26.35974	28.43511	–27.43541	19.38738
61.0	16.06011	16.72169	19.15613	21.61642	–37.75337	10.11754
62.0	12.57781	13.77202	14.76644	17.62330	–44.40106	4.52683
63.0	10.14690	11.89348	11.75231	15.02070	–49.25204	0.74416
64.0	8.30227	10.62590	9.50897	13.20751	–53.09343	–2.01487
65.0	6.81293	9.74383	7.73693	11.88840	–56.31437	–4.13631
66.0	5.55007	9.12486	6.27012	10.90205	–59.12997	–5.83170
67.0	4.43502	8.69788	5.00812	10.15369	–61.66951	–7.22611
68.0	3.41582	8.42000	3.88574	9.58491	–64.01608	–8.39758
69.0	2.45541	8.26520	2.85774	9.15872	–66.22634	–9.39658
70.0	1.52510	8.21856	1.89056	8.85186	–68.34115	–10.25629
71.0	0.60060	8.27321	0.95739	8.65051	–70.39186	–10.99834
72.0	–0.34077	8.42890	0.03488	8.54816	–72.40422	–11.63588
73.0	–1.32244	8.69180	–0.89926	8.54460	–74.40115	–12.17532
74.0	–2.37099	9.07518	–1.86863	8.64607	–76.40511	–12.61680
75.0	–3.51919	9.60135	–2.90063	8.86636	–78.44040	–12.95380
76.0	–4.81050	10.30539	–4.03011	9.22939	–80.53611	–13.17146
77.0	–6.30635	11.24188	–5.30493	9.77407	–82.73053	–13.24314
78.0	–8.09906	12.49745	–6.79574	10.56365	–85.07858	–13.12358
79.0	–10.33683	14.21547	–8.61476	11.70404	–87.66595	–12.73531
80.0	–13.27643	16.64854	–10.95523	13.38307	–90.63875	–11.93930
81.0	–17.40851	20.28365	–14.18527	15.96407	–94.27370	–10.46505
82.0	–23.80714	26.19157	–19.10936	20.24719	–99.17388	–7.71514
83.0	–35.37179	37.26874	–27.89760	28.39860	–106.96395	–2.07002

(continued 3) $\alpha = 2.0$

λ	A	B	C	D	G	H
84.0	–63.59085	65.00068	–49.14585	49.01043	–124.07597	12.89748
85.0	–251.38977	252.31009	–189.78509	189.01008	–230.51174	117.18452
86.0	137.26734	–136.84158	100.85191	–102.27322	–14.28858	–101.19630
87.0	54.53967	–54.61629	38.81841	–40.89628	–61.92005	–55.73595
88.0	33.97896	–34.56855	23.29233	–26.04069	–74.76077	–45.08462
89.0	24.48858	–25.60470	16.04472	–19.48134	–81.41295	–40.64515
90.0	18.92052	–20.57990	11.72689	–15.87368	–85.87938	–38.42015
91.0	15.18564	–17.40844	8.77487	–13.65826	–89.33141	–37.24416
92.0	12.45004	–15.26028	6.56370	–12.21514	–92.23802	–36.65478
93.0	10.31453	–13.74054	4.79351	–11.25010	–94.82513	–36.43349
94.0	8.56315	–12.63823	3.30137	–10.60668	–97.21646	–36.46502
95.0	7.06801	–11.83118	1.98995	–10.19509	–99.48692	–36.68429
96.0	5.74760	–11.24464	0.79642	–9.96128	–101.68598	–37.05335
97.0	4.54658	–10.83131	–0.32289	–9.87211	–103.84926	–37.55034
98.0	3.42499	–10.56086	–1.40058	–9.90760	–106.00468	–38.16384
99.0	2.35204	–10.41432	–2.46298	–10.05685	–108.17626	–38.89003
100.0	1.30232	–10.38087	–3.53319	–10.31586	–110.38664	–39.73124
105.0	–4.40813	–11.96372	–9.84954	–13.53937	–122.97129	–46.19020
110.0	–14.66833	–19.20989	–22.42997	–23.85883	–143.98357	–61.69825
115.0	–88.82345	–90.78011	–118.71196	–118.19943	–274.57551	–187.02396
120.0	32.95269	33.50107	41.27888	43.67497	–69.95588	22.82008
125.0	12.70746	15.94966	15.54392	19.96751	–108.24376	–10.13368
130.0	5.66757	12.11099	7.22266	14.05746	–123.64720	–19.91283
135.0	0.73862	11.42985	1.95204	11.98810	–134.84175	–24.88673
140.0	–4.38225	12.80999	–2.94593	11.99111	–145.07846	–27.62059
145.0	–12.06902	17.57178	–9.55325	14.77017	–156.30905	–27.96565
150.0	–32.70991	35.64546	–25.83019	27.68643	–174.70212	–20.53717
155.0	271.29164	–270.82783	202.72126	–204.10133	–8.44407	–197.59783
160.0	27.81547	–29.93136	18.40695	–23.16527	–153.55465	–63.47640
165.0	12.59454	–17.61761	6.12117	–14.68711	–169.06381	–59.51938
170.0	5.58546	–14.15983	–0.15171	–13.06597	–180.27381	–60.96774
175.0	0.26013	–13.60350	–5.49372	–13.97117	–191.22569	–64.76467
180.0	–5.39067	–15.21876	–11.77563	–17.20797	–203.79605	–71.22922
185.0	–13.71414	–20.38838	–21.80430	–24.85597	–221.45987	–83.28795
190.0	–33.87121	–37.83088	–47.45816	–48.45989	–259.68589	–116.15790
195.0	–794.29437	–795.72826	–1042.95978	–1042.05335	–1568.51269	–1419.73523
200.0	42.09765	43.17739	53.46263	56.26881	–136.79119	17.23023

VALUES OF THE FUNCTIONS A TO H

$\alpha = 2.1$

λ	A	B	C	D	G	H
0.2	1.52093	1.06925	0.18498	1.26656	0.44801	0.65932
0.4	1.43116	1.10821	0.11004	1.27757	0.17282	0.59824
0.6	1.33718	1.15055	0.03215	1.29048	–0.10458	0.53807
0.8	1.23857	1.19666	–0.04897	1.30555	–0.38437	0.47896
1.0	1.13480	1.24703	–0.13366	1.32308	–0.66680	0.42108
1.2	1.02527	1.30223	–0.22232	1.34342	–0.95212	0.36465
1.4	0.90927	1.36294	–0.31542	1.36699	–1.24065	0.30991
1.6	0.78598	1.42994	–0.41348	1.39428	–1.53274	0.25718
1.8	0.65440	1.50418	–0.51717	1.42588	–1.82882	0.20682
2.0	0.51334	1.58682	–0.62723	1.46252	–2.12939	0.15926
2.2	0.36137	1.67924	–0.74462	1.50506	–2.43507	0.11505
2.4	0.19673	1.78315	–0.87047	1.55460	–2.74661	0.07487
2.6	0.01725	1.90069	–1.00617	1.61246	–3.06490	0.03953
2.8	–0.17981	2.03453	–1.15348	1.68034	–3.39111	0.01011
3.0	–0.39790	2.18806	–1.31463	1.76040	–3.72665	–0.01204
3.2	–0.64149	2.36569	–1.49248	1.85541	–4.07339	–0.02517
3.4	–0.91642	2.57321	–1.69078	1.96906	–4.43374	–0.02696
3.6	–1.23056	2.81841	–1.91457	2.10629	–4.81094	–0.01427
3.8	–1.59468	3.11199	–2.17074	2.27390	–5.20941	0.01719
4.0	–2.02397	3.46907	–2.46900	2.48153	–5.63537	0.07355
4.2	–2.54059	3.91174	–2.82353	2.74322	–6.09790	0.16375
4.4	–3.17822	4.47355	–3.25580	3.08036	–6.61074	0.30137
4.6	–3.99053	5.20812	–3.80010	3.52712	–7.19577	0.50819
4.8	–5.06885	6.20666	–4.51470	4.14163	–7.89022	0.82124
5.0	–6.58208	7.63796	–5.50737	5.03154	–8.76285	1.30914
5.2	–8.88133	9.85303	–7.00208	6.42067	–9.95657	2.11461
5.4	–12.83670	13.72183	–9.55385	8.86387	–11.82541	3.59150
5.6	–21.35828	22.15432	–15.01882	14.21715	–15.55333	6.92354
5.8	–53.84521	54.54949	–35.77019	34.85350	–29.03225	20.00243
6.0	134.17392	–133.56421	84.12069	–85.15593	47.19217	–56.62640
6.2	31.50307	–30.99093	18.60281	–19.76032	5.15805	–15.00135
6.4	18.28106	–17.86966	10.13619	–11.41993	–0.48886	–9.76843
6.6	13.04265	–12.73536	6.76061	–8.17478	–2.88991	–7.78663
6.8	10.20877	–10.00918	4.91758	–6.46667	–4.31484	–6.78653
7.0	8.41560	–8.32755	3.73715	–5.42592	–5.31862	–6.21352
7.2	7.16597	–7.19353	2.90214	–4.73568	–6.10366	–5.86561
7.4	6.23525	–6.38279	2.26918	–4.25294	–6.76149	5.65169
7.6	5.50709	–5.77928	1.76396	–3.90378	–7.33964	–5.52473
7.8	4.91509	–5.31696	1.34405	–3.64617	–7.86531	–5.45805
8.0	4.41860	–4.95553	0.98339	–3.45454	–8.35525	–5.43548
8.2	3.99127	–4.66906	0.66505	–3.31246	–8.82041	–5.44670

(continued 1) $\alpha = 2.1$

λ	A	B	C	D	G	H
8.4	3.61523	−4.44015	0.37749	−3.20897	−9.26834	−5.48489
8.6	3.27788	−4.25670	0.11253	−3.13651	−9.70445	−5.54543
8.8	2.97007	−4.11011	−0.13586	−3.08978	−10.13278	−5.62514
9.0	2.68491	−3.99415	−0.37225	−3.06498	−10.55651	−5.72184
9.2	2.41710	−3.90421	−0.60023	−3.05938	−10.97818	−5.83406
9.4	2.16243	−3.83688	−0.82275	−3.07106	−11.39998	−5.96087
9.6	1.91749	−3.78962	−1.04224	−3.09867	−11.82381	−6.10175
9.8	1.67939	−3.76059	−1.26086	−3.14135	−12.25141	−6.25654
10.0	1.44567	−3.74847	−1.48053	−3.19858	−12.68446	−6.42534
11.0	0.27052	−3.92264	−2.65697	−3.70750	−14.99142	−7.50263
12.0	−1.10135	−4.52939	−4.15122	−4.68297	−17.74042	−9.11555
13.0	−3.05188	−5.84180	−6.40632	−6.49833	−21.46248	−11.75048
14.0	−6.63875	−8.86114	−10.72339	−10.41847	−27.76829	−16.99479
15.0	−17.66475	−19.35697	−24.32449	−23.64385	−45.65214	−33.82874
16.0	130.25055	129.07323	160.15466	161.20429	183.32556	196.19687
17.0	16.78820	16.12706	18.88079	20.30384	6.32147	20.24618
18.0	9.31136	9.18182	9.69776	11.50835	−6.05693	8.93341
19.0	6.36245	6.79358	6.16400	8.38588	−11.38954	4.68548
20.0	4.63542	5.67106	4.16310	6.83064	−14.80381	2.38261
21.0	3.39241	5.09396	2.78001	5.94041	−17.43687	0.89708
22.0	2.36826	4.81962	1.68998	5.40703	−19.68925	−0.15939
23.0	1.43589	4.75154	0.74231	5.10251	−21.74305	−0.95203
24.0	0.51623	4.85463	−0.15048	4.97209	−23.69912	−1.55733
25.0	−0.45634	5.12985	−1.05358	5.00029	−25.62629	−2.00699
26.0	−1.55526	5.61091	−2.03174	5.20235	−27.58532	−2.30200
27.0	−2.88615	6.37736	−3.17044	5.63123	−29.64821	−2.41093
28.0	−4.63381	7.59585	−4.61236	6.40696	−31.92731	−2.24970
29.0	−7.18494	9.63968	−6.65000	7.80529	−34.64553	−1.61633
30.0	−11.54891	13.50773	−10.04074	10.57047	−38.37246	0.04238
31.0	−21.53554	23.00110	−17.63551	17.54253	−45.24265	4.84725
32.0	−76.07347	77.04066	−58.54495	57.82232	−77.02584	34.55581
33.0	60.70332	−60.24699	43.59644	−44.96490	−1.84821	−42.71745
34.0	22.41351	−22.48801	14.82739	−16.86731	−24.56498	−22.12918
35.0	13.66834	−14.30169	8.14262	−10.88979	−30.77792	−18.09043
36.0	9.59885	−10.82818	4.94418	−8.44590	−34.39651	−16.70630
37.0	7.11438	−8.98765	2.91823	−7.23551	−37.15442	−16.26040
38.0	5.33983	−7.91832	1.40700	−6.61768	−39.54790	−16.27786
39.0	3.92712	−7.28908	0.14580	−6.34924	−41.78044	−16.58243
40.0	2.70487	−6.95090	−0.99933	−6.32456	−43.95954	−17.10248
41.0	1.57236	−6.83341	−2.11159	−6.49889	−46.15742	−17.81454
42.0	0.45836	−6.90802	−3.25531	−6.86195	−48.43625	−18.72496

(continued 2) $\alpha = 2.1$

λ	*A*	*B*	*C*	*D*	*G*	*H*
43.0	−0.69937	−7.17464	−4.49320	−7.42983	−50.86338	−19.86627
44.0	−1.96868	−7.66009	−5.90061	−8.24718	−53.52638	−21.30304
45.0	−3.43967	−8.42616	−7.58442	−9.39971	−56.55524	−23.14945
46.0	−5.25283	−9.59308	−9.71748	−11.04511	−60.16516	−25.60935
47.0	−7.65995	−11.39759	−12.61509	−13.48742	−64.75299	−29.07115
48.0	−11.18459	−14.35170	−16.93810	−17.37881	−71.15433	−34.36395
49.0	−17.15517	−19.77464	−24.36834	−24.39424	−81.49970	−43.61312
50.0	−30.27393	−32.36112	−40.86562	−40.48787	−103.34384	−64.36914
51.0	−88.27738	−89.84120	−114.27271	−113.49761	−197.34315	−157.28479
52.0	121.20422	120.16056	151.45140	152.62185	138.62597	179.76681
53.0	37.44588	36.92444	45.39009	46.95786	3.22544	45.45055
54.0	22.30255	22.31043	26.32423	28.29511	−21.87618	21.43790
55.0	15.77372	16.32311	18.18319	20.56684	−33.12443	11.28624
56.0	12.01940	13.12775	13.56404	16.37410	−39.90416	5.61377
57.0	9.49845	11.18892	10.51429	13.76876	−44.68972	1.94944
58.0	7.62612	9.92821	8.29414	12.01587	−48.41786	−0.63978
59.0	6.13012	9.08063	6.56023	10.77762	−51.52250	−2.58358
60.0	4.86489	8.50922	5.13017	9.87820	−54.23323	−4.10652
61.0	3.74383	8.13774	3.89674	9.21836	−56.68399	−5.33645
62.0	2.71024	7.92230	2.79131	8.73922	−58.95936	−6.35044
63.0	1.72330	7.83821	1.76614	8.40545	−61.11674	−7.19643
64.0	0.75030	7.87356	0.78498	8.19672	−63.19809	−7.90417
65.0	−0.23820	8.02633	−0.18262	8.10359	−65.23676	−8.49089
66.0	−1.27218	8.30357	−1.16538	8.12572	−67.26207	−8.96414
67.0	−2.38594	8.72242	−2.19392	8.27197	−69.30293	−9.32258
68.0	−3.62294	9.31318	−3.30484	8.56226	−71.39159	−9.55512
69.0	−5.04327	10.12562	−4.54683	9.03217	−73.56867	−9.63777
70.0	−6.73672	11.24122	−5.99082	9.74209	−75.89121	−9.52684
71.0	−8.84810	12.79797	−7.74955	10.79613	−78.44783	−9.14468
72.0	−11.63183	15.04455	−10.01959	12.38358	−81.39109	−8.34794
73.0	−15.58545	18.47359	−13.18371	14.88095	−85.01610	−6.84946
74.0	−21.83887	24.21064	−18.10949	19.15032	−89.98956	−3.98954
75.0	−33.63999	35.49967	−27.28864	27.67837	−98.22354	2.13753
76.0	−65.73271	67.08089	−52.02869	51.76801	−118.39061	20.19841
77.0	−588.70822	589.54202	−453.21214	452.29728	−427.24129	326.93804
78.0	88.87771	−88.56465	66.27555	−67.85278	−30.06043	−72.36447
79.0	41.64308	−41.86059	29.91859	−32.17078	−59.14258	−45.42006
80.0	27.06313	−27.82468	18.59999	−21.54437	−69.02802	−37.69438
81.0	19.82049	−21.14342	12.90371	−16.56240	−74.60755	−34.30287
82.0	15.38776	−17.29364	9.35675	−13.75726	−78.54585	−32.58768
83.0	12.32058	−14.83564	6.85027	−12.02609	−81.69477	−31.70458

(continued 3) $\alpha = 2.1$

λ	A	B	C	D	G	H
84.0	10.01371	–13.16954	4.91881	–10.91019	–84.41297	–31.30351
85.0	8.16767	–12.00202	3.33113	–10.18620	–86.87978	–31.21519
86.0	6.61604	–11.17392	1.95779	–9.73389	–89.19678	–31.34973
87.0	5.25763	–10.59267	0.71898	–9.48446	–91.42797	–31.65717
88.0	4.02590	–10.20224	–0.43897	–9.39759	–93.61789	–32.10998
89.0	2.87371	–9.96841	–1.55548	–9.45032	–95.80093	–32.69470
90.0	1.76490	–9.87115	–2.66237	–9.63138	–98.00670	–33.40789
91.0	0.66915	–9.90070	–3.78809	–9.93848	–100.26365	–34.25444
92.0	–0.44153	–10.05568	–4.96087	–10.37731	–102.60200	–35.24727
93.0	–1.59580	–10.34261	–6.21156	–10.96177	–105.05676	–36.40841
94.0	–2.82589	–10.77681	–7.57696	–11.71565	–107.67148	–37.77143
95.0	–4.17145	–11.38461	–9.10415	–12.67585	–110.50357	–39.38590
96.0	–5.68494	–12.20791	–10.85729	–13.89843	–113.63264	–41.32524
97.0	–7.44065	–13.31243	–12.92880	–15.46920	–117.17472	–43.70044
98.0	–9.55061	–14.80315	–15.45956	–17.52367	–121.30787	–46.68537
99.0	–12.19564	–16.85498	–18.67834	–20.28605	–126.32223	–50.56673
100.0	–15.69143	–19.77859	–22.98588	–24.15324	–132.72717	–55.85092
105.0	–402.51937	–403.92533	–514.91195	–514.01410	–763.79627	–681.42007
110.0	21.93521	23.18647	26.27033	29.21765	–79.27392	8.59229
115.0	8.89093	13.12389	10.40608	15.65562	–104.05206	–10.49902
120.0	2.93523	10.95791	3.76985	11.94786	–116.92307	–17.19787
125.0	–2.15862	11.45846	–1.31838	11.18487	–127.47420	–20.49582
130.0	–8.96372	14.97194	–7.40533	13.07778	–138.38408	–21.32528
135.0	–25.36889	28.62077	–20.78417	22.94370	–154.76411	–15.81788
140.0	596.52153	–595.86727	459.56249	–460.71445	210.77486	–391.97404
145.0	25.63361	–27.68420	17.33340	–21.93430	–137.28245	–54.70975
150.0	10.90746	–16.05114	5.16679	–13.71238	–152.83526	–50.58315
155.0	4.10232	–13.15237	–1.07799	–12.45014	–164.06890	–52.10016
160.0	–1.40918	–13.20853	–6.73976	–13.89046	–175.38423	–56.23971
165.0	–8.02714	–16.02791	–14.22946	–18.43960	–189.35343	–64.02343
170.0	–20.48679	–25.44467	–29.36150	–31.21497	–213.21884	–82.15453
175.0	–89.28128	–91.53160	–116.44386	–116.19934	–328.93578	–192.33667
180.0	55.28397	55.66013	68.63339	70.91410	–97.47786	44.60891
185.0	20.05548	23.18320	24.38065	28.79544	–158.54947	–10.89842
190.0	10.05480	16.28808	12.39872	19.22305	–178.38515	–24.95547
195.0	4.08778	14.11530	5.73408	15.50316	–191.30430	–31.68009
200.0	–1.01122	14.10752	0.50024	14.22149	–202.14698	–35.54570

$\alpha = 2.2$

λ	*A*	*B*	*C*	*D*	*G*	*H*
0.2	1.41024	1.03570	0.05804	1.14868	0.30240	0.52408
0.4	1.31153	1.08017	–0.02333	1.16357	0.02409	0.47108
0.6	1.20739	1.12905	–10.10848	1.18102	–0.25701	0.41938
0.8	1.09716	1.18299	–0.19785	1.20143	–0.54119	0.36923
1.0	0.98003	1.24274	–0.29197	1.22528	–0.82881	0.32091
1.2	0.85505	1.30921	–0.39147	1.25312	–1.12028	0.27478
1.4	0.72106	1.38351	–0.49709	1.28566	–1.41609	0.23125
1.6	0.57666	1.46698	–0.60975	1.32374	–1.71687	0.19086
1.8	0.42013	1.56130	–0.73058	1.36842	–2.02334	0.15427
2.0	0.24931	1.66856	–0.86097	1.42103	–2.33641	0.12230
2.2	0.06150	1.79140	–1.00267	1.48324	–2.65725	0.09601
2.4	–0.14678	1.93324	–1.15794	1.55722	–2.98731	0.07677
2.6	–0.38006	2.09853	–1.32970	1.64581	–3.32849	0.06636
2.8	–0.64437	2.29322	–1.52182	1.75280	–3.68330	0.06720
3.0	–0.94787	2.52539	–1.73957	1.88334	–4.05515	0.08256
3.2	–1.30196	2.80637	–1.99029	2.04467	–4.44877	0.11704
3.4	–1.72307	3.15246	–2.28453	2.24722	–4.87096	0.17730
3.6	–2.23567	3.58807	–2.63800	2.50660	–5.33182	0.27332
3.8	–2.87809	4.15138	–3.07530	2.84726	–5.84716	0.42072
4.0	–3.71373	4.90571	–3.63712	3.30974	–6.44310	0.64545
4.2	–4.85587	5.96420	–4.39608	3.96653	–7.16624	0.99394
4.4	–6.52852	7.55071	–5.49583	4.96110	–8.10870	1.55809
4.6	–9.24583	10.17926	–7.26597	6.62287	–9.48179	2.54902
4.8	–14.50688	15.34876	–10.66744	9.91259	–11.90154	4.58250
5.0	–29.32336	30.07073	–20.19469	19.32450	–18.24889	10.53921
5.2	–380.47745	381.12716	–245.31137	244.32200	–162.78766	154.68271
5.4	38.04382	–37.49512	22.92392	–24.03651	8.90525	–17.41039
5.6	18.77257	–18.32847	10.53824	–11.77841	0.72213	–9.63275
5.8	12.68381	–12.34813	6.60127	–7.97367	–2.04682	–7.27490
6.0	9.66726	–9.44411	4.63236	–6.14194	–3.55570	–6.18315
6.2	7.84515	–7.73892	3.42778	–5.07989	–4.57620	–5.58625
6.4	6.61001	–6.62542	2.59807	–4.39842	–5.35808	–5.23491
6.6	5.70579	–5.84791	1.97899	–3.93375	–6.00672	–5.02428
6.8	5.00575	–5.28007	1.48919	–3.60501	–6.57442	–4.90266
7.0	4.43995	–4.85239	1.08372	–3.36777	–7.09015	–4.84172
7.2	3.96653	–4.52351	0.73559	–3.19567	–7.57136	–4.82473
7.4	3.55886	–4.26736	0.42759	–3.07214	–8.02929	–4.84126
7.6	3.19914	–4.06672	0.14813	–2.98634	–8.47158	–4.88459
7.8	2.87493	–3.90988	–0.11097	–2.93093	–8.90371	–4.95024
8.0	2.57723	–3.78860	–0.35571	–2.90088	–9.32980	–5.03525
8.2	2.29928	–3.69703	–0.59065	–2.89268	–9.75307	–5.13769

(continued 1) $\alpha = 2.2$

λ	A	B	C	D	G	H
8.4	2.03588	−3.63093	−0.81943	−2.90389	−10.17622	−5.25634
8.6	1.78283	−3.58728	−1.04507	−2.93292	−10.60157	−5.39056
8.8	1.53668	−3.56394	−1.27015	−2.97875	−11.03123	−5.54018
9.0	1.29446	−3.55944	−1.49701	−3.04089	−11.46724	−5.70538
9.2	1.05351	−3.57290	−1.72785	−3.11926	−11.91159	−5.88672
9.4	0.81140	−3.60389	−1.96483	−3.21417	−12.36638	−6.08512
9.6	0.56575	−3.65245	−2.21014	−3.32633	−12.83386	−6.30185
9.8	0.31418	−3.71900	−2.46613	−3.45681	−13.31649	−6.53857
10.0	0.05420	−3.80442	−2.73533	−3.60713	−13.81707	−6.79742
11.0	−1.47976	−4.58118	−4.39877	−4.74886	−16.72603	−8.54325
12.0	−3.91105	−6.37717	−7.18598	−7.08987	−21.01965	−11.72016
13.0	−9.54860	−11.44262	−13.87886	−13.37405	−30.05898	−19.66448
14.0	−56.02194	−57.37254	−70.07561	−69.17687	−99.10368	−87.62090
15.0	21.81875	21.00637	24.39505	25.68869	14.46212	27.03709
16.0	9.86859	9.60791	10.03854	11.74093	−3.87098	9.80906
17.0	6.34885	6.67063	5.90535	8.04262	−9.80831	4.99846
18.0	4.47591	5.42934	3.77801	6.38979	−13.30639	2.65865
19.0	3.17837	4.83484	2.36319	5.50555	−15.92896	1.23820
20.0	2.11970	4.57972	1.25967	5.01060	−18.15730	0.27233
21.0	1.14800	4.55329	0.29271	4.76142	−20.19610	−0.42000
22.0	0.16837	4.72288	−0.63864	4.70431	−22.15709	−0.91456
23.0	−0.90434	5.10371	−1.61503	4.83511	−24.12080	−1.23364
24.0	−2.17863	5.76226	−2.72860	5.19500	−26.16897	−1.35733
25.0	−3.83650	6.85327	−4.12431	5.89588	−28.41961	−1.20775
26.0	−6.25961	8.73944	−6.09757	7.21008	−31.10326	−0.58236
27.0	−10.47510	12.43374	−9.43512	9.90714	−34.82400	1.10194
28.0	−20.65805	22.09987	−17.32458	17.16077	−42.01670	6.26311
29.0	−97.48963	98.40909	−76.05977	75.25270	−88.00009	50.20507
30.0	42.66866	−42.28645	30.73519	−32.20438	−7.71113	−32.14923
31.0	17.96904	−18.14848	11.75961	−13.92148	−23.37574	−18.58819
32.0	11.23575	−12.01163	6.48135	−9.37927	−28.60171	−15.51981
33.0	7.89873	−9.31798	3.78261	−7.47485	−31.87394	−14.47757
34.0	5.76839	−7.89274	1.98951	−6.55265	−34.47366	−14.20281
35.0	4.18289	−7.09304	0.59269	−6.12672	−36.79497	−14.33034
36.0	2.86622	−6.66813	−0.62436	−6.01175	−39.01043	−14.72617
37.0	1.67408	−6.50865	−1.78000	−6.13257	−41.22053	−15.34315
38.0	0.51260	−6.57101	−2.95773	−6.46782	−43.50369	−16.17971
39.0	−0.69663	−6.85244	−4.23529	−7.03329	−45.94133	−17.27001
40.0	−2.03915	−7.38720	−5.70628	−7.88347	−48.63995	−18.69069
41.0	−3.63320	−8.25941	−7.50885	−9.13045	−51.76417	−20.58650
42.0	−5.67649	−9.64293	−9.88179	−10.99483	−55.60497	−23.23450

(continued 2) $\alpha = 2.2$

λ	*A*	*B*	*C*	*D*	*G*	*H*
43.0	−8.56035	−11.91174	−13.30489	−13.94311	−60.75242	−27.21463
44.0	−13.22643	−15.99428	−18.93932	−19.12640	−68.63991	−33.95257
45.0	−22.72858	−24.93402	−30.55864	−30.31028	−83.93142	−48.10621
46.0	−55.97366	−57.62923	−71.53848	−70.86379	−135.52858	−98.57207
47.0	195.64401	194.53322	239.70295	240.80057	248.38244	286.46801
48.0	37.82340	37.25899	44.68844	46.21075	6.32598	45.54238
49.0	21.16013	21.15013	24.21157	26.16531	−19.91951	20.43332
50.0	14.57501	15.13383	16.19902	18.59587	−30.75084	10.74787
51.0	10.91224	12.06102	11.80444	14.66130	−37.10503	5.55307
52.0	8.48499	10.25219	8.94379	12.28328	−41.55760	2.27784
53.0	6.68632	9.10877	6.86853	10.71978	−45.03261	0.00321
54.0	5.24187	8.36623	5.24166	9.64149	−47.94499	−1.67977
55.0	4.00708	7.89192	3.88714	8.88168	−50.51031	−2.97935
56.0	2.89588	7.61461	2.70191	8.34890	−52.85334	−4.01123
57.0	1.85074	7.49579	1.61917	7.99125	−55.05333	−4.84299
58.0	0.82819	7.51694	0.59078	7.78014	−57.16536	−5.51439
59.0	−0.20962	7.67390	−0.42239	7.70283	−59.23183	−6.04716
60.0	−1.30083	7.97534	−1.45693	7.75943	−61.28971	−6.44949
61.0	−2.48982	8.44418	−2.55249	7.96305	−63.37604	−6.71708
62.0	−3.83494	9.12270	−3.75843	8.34320	−65.53406	−6.83132
63.0	−5.42123	10.08356	−5.14423	8.95403	−67.82167	−6.75321
64.0	−7.38479	11.45307	−6.81882	9.89236	−70.32704	−6.40961
65.0	−9.96541	13.46304	−8.97199	11.33811	−73.20119	−5.66167
66.0	−13.63728	16.58100	−11.97702	13.65632	−76.73863	−4.22508
67.0	−19.49979	21.90060	−16.69715	17.70312	−81.61779	−1.43022
68.0	−30.83467	32.69847	−25.70645	26.04627	−89.85818	4.73461
69.0	−63.91958	65.24760	−51.76787	51.44294	−111.46745	24.26987
70.0	−7190.37600	7191.16488	−5639.12127	5638.12738	−4493.19891	4403.92272
71.0	68.08548	−67.84304	51.47735	−53.14953	−32.87987	−58.48820
72.0	34.06873	−34.38487	24.67987	−27.04544	−55.07307	−38.40499
73.0	22.54157	−23.43291	15.51060	−18.59032	−63.45036	−32.16359
74.0	16.58822	−18.07641	10.70623	−14.52707	−68.41179	−29.37167
75.0	12.84819	−14.96043	7.63082	−12.22666	−72.02552	−27.96961
76.0	10.20240	−12.97219	5.40556	−10.81809	−74.98237	−27.27637
77.0	8.16970	−11.63791	3.65153	−9.93160	−77.58148	−27.00410
78.0	6.50720	−10.72338	2.17632	−9.38557	−79.97619	−27.01291
79.0	5.07713	−10.10140	0.86949	−9.08268	−82.25610	−27.22948
80.0	3.79349	−9.69893	−0.33937	−8.96864	−84.48009	−27.61498
81.0	2.59747	−9.47349	−1.50009	−9.01388	−86.69189	−28.15085
82.0	1.44482	−9.40169	−2.65215	−9.20487	−88.92841	−28.83243
83.0	0.29844	−9.47369	−3.83091	−9.54013	−91.22512	−29.66633

(continued 3) $\alpha = 2.2$

λ	A	B	C	D	G	H
84.0	–0.87672	–9.69087	–5.07219	–10.02909	–93.62049	–30.67034
85.0	–2.11801	–10.06575	–6.41671	–10.69302	–96.16068	–31.87532
86.0	–3.46996	–10.62411	–7.91552	–11.56825	–98.90575	–33.32983
87.0	–4.99111	–11.40996	–9.63803	–12.71281	–101.93934	–35.10859
88.0	–6.76505	–12.49548	–11.68553	–14.21902	–105.38529	–37.32837
89.0	–8.92084	–14.00050	–14.21628	–16.23792	–109.43862	–40.17855
90.0	–11.67425	–16.13324	–17.49670	–19.03001	–114.42911	–43.98424
91.0	–15.42152	–19.28364	–22.01782	–23.08135	–120.96582	–49.35065
92.0	–20.98570	–24.26936	–28.80128	–29.40937	–130.31752	–57.54321
93.0	–30.41415	–33.13309	–40.39047	–40.55378	–145.64428	–71.71910
94.0	–50.62558	–52.78935	–65.38300	–65.10891	–177.63046	–102.56011
95.0	–129.28059	–130.89497	–163.01060	–162.30352	–299.88856	–223.67641
96.0	279.93112	278.86389	345.68190	346.82033	331.38737	408.74020
97.0	69.01817	68.49920	83.68373	85.25452	4.84932	83.34362
98.0	39.53489	39.56859	47.16726	49.17402	–41.45940	38.17937
99.0	27.58130	28.17537	32.43836	34.88731	–60.68918	20.09906
100.0	20.98175	22.14728	24.36591	27.26591	–71.64437	10.30043
105.0	7.83828	12.16333	8.74602	14.14160	–95.61693	–7.69821
110.0	1.88530	10.39088	2.29196	10.99259	–108.01636	–13.48201
115.0	–3.64982	11.55301	–3.07544	10.92280	–118.61619	–15.88868
120.0	–12.62529	17.32631	–10.90151	14.76617	–130.83386	–14.87061
125.0	–54.57717	56.49885	–44.72624	45.13296	–163.50586	7.25302
130.0	40.93387	–41.76279	30.22552	–33.24204	–110.09891	–56.66043
135.0	13.33197	–17.21506	7.69164	–14.51016	–133.91646	–43.82193
140.0	5.14526	–12.83652	0.34875	–11.90878	–145.92763	–43.96049
145.0	–0.67085	–12.46723	–5.49087	–12.85177	–157.21319	–47.36703
150.0	–7.44238	–15.17879	–13.00336	–17.14878	–170.96420	–54.56878
155.0	–20.97626	–25.53981	–29.13931	–30.77045	–195.61416	–73.22373
160.0	–159.18929	–160.93939	–200.62760	–200.02794	–413.79827	–285.63643
165.0	36.84747	37.86120	44.13281	46.92515	–113.61295	20.29106
170.0	14.82964	18.81886	17.38433	22.53849	–151.51576	–11.73445
175.0	6.76955	14.26713	8.13790	16.07797	–167.52707	–21.53094
180.0	1.18855	13.26548	2.21984	13.79737	–179.20177	–26.31377
185.0	–4.40183	14.48418	–3.20771	13.77948	–189.86650	–28.67807
190.0	–12.25925	18.96613	–10.21244	16.57918	–201.50520	–28.52395
195.0	–30.27528	34.07288	–25.18452	27.92653	–219.34434	–21.44704
200.0	–382.65874	383.71648	–305.81037	305.13826	–448.22733	196.93568

VALUES OF THE FUNCTIONS *A* TO *H*

$\alpha = 2.3$

λ	*A*	*B*	*C*	*D*	*G*	*H*
0.2	1.30602	1.00672	–0.05391	1.04647	0.18211	0.41425
0.4	1.19737	1.05750	–0.14230	1.06567	–0.09996	0.36899
0.6	1.08172	1.11409	–0.23554	1.08830	–0.38554	0.32556
0.8	0.95803	1.17747	–0.33431	1.11497	–0.67508	0.28434
1.0	0.82506	1.24881	–0.43940	1.14642	–0.96913	0.24578
1.2	0.68128	1.32959	–0.55184	1.18358	–1.26835	0.21047
1.4	0.52477	1.42165	–0.67285	1.22761	–1.57355	0.17913
1.6	0.35314	1.52733	–0.80402	1.28000	–1.88578	0.15269
1.8	0.16327	1.64965	–0.94734	1.34267	–2.20634	0.13235
2.0	–0.04886	1.79256	–1.10544	1.41814	–2.53695	0.11972
2.2	–0.28863	1.96135	–1.28179	1.50977	–2.87985	0.11692
2.4	–0.56336	2.16324	–1.48111	1.62218	–3.23812	0.12689
2.6	–0.88329	2.40837	–1.71002	1.76185	–3.61604	0.15376
2.8	–1.26319	2.71140	–1.97802	1.93818	–4.01974	0.20353
3.0	–1.72523	3.09439	–2.29940	2.16528	–4.45844	0.28522
3.2	–2.30426	3.59204	–2.69655	2.46542	–4.94657	0.41310
3.4	–3.05840	4.26234	–3.20685	2.87579	–5.50816	0.61100
3.6	–4.09244	5.20993	–3.89763	3.46356	–6.18652	0.92201
3.8	–5.61652	6.64478	–4.90389	4.36353	–7.06839	1.43265
4.0	–8.12312	9.05920	–6.54195	5.89179	–8.35695	2.34586
4.2	–13.10138	13.94212	–9.76817	9.00449	–10.66646	4.27559
4.4	–28.14243	28.88446	–19.45760	18.57641	–17.12770	10.35235
4.6	1209.79193	–1209.15222	775.50538	–776.50833	493.15292	–500.31777
4.8	29.49953	–28.96600	17.49633	–18.62564	6.12280	–13.68260
5.0	15.48784	–15.06465	8.46473	–9.72528	0.08023	–8.04075
5.2	10.69045	–10.38206	5.34925	–6.74633	–2.16449	–6.20300
5.4	8.23564	–8.04687	3.73683	–5.27613	–3.44552	–5.33566
5.6	6.72278	–6.65883	2.72788	–4.41555	–4.34068	–4.86146
5.8	5.68111	–5.74761	2.01997	–3.86267	–5.04435	–4.58663
6.0	4.90782	–5.11089	1.48273	–3.48769	–5.64039	–4.42796
6.2	4.30105	–4.64735	1.05060	–3.22567	–6.17108	–4.34393
6.4	3.80394	–4.30072	0.68689	–3.04066	–6.66017	–4.31163
6.6	3.38214	–4.03734	0.36937	–2.91122	–7.12217	–4.31752
6.8	3.01362	–3.83596	0.08367	–2.82389	–7.56653	–4.35320
7.0	2.68347	–3.68253	–0.18001	–2.76991	–7.99985	–4.41332
7.2	2.38117	–3.56754	–0.42871	–2.74342	–8.42694	–4.49444
7.4	2.09898	–3.48441	–0.66770	2.74052	–8.85158	–4.59440
7.6	1.83098	–3.42852	–0.90114	–2.75858	–9.27685	–4.71196
7.8	1.57245	–3.39670	–1.13248	–2.79599	–9.70545	–4.84655
8.0	1.31949	–3.38681	–1.36471	–2.85184	–10.13987	–4.99816
8.2	1.06868	–3.39753	–1.60057	–2.92586	–10.58254	–5.16727

(continued 1) $\alpha = 2.3$

λ	A	B	C	D	G	H
8.4	0.81694	–3.42824	–1.84272	–3.01831	–11.03594	–5.35484
8.6	0.56132	–3.47890	–2.09385	–3.12994	–11.50274	–5.56228
8.8	0.29884	–3.55006	–2.35679	–3.26204	–11.98588	–5.79152
9.0	0.02640	–3.64281	–2.63470	–3.41646	–12.48871	–6.04510
9.2	–0.25938	–3.75893	–2.93112	–3.59572	–13.01516	–6.32627
9.4	–0.56238	–3.90087	–3.25026	–3.80311	–13.56994	–6.63916
9.6	–0.88712	–4.07201	–3.59714	–4.04291	–14.15879	–6.98902
9.8	–1.23907	–4.27682	–3.97796	–4.32068	–14.78887	–7.38259
10.0	–1.62502	–4.52126	–4.40052	–4.64366	–15.46923	–7.82857
11.0	–4.49097	–6.74107	–7.63658	–7.41917	–20.20846	–11.41829
12.0	–13.05977	–14.72445	–17.62587	–16.98292	–32.93404	–23.01370
13.0	94.29328	93.19160	109.27449	110.33342	115.99597	127.04409
14.0	12.87019	12.33561	13.23722	14.72069	1.64712	13.83312
15.0	7.12813	7.18695	6.57896	8.51121	–7.14584	6.19934
16.0	4.76286	5.46363	3.91679	6.33836	–11.21248	3.32521
17.0	3.29400	4.71129	2.32709	5.29807	–14.00130	1.77694
18.0	2.15623	4.39892	1.14941	4.75605	–16.28684	0.80004
19.0	1.13321	4.35974	0.13843	4.50511	–18.35102	0.14198
20.0	0.10150	4.54775	–0.83586	4.47515	–20.33518	–0.29225
21.0	–1.04811	4.98499	–1.87580	4.66552	–22.33962	–0.52408
22.0	–2.46294	5.76870	–3.10589	5.14148	–24.47293	–0.51489
23.0	–4.41751	7.14182	–4.74562	6.08520	–26.91312	–0.13085
24.0	–7.58431	9.75524	–7.32152	7.99760	–30.07418	1.01168
25.0	–14.32580	15.95528	–12.67115	12.69708	–35.38680	4.32064
26.0	–43.74301	44.82975	–35.62790	35.00131	–54.37134	21.29769
27.0	59.62727	–59.09641	44.51264	–45.80830	6.68822	–41.79013
28.0	18.64561	–18.69555	12.56762	–14.56298	–19.28631	–17.88150
29.0	10.94978	–11.61813	6.45829	–9.19925	–25.21143	–14.07826
30.0	7.46454	–8.80377	3.60652	–7.15682	–28.62140	–12.86763
31.0	5.31437	–7.39535	1.77569	–6.22129	–31.25756	–12.53506
32.0	3.72913	–6.64689	0.36266	–5.81866	–33.59447	–12.64114
33.0	2.40494	–6.28769	–0.87568	–5.74591	–35.82952	–13.03686
34.0	1.18488	–6.20861	–2.07145	–5.92857	–38.07775	–13.67687
35.0	–0.03733	–6.37549	–3.32276	–6.35572	–40.43416	–14.57149
36.0	–1.36003	–6.80525	–4.73085	–7.06536	–43.00742	–15.78027
37.0	–2.90713	–7.57103	–6.43450	–8.15683	–45.95673	–17.43180
38.0	–4.87960	–8.84020	–8.66933	–9.83965	–49.56117	–19.78486
39.0	–7.67851	–10.99043	–11.91496	–12.57523	–54.39645	–23.40100
40.0	–12.30421	–15.00489	–17.37658	–17.55541	–61.91613	–29.72339
41.0	–22.26004	–24.37375	–29.28508	–29.00079	–77.23089	–43.85468
42.0	–65.02811	–66.56842	–80.84954	–80.11211	–140.47122	–105.91885

(continued 2) $\alpha = 2.3$

λ	A	B	C	D	G	H
43.0	95.62224	94.65090	113.41029	114.59805	93.35945	129.08630
44.0	28.95103	28.55249	32.95865	34.60043	−4.79069	32.11405
45.0	17.13085	17.31692	18.79510	20.90086	−22.83171	15.25928
46.0	11.99049	12.78110	12.70775	15.29387	−31.10973	8.18095
47.0	8.98035	10.40413	9.20031	12.29009	−36.26543	4.24388
48.0	6.90646	9.00196	6.83198	10.45657	−40.03836	1.71475
49.0	5.31504	8.13241	5.05676	9.25658	−43.08643	−0.05697
50.0	3.99243	7.59610	3.61938	8.44622	−45.71580	−1.36837
51.0	2.82108	7.29352	2.38151	7.90156	−48.09068	−2.37218
52.0	1.72664	7.17392	1.25810	7.55639	−50.30998	−3.15224
53.0	0.65452	7.21450	0.18970	7.37668	−52.44035	−3.75490
54.0	−0.44271	7.41214	−0.87194	7.34958	−54.53287	−4.20293
55.0	−1.61435	7.78156	−1.97331	7.47979	−56.63324	−4.50113
56.0	−2.92088	8.35865	−3.16779	7.79098	−58.79023	−4.63665
57.0	−4.44769	9.21068	−4.52738	8.33321	−61.06588	−4.57393
58.0	−6.33040	10.45948	−6.16334	9.20109	−63.55259	−4.24075
59.0	−8.81018	12.33535	−8.27074	10.57656	−66.40902	−3.49429
60.0	−12.37491	15.31736	−11.24154	12.84092	−69.95079	−2.03185
61.0	−18.20169	20.57520	−16.01854	16.92798	−74.93368	0.89150
62.0	−30.07137	31.88323	−25.62651	25.85469	−83.77811	7.68670
63.0	−70.60573	71.85738	−58.15663	57.70511	−110.95254	32.81359
64.0	294.39476	−293.70744	233.60671	−234.74307	121.19988	−201.39282
65.0	49.43565	−49.32223	37.60100	−39.43396	−36.70498	−45.55640
66.0	27.00642	−27.48202	19.54269	−22.09077	−52.31444	−32.03803
67.0	18.34716	−19.43284	12.49135	−15.78026	−59.12738	−27.34754
68.0	13.60588	−15.32923	8.56729	−12.63064	−63.44662	−25.19175
69.0	10.50880	−12.90479	5.95069	−10.83107	−66.72925	−24.12442
70.0	8.24645	−11.35865	3.99245	−9.74292	−69.49653	−23.63705
71.0	6.45582	−10.33815	2.40014	−9.08631	−71.98490	−23.50841
72.0	4.94726	−9.66625	1.01948	−8.72229	−74.32092	−23.63060
73.0	3.60914	−9.24716	−0.24208	−8.57756	−76.58262	−23.94897
74.0	2.36820	−9.02790	−1.44734	−8.61395	−78.82531	−24.43815
75.0	1.17050	−8.98109	−2.64498	−8.81510	−81.09465	−25.09176
76.0	−0.02924	−9.09715	−3.87860	−9.18076	−83.43453	−25.91861
77.0	−1.27484	−9.38148	−5.19351	−9.72551	−85.89363	−26.94282
78.0	−2.61527	−9.85528	−6.64343	−10.48086	−88.53263	−28.20730
79.0	−4.11265	−10.55998	−8.29960	−11.50148	−91.43469	−29.78197
80.0	−5.85481	−11.56762	−10.26549	−12.87825	−94.72296	−31.77989
81.0	−7.97826	−13.00237	−12.70448	−14.76464	−98.59400	−34.38971
82.0	−10.71601	−15.08736	−15.89776	−17.43397	−103.38800	−37.94530
83.0	−14.51085	−18.25729	−20.38193	−21.41633	−109.75589	−43.09239

(continued 3) $\alpha = 2.3$

λ	A	B	C	D	G	H
84.0	−20.33222	−23.47483	−27.33406	−27.88338	−119.12741	−51.25640
85.0	−30.80948	−33.36349	−39.94920	−40.02550	−135.38557	−66.31659
86.0	−56.48818	−58.46372	−71.04567	−70.65689	−174.11168	−103.85096
87.0	−236.88726	−238.28978	−290.22407	−289.37443	−441.47433	−370.02511
88.0	117.66843	116.83778	140.89728	142.20702	81.67821	154.31542
89.0	47.91629	47.66051	56.22336	57.99580	−22.17830	51.64915
90.0	30.06295	30.38916	34.63985	36.88088	−49.33989	25.68265
91.0	21.71356	22.63306	24.61175	27.33061	−62.45272	13.77249
92.0	16.76023	18.28866	18.71509	21.92452	−70.54077	6.89750
93.0	13.39991	15.55763	14.75899	18.47555	−76.26650	2.39825
94.0	10.90864	13.72113	11.86446	16.10883	−80.69761	−0.78958
95.0	8.93810	12.43668	9.60928	14.40683	−84.34640	−3.17453
96.0	7.29880	11.52144	7.76444	13.14593	−87.49013	−5.02952
97.0	5.87759	10.87005	6.19401	12.19647	−90.29279	−6.51347
98.0	4.60146	10.41879	4.81112	11.47907	−92.85842	−7.72437
99.0	3.41980	10.12827	3.55654	10.94359	−95.25719	−8.72507
100.0	2.29482	9.97456	2.38718	10.55808	−97.53920	−9.55666
105.0	−3.48401	11.07288	−3.23521	10.48791	−108.34387	−11.78976
110.0	−13.43403	17.70234	−11.94994	15.16249	−12130986	−9.85378
115.0	−99.99407	101.37906	−82.44973	82.15265	−184.06868	42.51872
120.0	25.87749	−27.41212	18.61019	−22.46241	−108.27950	−43.71380
125.0	9.22324	−14.14714	4.46333	−12.44380	−125.65442	−37.48156
130.0	2.29377	−11.76422	−2.06780	−11.45165	−137.16904	−38.82932
135.0	−3.92321	−12.90720	−8.60208	−13.88494	−149.39110	−43.64791
140.0	−13.91163	−19.30303	−20.03679	−22.41427	−167.83053	−55.64879
145.0	−61.85435	−64.23112	−77.83835	−77.77617	−242.86438	−124.61938
150.0	44.30540	44.80800	52.10581	54.49979	−89.15240	35.06985
155.0	14.98357	18.56068	17.00425	21.88957	−136.51727	−6.18734
160.0	6.36574	13.60237	7.24914	15.10127	−152.90111	−16.07707
165.0	0.56877	12.72590	1.18754	13.03017	−164.57637	−20.42718
170.0	−5.49507	14.43720	−4.61843	13.53119	−175.46889	−22.11475
175.0	−15.18919	20.76178	−13.16988	17.97515	−188.34536	−20.35965
180.0	−48.40157	51.02104	−40.71724	41.92206	−216.53033	−2.67868
185.0	89.21414	−89.45653	70.34609	−72.63101	−132.23860	−97.34012
190.0	22.45845	−25.73414	15.56785	−21.55211	−182.53704	−57.66761
195.0	10.17896	−16.96569	4.88814	−15.15487	−197.17789	−54.36446
200.0	3.34165	−14.60529	−1.57398	−14.15378	−208.63810	−55.68003

VALUES OF THE FUNCTIONS *A* TO *H*

$\alpha = 2.4$

λ	*A*	*B*	*C*	*D*	*G*	*H*
0.2	1.20721	0.98193	–0.15342	0.95745	0.08186	0.32455
0.4	1.08747	1.04000	–0.24950	0.98145	–0.20457	0.28678
0.6	0.95865	1.10573	–0.35181	1.01002	–0.49543	0.25154
0.8	0.81919	1.18059	–0.46136	1.04410	–0.79140	0.21940
1.0	0.66715	1.26646	–0.57941	1.08485	–1.09331	0.19110
1.2	0.50004	1.36573	–0.70759	1.13380	–1.40225	0.16758
1.4	0.31462	1.48154	–0.84798	1.19294	–1.71956	0.15011
1.6	0.10665	1.61805	–1.00335	1.26492	–2.04705	0.14033
1.8	–0.12964	1.78092	–1.17743	1.35332	–2.38715	0.14054
2.0	–0.40222	1.97800	–1.37535	1.46318	–2.74319	0.15392
2.2	–0.72249	2.22057	–1.60446	1.60168	–3.11992	0.18505
2.4	–1.10732	2.52535	–1.87562	1.77953	–3.52434	0.24075
2.6	–1.58278	2.91829	–2.20560	2.01332	–3.96725	0.33164
2.8	–2.19153	3.44186	–2.62180	2.33029	–4.46630	0.47513
3.0	–3.00840	4.17074	–3.17229	2.77828	–5.05236	0.70187
3.2	–4.17800	5.24933	–3.94988	3.44990	–5.78504	1.07121
3.4	–6.02076	6.99785	–5.16021	4.55053	–6.79634	1.71488
3.6	–9.41730	10.29671	–7.36832	6.64496	–8.44883	2.99514
3.8	–17.98437	18.76238	–12.89466	12.05334	–12.23132	6.40045
4.0	–85.69977	86.37240	–56.36976	55.40587	–40.36174	34.14838
4.2	36.38625	–35.82331	21.92458	–23.01600	9.62995	–16.23152
4.4	15.81839	–15.36979	8.69548	–9.91979	0.90170	–7.89767
4.6	10.35310	–10.02387	5.15438	–6.51739	–1.61256	–5.78452
4.8	7.78463	–7.58022	3.47035	–4.97835	–2.93703	–4.86841
5.0	6.26677	–6.19313	2.45879	–4.11865	–3.83184	–4.38987
5.2	5.24583	–5.30943	1.76434	–3.58351	–4.52517	–4.12144
5.4	4.49783	–4.70576	1.24313	–3.22981	–5.10947	–3.97146
5.6	3.91476	–4.27480	0.82571	–2.98885	–5.62952	–3.89608
5.8	3.43797	–3.95869	0.47423	–2.82371	–6.10983	–3.87183
6.0	3.03278	–3.72361	0.16621	–2.71294	–6.56508	–3.88521
6.2	2.67718	–3.54862	–0.11271	–2.64335	–7.00481	–3.92806
6.4	2.35644	–3.42015	–0.37231	–2.60653	–7.43561	–3.99534
6.6	2.06018	–3.32919	–0.61959	–2.59706	–7.86240	–4.08396
6.8	1.78074	–3.26968	–0.85988	–2.61144	–8.28906	–4.19216
7.0	1.51217	–3.23753	–1.09748	–2.64755	–8.71891	–4.31911
7.2	1.24962	–3.23011	–1.33604	–2.70424	–9.15494	–4.46474
7.4	0.98895	–3.24590	–1.57893	–2.78121	–9.60007	1.62961
7.6	0.72637	–3.28425	–1.82935	–2.87886	–10.05732	–4.81492
7.8	0.45824	–3.34533	–2.09063	–2.99825	–10.52990	–5.02245
8.0	0.18087	–3.43007	–2.36633	–3.14115	–11.02149	–5.25472
8.2	–0.10975	–3.54019	–2.66045	–3.31015	–11.53634	–5.51505

(continued 1) α = 2.4

λ	A	B	C	D	G	H
8.4	–0.41807	–3.67834	–2.97772	–3.50875	–12.07957	–5.80780
8.6	–0.74939	–3.84827	–3.32379	–3.74166	–12.65749	–6.13862
8.8	–1.11010	–4.05510	–3.70575	–4.01511	–13.27801	–6.51494
9.0	–1.50826	–4.30582	–4.13262	–4.33743	–13.95138	–6.94651
9.2	–1.95435	–4.60998	–4.61622	–4.71985	–14.69107	–7.44643
9.4	–2.46234	–4.98076	–5.17245	–5.17778	–15.51530	–8.03257
9.6	–3.05148	–5.43675	–5.82337	–5.73279	–16.44933	–8.72995
9.8	–3.74927	–6.00483	–6.60048	–6.41602	–17.52943	–9.57452
10.0	–4.59644	–6.72521	–7.55062	–7.27397	–18.80947	–10.61997
11.0	–16.07604	–17.59995	–20.65967	–19.93747	–34.84889	–25.49377
12.0	32.96635	32.02838	35.98149	37.14338	29.50241	40.02434
13.0	9.65011	9.31041	9.22039	10.83773	–2.27587	9.42920
14.0	5.67975	5.97784	4.76250	6.87069	–8.34234	4.57652
15.0	3.77826	4.78281	2.69997	5.35628	–11.63866	2.54116
16.0	2.47793	4.29461	1.34823	4.63834	–14.10111	1.40875
17.0	1.38268	4.17020	0.26063	4.31152	–16.23684	0.70440
18.0	0.31257	4.31362	–0.75480	4.24972	–18.25502	0.27061
19.0	–0.86919	4.73428	–1.82897	4.43703	–20.28560	0.06798
20.0	–2.33849	5.54025	–3.11290	4.94615	–22.46187	0.13948
21.0	–4.43559	7.03128	–4.88222	6.00597	–25.00970	0.65700
22.0	–8.08254	10.10273	–7.86888	8.31754	–28.51191	2.16760
23.0	–17.30480	18.76053	–15.25068	15.03602	–35.47852	7.15501
24.0	–131.07431	131.96069	–105.07386	104.18909	–107.45550	77.14362
25.0	28.08337	–27.78581	20.31228	–21.89095	–9.73681	–22.59247
26.0	12.92723	–13.25284	8.23779	–10.55163	–20.41991	–13.97634
27.0	8.15750	–9.15747	4.34280	–7.45289	–24.66567	–11.87041
28.0	5.60756	–7.35375	2.18358	–6.17537	–27.55998	–11.21753
29.0	3.86066	–6.45209	0.63764	–5.62866	–29.99426	–11.16416
30.0	2.45694	–6.03048	–0.66508	–5.48746	–32.26911	–11.46260
31.0	1.18549	–5.93373	–1.90186	–5.64033	–34.53734	–12.03796
32.0	–0.08699	–6.11535	–3.19495	–6.06777	–36.91559	–12.89510
33.0	–1.48227	–6.59810	–4.66892	–6.81623	–39.53584	–14.10441
34.0	–3.16072	–7.48290	–6.50179	–8.01684	–42.59953	–15.83029
35.0	–5.40520	–9.01430	–9.02073	–9.96655	–46.48785	–18.43012
36.0	–8.85427	–11.80469	–12.97630	–13.39521	–52.08494	–22.77165
37.0	–15.45132	–17.77843	–20.66455	–20.58384	–62.10501	–31.55719
38.0	–35.62665	–37.35117	–44.42588	–43.86122	–91.14947	–59.37894
39.0	388.98298	387.85231	458.17106	459.21353	502.69912	535.68813
40.0	33.00202	32.46700	37.01363	38.53616	3.37289	37.58279
41.0	17.45983	17.53215	18.73466	20.74745	–19.18368	16.25595
42.0	11.70660	12.40795	12.04367	14.56493	–28.02392	8.66070

(continued 2) $\alpha = 2.4$

λ	A	B	C	D	G	H
43.0	8.54531	9.90813	8.42569	11.48232	−33.22280	4.72907
44.0	6.43072	8.49973	6.05438	9.68320	−36.93980	2.30961
45.0	4.82722	7.66190	4.29865	8.54843	−39.91951	0.66744
46.0	3.49518	7.17353	2.87836	7.81292	−42.49064	−0.51398
47.0	2.30589	6.93006	1.64568	7.34844	−44.82453	−1.39002
48.0	1.17747	6.88236	0.50977	7.09080	−47.02371	−2.04159
49.0	0.04720	7.01373	−0.59508	7.01170	−49.15906	−2.50909
50.0	−1.14425	7.33229	−1.72660	7.10830	−51.28873	−2.80603
51.0	−2.46595	7.87247	−2.94742	7.40219	−53.47128	−2.92244
52.0	−4.01420	8.70601	−4.34044	7.94724	−55.77954	−2.81986
53.0	−5.94446	9.97034	−6.03554	8.85193	−58.32318	−2.41392
54.0	−8.54416	11.93906	−8.26920	10.33641	−61.29772	−1.52850
55.0	−12.43306	15.22093	−11.54791	12.89413	−65.11779	0.23564
56.0	−19.27132	21.46702	−17.22461	17.86723	−70.88659	3.96891
57.0	−35.59275	37.20325	−30.61882	30.56591	−82.93560	13.99167
58.0	−140.79581	141.82092	−116.35674	115.60786	−153.86762	82.89671
59.0	80.73960	−80.30685	63.84640	−65.29973	−8.34378	−64.66447
60.0	31.87384	−32.04721	23.95894	−26.13329	−41.96172	−33.10377
61.0	19.68816	−20.48850	13.92248	−16.84281	−51.28787	−25.86479
62.0	13.95094	−15.40684	9.12911	−12.82961	−56.35306	−22.92763
63.0	10.47960	−12.62839	6.17284	−10.69811	−59.93150	−21.53049
64.0	8.05306	−10.94240	4.05788	−9.46479	−62.83600	−20.87515
65.0	6.18136	−9.87132	2.38308	−8.74330	−65.39587	−20.64995
66.0	4.62608	−9.19226	0.95147	−8.35513	−67.77490	−20.71307
67.0	3.25330	−8.79110	−0.34969	−8.21115	−70.06930	−20.99640
68.0	1.97735	−8.60813	−1.59504	−8.26765	−72.34567	−21.47022
69.0	0.73518	−8.61542	−2.84251	−8.50852	−74.65904	−22.12929
70.0	−0.52753	−8.80734	−4.14551	−8.93836	−77.06383	−22.98831
71.0	−1.86645	−9.19830	−5.56257	−9.58212	−79.62344	−24.08334
72.0	−3.34865	−9.82596	−7.16805	−10.49023	−82.42195	−25.47897
73.0	−5.06710	−10.76164	−9.06864	−11.75172	−85.58312	−27.28460
74.0	−7.16677	−12.13385	−11.43394	−13.52284	−89.30652	−29.68890
75.0	−9.9004	−14.18221	−14.56224	−16.09144	−93.94555	−33.03680
76.0	−13.76392	−17.39347	−19.04375	−20.03939	−100.20183	−38.02310
77.0	−19.91054	−22.91132	−26.24955	−26.73088	−109.70758	−46.27442
78.0	−31.78907	−34.17777	−40.28767	−40.26812	−127.35606	−62.67927
79.0	−66.65686	−68.44394	−81.71758	−81.20546	−177.64114	−111.72734
80.0	35193.1504	35191.9600	41927.3803	41928.3812	49871.6127	49938.7601
81.0	70.02039	69.42743	81.34345	82.83384	14.78035	83.16309
82.0	35.35150	35.36139	40.15038	42.13495	−35.22256	34.39901
83.0	23.50645	24.12985	26.15181	28.63949	−52.82406	18.04385

(continued 3) $\alpha = 2.4$

λ	A	B	C	D	G	H
84.0	17.36822	18.62108	18.95526	21.95933	–62.31916	9.80612
85.0	13.50562	15.40955	14.47382	18.01220	–68.57620	4.82128
86.0	10.77191	13.35474	11.34242	15.43811	–73.22062	1.46809
87.0	8.67314	11.96978	8.97385	13.65569	–76.95094	–0.94719
88.0	6.95985	11.01339	7.07244	12.37596	–80.11818	–2.77008
89.0	5.49080	10.35401	5.47168	11.44039	–82.91927	–4.19093
90.0	4.17813	9.91558	4.06909	10.75619	–85.47450	–5.32211
91.0	2.96237	9.65312	2.79647	10.26709	–87.86291	–6.23282
92.0	1.79944	9.54089	1.60470	9.93904	–90.14026	–6.96645
93.0	0.65332	9.56642	0.45521	9.75283	–92.34893	–7.54944
94.0	–0.50889	9.72779	–0.68549	9.70046	–94.52411	–7.99601
95.0	–1.72089	10.03300	–1.84984	9.78377	–96.69814	–8.31031
96.0	–3.02123	10.50115	–3.07317	10.01489	–98.90435	–8.48664
97.0	–4.45855	11.16580	–4.39825	10.41864	–101.18124	–8.50789
98.0	–6.09960	12.08178	–5.88219	11.03792	–103.57841	–8.34139
99.0	–8.04313	13.33821	–7.60800	11.94427	–106.16627	–7.93041
100.0	–10.44648	15.08450	–9.70659	13.25914	–109.05423	–7.17680
105.0	–67.62927	69.22671	–57.30990	57.23465	–153.96913	27.41194
110.0	24.60191	–26.04340	17.99771	–21.70053	–97.77557	–39.10958
115.0	8.10199	–13.11959	3.75926	–11.73246	–115.36559	–32.61671
120.0	1.20564	–11.23148	–2.84623	–11.10894	–126.99571	–34.12937
125.0	–5.65300	–13.27585	–10.14595	–14.44466	–140.06321	–39.76453
130.0	–20.00716	–24.13668	–26.64664	–28.06689	–164.32530	–57.48632
135.0	618.44441	617.36068	736.64232	737.73382	742.91842	855.99739
140.0	21.77921	23.75982	24.40238	28.02286	–112.57456	6.75979
145.0	8.66588	14.15386	9.38293	15.89985	–135.06695	–9.17227
150.0	2.10461	12.19208	2.39126	12.70829	–147.80324	–14.60050
155.0	–3.96211	13.29558	–3.53706	12.70577	–158.87851	–16.61002
160.0	–13.16296	18.86495	–11.81150	16.63997	–171.60090	–15.07271
165.0	–45.15866	47.75947	–38.83145	39.95270	–199.70263	2.59963
170.0	71.17448	–71.56289	56.77982	–59.23265	–126.41134	–80.96274
175.0	18.82149	–22.41483	12.91192	–19.19744	–168.46419	–49.49097
180.0	7.93749	–15.34878	3.20932	–14.06123	–182.41143	–47.00302
185.0	1.32657	–13.86238	–3.20292	–13.77871	–193.92807	–48.81615
190.0	–5.21773	–15.37628	–10.10264	–16.51842	–206.49669	–53.77618
195.0	–15.30289	–21.70545	–21.47409	–24.78283	–224.60965	–65.15057
200.0	–48.11440	–51.32514	–60.13714	–60.80462	–275.45724	–109.64417

$a = 2.5$

λ	A	B	C	D	G	H
0.2	1.11289	0.96105	–0.24255	0.87962	–0.00241	0.25097
0.4	0.98069	1.02759	–0.34709	0.90900	–0.29378	0.22053
0.6	0.83669	1.10422	–0.45963	0.94446	–0.59077	0.19353
0.8	0.67852	1.19322	–0.58169	0.98741	–0.89437	0.17084
1.0	0.50310	1.29756	–0.71528	1.03974	–1.20589	0.15366
1.2	0.30633	1.42124	–0.86305	1.10397	–1.52706	0.14354
1.4	0.08263	1.56972	–1.02859	1.18356	–1.86022	0.14268
1.6	–0.17581	1.75065	–1.21694	1.28338	–2.20861	0.15415
1.8	–0.48024	1.97516	–1.43532	1.41050	–2.57690	0.18244
2.0	–0.84758	2.25999	–1.69457	1.57557	–2.97207	0.23430
2.2	–1.30439	2.63152	–2.01174	1.79544	–3.40505	0.32046
2.4	–1.89504	3.13394	–2.41522	2.09830	–3.89405	0.45888
2.6	–2.69965	3.84717	–2.95630	2.53517	–4.47192	0.68213
2.8	–3.87934	4.93207	–3.73801	3.20885	–5.20458	1.05584
3.0	–5.81320	6.76748	–5.00275	4.36143	–6.24706	1.73472
3.2	–9.65628	10.50817	–7.48896	6.73102	–8.07147	3.19051
3.4	–21.40798	22.15321	–15.03175	14.15239	–13.12947	7.87445
3.6	470.28826	–469.65429	299.24809	–300.25408	187.53297	–193.16790
3.8	22.03093	–21.51324	12.67120	–13.80948	4.10262	–10.12387
4.0	11.78831	–11.39235	6.08924	–7.36596	–0.34545	–6.06904
4.2	8.21158	–7.94332	3.76687	–5.18874	–2.07230	–4.74301
4.4	6.35590	–6.22185	2.54302	–4.11740	–3.09877	–4.12565
4.6	5.19568	–5.20300	1.76193	–3.49689	–3.84423	–3.79840
4.8	4.38353	–4.54011	1.20136	–3.10580	–4.45094	–3.61994
5.0	3.76901	–4.08357	0.76494	–2.84869	–4.98074	–3.52947
5.2	3.27620	–3.75842	0.40389	–2.67785	–5.46507	–3.49678
5.4	2.86243	–3.52309	0.09068	–2.56702	–5.92184	–3.50535
5.6	2.50174	–3.35290	–0.19161	–2.50070	–6.36221	–3.54565
5.8	2.17719	–3.23241	–0.45417	–2.46940	–6.79375	–3.61201
6.0	1.87711	–3.15172	–0.70490	–2.46720	–7.22199	–3.70113
6.2	1.59296	–3.10436	–0.94976	–2.49049	–7.65132	–3.81127
6.4	1.31812	–3.08620	–1.19358	–2.53722	–8.08552	–3.94180
6.6	1.04717	–3.09479	–1.44058	–2.60655	–8.52810	–4.09301
6.8	0.77532	–3.12898	–1.69469	–2.69863	–8.98256	–4.26598
7.0	0.49808	–3.18870	–1.95987	–2.81449	–9.45266	–4.46258
7.2	0.21094	–3.27493	–2.24034	–2.95608	–9.94261	–4.68557
7.4	–0.09096	–3.38969	–2.54088	–3.12639	–10.45734	–4.93873
7.6	–0.41311	–3.53623	–2.86714	–3.32960	–11.00289	–5.22715
7.8	–0.76208	–3.71926	–3.22607	–3.57147	–11.58678	–5.55759
8.0	–1.14603	–3.94543	–3.62650	–3.85986	–12.21873	–5.93914
8.2	–1.57550	–4.22400	–4.08000	–4.20552	–12.91158	–6.38412

(continued 1) $\alpha = 2.5$

λ	A	B	C	D	G	H
8.4	−2.06463	−4.56806	−4.60224	−4.62343	−13.68282	−6.90956
8.6	−2.63302	−4.99629	−5.21508	−5.13487	−14.55694	−7.53957
8.8	−3.30894	−5.53620	−5.95021	−5.77101	−15.56951	−8.30940
9.0	−4.13497	−6.22967	−6.85540	−6.57919	−16.77427	−9.27247
9.2	−5.17852	−7.14353	−8.00650	−7.63485	−18.25654	−10.51389
9.4	−6.55350	−8.39117	−9.53168	−9.06585	−20.16068	−12.17775
9.6	−8.46889	−10.18110	−11.66630	−11.10720	−22.75175	−14.52891
9.8	−11.35473	−12.94293	−14.89471	−14.24300	−26.57537	−18.11281
10.0	−16.26004	−17.72530	−20.39869	−19.65475	−32.96286	−24.26059
11.0	24.33548	23.48036	25.44426	26.65143	17.74864	27.65480
12.0	8.15275	7.92634	7.32016	9.01249	−3.60400	7.52812
13.0	4.83714	5.29167	3.69809	5.92217	−8.61281	3.78562
14.0	3.13499	4.36063	1.90654	4.73793	−11.54838	2.17901
15.0	1.89613	4.03462	0.65858	4.21318	−13.85369	1.29699
16.0	0.78152	4.05457	−0.41439	4.04264	−15.93351	0.78501
17.0	−0.39252	4.37572	−1.49641	4.15288	−17.97698	0.54209
18.0	−1.81576	5.08038	−2.75665	4.59193	−20.13910	0.58662
19.0	−3.82958	6.45141	−4.47749	5.57610	−22.65639	1.08056
20.0	−7.36778	9.38651	−7.41118	7.81686	−26.13682	2.58574
21.0	−16.80546	18.23571	−15.05878	14.78679	−33.38340	7.87290
22.0	−318.67828	319.51485	−256.44344	255.48650	−227.45168	199.97189
23.0	22.12104	−21.90130	15.83660	−17.50632	−10.96808	−18.51438
24.0	10.79736	−11.23611	6.66647	−9.09806	−19.44916	−12.09382
25.0	6.84925	−8.00958	3.37942	−6.64675	−23.24197	−10.44796
26.0	4.61758	−6.58993	1.44740	−5.65595	−25.97128	−9.98859
27.0	3.01053	−5.92295	−0.00899	−5.28981	−28.34781	−10.05547
28.0	1.64983	−5.68560	−1.30207	−5.30012	−30.63274	−10.46177
29.0	0.34315	−5.77236	−2.60122	−5.61818	−32.97963	−11.17073
30.0	−1.05706	−6.18216	−4.05086	−6.26959	−35.53465	−12.23354
31.0	−2.72761	−7.00873	−5.84120	−7.37816	−38.50659	−13.80589
32.0	−4.97571	−8.50865	−8.31973	−9.25086	−42.29151	−16.25143
33.0	−8.52104	−11.36837	−12.31591	−12.69054	−47.84751	−20.50693
34.0	−15.75905	−17.95997	−20.60634	−20.45504	−58.39446	−29.77721
35.0	−43.26601	−44.84205	−52.43061	−51.76975	−96.26434	−66.38283
36.0	94.50985	93.55150	107.61383	108.77943	88.59754	119.74022
37.0	24.14675	23.81154	26.03803	27.71381	−7.02787	25.38120
38.0	13.86978	14.17526	14.21642	16.41766	−21.67612	12.01259
39.0	9.49616	10.47245	9.25243	12.00468	−28.35967	6.63029
40.0	6.91442	8.60570	6.37561	9.71594	−32.61441	3.70790
41.0	5.09295	7.56014	4.39122	8.37050	−35.82492	1.87216
42.0	3.64451	6.96964	2.85330	7.53976	−38.50598	0.62261

(continued 2) $\alpha = 2.5$

λ	A	B	C	D	G	H
43.0	2.38387	6.67681	1.55149	7.03634	−40.89793	−0.26213
44.0	1.20251	6.61162	0.36627	6.77250	−43.13420	−0.88925
45.0	0.02145	6.75056	−0.78478	6.71166	−45.30280	−1.30887
46.0	−1.23297	7.10398	−1.97313	6.85179	−47.47509	−1.53496
47.0	−2.64873	7.71833	−3.27844	7.22436	−49.72499	−1.54957
48.0	−4.35533	8.69448	−4.81259	7.90796	−52.15056	−1.29327
49.0	−6.57895	10.23790	−6.76604	9.06903	−54.91332	−0.63249
50.0	−9.78539	12.79871	−9.52658	11.07712	−58.33695	0.73532
51.0	−15.15413	17.54397	−14.07157	14.89517	−63.23202	3.60393
52.0	−26.87333	28.65148	−23.86584	23.97596	−72.46556	10.82638
53.0	−79.06716	80.23634	−67.12744	66.52697	−109.37223	45.72500
54.0	103.52465	−102.97017	83.74885	−85.06684	14.24805	−79.91172
55.0	32.13552	−32.20977	24.61133	−26.66346	−35.79047	−31.90931
56.0	18.87861	−19.60423	13.53800	−16.35095	−46.09050	−23.67680
57.0	13.05783	−14.46690	8.60782	−12.21925	−51.31751	−20.56157
58.0	9.63249	−11.76788	5.65088	−10.11109	−54.92163	−19.12820
59.0	7.26232	−10.17977	3.55676	−8.93108	−57.82363	−18.47352
60.0	5.43369	−9.20483	1.89802	−8.27034	−60.38013	−18.26261
61.0	3.90240	−8.61917	0.46928	−7.94725	−62.76502	−18.34936
62.0	2.53180	−8.31289	−0.84705	−7.87549	−65.08087	−18.66775
63.0	1.23250	−8.23294	−2.13112	−8.01745	−67.40126	−19.19396
64.0	−0.06515	−8.36079	−3.44928	−8.36654	−69.79129	−19.93280
65.0	−1.42730	−8.70490	−4.86901	−8.94255	−72.32173	−20.91558
66.0	−2.93056	−9.30239	−6.47316	−9.79569	−75.08389	−22.20661
67.0	−4.68046	−10.23071	−8.38036	−11.02133	−78.21257	−23.92148
68.0	−6.84472	−11.63687	−10.78329	−12.79503	−81.93020	−26.26843
69.0	−9.72728	−13.80904	−14.03475	−15.45654	−86.64728	−29.64718
70.0	−13.96534	−17.37201	−18.87802	−19.73889	−93.23152	−34.91694
71.0	−21.19143	−23.94826	−27.22113	−27.54174	−103.91348	−44.30136
72.0	−37.28640	−39.41014	−45.94194	−45.73596	−126.73762	−65.83909
73.0	−112.32184	−113.82188	−133.61246	−132.88741	−230.21983	−168.04093
74.0	131.47026	130.59125	151.67758	152.91973	102.58286	166.04075
75.0	42.57695	42.32262	47.79613	49.55867	−19.86275	44.87703
76.0	25.37432	25.75456	27.77922	30.07064	−44.20319	21.82576
77.0	17.83947	18.87057	19.07420	21.90830	−55.30891	12.02090
78.0	13.47136	15.17639	14.07732	17.47357	−62.07372	6.57338
79.0	10.52154	12.93111	10.74456	14.72871	−66.88766	3.09838
80.0	8.31987	11.47316	8.29343	12.89837	−70.66487	0.68773
81.0	6.55199	10.49823	6.35786	11.62489	−73.83217	−1.07836
82.0	5.04849	9.84905	4.74173	10.72229	−76.61729	−2.41917
83.0	3.70747	9.43868	3.32830	10.08634	−79.15469	−3.45869

(continued 3) $\alpha = 2.5$

λ	A	B	C	D	G	H
84.0	2.46098	9.21813	2.04128	9.65657	−81.53098	−4.27030
85.0	1.25860	9.16147	0.82577	9.39859	−83.80696	−4.89768
86.0	0.05803	9.25894	−0.36229	9.29552	−86.02963	−5.36508
87.0	−1.18129	9.51427	−1.56296	9.34435	−88.23969	−5.68221
88.0	−2.50391	9.94513	−2.81798	9.55537	−90.47727	−5.84572
89.0	−3.96507	10.58658	−4.17691	9.95453	−92.78777	−5.83783
90.0	−5.64088	11.49920	−5.70599	10.58992	−95.22961	−5.62138
91.0	−7.64619	12.78557	−7.50331	11.54526	−97.88721	−5.12871
92.0	−10.17018	14.62498	−9.72875	12.96885	−100.89623	−4.23777
93.0	−13.55633	17.35269	−12.67092	15.13970	−104.49963	−2.71680
94.0	−18.50972	21.66682	−16.92068	18.64049	−109.19292	−0.07986
95.0	−26.75707	29.28801	−23.92312	24.90922	−116.18585	4.87449
96.0	−43.97518	45.88758	−38.42677	38.68802	−129.45079	16.11113
97.0	−106.29887	107.59533	−90.65276	90.19212	−174.26327	58.89864
98.0	317.22616	−316.54785	263.44594	−264.63117	120.74135	−238.13415
99.0	65.07304	−65.01985	52.46008	−54.37812	−56.84721	−62.58354
100.0	36.19628	−36.77996	28.20271	−30.86743	−78.27331	−43.21172
105.0	9.26879	−13.40681	5.08394	−11.91698	−103.37791	−28.88977
110.0	1.67913	−10.77116	−2.09916	−10.54553	−115.43317	−29.54659
115.0	−5.39786	−12.73671	−9.54352	−13.71101	−128.51650	−34.80634
120.0	−21.53834	−25.27163	−27.87773	−29.02434	−154.59088	−54.10808
125.0	107.75084	107.17056	124.35760	125.85519	19.41629	126.35819
130.0	16.23293	18.90715	17.51385	21.74310	−110.56613	2.91022
135.0	5.95336	12.53936	6.10348	13.61835	−128.47580	−7.99818
140.0	−0.33517	11.82406	−0.34807	11.85011	−140.33492	−11.68024
145.0	−7.57532	14.68713	−7.16319	13.52719	−151.98751	−11.95320
150.0	−24.16140	27.85900	−21.67535	24.03061	−169.92333	−4.61848
155.0	301.58811	−301.00847	251.64572	−252.95229	54.16505	−238.90174
160.0	22.36607	−25.02978	16.37557	−21.49210	−149.31938	−45.78711
165.0	8.75088	−15.23138	4.29153	−13.89245	−165.29565	−40.97388
170.0	1.66923	−13.29785	−2.52322	−13.12708	−177.10452	−42.16770
175.0	−5.13295	−14.81947	−9.63468	−15.79740	−189.78755	−46.87406
180.0	−16.21655	−22.04561	−22.01618	−24.93351	−208.86441	−58.97897
185.0	−66.12250	−68.67624	−80.11997	−80.28057	−281.75676	−125.30957
190.0	54.92487	55.50159	62.65765	65.13291	−118.59185	44.31628
195.0	17.88910	21.78645	19.68532	24.95761	−173.69596	−4.19034
200.0	7.77174	15.57277	8.44424	17.00569	−191.42777	−14.90928

$\alpha = 2.6$

λ	*A*	*B*	*C*	*D*	*G*	*H*
0.2	1.02222	0.94389	–0.32300	0.81138	–0.07381	0.19039
0.4	0.87593	1.02029	–0.43687	0.84683	–0.37075	0.16720
0.6	0.71425	1.11006	–0.56103	0.89033	–0.67479	0.14864
0.8	0.53356	1.21669	–0.69783	0.94409	–0.98744	0.13607
1.0	0.32890	1.34500	–0.85042	1.01113	–1.31076	0.13138
1.2	0.09338	1.50173	–1.02324	1.09569	–1.64761	0.13722
1.4	–0.18299	1.69671	–1.22268	1.20399	–2.00207	0.15750
1.6	–0.51520	1.94474	–1.45829	1.34537	–2.38029	0.19811
1.8	–0.92685	2.26920	–1.74512	1.53466	–2.79186	0.26842
2.0	–1.45740	2.70936	–2.10830	1.79675	–3.25281	0.38418
2.2	–2.17836	3.33645	–2.59331	2.17684	–3.79211	0.57405
2.4	–3.23404	4.29451	–3.29191	2.76638	–4.46812	0.89607
2.6	–4.96621	5.92503	–4.42132	3.78225	–5.41893	1.48794
2.8	–8.42832	9.28107	–6.65089	5.89344	–7.06990	2.77460
3.0	–19.25157	19.99349	–13.55841	12.67729	–11.69419	7.02877
3.2	239.57554	–238.94966	150.85031	–151.86084	92.52282	–97.56472
3.4	18.80146	–18.29731	10.55162	–11.69786	3.14792	–8.57324
3.6	10.25657	–9.88041	5.08742	–6.37626	–0.56398	–5.25238
3.8	7.19970	–6.95838	3.10825	–4.54727	–2.06355	–4.15222
4.0	5.59206	–5.49319	2.04799	–3.64558	–2.98132	–3.64310
4.2	4.57488	–4.62686	1.36082	–3.12626	–3.66428	–3.37904
4.4	3.85405	–4.06624	0.85968	–2.80329	–4.23154	–3.24209
4.6	3.30127	–3.68413	0.46279	–2.59610	–4.73532	–3.18134
4.8	2.85141	–3.41667	0.12846	–2.46443	–5.20244	–3.17153
5.0	2.46762	–3.22851	–0.16707	–2.38616	–5.64840	–3.19901
5.2	2.12722	–3.09876	–0.43862	–2.34843	–6.08306	–3.25606
5.4	1.81524	–3.01455	–0.69621	–2.34356	–6.51333	–3.33833
5.6	1.52110	–2.96783	–0.94714	–2.36702	–6.94452	–3.44361
5.8	1.23678	–2.95366	–1.19724	–2.41638	–7.38112	–3.57112
6.0	0.95578	–2.96927	–1.45151	–2.49076	–7.82727	–3.72128
6.2	0.67237	–3.01354	–1.71469	–2.59052	–8.28716	–3.89551
6.4	0.38107	–3.08681	–1.99158	–2.71721	–8.76537	–4.09627
6.6	0.07626	–3.19074	–2.28750	–2.87361	–9.26713	–4.32721
6.8	–0.24830	–3.32849	–2.60860	–3.06392	–9.79880	–4.59342
7.0	–0.60000	–3.50498	–2.96245	–3.29411	–10.36834	–4.90185
7.2	–0.98802	–3.72741	–3.35871	–3.57260	–10.98612	–5.26207
7.4	–1.42433	–4.00614	–3.81022	–3.91118	–11.66605	–5.68732
7.6	–1.92522	–4.35612	–4.33463	–4.32663	–12.42741	–6.19631
7.8	–2.51383	–4.79934	–4.95719	–4.84348	–13.29786	–6.81623
8.0	–3.22444	–5.36917	–5.71549	–5.49871	–14.31872	–7.58800
8.2	–4.11059	–6.11831	–6.66837	–6.35063	–15.55473	–8.57603

(continued 1) $\alpha = 2.6$

λ	A	B	C	D	G	H
8.4	−5.26109	−7.13487	−7.91361	−7.49653	−17.11361	−9.88772
8.6	−6.83534	−8.57761	−9.62679	−9.11161	−19.18886	−11.71630
8.8	−9.15129	−10.76396	−12.15839	−11.54598	−22.16685	−14.44788
9.0	−12.94893	−14.43340	−16.32419	−15.61509	−26.94695	−18.98164
9.2	−20.43940	−21.79663	−24.56208	−23.75651	−36.21710	−28.00528
9.4	−42.57970	−43.81024	−48.95300	−48.05091	−63.29751	−54.83884
9.6	−4008.61985	−4009.72387	−4421.83346	−4420.83456	−4884.96234	−4876.25622
9.8	46.97702	45.99975	49.83165	50.92795	45.24616	54.20032
10.0	24.08309	23.23310	24.60924	25.80374	17.24879	26.45195
11.0	7.51183	7.31775	6.43906	8.14492	−3.72148	6.74602
12.0	4.36230	4.88696	3.07362	5.34737	−8.36205	3.41919
13.0	2.72590	4.08037	1.39058	4.32607	−11.13546	2.03800
14.0	1.49223	3.85671	0.17625	3.92219	−13.37152	1.31633
15.0	0.32406	3.99410	−0.92391	3.87373	−15.44741	0.95141
16.0	−0.99429	4.49657	−2.11525	4.15256	−17.56340	0.88598
17.0	−2.75750	5.55923	−3.65063	4.90162	−19.94011	1.21927
18.0	−5.68135	7.84293	−6.11694	6.64704	−23.06114	2.36770
19.0	−12.76785	14.31475	−11.94713	11.78302	−28.89980	6.26162
20.0	−87.51891	88.45089	−72.50024	71.63997	−78.99550	54.40844
21.0	22.92235	−22.62772	16.71039	−18.29366	−7.97987	−18.58802
22.0	10.33756	−10.72505	6.41927	−8.77756	−17.44258	−11.16654
23.0	6.37070	−7.51102	3.08519	−6.29990	−21.29318	−9.45085
24.0	4.17492	−6.17244	1.16547	−5.35623	−24.02040	−8.99566
25.0	2.58414	−5.59101	−0.29080	−5.05005	−26.40219	−9.08491
26.0	1.20628	−5.44801	−1.61282	−5.13571	−28.71863	−9.53377
27.0	−0.16653	−5.65734	−2.98889	−5.56378	−31.14499	−10.32525
28.0	−1.71716	−6.25746	−4.60358	−6.40201	−33.87242	−11.55597
29.0	−3.71174	−7.43447	−6.74716	−7.87590	−37.22120	−13.49380
30.0	−6.72266	−9.71200	−10.06437	−10.59070	−41.92014	−16.83554
31.0	−12.52973	−14.83758	−16.57979	−16.54478	−50.27394	−23.86467
32.0	−31.71147	−33.36655	−38.35450	−37.78040	−76.03533	−48.31856
33.0	137.69650	136.68362	154.95678	156.06254	143.50915	172.52833
34.0	23.77353	23.40785	25.12646	26.76924	−5.49224	24.83474
35.0	13.06673	13.36803	13.01763	15.21493	−20.22896	11.42125
36.0	8.75711	9.76052	8.20968	10.99174	−26.63477	6.36462
37.0	6.25051	8.00860	5.46562	8.87718	−30.68030	3.70632
38.0	4.47849	7.06516	3.57015	7.67374	−33.74886	2.07786
39.0	3.05163	6.56854	2.08338	6.96471	−36.33864	1.00043
40.0	1.78371	6.37058	0.79868	6.57535	−38.68047	0.26990
41.0	0.56122	6.41289	−0.40506	6.43070	−40.90547	−0.20619
42.0	−0.70675	6.68859	−1.61884	6.51022	−43.10680	−0.46196

(continued 2) $\alpha = 2.6$

λ	A	B	C	D	G	H
43.0	–2.11843	7.23574	–2.93408	6.83671	–45.37146	–0.48863
44.0	–3.81247	8.15358	–4.47300	7.48744	–47.80874	–0.22824
45.0	–6.03143	9.65735	–6.44314	8.63874	–50.59739	0.46238
46.0	–9.29055	12.24228	–9.27960	10.70288	–54.10573	1.92549
47.0	–14.97844	17.28155	–14.14918	14.82906	–59.31491	5.12210
48.0	–28.71061	30.37803	–25.76296	25.71389	–70.17781	13.98859
49.0	–129.43267	130.46635	–110.32611	109.54998	–142.21576	84.03191
50.0	58.86009	–58.46844	47.47599	–48.98900	–11.00885	–49.18124
51.0	24.33647	–24.60528	18.42201	–20.69334	–36.50233	–25.71897
52.0	15.09943	–16.05789	10.56852	–13.63197	–44.22261	–20.06901
53.0	10.59386	–12.28321	6.67622	–10.57943	–48.62864	–17.78831
54.0	7.77398	–10.24975	4.18873	–8.99572	–51.86674	–16.74938
55.0	5.72675	–9.06203	2.33762	–8.13263	–54.58446	–16.32791
56.0	4.07667	–8.36721	0.80450	–7.69781	–57.05283	–16.28279
57.0	2.63364	–8.00535	–0.57476	–7.56180	–59.41491	–16.51085
58.0	1.28204	–7.90259	–1.90362	–7.66919	–61.76493	–16.97311
59.0	–0.06400	–8.03312	–3.26344	–8.00769	–64.18190	–17.67003
60.0	–1.48474	–8.40688	–4.73562	–8.59970	–66.75024	–18.63655
61.0	–3.07490	–9.07186	–6.42180	–9.50768	–69.58080	–19.95065
62.0	–4.97096	–10.13204	–8.47395	–10.85633	–72.84413	–21.76001
63.0	–7.40481	–11.79583	–11.15518	–12.88905	–76.83986	–24.34773
64.0	–10.83692	–14.50599	–14.99244	–16.11796	–82.17189	–28.30525
65.0	–16.37088	–19.35232	–21.25320	–21.79890	–90.29651	–35.07910
66.0	–27.54760	–29.86450	–34.00919	–33.99422	–105.89643	–49.34411
67.0	–65.61761	–67.28360	–77.70582	–77.14132	–157.09268	–99.21458
68.0	269.72904	268.70875	308.17919	309.28914	285.90568	345.10642
69.0	46.18648	45.81448	51.11215	52.77004	–10.75479	49.77090
70.0	25.31923	25.60568	27.20639	29.42108	–39.18278	22.67563
71.0	17.20365	18.16645	17.97284	20.75978	–50.72949	12.47493
72.0	12.72313	14.38847	12.92492	16.30655	–57.45772	7.11191
73.0	9.76787	12.17117	9.63692	13.64346	–62.15689	3.80378
74.0	7.58501	10.77236	7.24434	11.91508	–65.82011	1.56506
75.0	5.83570	9.86594	5.35929	10.74433	–68.89200	–0.03968
76.0	4.34160	9.28932	3.77899	9.94177	–71.60382	–1.23040
77.0	2.99622	8.95597	2.38396	9.40485	–74.09047	–2.12761
78.0	1.72776	8.82010	1.09551	9.07696	–76.43912	–2.79978
79.0	0.48084	8.86120	–0.14480	8.92924	–78.71256	–3.28458
80.0	–0.79429	9.07771	–1.38692	8.95264	–80.96207	–3.59861
81.0	–2.15005	9.48600	–2.68080	9.15560	–83.23631	–3.74093
82.0	–3.65062	10.12407	–4.08485	9.56642	–85.58941	–3.69210
83.0	–5.38527	11.06150	–5.67782	10.24119	–88.09161	–3.40791

(continued 3) $\alpha = 2.6$

λ	A	B	C	D	G	H
84.0	–7.49243	12.42157	–7.57948	11.28227	–90.84721	–2.80412
85.0	–10.21030	14.43045	–9.99384	12.87979	–94.03146	–1.72142
86.0	–13.99923	17.53867	–13.31313	15.41471	–97.97900	0.16181
87.0	–19.89610	22.77481	–18.41870	19.75882	–103.43935	3.58413
88.0	–30.85417	33.08499	–27.81837	28.41171	–112.54242	10.66594
89.0	–60.04574	61.63504	–52.69381	52.54760	–134.78035	30.89083
90.0	–461.69687	462.64503	–393.71404	392.82860	–425.36295	319.46062
91.0	85.67022	–85.36860	70.79429	–72.42527	–32.21157	–75.71094
92.0	39.38779	–39.74392	31.40936	–33.79891	–66.76834	–43.18949
93.0	25.31755	–26.34865	19.36355	–22.53162	–78.12269	–33.89358
94.0	18.31916	–20.04890	13.31666	–17.29066	–84.38939	–29.71703
95.0	14.00250	–16.46170	9.54127	–14.35682	–88.73354	–27.50424
96.0	10.97843	–14.20605	6.85702	–12.55917	–92.15847	–26.26273
97.0	8.66633	–12.71086	4.76963	–11.41441	–95.08539	–25.58397
98.0	6.77812	–11.69946	3.03285	–10.68943	–97.72533	–25.27211
99.0	5.15206	–11.02403	1.50724	–10.26091	–100.19912	–25.22491
100.0	3.68741	–10.60131	0.10464	–10.06083	–102.58464	–25.38771
105.0	–3.25263	–11.33580	–6.96042	–11.85439	–114.99199	–29.00813
110.0	–16.29029	–20.42535	–21.41076	–22.94775	–136.21867	–43.07156
115.0	181.89050	181.05913	207.87279	209.14790	123.84021	223.68984
120.0	15.64588	18.18952	16.46736	20.61816	–101.74239	4.86631
125.0	5.20780	11.87616	5.03902	12.70707	–119.46123	–5.54445
130.0	–1.28099	11.56227	–1.52256	11.41372	–131.29913	–8.57885
135.0	–9.70013	15.70973	–9.35157	14.30904	–143.77991	–7.59872
140.0	–38.88008	41.47533	–34.77124	35.78403	–171.12590	9.38750
145.0	49.11568	–49.76613	39.84089	–42.57908	–113.06833	–58.80735
150.0	13.42253	–17.64914	8.83614	–15.70827	–145.21146	–37.24529
155.0	4.31021	–13.10898	0.36619	–12.52463	–158.29670	–36.07441
160.0	–2.47090	–13.34448	–6.48489	–13.78710	–170.42900	–39.12644
165.0	–11.82204	–18.35883	–16.64309	–20.24121	–186.67167	–47.89667
170.0	–43.66176	–46.70355	–52.93574	–53.54737	–233.24608	–87.61042
175.0	64.83643	65.05161	73.02530	75.19804	–92.21736	60.10738
180.0	17.53303	21.18320	18.83944	23.95002	–159.49270	–0.34661
185.0	7.07207	14.80882	7.35653	15.96344	–177.30181	–10.85564
190.0	0.41011	13.71821	0.48949	13.86429	–189.58237	–14.74721
195.0	–6.61937	15.82251	–6.26781	14.92480	–201.27824	–15.44294
200.0	–18.81496	24.23950	–17.27544	21.56131	–216.41365	–11.16646

$\alpha = 2.7$

λ	*A*	*B*	*C*	*D*	*G*	*H*
0.2	0.93446	0.93033	−0.39615	0.75142	−0.13479	0.14039
0.4	0.77213	1.01830	−0.52037	0.79375	−0.43794	0.12447
0.6	0.58963	1.12403	−0.65789	0.84678	−0.75009	0.11475
0.8	0.38135	1.25300	−0.81229	0.91391	−1.07355	0.11332
1.0	0.13923	1.41306	−0.98868	1.00006	−1.41158	0.12324
1.2	−0.14870	1.61599	−1.19465	1.11260	−1.76901	0.14911
1.4	−0.50108	1.88018	−1.44200	1.26308	−2.15334	0.19814
1.6	−0.94859	2.23608	−1.75011	1.47060	−2.57683	0.28231
1.8	−1.54571	2.73788	−2.15337	1.76926	−3.06122	0.42303
2.0	−2.39939	3.49221	−2.71923	2.22615	−3.64909	0.66250
2.2	−3.75254	4.74164	−3.60079	2.99401	−4.43695	1.09685
2.4	−6.30124	7.18185	−5.23665	4.51101	−5.70123	2.00202
2.6	−13.17826	13.94516	−9.60045	8.75030	−8.68534	4.62091
2.8	−115.53178	116.17925	−74.11249	73.13164	−49.56222	45.12586
3.0	20.60184	−20.08010	11.61259	−12.73097	4.20348	−9.01914
3.2	10.07058	−9.68153	4.94224	−6.20566	−0.23850	−4.96460
3.4	6.84148	−6.59284	2.87018	−4.28699	−1.78567	−3.81390
3.6	5.23205	−5.13240	1.81644	−3.39593	−2.69349	−3.31259
3.8	4.23858	−4.29751	1.14848	−2.90103	−3.36071	−3.06313
4.0	3.54231	−3.77062	0.66522	−2.60249	−3.91426	−2.93993
4.2	3.00997	−3.41981	0.28240	−2.41755	−4.40759	−2.89119
4.4	2.57562	−3.18082	−0.04190	−2.30607	−4.86750	−2.89202
4.6	2.20257	−3.01891	−0.33124	−2.24661	−5.30931	−2.92939
4.8	1.86842	−2.91402	−0.60026	−2.22706	−5.74282	−2.99624
5.0	1.55832	−2.85417	−0.85897	−2.24051	−6.17504	−3.08891
5.2	1.26162	−2.83221	−1.11486	−2.28324	−6.61154	−3.20591
5.4	0.97002	−2.84416	−1.37415	−2.35372	−7.05729	−3.34731
5.6	0.67643	−2.88836	−1.64253	−2.45211	−7.51720	−3.51453
5.8	0.37420	−2.96505	−1.92574	−2.58011	−7.99657	−3.71033
6.0	0.05648	−3.07626	−2.23010	−2.74101	−8.50155	−3.93893
6.2	−0.28442	−3.22600	−2.56310	−2.93995	−9.03970	−4.20639
6.4	−0.65775	−3.42065	−2.93409	−3.18445	−9.62072	−4.52127
6.6	−1.07537	−3.66976	−3.35537	−3.48537	−10.25758	−4.89561
6.8	−1.55330	−3.98754	−3.84385	−3.85843	−10.96827	−5.34665
7.0	−2.11428	−4.39519	−4.42374	−4.32690	−11.77876	−5.89973
7.2	−2.79220	−4.92537	−5.13141	−4.92631	−12.72815	−6.59344
7.4	−3.64052	−5.63046	−6.02435	−5.71350	−13.87840	−7.48929
7.6	−4.74921	−6.59956	−7.19958	−6.78490	−15.33420	−8.69160
7.8	−6.28314	−7.99674	−8.83504	−8.31792	−17.28766	−10.39213
8.0	−8.58122	−10.16026	−11.29671	−10.67812	−20.13406	−12.98588
8.2	−12.46972	−13.91575	−15.47706	−14.75755	−24.83745	−17.43660

(continued 1) $\alpha = 2.7$

λ	A	B	C	D	G	H
8.4	−20.63223	−21.94627	−24.27504	−23.45479	−34.52944	−26.87567
8.6	−49.53501	−50.71760	−55.47886	−54.55773	−68.42638	−60.51919
8.8	182.04771	180.99651	194.71642	195.73890	201.66866	209.82997
9.0	34.23210	33.31265	35.05200	36.17660	28.99710	37.41347
9.2	19.42959	18.64267	19.07993	20.30772	11.55466	20.22723
9.4	13.76219	13.10897	12.97636	14.30869	4.77309	13.70321
9.6	10.73687	10.21892	9.72701	11.16552	1.07499	10.26422
9.8	8.83476	8.45403	7.69114	9.23774	−1.31208	8.13805
10.0	7.51343	7.27227	6.28282	7.93974	−3.02101	6.69201
11.0	4.16297	4.66943	2.76335	5.01625	−7.75635	3.30979
12.0	2.50291	3.88539	1.08417	4.04274	−10.49112	2.02082
13.0	1.24222	3.71719	−0.13719	3.70735	−12.70911	1.39812
14.0	0.00561	3.94594	−1.28519	3.75251	−14.80877	1.14640
15.0	−1.47800	4.62641	−2.60957	4.19413	−17.02479	1.24760
16.0	−3.67486	6.12480	−4.50431	5.31791	−19.69332	1.92063
17.0	−8.10617	9.90702	−8.21946	8.31388	−23.84356	4.12957
18.0	−27.33861	28.50355	−24.02640	23.41399	−37.87218	16.22811
19.0	35.17695	−34.66296	26.94821	−28.28608	2.65322	−26.25017
20.0	11.43034	−11.60897	7.44224	−9.55361	−14.40987	−11.19527
21.0	6.57489	−7.51754	3.35787	−6.32384	−18.88727	−8.81813
22.0	4.17677	−5.99319	1.26351	−5.20802	−21.75852	−8.18704
23.0	2.51468	−5.36977	−0.25539	−4.85329	−24.18859	−8.20609
24.0	1.09151	−5.23715	−1.61817	−4.93783	−26.53489	−8.62718
25.0	−0.34141	−5.49713	−3.05093	−5.40429	−29.00785	−9.42914
26.0	−2.01007	−6.21331	−4.78243	−6.34876	−31.84390	−10.73999
27.0	−4.28008	−7.66409	−7.20963	−8.09688	−35.47757	−12.93592
28.0	−8.07183	−10.71770	−11.35734	−11.63075	−41.05227	−17.12536
29.0	−17.16883	−19.12329	−21.46566	−21.16231	−53.32141	−28.03861
30.0	−96.93287	−98.21774	−110.87087	−110.00731	−154.56972	−127.94354
31.0	33.30993	32.69289	35.36350	36.78728	8.58341	36.55423
32.0	14.64543	14.71246	14.51615	16.51518	−15.73802	13.59115
33.0	9.17849	9.96403	8.48140	11.08589	−23.43519	7.27878
34.0	6.33967	7.89850	5.40253	8.65982	−27.80964	4.32996
35.0	4.43869	6.85049	3.38641	7.36482	−30.98268	2.64112
36.0	2.94698	6.32422	1.84448	6.64011	−33.61088	1.57917
37.0	1.63217	6.13346	0.52233	6.27036	−35.97433	0.89717
38.0	0.35804	6.21089	−0.72345	6.17061	−38.22596	0.49175
39.0	−0.98587	6.55558	−2.00185	6.32485	−40.47536	0.33224
40.0	−2.52662	7.22797	−3.42983	6.77526	−42.83204	0.44423
41.0	−4.46159	8.38257	−5.18084	7.64719	−45.44986	0.92760
42.0	−7.18061	10.37984	−7.58993	9.24268	−48.61782	2.03428

(continued 2) $\alpha = 2.7$

λ	*A*	*B*	*C*	*D*	*G*	*H*
43.0	–11.66405	14.17869	–11.49251	12.37307	–53.04864	4.44993
44.0	–21.44020	23.29069	–19.88699	20.01795	–61.29104	10.70213
45.0	–67.10479	68.29762	–58.75622	58.14448	–95.41098	42.83918
46.0	75.69100	–75.16191	62.40793	–63.76963	6.36298	–60.92633
47.0	24.89841	–25.05136	19.18016	–21.31288	–31.46267	–25.11643
48.0	14.68469	–15.55082	10.40602	–13.34531	–40.03668	–18.59875
49.0	10.04477	–11.66966	6.35829	–10.15601	–44.60473	–16.14604
50.0	7.22046	–9.66688	3.84321	–8.57069	–47.88162	–15.06554
51.0	5.18798	–8.54040	1.98838	–7.74147	–50.61162	–14.64072
52.0	3.54514	–7.91654	0.44819	–7.35508	–53.09278	–14.61013
53.0	2.09156	–7.63418	–0.95304	–7.28026	–55.48061	–14.86861
54.0	0.70331	–7.62573	–2.32856	–7.46760	–57.88030	–15.38325
55.0	–0.71760	–7.87795	–3.77359	–7.91788	–60.38634	–16.16834
56.0	–2.27315	–8.42347	–5.39384	–8.67790	–63.11014	–17.28523
57.0	–4.10041	–9.35359	–7.33811	–9.85757	–66.21476	–18.86398
58.0	–6.42636	–10.86353	–9.85906	–11.68259	–69.98265	–21.16412
59.0	–9.70194	–13.38120	–13.46435	–14.64104	–74.98620	–24.74140
60.0	–15.02790	–17.98986	–19.39863	–19.96271	–82.63349	–30.99124
61.0	–26.08527	–28.35670	–31.83053	–31.80443	–97.64586	–44.62499
62.0	–67.81792	–69.41404	–79.01960	–78.41593	–152.03977	–97.65071
63.0	157.73928	156.81335	176.62740	177.80465	136.67611	192.43029
64.0	37.96396	37.71246	41.01611	42.77094	–17.89976	39.22324
65.0	21.54183	21.97819	22.50471	24.84898	–39.80105	18.70124
66.0	14.75634	15.90348	14.91451	17.86817	–49.32615	10.57282
67.0	10.88364	12.77482	10.62886	14.22077	–55.10342	6.21725
68.0	8.26111	10.94142	7.76565	12.03481	–59.26387	3.51220
69.0	6.27518	9.80380	5.63167	10.62916	–62.59127	1.68431
70.0	4.64238	9.09590	3.90811	9.69997	–65.44475	0.38728
71.0	3.20890	8.68617	2.42376	9.09517	–68.01568	–0.55382
72.0	1.87853	8.50774	1.07361	8.73498	–70.41960	–1.23282
73.0	0.58116	8.53055	–0.21630	8.57987	–72.73559	–1.69920
74.0	–0.74402	8.75030	–1.50713	8.61726	–75.02634	–1.97386
75.0	–2.16061	9.18627	–2.85966	8.85767	–77.35096	–2.05461
76.0	–3.74788	9.88632	–4.34635	9.33799	–79.77663	–1.91458
77.0	–5.62045	10.94200	–6.06894	10.13381	–82.39444	–1.49230
78.0	–7.96632	12.52379	–8.19151	11.38937	–85.34823	–0.66643
79.0	–11.13443	14.96688	–11.01610	13.39112	–88.89991	0.80727
80.0	–15.87513	19.01050	–15.18964	16.77337	–93.60770	3.47271
81.0	–24.18162	26.63844	–22.42788	23.24121	–100.94885	8.79507
82.0	–43.74950	45.53813	–39.35134	39.40590	–116.61930	22.45977
83.0	–160.25523	161.37859	–139.65618	138.95516	–204.01041	107.84861

(continued 3) $\alpha = 2.7$

λ	A	B	C	D	G	H
84.0	106.60370	−106.14973	89.79171	−91.25314	−7.76351	−90.40620
85.0	40.46451	−40.69104	32.80766	−35.04223	−57.89461	−42.29761
86.0	24.73621	−25.66156	19.18221	−22.21085	−70.73352	−31.50511
87.0	17.46102	−19.11123	12.82408	−16.67649	−77.32560	−26.99332
88.0	13.11880	−15.52853	8.98388	−13.69955	−81.75742	−24.68681
89.0	10.12459	−13.33842	6.29693	−11.92666	−85.20437	−23.42301
90.0	7.84914	−11.92343	4.22035	−10.82830	−88.13525	−22.74832
91.0	5.98962	−10.99510	2.49167	−10.15834	−90.77811	−22.45328
92.0	4.37869	−10.40404	0.96447	−9.79077	−93.26172	−22.43235
93.0	2.91228	−10.06923	−0.45397	−9.65910	−95.66945	−22.63194
94.0	1.51749	−9.94806	−1.83056	−9.73087	−98.06364	−23.02893
95.0	0.13585	−10.02302	−3.22125	−9.99670	−100.49886	−23.62169
96.0	−1.28717	−10.29653	−4.68085	−10.46655	−103.03125	−24.42776
97.0	−2.81104	−10.79100	−6.27187	−11.17106	−105.72748	−25.48643
98.0	−4.51064	−11.55410	−8.07570	−12.16822	−108.67689	−26.86687
99.0	−6.49236	−12.67169	−10.21049	−13.55851	−112.01140	−28.68583
100.0	−8.92417	−14.29584	−12.86512	−15.51712	−115.94396	−31.14447
105.0	−78.55922	−80.36609	−91.57876	−91.15652	−210.09670	−118.27546
110.0	19.92843	21.54808	20.91133	24.29211	−86.78977	11.95648
115.0	6.15477	11.84403	5.80011	12.69809	−108.54341	−2.38686
120.0	−0.71963	11.04096	−1.19241	10.95614	−120.82779	−5.75999
125.0	−9.34745	15.16302	−9.27116	13.90727	−133.56190	−4.58172
130.0	−44.83148	47.11443	−40.45405	41.06851	−166.13580	17.69330
135.0	34.79541	−35.89946	27.97463	−31.21745	−112.50550	−46.05038
140.0	10.28137	−15.23919	6.23682	−13.86981	−136.96023	−32.31738
145.0	2.11250	−12.31110	−1.55033	−12.05416	−149.56265	−32.23957
150.0	−5.26416	−13.96629	−9.17289	−14.72400	−162.61618	−36.63543
155.0	−19.39685	−24.07686	−24.72866	−26.79644	−184.91288	−51.54502
160.0	−855.67152	−856.90571	−977.30450	−976.38646	−1275.13775	−1134.81221
165.0	24.26223	26.47117	25.96642	29.86951	−136.41555	10.86824
170.0	9.12120	15.28715	9.24959	16.58475	−160.04516	−5.41422
175.0	1.73623	13.18753	1.55179	13.47262	−173.24014	−10.26021
180.0	−5.29635	14.75266	−5.29484	14.08707	−185.07485	−11.24985
185.0	−17.10816	22.55818	−16.09129	20.31618	−200.11187	−7.14031
190.0	−83.63741	85.63191	−74.30312	74.58785	−256.21635	38.75094
195.0	43.30604	−44.70564	35.40967	−38.99569	−166.56860	−61.03169
200.0	14.77680	−19.92685	10.13720	−18.00103	−194.13175	−44.06833

$\alpha = 2.8$

λ	*A*	*B*	*C*	*D*	*G*	*H*
0.2	0.84889	0.92030	–0.46317	0.69870	–0.18731	0.09908
0.4	0.66816	1.02193	–0.59895	0.74889	–0.49735	0.09054
0.6	0.46084	1.14732	–0.75199	0.81333	–0.81886	0.09027
0.8	0.21811	1.30508	–0.92783	0.89733	–1.15535	0.10153
1.0	–0.07344	1.50837	–1.13489	1.00906	–1.51212	0.12936
1.2	–0.43533	1.77842	–1.38664	1.16168	–1.89764	0.18189
1.4	–0.90438	2.15176	–1.70609	1.37789	–2.32631	0.27322
1.6	–1.54943	2.69689	–2.13624	1.70031	–2.82502	0.42982
1.8	–2.51601	3.55894	–2.76787	2.21932	–3.45044	0.70793
2.0	–4.17476	5.10811	–3.83208	3.16555	–4.34670	1.25119
2.2	–7.83246	8.65068	–6.14344	5.35304	–6.02161	2.56684
2.4	–23.76541	24.46238	–16.10680	15.18603	–12.46715	8.64623
2.6	36.24010	–35.67114	21.25711	–22.31545	10.58252	–14.77724
2.8	11.44129	–11.00789	5.76477	–6.96865	0.68756	–5.26455
3.0	7.07746	–6.78802	3.00665	–4.36500	–1.27279	–3.69593
3.2	5.20463	–5.06857	1.79889	–3.32172	–2.26915	–3.10193
3.4	4.12693	–4.15487	1.08428	–2.78284	–2.96070	–2.82468
3.6	3.40009	–3.60408	0.58560	–2.47268	–3.52055	–2.69268
3.8	2.85618	–3.24994	0.19783	–2.28797	–4.01454	–2.64197
4.0	2.41719	–3.01646	–0.12810	–2.18179	–4.47374	–2.64376
4.2	2.04149	–2.86445	–0.41868	–2.13024	–4.91521	–2.68372
4.4	1.70433	–2.77213	–0.68999	–2.12040	–5.34967	–2.75447
4.6	1.38951	–2.72701	–0.95292	–2.14528	–5.78478	–2.85250
4.8	1.08542	–2.72206	–1.21573	–2.20149	–6.22673	–2.97677
5.0	0.78280	–2.75385	–1.48547	–2.28816	–6.68128	–3.12808
5.2	0.47346	–2.82169	–1.76887	–2.40648	–7.15437	–3.30889
5.4	0.14927	–2.92722	–2.07316	–2.55962	–7.65275	–3.52343
5.6	–0.19871	–3.07457	–2.40674	–2.75298	–8.18470	–3.77807
5.8	–0.58118	–3.27080	–2.78019	–2.99484	–8.76091	–4.08209
6.0	–1.01203	–3.52695	–3.20758	–3.29749	–9.39593	–4.44888
6.2	–1.51034	–3.85986	–3.70868	–3.67926	–10.11038	–4.89818
6.4	–2.10399	–4.29558	–4.31262	–4.16815	–10.93489	–5.45988
6.6	–2.83605	–4.87572	–5.06472	–4.80853	–11.91729	–6.18121
6.8	–3.77769	–5.67020	–6.04013	–5.67476	–13.13701	–7.14109
7.0	–5.05633	–6.80542	–7.37364	–6.90097	–14.73673	–8.48176
7.2	–6.92609	–8.53459	–9.33429	–8.75559	–17.00137	–10.48776
7.4	–9.97861	–11.44858	–12.54874	–11.86479	–20.59478	–13.82257
7.6	–15.98483	–17.31765	–18.89312	–18.10424	–27.50453	–20.47350
7.8	–33.76049	–34.95690	–37.70727	–36.81336	–47.62582	–40.33541
8.0	–6678.52038	–6679.58081	–7077.58407	–7076.58483	–7506.32189	–7498.77160
8.2	37.37144	36.44782	37.69068	38.79650	31.87865	39.69049

(continued 1) $\alpha = 2.8$

λ	A	B	C	D	G	H
8.4	19.30902	18.52281	18.57192	19.78535	11.43436	19.50879
8.6	13.24783	12.60036	12.16796	13.49056	4.46099	12.79958
8.8	10.17398	9.66706	8.92897	10.36265	0.84088	9.44548
9.0	8.29220	7.92809	6.95300	8.50001	−1.44098	7.43171
9.2	7.00496	6.78641	5.60705	7.27002	−3.05517	6.08801
9.4	6.05629	5.98654	4.62002	6.40192	−4.28887	5.12746
9.6	5.31817	5.40095	3.85633	5.76055	−5.28567	4.40678
9.8	4.71935	4.95894	3.24058	5.27091	−6.12537	3.84650
10.0	4.21701	4.61825	2.72747	4.88815	−6.85594	3.39900
11.0	2.44539	3.75086	0.95547	3.85018	−9.65960	2.08046
12.0	1.13637	3.59050	−0.30018	3.53408	−11.89388	1.50332
13.0	−0.16737	3.87670	−1.50038	3.64011	−14.02784	1.33058
14.0	−1.80932	4.71027	−2.95651	4.22231	−16.34829	1.57618
15.0	−4.49234	6.68595	−5.26042	5.75771	−19.35650	2.61940
16.0	−11.33483	12.86298	−10.98895	10.76007	−25.18353	6.52408
17.0	−228.28874	229.15385	−189.86986	188.91486	−173.70558	153.12105
18.0	15.82757	−15.65561	11.20696	−12.92323	−9.11451	−13.43595
19.0	7.64764	−8.23218	4.35880	−6.90773	−15.88351	−8.71323
20.0	4.63669	−6.08190	1.75469	−5.25246	−19.17228	−7.60026
21.0	2.78487	−5.25273	0.08216	−4.70862	−21.71941	−7.42857
22.0	1.28852	−5.03270	−1.33389	−4.70228	−24.09540	−7.73935
23.0	−0.18149	−5.25430	−2.78738	−5.11808	−26.56753	−8.46252
24.0	−1.89197	−5.97212	−4.54333	−6.04609	−29.40410	−9.72218
25.0	−4.27276	−7.50867	−7.06178	−7.85781	−33.10112	−11.94131
26.0	−8.49711	−10.97554	−11.63083	−11.79045	−39.07519	−16.49445
27.0	−20.49172	−22.25987	−24.79563	−24.35636	−54.55592	−30.58405
28.0	522.05550	520.97891	574.95702	575.98133	607.40268	632.75557
29.0	21.93555	21.55497	22.27569	23.89054	−4.39487	22.34513
30.0	11.28133	11.62268	10.59081	12.81979	−18.24055	9.90789
31.0	7.29038	8.40224	6.27601	9.16189	−23.93556	5.65892
32.0	4.98217	6.93991	3.82983	7.43823	−27.55761	3.53993
33.0	3.31253	6.22638	2.10255	6.52885	−30.37375	2.30935
34.0	1.90947	5.93867	0.68912	6.07068	−32.82645	1.56066
35.0	0.58534	5.96322	−0.60870	5.92901	−35.12695	1.13710
36.0	−0.79619	6.28390	−1.92674	6.07115	−37.41263	0.98978
37.0	−2.38410	6.96355	−3.40353	6.54041	−39.81420	1.14411
38.0	−4.41234	8.18302	−5.24649	7.48550	−42.51698	1.72052
39.0	−7.36948	10.39559	−7.87949	9.29125	−45.88976	3.04462
40.0	−12.61979	14.94033	−12.47711	13.10435	−50.94456	6.09754
41.0	−26.28359	27.91811	−24.29708	24.16100	−62.19874	15.37280
42.0	−234.68568	235.63751	−203.33333	202.43709	−217.03681	168.23490

(continued 2) $\alpha = 2.8$

λ	A	B	C	D	G	H
43.0	38.04966	–37.79208	30.76777	–32.43749	–17.12755	–33.66572
44.0	17.69658	–18.15963	13.19837	–15.67143	–33.32480	–19.49355
45.0	11.19669	–12.42291	7.51854	–10.84277	–39.31926	–15.57795
46.0	7.76732	–9.81820	4.46710	–8.71144	–43.06594	–13.98745
47.0	5.48189	–8.44261	2.38668	–7.64654	–45.99139	–13.32502
48.0	3.71560	–7.70274	0.73676	–7.14257	–48.56425	–13.16113
49.0	2.19214	–7.36591	–0.72557	–7.00538	–50.99939	–13.33535
50.0	0.75416	–7.33867	–2.14368	–7.16296	–53.42913	–13.79507
51.0	–0.71678	–7.60117	–3.63199	–7.61100	–55.96597	–14.55081
52.0	–2.34308	–8.19287	–5.31657	–8.40433	–58.74152	–15.67303
53.0	–4.29493	–9.23080	–7.38066	–9.68054	–61.95465	–17.32136
54.0	–6.87273	–10.97929	–10.15525	–11.73961	–65.97114	–19.83495
55.0	–10.73285	–14.06893	–14.37036	–15.28943	–71.60379	–24.00756
56.0	–17.74052	–20.34554	–22.10750	–22.39479	–81.17606	–32.14830
57.0	–36.37196	–38.26996	–42.83503	–42.51082	–105.26515	–54.82296
58.0	–352.50428	–353.70634	–395.95765	–395.03106	–500.73579	–448.88669
59.0	50.51466	50.00897	54.41676	55.94655	1.52892	54.78608
60.0	23.80791	24.01009	24.66610	26.80949	–32.64201	22.03255
61.0	15.32081	16.25356	15.27460	18.05166	–44.06367	12.04605
62.0	10.94191	12.64005	10.47724	13.91858	–50.34873	7.22303
63.0	8.12537	10.63762	7.43132	11.57960	–54.67182	4.39932
64.0	6.05145	9.44319	5.22305	10.13540	–58.05209	2.56834
65.0	4.36994	8.72748	3.46375	9.21548	–60.92159	1.31387
66.0	2.89976	8.33665	1.95443	8.64457	–63.49966	0.43717
67.0	1.53020	8.19692	0.57599	8.33569	–65.91543	–0.16295
68.0	0.18020	8.27913	–0.75574	8.24983	–68.25736	–0.53562
69.0	–1.22332	8.58618	–2.11302	8.38086	–70.59803	–0.69587
70.0	–2.76222	9.15290	–3.57291	8.75328	–73.01097	–0.62888
71.0	–4.54769	10.05753	–5.23626	9.43111	–75.58855	–0.28346
72.0	–6.75695	11.45325	–7.26030	10.54473	–78.47092	0.45055
73.0	–9.71587	13.64771	–9.93093	12.35967	–81.90948	1.80178
74.0	–14.12629	17.32841	–13.86085	15.47261	–86.43951	4.28761
75.0	–21.87128	24.36666	–20.69065	21.51097	–93.49073	9.32302
76.0	–40.39960	42.20116	–36.90539	36.94857	–108.70948	22.54179
77.0	–162.54884	163.66050	–143.31739	142.58766	–202.44073	114.27761
78.0	88.96646	–88.54927	75.52906	–77.03699	–13.04908	–77.11558
79.0	35.28871	–35.57894	28.71616	31.01693	–54.90456	–37.27820
80.0	21.73478	–22.75406	16.82653	–19.94455	–66.36400	–27.86439
81.0	15.32322	–17.10266	11.15003	–15.12029	–72.41965	–23.89373
82.0	11.43080	–14.01218	7.66098	–12.53047	–76.57722	–21.87398
83.0	8.70163	–12.13933	5.17759	–11.00739	–79.86717	–20.79049

(continued 3) $\alpha = 2.8$

λ	A	B	C	D	G	H
84.0	6.58995	−10.95362	3.22284	−10.09120	−82.70688	−20.24514
85.0	4.82918	−10.20766	1.56230	−9.56898	−85.30312	−20.05526
86.0	3.26864	−9.77551	0.06168	−9.33419	−87.77628	−20.13164
87.0	1.81071	−9.59244	−1.36812	−9.33464	−90.20853	−20.43347
88.0	0.38206	−9.63012	−2.79659	−9.55132	−92.66675	−20.95055
89.0	−1.08272	−9.88684	−4.28869	−9.99077	−95.21645	−21.69758
90.0	−2.65355	−10.38631	−5.91707	−10.68539	−97.93410	−22.71598
91.0	−4.41887	−11.18392	−7.77680	−11.70150	−100.92277	−24.08377
92.0	−6.50762	−12.38419	−10.00951	−13.15945	−104.33871	−25.93870
93.0	−9.13226	−14.18100	−12.85126	−15.27915	−108.44478	−28.52955
94.0	−12.68826	−16.95523	−16.74400	−18.48982	−113.73531	−32.33957
95.0	−18.02787	−21.54737	−22.64261	−23.73611	−121.28053	−38.43007
96.0	−27.42083	−30.21734	−33.09284	−33.55518	−133.93606	−49.64921
97.0	−49.67128	−51.76082	−57.97313	−57.81815	−162.78607	−77.07474
98.0	−184.74816	−186.13920	−209.46234	−208.69725	−333.71078	−246.58112
99.0	118.72143	118.02735	131.16142	132.53544	47.58700	136.13407
100.0	45.54280	45.55090	49.13137	51.11902	−45.39411	44.57464
105.0	9.27648	13.12377	8.90992	14.25505	−95.15080	2.18123
110.0	1.25342	10.46117	0.56748	10.60457	−108.97156	−3.10567
115.0	−6.59399	13.09116	−6.94886	12.26331	−121.31429	−2.98678
120.0	−31.14536	33.86550	−28.91909	29.99126	−146.11865	11.40520
125.0	38.47036	−39.25211	31.61996	−34.48278	−98.62032	−46.16229
130.0	9.91260	−14.66394	6.11424	−13.43921	−126.55120	−28.89477
135.0	1.55701	−11.84100	−1.89255	−11.65318	−139.37509	−28.71284
140.0	−6.37624	−14.13808	−10.11441	−14.93561	−153.04516	−33.62718
145.0	−25.53860	−29.30115	−31.16873	−32.48311	−181.32536	−54.40545
150.0	86.19390	85.95666	94.91992	96.69866	−44.18259	89.87813
155.0	15.86075	19.31335	16.27892	21.29697	−137.25931	4.07211
160.0	5.31465	13.31497	4.97847	13.99078	−154.51350	−5.24704
165.0	−1.70687	13.07979	−2.08371	12.89087	−166.76003	−7.82845
170.0	−10.80680	17.52408	−10.65313	16.22291	−179.97210	−6.15568
175.0	−39.75339	42.74845	−36.54895	37.93056	−208.28161	11.79388
180.0	66.19025	−66.69265	56.11960	−58.67409	−132.37006	−74.19493
185.0	16.55259	−20.83079	12.03325	−18.83788	−176.67378	−40.32277
190.0	5.83466	−14.79770	2.03555	−14.11434	−191.14846	−37.43318
195.0	−1.45504	−14.37204	−5.21928	−14.58774	−203.51651	−39.41385
200.0	−10.37845	−18.33085	−14.65402	−19.65788	−218.63821	−46.27158

$\alpha = 2.9$

λ	*A*	*B*	*C*	*D*	*G*	*H*
0.2	0.76484	0.91384	−0.52506	0.65236	−0.23289	0.06493
0.4	0.56283	1.03172	−0.67383	0.71161	−0.55060	0.06400
0.6	0.32546	1.18162	−0.84515	0.78992	−0.88299	0.07412
0.8	0.03870	1.37729	−1.04772	0.89570	−1.23549	0.10039
1.0	−0.32044	1.64140	−1.29575	1.04282	−1.61691	0.15128
1.2	−0.79250	2.01410	−1.61426	1.25591	−2.04273	0.24188
1.4	−1.45633	2.57384	−2.05188	1.58317	−2.54296	0.40173
1.6	−2.48882	3.49704	−2.71753	2.13305	−3.18474	0.69748
1.8	−4.38655	5.27973	−3.91688	3.21064	−4.15626	1.31673
2.0	−9.28627	10.05806	−6.96482	6.13018	−6.26673	3.06800
2.2	−59.95205	60.59535	−38.17431	37.20389	−25.70670	22.14108
2.4	17.47159	−16.96468	9.42806	−10.54251	3.34455	−7.28565
2.6	8.26033	−7.89870	3.72181	−4.98954	−0.40714	−3.91908
2.8	5.57559	−5.36927	2.02910	−3.46055	−1.69163	−3.03060
3.0	4.24573	−4.20609	1.16749	−2.77453	−2.46735	−2.66323
3.2	3.41737	−3.55743	0.61158	−2.40773	−3.05797	−2.49506
3.4	2.82602	−3.16070	0.19818	−2.19901	−3.56433	−2.42735
3.6	2.36218	−2.90878	−0.14058	−2.08293	−4.02869	−2.42037
3.8	1.97170	−2.75043	−0.43862	−2.02858	−4.47260	−2.45573
4.0	1.62406	−2.65869	−0.71552	−2.02013	−4.90888	−2.52448
4.2	1.30001	−2.61881	−0.98407	−2.04945	−5.34642	−2.62259
4.4	0.98595	−2.62294	−1.25386	−2.11292	−5.79237	−2.74911
4.6	0.67110	−2.66770	−1.53312	−2.21003	−6.25346	−2.90534
4.8	0.34567	−2.75306	−1.82993	−2.34282	−6.73680	−3.09470
5.0	−0.00036	−2.88211	−2.15327	−2.51596	−7.25084	−3.32294
5.2	−0.37858	−3.06129	−2.51410	−2.73724	−7.80630	−3.59884
5.4	−0.80367	−3.30146	−2.92688	−3.01875	−8.41772	−3.93545
5.6	−1.29579	−3.61981	−3.41197	−3.37899	−9.10590	−4.35240
5.8	−1.88458	−4.04364	−3.99970	−3.84684	−9.90207	−4.88002
6.0	−2.61662	−4.61768	−4.73812	−4.46918	−10.85598	−5.56730
6.2	−3.57075	−5.41923	−5.70892	−5.32672	−12.05236	−6.49836
6.4	−4.89314	−6.59322	−7.06385	−6.57041	−13.64888	−7.83036
6.6	−6.89031	−8.44507	−9.12145	−8.51810	−15.97665	−9.89394
6.8	−10.33425	−11.74587	−12.68435	−11.97181	−19.87022	−13.52326
7.0	−17.89155	−19.16138	−20.52543	−19.70386	−28.21306	−21.60140
7.2	−49.18811	−50.31678	−53.05378	−52.12290	−62.22878	−55.35167
7.4	87.65486	86.66740	89.26814	90.30912	85.58458	92.72825
7.6	24.75912	23.91353	23.87871	25.03098	17.39591	24.80754
7.8	14.83372	14.13123	13.57368	14.83887	6.49025	14.17155
8.0	10.73914	10.18160	9.33214	10.71229	1.89013	9.84311
8.2	8.47247	8.06225	6.99145	8.48901	−0.73318	7.49382

(continued 1) $\alpha = 2.9$

λ	A	B	C	D	G	H
8.4	7.01161	6.75167	5.48882	7.10667	–2.48482	6.01885
8.6	5.97597	5.86985	4.42854	6.17001	–3.77625	5.00709
8.8	5.19132	5.24317	3.62951	5.49840	–4.79574	4.27065
9.0	4.56652	4.78111	2.99702	4.99762	–5.64171	3.71150
9.2	4.04915	4.43196	2.47666	4.61380	–6.37067	3.27354
9.4	3.60686	4.16409	2.03488	4.31399	–7.01747	2.92243
9.6	3.21850	3.95718	1.64976	4.07692	–7.60480	2.63599
9.8	2.86960	3.79761	1.30637	3.88836	–8.14813	2.39933
10.0	2.54979	3.67605	0.99404	3.73848	–8.65844	2.20214
11.0	1.17007	3.46614	–0.32255	3.38511	–10.94060	1.61266
12.0	–0.18756	376438	–1.56768	3.51031	–13.10764	1.47856
13.0	–1.94692	4.69459	–3.12435	4.18484	–15.51016	1.82540
14.0	–5.06926	7.09258	–5.80361	6.08890	–18.83521	3.20377
15.0	–15.36175	16.69882	–14.43349	13.98065	–27.09838	9.55602
16.0	48.72384	–48.07964	38.68026	–39.88136	15.89300	–35.36037
17.0	10.67231	–10.76607	6.99499	–8.99555	–11.52165	–9.92866
18.0	5.74650	–6.66543	2.79714	–5.69370	–16.13074	–7.40930
19.0	3.42968	–5.31713	0.74545	–4.69538	–18.97958	–6.82288
20.0	1.78733	–4.87328	–0.77718	–4.47768	–21.42277	–6.91722
21.0	0.28833	–4.95436	–2.23127	–4.74511	–23.86398	–7.46833
22.0	–1.38392	–5.54553	–3.91883	–5.52445	–26.59659	–8.54108
23.0	–3.66138	–6.92732	–6.29100	–7.14106	–30.09599	–10.50631
24.0	–7.68570	–10.16037	–10.58161	–10.76138	–35.69825	–14.64404
25.0	–19.43980	–21.17821	–23.30535	–22.85888	–50.49869	–28.01518
26.0	208.61639	207.59361	225.34557	226.40300	219.58017	243.48188
27.0	18.87489	18.57463	18.62239	20.29872	–6.67073	18.65764
28.0	9.87909	10.33389	8.90359	11.22858	–18.19648	8.58561
29.0	6.34105	7.61119	5.13936	8.16650	–23.22679	5.05637
30.0	4.21879	6.39859	2.92812	6.74019	–26.55579	3.30103
31.0	2.62405	5.85421	1.30695	6.02677	–29.22844	2.30920
32.0	1.22444	5.71460	–0.07868	5.73139	–31.62448	1.75279
33.0	–0.16617	5.90475	–1.41923	5.75984	–33.94075	1.51836
34.0	–1.71234	6.45574	–2.87225	6.12416	–36.33006	1.60009
35.0	–3.64425	7.52145	–4.64581	6.94924	–38.98210	2.09699
36.0	–6.42436	9.51656	–7.14628	8.58934	–42.25507	3.31042
37.0	–11.34210	13.69796	–11.49545	12.13073	–47.12574	6.17837
38.0	–24.31781	25.96181	–22.83005	22.68372	–58.05156	15.12944
39.0	–292.21986	293.15681	–255.23245	254.30914	–260.68362	215.79115
40.0	32.88565	–32.66875	26.60947	–28.32459	–17.37375	–29.50601
41.0	15.59818	–16.13206	11.52993	–14.07122	–31.55361	–17.35180
42.0	9.85385	–11.18883	6.45415	–9.87747	–37.06528	–13.92779

(continued 2) $\alpha = 2.9$

λ	*A*	*B*	*C*	*D*	*G*	*H*
43.0	6.74411	–8.95405	3.65406	–8.04116	–40.61342	–12.55782
44.0	4.61637	–7.80544	1.69299	–7.15906	–43.44781	–12.02894
45.0	2.92165	–7.23575	0.08997	–6.79618	–45.99096	–11.96934
46.0	1.40714	–7.05286	–1.38138	–6.79304	–48.44704	–12.24860
47.0	–0.08428	–7.19332	–2.86835	–7.10568	–50.95596	–12.84076
48.0	–1.69100	–7.67319	–4.50916	–7.76622	–53.65768	–13.79376
49.0	–3.58798	–8.59359	–6.48811	–8.89490	–56.74794	–15.24778
50.0	–6.07385	–10.20537	–9.12895	–10.77404	–60.57899	–17.51909
51.0	–9.80293	–13.12999	–13.14962	–14.09303	–65.93939	–21.37153
52.0	–16.68817	–19.25643	–20.65744	–20.93836	–75.15126	–29.10909
53.0	–36.10333	–37.93983	–41.99026	–41.63174	–99.61624	–52.11936
54.0	–2098.92698	–2100.04338	–2317.37626	–2316.38793	–2610.49279	–2561.54930
55.0	40.80492	40.41066	43.01005	44.63053	–7.72271	42.67020
56.0	20.30153	20.64459	20.47649	22.74291	–33.51231	18.34236
57.0	13.20691	14.31603	12.73692	15.67492	–42.98025	10.35907
58.0	9.39752	11.31643	8.62591	12.27429	–48.44143	6.41695
59.0	6.87107	9.66116	5.93675	10.34981	–52.32829	4.09715
60.0	4.95590	8.70051	3.93103	9.18232	–55.45321	2.60423
61.0	3.35530	8.16632	2.28464	8.47283	–58.17089	1.60554
62.0	1.90867	7.93684	0.82476	8.08268	–60.66812	0.94431
63.0	0.50982	7.96109	–0.55954	7.94947	–63.06188	0.54627
64.0	–0.92967	8.23271	–1.95665	8.05700	–65.44158	0.38550
65.0	–2.50451	8.78555	–3.45669	8.42825	–67.89400	0.47471
66.0	–4.34256	9.70625	–5.17673	9.13483	–70.52722	0.87181
67.0	–6.65241	11.17415	–7.30355	10.33125	–73.50916	1.70912
68.0	–9.83625	13.56977	–10.19343	12.34986	–77.15595	3.27637
69.0	–14.82936	17.81157	–14.67108	15.99682	–82.19523	6.28071
70.0	–24.49550	26.74966	–23.25751	23.77807	–90.84663	12.92513
71.0	–54.11664	55.65436	–49.39885	49.12686	–114.94127	35.02654
72.0	645.24863	–644.42624	565.80955	–566.87306	425.21260	–507.11963
73.0	48.54827	–48.45011	40.76044	–42.62552	–37.81388	–46.09679
74.0	25.13519	–25.78028	20.07738	–22.76523	–57.11028	–28.82770
75.0	16.58843	–18.00656	12.46993	–16.01372	–64.90688	–23.09529
76.0	11.95836	–14.19139	8.30343	–12.74966	–69.68161	–20.43643
77.0	8.91110	–12.01506	5.52295	–10.93384	–73.24414	–19.05885
78.0	6.64178	–10.69001	3.41842	–9.87535	–76.22139	–18.35685
79.0	4.79285	–9.88058	1.67278	–9.28136	–78.89505	–18.07560
80.0	3.17477	–9.42588	0.11598	–9.01362	–81.41835	–18.09709
81.0	1.66991	–9.24707	–1.35988	–9.00719	–83.89115	–18.36908
82.0	0.19164	–9.31189	–2.83725	–9.23988	–86.39284	–18.87895
83.0	–1.33781	–9.62169	–4.39357	–9.72195	–89.00162	–19.64574

(continued 3) $\alpha = 2.9$

λ	A	B	C	D	G	H
84.0	−3.00444	−10.21094	−6.11829	−10.49747	−91.81141	−20.72348
85.0	−4.92306	−11.15820	−8.13451	−11.65767	−94.95504	−22.21694
86.0	−7.27294	−12.61632	−10.63793	−13.37496	−98.64674	−24.31987
87.0	−10.37574	−14.88696	−13.98279	−15.98602	−103.27716	−27.40744
88.0	−14.90164	−18.62469	−18.91013	−20.21813	−109.66611	−32.28736
89.0	−22.55749	−25.52374	−27.31041	−27.95063	−119.90770	−41.04394
90.0	−39.43467	−41.66487	−45.93413	−45.92468	−141.48320	−61.15023
91.0	−115.82056	−117.32621	−130.51912	−129.86995	−236.17178	−154.37823
92.0	145.73979	144.95562	159.47721	160.76360	84.32750	167.57966
93.0	45.66387	45.60615	48.63403	50.56221	−39.46581	45.24908
94.0	26.90626	27.58795	27.92592	30.50752	−63.35164	22.83645
95.0	18.72822	20.17060	18.94658	22.20058	−74.23520	13.44302
96.0	13.98340	16.21677	13.77592	17.72929	−80.89426	8.29808
97.0	10.76713	13.83197	10.30396	14.99265	−85.66653	5.07186
98.0	8.35194	12.30055	7.72568	13.19611	−89.44315	2.88250
99.0	6.39640	11.29537	5.66424	11.97548	−92.64036	1.32492
100.0	4.71554	10.64909	3.91657	11.14328	−95.48153	0.18958
105.0	−2.62788	10.84764	−3.38273	10.52033	−107.85922	−1.79925
110.0	−16.00975	19.91838	−15.71487	18.07007	−124.34510	3.79910
115.0	95.93495	−95.66345	83.00952	−84.69121	−42.40329	−88.20561
120.0	12.20236	−15.84482	8.41957	−14.44743	−113.97603	−27.04132
125.0	2.52895	−11.48424	−0.74612	−11.18322	−127.78880	−25.36624
130.0	−5.44562	−13.28449	−8.91147	−13.87773	−141.27570	−29.52697
135.0	−24.59753	−28.25742	−29.72165	−30.98279	−169.01201	−49.49963
140.0	64.94438	64.93399	70.17498	72.15177	−62.68704	64.17678
145.0	13.25061	17.12408	13.17230	18.59715	−130.66749	3.73610
150.0	3.61720	12.49133	3.02545	12.89202	−146.47761	−3.65016
155.0	−3.73628	13.28300	−4.23917	12.86438	−158.77511	−5.03899
160.0	−15.88591	21.03365	−15.52603	19.26151	−174.38043	−0.45120
165.0	−147.66641	149.11560	−133.04710	132.67042	−284.30630	99.32031
170.0	26.85607	−29.10302	21.63138	−26.11857	−152.33912	−42.80044
175.0	9.13347	−15.70428	5.39799	−14.69470	−172.33381	−33.73780
180.0	1.01528	−13.60189	−2.49025	−13.49892	−185.12378	−33.97535
185.0	−7.18212	−16.04736	−10.96743	−16.86267	−199.01405	−38.84124
190.0	−24.32978	−28.98431	−29.57583	−31.72662	−224.33016	−56.34897
195.0	705.20934	704.23657	782.52905	783.65348	674.56251	849.93595
200.0	25.27040	28.01294	26.43411	30.86471	−171.34043	11.45353

$\alpha = 3.0$

λ	*A*	*B*	*C*	*D*	*G*	*H*
0.2	0.68165	0.91103	–0.58267	0.61169	–0.27279	0.03676
0.4	0.45479	1.04841	–0.74617	0.68149	–0.59904	0.04381
0.6	0.18043	1.22936	–0.93938	0.77696	–0.94423	0.06567
0.8	–0.16421	1.47627	–1.17620	0.91162	–1.31686	0.11042
1.0	–0.61999	1.82958	–1.48149	1.10988	–1.73209	0.19280
1.2	–1.26844	2.37033	–1.90477	1.42078	–2.22007	0.34243
1.4	–2.29950	3.28791	–2.56131	1.95901	–2.85078	0.62872
1.6	–4.28254	5.15102	–3.79642	3.06919	–3.83370	1.26045
1.8	–10.06388	10.80523	–7.33572	6.47621	–6.21509	3.28310
2.0	29884.4909	–29883.8842	18115.0924	–18116.0924	10979.5965	–10982.8961
2.2	12.48981	–12.02796	6.26183	–7.41175	1.55056	–5.22650
2.4	6.69961	–6.39240	2.71103	–4.02141	–0.84399	–3.21951
2.6	4.69184	–4.55107	1.45060	–2.93350	–1.85274	–2.61070
2.8	3.62419	–3.66340	0.75711	–2.42640	–2.52080	–2.35674
3.0	2.92709	–3.16199	0.28483	–2.15657	–3.05824	–2.24985
3.2	2.40970	–2.85860	–0.08248	–2.01044	–3.53607	–2.22170
3.4	1.98913	–2.67359	–0.39574	–1.94040	–3.98587	–2.24414
3.6	1.62270	–2.56837	–0.68169	–1.92388	–4.42472	–2.30434
3.8	1.28520	–2.52292	–0.95670	–1.94984	–4.86364	–2.39669
4.0	0.95970	–2.52703	–1.23244	–2.01346	–5.31116	–2.51970
4.2	0.63310	–2.57647	–1.51869	–2.11401	–5.77507	–2.67478
4.4	0.29367	–2.67140	–1.82506	–2.25405	–6.26375	–2.86602
4.6	–0.07082	–2.81598	–2.16238	–2.43949	–6.78725	–3.10046
4.8	–0.47489	–3.01895	–2.54426	–2.68044	–7.35875	–3.38915
5.0	–0.93767	–3.29516	–2.98941	–2.99295	–7.99683	–3.74902
5.2	–1.48678	–3.66896	–3.52536	–3.40256	–8.72919	–4.20655
5.4	–2.16518	–4.18078	–4.19547	–3.95108	–9.59984	–4.80477
5.6	–3.04538	–4.90114	–5.07343	–4.71093	–10.68378	–5.61789
5.8	–4.26192	–5.96294	–6.29633	–5.81817	–12.12086	–6.78511
6.0	–6.09787	–7.64789	–8.15307	–7.56085	–14.20600	–8.60080
6.2	–9.27110	–10.67270	–11.37669	–10.67126	–17.68786	–11.81311
6.4	–16.29563	–17.55043	–18.53499	–17.71655	–25.18921	–19.04438
6.6	–46.51786	–47.62657	–49.38934	–48.45749	–56.89534	–50.47950
6.8	72.88042	71.91788	72.58685	73.63304	67.50821	74.19636
7.0	21.89876	21.08322	20.52687	21.68885	14.14111	21.10325
7.2	13.27189	12.60487	11.73029	13.01000	4.96568	12.20386
7.4	9.65623	9.13992	8.05231	9.45221	1.01872	8.53534
7.6	7.63578	7.27308	6.00377	7.52680	–1.26376	6.53410
7.8	6.32342	6.11789	4.67857	6.32820	–2.80723	5.27506
8.0	5.38599	5.34190	3.73650	5.51674	–3.95913	4.41121
8.2	4.67014	4.79251	3.02102	4.93645	–4.87930	3.78316

(continued 1) $\alpha = 3.0$

λ	A	B	C	D	G	H
8.4	4.09531	4.38996	2.44992	4.50577	–5.65164	3.30752
8.6	3.61501	4.08860	1.97581	4.17797	–6.32449	2.93651
8.8	3.20041	3.86056	1.56933	3.92447	–6.92775	2.64083
9.0	2.83256	3.68793	1.21126	3.72689	–7.48098	2.40163
9.2	2.49839	3.55883	0.88836	3.57294	–7.99761	2.20626
9.4	2.18848	3.46514	0.59113	3.45423	–8.48723	2.04604
9.6	1.89573	3.40130	0.31246	3.36490	–8.95698	1.91486
9.8	1.61456	3.36350	0.04685	3.30089	–9.41235	1.80844
10.0	1.34040	3.34919	–0.21021	3.25941	–9.85776	1.72375
11.0	–0.05498	3.60311	–1.48969	3.35258	–12.05280	1.58010
12.0	–1.86260	4.54426	–3.09052	4.04602	–14.49238	1.96378
13.0	–5.23563	7.16887	–5.99041	6.15663	–18.00817	3.54168
14.0	–19.17380	20.39762	–17.71413	17.12782	–28.89332	12.52311
15.0	25.92342	–25.42213	19.86946	–21.22580	1.40283	–19.69903
16.0	8.37184	–8.65211	5.11440	–7.30657	–12.02838	–8.26697
17.0	4.62831	–5.80218	1.87801	–5.02821	–15.85426	–6.57342
18.0	2.61391	–4.86913	0.06208	–4.37356	–18.51923	–6.26072
19.0	1.03174	–4.68274	–1.43179	–4.38023	–20.95699	–6.54122
20.0	–0.57103	–5.03472	–3.01116	–4.90355	–23.53955	–7.32556
21.0	–2.60075	–6.07693	–5.08259	–6.13739	–26.67869	–8.84780
22.0	–5.93660	–8.56727	–8.57740	–8.91162	–31.36422	–12.01286
23.0	–14.52458	–16.38380	–17.73288	–17.40659	–42.14814	–21.32495
24.0	–308.14518	–309.26293	–333.55199	–332.58807	–382.86688	–360.58918
25.0	19.73446	19.36190	19.28285	20.89005	–4.19990	19.53923
26.0	9.63485	10.04168	8.49747	10.77970	–16.73962	8.49020
27.0	6.00946	7.26270	4.68338	7.70070	–21.77387	5.00035
28.0	3.88275	6.09034	2.49178	6.33976	–25.05354	3.34959
29.0	2.27965	5.60694	0.87950	5.70373	–27.69588	2.46429
30.0	0.84541	5.54680	–0.52612	5.49769	–30.09393	2.01926
31.0	–0.62813	5.85337	–1.93371	5.64551	–32.45880	1.91924
32.0	–2.35115	6.60371	–3.54051	6.20114	–34.97741	2.19694
33.0	–4.67422	8.07279	–5.66076	7.39748	–37.93283	3.04786
34.0	–8.46167	11.07748	–9.05533	9.94426	–41.99580	5.08769
35.0	–17.03678	18.90823	–16.63466	16.71654	–49.71587	10.82781
36.0	–70.26624	71.40655	–63.27565	62.56402	–91.60522	50.75106
37.0	41.78756	–41.38677	34.66743	–36.18239	–7.01776	–35.81394
38.0	16.41600	–16.78402	12.39079	–14.74155	–27.59931	–17.24592
39.0	9.86237	–11.05111	6.56961	–9.81320	–33.79256	–13.12884
40.0	6.57171	–8.66022	3.59392	–7.81691	–37.50630	–11.58604
41.0	4.38735	–7.49014	1.57337	–6.90089	–40.39830	–11.00175
42.0	2.66244	–6.94371	–0.06327	–6.54809	–42.97404	–10.92949

(continued 2) $\alpha = 3.0$

λ	A	B	C	D	G	H
43.0	1.11195	−6.81051	−1.57332	−6.58245	−45.46738	−11.22396
44.0	−0.44251	−7.03188	−3.12573	−6.96554	−48.04015	−11.86558
45.0	−2.16904	−7.64752	−4.88994	−7.75446	−50.86498	−12.92870
46.0	−4.30500	−8.81851	−7.11650	−9.13300	−54.20782	−14.62085
47.0	−7.31311	−10.95859	−10.30460	−11.55749	−58.60827	−17.44412
48.0	−12.40297	−15.24377	−15.76935	−16.31360	−65.49693	−22.80331
49.0	−24.39425	−26.46893	−28.76251	−28.63140	−80.59698	−36.40257
50.0	−108.30929	−109.63686	−120.16977	−119.37937	−181.18656	−135.50494
51.0	52.89248	52.30979	55.65013	57.09849	9.55598	56.72416
52.0	21.66223	21.83783	21.68057	23.79933	−28.41420	20.25207
53.0	13.33947	14.30267	12.68755	15.50320	−39.15239	11.03587
54.0	9.23005	11.02764	8.29243	11.84691	−44.87392	6.87393
55.0	6.60601	9.30552	5.52327	9.87690	−48.81692	4.54437
56.0	4.64989	8.34496	3.49155	8.72777	−51.94771	3.10139
57.0	3.02120	7.84019	1.82957	8.06266	−54.66405	2.17458
58.0	1.54045	7.66028	0.34656	7.73390	−57.16952	1.59861
59.0	0.08790	7.75689	−1.08077	7.68157	−59.59216	1.30173
60.0	−1.44151	8.13570	−2.55579	7.90066	−62.03471	1.26855
61.0	−3.17077	8.85675	−4.19408	8.43814	−64.60674	1.53381
62.0	−5.28471	10.06262	−6.16409	9.41636	−67.46254	2.20156
63.0	−8.12393	12.06403	−8.77134	11.10827	−70.87676	3.51091
64.0	−12.45472	15.60505	−12.69872	14.17251	−75.45840	6.04440
65.0	−20.52980	22.92096	−19.94884	20.59268	−82.98812	11.56196
66.0	−43.33352	44.98169	−40.27952	40.11086	−102.13443	28.71491
67.0	−1683.09310	1684.00177	−1496.79579	1495.81819	−1396.90548	1321.49633
68.0	48.70427	−48.54354	41.31539	−43.11140	−31.83940	−45.57018
69.0	23.95957	−24.56717	19.25761	−21.89454	−52.52318	−26.91159
70.0	15.50258	−16.91165	11.66303	−15.17736	−60.36461	−21.13552
71.0	11.01131	−13.26942	7.58577	−12.02977	−65.08750	−18.53541
72.0	8.06856	−11.24044	4.87706	−10.32181	−68.60233	−17.22140
73.0	5.86790	−10.03969	2.81842	−9.35844	−71.54970	−16.57856
74.0	4.05594	−9.34154	1.09310	−8.85331	−74.21409	−16.35573
75.0	2.44456	−8.99531	−0.46990	−8.67646	−76.75165	−16.44172
76.0	0.91337	−8.93309	−1.98295	−8.77304	−79.26815	−16.79378
77.0	−0.63259	−9.13657	−3.53835	−9.13473	−81.85431	−17.41325
78.0	−2.28825	−9.62815	−5.23262	−9.79362	−84.60920	−18.34200
79.0	−4.17304	−10.47842	−7.19157	−10.83215	−87.66623	−19.67486
80.0	−6.46929	−11.83531	−9.61147	−12.41601	−91.23732	−21.59663
81.0	−9.50482	−14.00141	−12.84899	−14.87947	−95.71107	−24.47596
82.0	−13.97607	−17.65394	−17.66546	−18.96675	−101.92012	−29.13032
83.0	−21.72716	−24.62174	−26.08045	−26.68387	−112.07654	−37.75973

(continued 3) $\alpha = 3.0$

λ	A	B	C	D	G	H
84.0	−39.89113	−42.02527	−45.91056	−45.83622	−134.74627	−58.92014
85.0	−149.00978	−150.39539	−165.42940	−164.68767	−266.67712	−189.35073
86.0	97.02248	96.38340	104.30081	105.70838	28.01301	106.83851
87.0	37.12280	37.23773	38.72682	40.80711	−44.79341	35.53753
88.0	22.69933	23.58529	22.99686	25.76527	−62.96876	18.88154
89.0	15.96448	17.64857	15.69635	19.17728	−71.90298	11.48865
90.0	11.90093	14.42142	11.32739	15.55521	−77.62080	7.34301
91.0	9.06265	12.47072	8.30640	13.32701	−81.85669	4.72047
92.0	6.87302	11.23532	6.00299	11.87610	−85.30016	2.94386
93.0	5.05232	10.45471	4.11251	10.91502	−88.28336	1.69638
94.0	3.44342	9.99633	2.46524	10.29601	−90.99010	0.81385
95.0	1.94537	9.79167	0.95379	9.94069	−93.53514	0.20745
96.0	0.48307	9.80983	−0.49967	9.81073	−95.99982	−0.16882
97.0	−1.00949	10.04697	−1.96124	9.89565	−98.45095	−0.33205
98.0	−2.60184	10.52452	−3.49793	10.20978	−100.95380	−0.27476
99.0	−4.38066	11.29476	−5.19074	10.79640	−103.58459	0.03798
100.0	−6.47070	12.45703	−7.15379	11.74177	−106.44821	0.68047
105.0	−46.80521	48.80563	−43.63107	43.84528	−144.54022	28.44907
110.0	20.32857	−22.14065	16.02120	−19.99303	−96.64203	−29.56932
115.0	5.11092	−11.67350	1.92549	−11.11569	−114.80699	−22.66070
120.0	−2.80512	−11.78223	−5.97607	−12.03168	−127.80147	−24.87888
125.0	−16.95421	−21.29674	−21.05291	−22.95908	−148.97219	−37.80856
130.0	117.00709	116.47560	126.54754	128.05614	8.55147	127.30080
135.0	13.85222	17.28317	13.58802	18.64967	−120.24902	6.21217
140.0	3.52404	12.07933	2.75436	12.41351	−136.71633	−1.60440
145.0	−4.15795	13.07355	−4.81467	12.63138	−149.27583	−2.64927
150.0	−18.41887	22.89938	−18.06339	21.00380	−166.68543	3.87493
155.0	388.98602	−388.29061	347.26631	−348.48877	155.81175	−328.72464
160.0	18.58734	−21.81140	14.37611	−19.91042	−148.47303	−34.70729
165.0	5.78803	−13.88223	2.39252	−13.28567	−164.79877	−29.80213
170.0	−2.04107	−13.56325	−5.39791	−13.76015	−177.65588	−31.56455
175.0	−12.73138	−19.21967	−16.65391	−20.49499	−194.61127	−39.93346
180.0	−63.08692	−65..59196	−71.67264	−71.93485	−259.82810	−97.40887
185.0	39.32840	40.59398	41.42591	44.55287	−140.03499	29.95638
190.0	12.41842	17.85385	12.22855	19.10483	−176.81714	1.07128
195.0	3.26511	14.11606	2.69885	14.44581	−191.84085	−5.04594
200.0	−4.38004	15.09578	−4.84894	14.65655	−204.39351	−6.09359

$\alpha = 3.1$

λ	A	B	C	D	G	H
0.2	0.59862	0.91205	−0.63675	0.57615	−0.30803	0.01359
0.4	0.34242	1.07307	−0.81712	0.65837	−0.64385	0.02919
0.6	0.02160	1.29406	−1.03707	0.77548	−1.00433	0.06472
0.8	−0.40173	1.61245	−1.31930	0.94971	−1.40309	0.13333
1.0	−1.00387	2.10395	−1.70941	1.22608	−1.86744	0.26174
1.2	−1.96490	2.94798	−2.31488	1.71141	−2.46158	0.51350
1.4	−3.83772	4.69670	−3.46544	2.73467	−3.38223	1.08455
1.6	−9.55478	10.28165	−6.90925	6.04312	−5.67161	3.01622
1.8	142.61415	−142.02844	83.75761	−84.76823	48.13451	−51.15680
2.0	10.48877	−10.05456	4.96005	−6.12554	0.92401	−4.32366
2.2	5.83729	−5.56641	2.14455	−3.47681	−0.99767	−2.79130
2.4	4.13468	−4.04077	1.08385	−2.59662	−1.87630	−2.31581
2.6	3.20130	−3.30027	0.47825	−2.18751	−2.48708	−2.12426
2.8	2.57556	−2.88608	0.05203	−1.97655	−2.99493	−2.05450
3.0	2.09873	−2.64291	−0.29015	−1.87187	−3.45740	−2.05248
3.2	1.70036	−2.50471	−0.59118	−1.83497	−3.90095	−2.09614
3.4	1.34320	−2.43989	−0.87443	−1.84818	−4.34068	−2.17610
3.6	1.00428	−2.43286	−1.15512	−1.90370	−4.78720	−2.28921
3.8	0.66700	−2.47693	−1.44516	−1.9954	−5.24965	−2.43631
4.0	0.31713	−2.57142	−1.75581	−2.13805	−5.73751	−2.62155
4.2	−0.05983	−2.72081	−2.09962	−2.32576	−6.26217	−2.85271
4.4	−0.48119	−2.93541	−2.49255	−2.57443	−6.83881	−3.14244
4.6	−0.97028	−3.23349	−2.95699	−2.90341	−7.48931	−3.51077
4.8	−1.56191	−3.64607	−3.52715	−3.34462	−8.24756	−3.99018
5.0	−2.31286	−4.22709	−4.25948	−3.95274	−9.16988	−4.63595
5.2	−3.32513	−5.07625	−5.25592	−4.82834	−10.35837	−5.54926
5.4	−4.80476	−6.39779	−6.72300	−6.17679	−12.02015	−6.93655
5.6	−7.24389	−8.68234	−9.15450	−8.49093	−14.65079	−9.29276
5.8	−12.18725	−13.47335	−14.10103	−13.32055	−19.80686	−14.17391
6.0	−28.30331	−29.43819	−30.26507	−29.36740	−36.22505	−30.31623
6.2	196.42042	195.43660	195.36018	196.37597	190.09957	196.28569
6.4	24.28054	23.44852	22.55586	23.69131	16.42275	22.88802
6.6	13.45112	12.77250	11.69803	12.95531	5.33117	12.07789
6.8	9.46823	8.94540	7.71356	9.09539	1.14004	8.17093
7.0	7.36064	6.99683	5.61171	7.12142	−1.16099	6.15725
7.2	6.03074	5.82997	4.29063	5.93218	−2.67802	4.93122
7.4	5.09650	5.06364	3.36690	5.14488	−3.79596	4.10845
7.6	4.38983	4.53066	2.67186	4.59157	−4.68403	3.52028
7.8	3.82508	4.14633	2.11961	4.18712	−5.42840	3.08115
8.0	3.35382	3.86327	1.66165	3.88386	−6.07769	2.74312
8.2	2.94649	3.65308	1.26840	3.65317	−6.66150	2.47737

(continued 1) $\alpha = 3.1$

λ	A	B	C	D	G	H
8.4	2.58384	3.49784	0.92067	3.47692	–7.19901	2.26561
8.6	2.25266	3.38583	0.60533	3.34321	–7.70329	2.09577
8.8	1.94343	3.30923	0.31295	3.24405	–8.18367	1.95971
9.0	1.64891	3.26281	0.03647	3.17400	–8.64713	1.85182
9.2	1.36334	3.24316	–0.22971	3.12943	–9.09910	1.76830
9.4	1.08188	3.24821	–0.49022	3.10803	–9.54404	1.70663
9.6	0.80020	3.27697	–0.74911	3.10853	–9.98579	1.66528
9.8	0.51420	3.32938	–1.01017	3.13051	–10.42788	1.64357
10.0	0.21975	3.40625	–1.27713	3.17440	–10.87370	1.64157
11.0	–1.56040	4.26060	–2.85894	3.80206	–13.29837	1.98748
12.0	–4.89195	6.81171	–5.73412	5.86616	–16.79954	3.55046
13.0	–20.21809	21.40390	–18.67793	18.04188	–28.75252	13.60446
14.0	20.30847	–19.87235	15.26172	–16.68662	–1.35195	–15.72119
15.0	7.19752	–7.58148	4.16274	–6.45398	–11.77180	–7.31146
16.0	3.94253	–5.28220	1.32092	–4.62457	–15.27764	–5.97292
17.0	2.05580	–4.58542	–0.39987	–4.16655	–17.87164	–5.81376
18.0	0.45785	–4.58920	–1.92555	–4.34242	–20.35200	–6.23571
19.0	–1.31657	–5.20684	–3.68891	–5.12155	–23.12692	–7.25782
20.0	–3.87882	–6.83199	–6.31544	–6.94650	–26.84082	–9.37080
21.0	–9.23097	–11.36018	–11.92135	–11.84381	–33.73309	–14.73854
22.0	–39.59728	–40.95350	–44.12245	–43.37678	–68.89971	–48.41262
23.0	27.09443	26.50340	26.83889	28.24896	5.58526	27.56418
24.0	10.55871	10.76158	9.33731	11.43940	–13.95748	9.53950
25.0	6.22822	7.29130	4.81385	7.66817	–19.70099	5.36855
26.0	3.91943	5.95523	2.44859	6.15571	–23.14215	3.85972
27.0	2.24418	5.43066	0.77211	5.49003	–25.83755	2.69684
28.0	0.75958	5.37872	–0.67681	5.30136	–28.26904	2.28851
29.0	–0.78196	5.73591	–2.14462	5.50547	–30.68383	2.26039
30.0	–2.63993	6.61184	–3.87359	6.18527	–33.30991	2.67106
31.0	–5.29354	8.40291	–6.29367	7.68264	–36.53427	3.81384
32.0	–10.13084	12.44588	–10.63306	11.17121	–41.44456	6.71828
33.0	–24.66018	26.21359	–23.51376	23.23509	–53.88160	17.18484
34.0	212.50787	–211.71121	185.47959	–186.57073	129.26570	–167.92926
35.0	21.96667	–21.94665	17.43962	–19.36542	–19.94905	–20.70600
36.0	11.39589	–12.19774	8.04079	–10.85061	–29.32511	–13.37530
37.0	7.22983	–8.92807	4.28001	–8.05472	–33.73952	–11.09395
38.0	4.74580	–7.45262	1.99054	–6.85152	–36.86927	–10.22860
39.0	2.89396	–6.77433	0.24175	–6.36725	–39.54030	–10.01458
40.0	1.28005	–6.57891	–1.32176	–6.33274	–42.07436	–10.22306
41.0	–0.31696	–6.77613	–2.90770	–6.68127	–44.66843	–10.80911
42.0	–2.09341	–7.40315	–4.71213	–7.46859	–47.52012	–11.84664

(continued 2) $\alpha = 3.1$

λ	A	B	C	D	G	H
43.0	−4.32771	−8.64765	−7.02640	−8.90604	−50.93578	−13.57142
44.0	−7.59340	−11.02663	−10.46348	−11.55670	−55.57159	−16.59565
45.0	−13.55835	−16.16990	−16.81845	−17.18206	−63.36024	−22.82234
46.0	−30.65895	−32.48608	−35.18718	−34.85345	−84.10915	−42.03723
47.0	1192.77330	1191.71533	1284.34847	1285.36676	1338.06109	1381.65638
48.0	32.05199	31.76711	32.59058	34.29763	−12.89128	32.23168
49.0	16.22006	16.73053	15.58981	18.00616	−32.16494	14.50464
50.0	10.47724	11.82478	9.47357	12.63708	−39.69652	8.55400
51.0	7.27775	9.52648	6.10589	10.07441	−44.25877	5.62489
52.0	5.06816	8.31013	3.81397	8.67031	−47.65347	3.93780
53.0	3.31174	7.67613	2.02237	7.88256	−50.49819	2.90490
54.0	1.75870	7.42698	0.46649	7.49337	−53.07400	2.28702
55.0	0.25740	7.48909	−1.01008	7.41621	−55.54320	1.98443
56.0	−1.31718	7.86109	−2.53086	7.63829	−58.02887	1.97322
57.0	−3.10866	8.60969	−4.23152	8.21235	−60.66013	2.29302
58.0	−5.33787	9.90564	−6.31446	9.28674	−63.62322	3.07178
59.0	−8.43441	12.14391	−9.16767	11.21208	−67.26909	4.62017
60.0	−13.45575	16.35661	−13.74074	14.91062	−72.45091	7.76179
61.0	−24.06196	26.18408	−23.31411	23.64148	−82.10918	15.41417
62.0	−68.11499	69.47192	−62.84767	62.34689	−118.60428	49.91860
63.0	107.27825	−106.68754	94.19557	−95.52589	20.99117	−91.66936
64.0	30.80031	−30.99082	25.61430	−27.79066	−41.52657	−31.16225
65.0	17.68735	−18.68848	13.79160	−16.84611	−53.20393	−21.53058
66.0	11.95417	−13.81129	8.57518	−12.55723	−58.96829	−17.86567
67.0	8.53871	−11.31601	5.42874	−10.40810	−62.88512	−16.12415
68.0	6.12118	−9.90622	3.16782	−9.23963	−66.01782	−15.27009
69.0	4.19585	−9.10670	1.33652	−8.62898	−68.77791	−14.92786
70.0	2.51634	−8.71272	−0.28977	−8.39669	−71.37086	−14.94093
71.0	0.93459	−8.63551	−1.84932	−8.46885	−73.92659	−15.24918
72.0	−0.66205	−8.85105	−3.45139	−8.83229	−76.55207	−15.84990
73.0	−2.38574	−9.38675	−5.20964	−9.52476	−79.36342	−16.79120
74.0	−4.38164	−10.33205	−7.27629	−10.64856	−82.52103	−18.18876
75.0	−6.88326	−11.88197	−9.90099	−12.41884	−86.29251	−20.27924
76.0	−10.34294	−14.46099	−13.57185	−15.29874	−91.20495	−23.56739
77.0	−15.83234	−19.11967	−19.44939	−20.42988	−98.51554	−29.29345
78.0	−26.76066	−29.25053	−31.23036	−31.49404	−112.23311	−41.45281
79.0	−63.26067	−64.97245	−70.75357	−70.31757	−156.04730	−83.72393
80.0	285.64398	284.70311	307.81063	308.94013	253.68037	327.54149
81.0	45.67127	45.50546	47.55899	49.38598	−29.58125	45.82146
82.0	24.69800	25.32255	24.87990	27.41842	−55.12222	21.83491
83.0	16.54289	17.98473	16.10833	19.38285	−65.57416	12.95966

(continued 3) $\alpha = 3.1$

λ	A	B	C	D	G	H
84.0	12.00353	14.30243	11.26267	15.30925	−71.76407	8.37912
85.0	8.96818	12.17873	8.05342	12.92147	−76.17438	5.62291
86.0	6.68380	10.87853	5.66537	11.42047	−79.68795	3.82271
87.0	4.80896	10.08295	3.73018	10.45823	−82.70251	2.59915
88.0	3.15991	9.63757	2.05118	9.86457	−85.42896	1.76519
89.0	1.62116	9.46650	0.50673	9.55355	−87.99603	1.22434
90.0	0.10666	9.53909	−0.99147	9.48684	−90.49491	0.93046
91.0	−1.46155	9.85786	−2.52064	9.65983	−93.00277	0.87193
92.0	−3.17025	10.45839	−4.16376	10.09997	−95.59955	1.06842
93.0	−5.13572	11.42076	−6.02905	10.87618	−98.38646	1.57854
94.0	−7.53971	12.90012	−8.28293	12.12605	−101.51631	2.52380
95.0	−10.70826	15.20229	−11.22145	14.12367	−105.25822	4.14952
96.0	−15.32020	18.99015	−15.45856	17.46568	−110.16777	6.99248
97.0	−23.10198	25.97706	−22.55327	23.69691	−117.65296	12.44526
98.0	−40.19841	42.29674	−38.05114	38.35084	−132.71873	25.49973
99.0	−116.88658	118.21636	−107.31876	106.78332	−196.30093	87.08006
100.0	149.87033	−149.31008	133.30848	−134.68025	19.73891	−130.96306
105.0	10.51857	−14.15196	7.18116	−13.11226	−99.51010	−22.14279
110.0	0.95270	−10.67424	−2.01594	−10.53718	−113.44430	−20.88309
115.0	−8.80465	−14.68749	−12.10021	−15.43420	−128.95622	−27.14144
120.0	−88.24672	−90.01548	−98.15234	−97.76798	−227.25686	−117.51221
125.0	18.71723	20.90851	18.58448	22.55109	−104.94254	12.61176
130.0	4.98873	12.07056	4.09445	12.48813	−125.32396	0.80699
135.0	−2.88961	12.24083	−3.74409	11.93860	−138.20901	−0.90785
140.0	−16.23182	20.85653	−16.26135	19.31085	−155.03769	4.84851
145.0	359.65847	−358.95746	323.87467	−325.09775	147.65834	−307.97187
150.0	16.62812	−19.99879	12.77145	−18.42967	−139.57777	−31.03852
155.0	4.52835	−13.10886	1.33184	−12.66639	−155.48259	−26.79006
160.0	−3.58197	−13.61789	−6.80821	−13.91888	−168.74098	−29.23854
165.0	−17.24452	−22.44732	−21.25427	−23.98407	−189.10278	−41.06034
170.0	−649.34954	−650.57157	−709.75287	−708.87281	−944.11291	−788.22797
175.0	22.08795	24.84358	22.33699	26.82564	−150.97878	12.74902
180.0	7.30090	14.79849	6.65394	15.44613	−172.69866	−0.49618
185.0	−0.70967	13.68942	−1.44832	13.60886	−185.97922	−3.52052
190.0	−10.17411	17.70838	−10.53704	16.76172	−199.79189	−1.65002
195.0	−38.35056	41.68657	−36.50245	38.15002	−228.80471	16.90089
200.0	69.83386	−70.37775	61.36989	−63.95313	−145.34768	−76.64139

$\alpha = 3.2$

λ	A	B	C	D	G	H
0.2	0.51503	0.91717	–0.68798	0.54531	–0.33944	–0.00533
0.4	0.22376	1.10718	–0.88788	0.64232	–0.68607	0.01959
0.6	–0.15679	1.38098	–1.14126	0.78737	–1.06520	0.07163
0.8	–0.69181	1.80316	–1.48639	1.01812	–1.49935	0.17269
1.0	–1.53363	2.52529	–2.01253	1.42310	–2.04089	0.37442
1.2	–3.14168	4.00592	–2.98863	2.27038	–2.84734	0.83347
1.4	–7.84980	8.57786	–5.78017	4.92443	–4.71771	2.34783
1.6	436.45277	–435.87089	254.09305	–255.09618	147.06614	–149.80182
1.8	9.82455	–9.40035	4.47759	–5.63950	0.80155	–3.91431
2.0	5.37256	–5.11936	1.82831	–3.16217	–0.99302	–2.50984
2.2	3.78262	–3.71595	0.84995	–2.37114	–1.81339	–2.09480
2.4	2.91137	–3.04956	0.28812	–2.01476	–2.39393	–1.93750
2.6	2.32163	–2.68650	–0.11364	–1.84001	–2.88538	–1.89063
2.8	1.86502	–2.48282	–0.44306	–1.76357	–3.33993	–1.90631
3.0	1.47571	–2.37852	–0.73981	–1.75148	–3.78198	–1.96582
3.2	1.11832	–2.34584	–1.02615	–1.78905	–4.22605	–2.06208
3.4	0.77008	–2.37234	–1.31749	–1.87104	–4.68325	–2.19405
3.6	0.41325	–2.45458	–1.62710	–1.99831	–5.16417	–2.36507
3.8	0.03085	–2.59596	–1.96903	–2.17693	–5.68111	–2.58271
4.0	–0.39671	–2.80695	–2.36080	–2.41904	–6.25030	–2.86004
4.2	–0.89576	–3.10752	–2.82715	–2.74563	–6.89543	–3.21851
4.4	–1.50623	–3.53298	–3.40661	–3.19246	–7.65398	–3.693395
4.6	–2.29531	–4.14706	–4.16484	–3.82311	–8.59040	–4.34954
4.8	–3.38932	–5.07341	–5.22602	–4.76017	–9.82699	–5.30660
5.0	–5.06123	–6.58290	–6.85927	–6.27144	–11.62920	–6.82976
5.2	–8.03689	–9.39964	–9.78085	–9.07212	–14.70397	–9.62527
5.4	–15.11203	–16.31792	–16.75061	–15.92111	–21.77578	–16.41700
5.6	–56.97112	–58.02096	–58.06144	–57.11049	–62.75105	–57.11082
5.8	40.55994	39.66648	38.23949	39.31333	32.13006	38.05363
6.0	15.97960	15.24387	13.98602	15.18489	7.99416	14.20348
6.2	10.23146	9.65578	8.32426	9.65100	2.21277	8.71075
6.4	7.61638	7.20405	5.75543	7.21359	–0.51514	6.27494
6.6	6.08882	5.84409	4.26018	5.85402	–2.18309	4.90309
6.8	5.06420	4.99231	3.26150	4.99604	–3.36065	4.02622
7.0	4.31193	4.41917	2.53186	4.41294	–4.27239	3.42042
7.2	3.72248	4.01627	1.96321	3.99757	–5.02487	2.97984
7.4	3.23692	3.72594	1.49750	3.69290	–5.67532	2.64808
7.6	2.82056	3.51487	1.10060	3.46591	–6.25737	2.39244
7.8	2.45140	3.36266	0.75094	3.29633	–6.79219	2.19282
8.0	2.11467	3.25634	0.43405	3.17120	–7.29395	2.03629
8.2	1.79980	3.18746	0.13968	3.08202	–7.77264	1.91434

(continued 1) $\alpha = 3.2$

λ	A	B	C	D	G	H
8.4	1.49885	3.15059	−0.13982	3.02323	−8.23566	1.82130
8.6	1.20545	3.14231	−0.41052	2.99124	−8.68881	1.75344
8.8	0.91416	3.16081	−0.67751	2.98396	−9.13691	1.70850
9.0	0.62002	3.20549	−0.94538	3.00050	−9.58421	1.68530
9.2	0.31818	3.27693	−1.21852	3.04101	−10.03472	1.68366
9.4	0.00351	3.37683	−1.50148	3.10668	−10.49250	1.70434
9.6	−0.32965	3.50816	−1.79925	3.19981	−10.96197	1.74907
9.8	−0.68804	3.67541	−2.11764	3.32402	−11.44818	1.82074
10.0	−1.07995	3.88508	−2.46378	3.48463	−11.95727	1.92370
11.0	−4.09992	6.08091	−5.08407	5.26114	−15.24819	3.26460
12.0	−17.09100	18.31168	−16.10766	15.49972	−25.62092	11.74087
13.0	19.43744	−18.98971	14.60246	−16.01364	−0.82258	−14.97800
14.0	6.65643	−7.05885	3.74035	−6.03912	−11.07540	−6.73753
15.0	3.54910	−4.95667	1.01378	−4.36524	−14.48994	−5.51105
16.0	1.69749	−4.38764	−0.68495	−4.01239	−17.07012	−5.42867
17.0	0.04676	−4.53218	−2.26921	−4.31404	−19.61131	−5.95402
18.0	−1.93725	−5.40797	−4.24659	−5.34232	−22.60162	−7.20804
19.0	−5.21728	−7.77402	−7.60768	−7.91590	−27.06423	−10.06880
20.0	−14.66108	−16.39704	−17.45784	−17.05464	−38.35581	−19.82056
21.0	111.26305	110.31574	114.95131	116.04183	99.85545	119.91274
22.0	13.69142	13.54620	12.47768	14.27052	−8.78284	12.81235
23.0	7.13470	7.84767	5.65903	8.20626	−16.88981	6.29134
24.0	4.34554	6.02237	2.80740	6.20454	−20.82080	4.03249
25.0	2.51016	5.32522	0.97147	5.37464	−23.67240	2.99269
26.0	0.95981	5.19676	−0.54254	5.11950	−26.16574	2.53646
27.0	−0.61867	5.52068	−2.04758	5.30070	−28.61536	2.50394
28.0	−2.52440	6.40484	−3.82467	5.99823	−31.28709	2.94962
29.0	−5.31563	8.30727	−6.37743	7.61031	−34.63646	4.21820
30.0	−10.76188	12.93755	−11.28113	11.64911	−40.06665	7.64645
31.0	−31.38899	32.78077	−29.65535	29.19125	−57.44939	23.06080
32.0	61.51816	−60.91002	52.72685	−54.02384	14.58438	−50.94333
33.0	16.41484	−16.61906	12.62697	−14.78832	−22.08333	−16.28047
34.0	9.13719	−10.21314	6.08999	−9.17976	−28.97159	−11.46613
35.0	5.80441	−7.84868	3.04472	−7.16658	−32.77089	−9.85200
36.0	3.63619	−6.79583	1.01927	−6.33068	−35.67976	−9.29692
37.0	1.89731	−6.39461	−0.64560	−6.09308	−38.29044	−9.29425
38.0	0.26743	−6.44685	−2.24520	−6.28892	−40.87674	−9.70942
39.0	−1.48149	−6.94247	−4.00151	−6.92886	−43.65665	−10.57283
40.0	−3.63047	−8.03660	−6.20328	−8.18988	−46.92839	−12.08352
41.0	−6.72996	−10.20566	−9.43160	−10.58735	−51.30656	−14.79769
42.0	−12.37554	−14.99753	−15.38599	−15.77836	−58.60210	−20.48860

(continued 2) $\alpha = 3.2$

λ	*A*	*B*	*C*	*D*	*G*	*H*
43.0	–28.86291	–30.67427	–32.92165	–32.58822	–78.26802	–38.58264
44.0	459.70429	458.68653	488.84283	489.88773	477.93428	519.17947
45.0	28.19682	27.97812	28.14264	29.90474	–14.94873	27.86219
46.0	14.45349	15.06087	13.53686	16.04120	–31.48609	12.91374
47.0	9.29772	10.78138	8.10474	11.39721	–38.22425	7.80636
48.0	6.35390	8.79151	5.04069	9.19184	–42.42823	5.29726
49.0	4.26791	7.77249	2.90168	8.01393	–45.63624	3.87693
50.0	2.55995	7.29345	1.17947	7.39933	–48.38731	3.04606
51.0	0.99595	7.19248	–0.36992	7.16917	–50.93658	2.60827
52.0	–0.58135	7.42616	–1.90498	7.26767	–53.44478	2.49547
53.0	–2.32517	8.03450	–3.57336	7.72154	–56.05519	2.72180
54.0	–4.45010	9.16255	–5.57449	8.65598	–58.95392	3.39408
55.0	–7.35823	11.16663	–8.27508	10.38878	–62.47601	4.79511
56.0	–12.02948	14.99443	–12.56226	13.77265	–67.42503	7.69166
57.0	–21.86413	24.02195	–21.50711	21.85273	–76.57508	14.83013
58.0	–63.26219	64.62960	–58.93987	58.43824	–111.43701	47.69887
59.0	94.26236	–93.68588	83.17403	–84.52376	15.75899	–81.49160
60.0	27.73431	–27.96565	23.05707	–25.27336	–39.57975	–28.16745
61.0	15.92998	–17.00304	12.33046	–15.44996	–50.34226	–19.45917
62.0	10.69172	–12.65944	7.52563	–11.60545	–55.76487	–16.15238
63.0	7.52211	–10.46038	4.58128	–9.70313	–59.51541	–14.60566
64.0	5.23637	–9.25033	2.42552	–8.70254	–62.56405	–13.88259
65.0	3.37433	–8.60854	0.63959	–8.22729	–65.29239	–13.64708
66.0	1.70497	–8.35981	–0.98979	–8.12404	–67.89809	–13.76587
67.0	0.07980	–8.43884	–2.60381	–8.34100	–70.51629	–14.19989
68.0	–1.62859	–8.84928	–4.32874	–8.89272	–73.27306	–14.97439
69.0	–3.56915	–9.66307	–6.31804	–9.86314	–76.32732	–16.18553
70.0	–5.97009	–11.05744	–8.81255	–11.44709	–79.93397	–18.04666
71.0	–9.26724	–13.43268	–12.27759	–14.07778	–84.59026	–21.02606
72.0	–14.49816	–17.80024	–17.82593	–18.84441	–91.49003	–26.29624
73.0	–25.03343	–27.51051	–29.07801	–29.34919	–104.52241	–37.72989
74.0	–62.08131	–63.75528	–68.82381	–68.36716	–148.17725	–79.80336
75.0	197.01264	196.13420	209.67708	210.85500	150.16954	220.11920
76.0	39.57332	39.49610	40.55261	42.45727	–32.52195	39.00864
77.0	21.80557	22.54837	21.52698	24.17571	–53.90911	19.21842
78.0	14.62964	16.22509	13.88637	17.30907	–63.05909	11.69279
79.0	10.54197	13.03827	9.56839	13.80909	–68.63488	7.78153
80.0	7.75199	11.21574	6.65035	11.76979	–72.70133	5.43502
81.0	5.60728	10.12796	4.43302	10.51274	–76.00814	3.92245
82.0	3.80551	9.50231	2.59399	9.74252	–78.89954	2.92412
83.0	2.17836	9.21053	0.95565	9.31797	–81.56338	2.28530

(continued 3) $\alpha = 3.2$

λ	A	B	C	D	G	H
84.0	0.61321	9.19634	−0.59834	9.17395	−84.12047	1.93178
85.0	−0.98299	9.44926	−2.16117	9.29238	−86.66474	1.83790
86.0	−2.70746	9.99920	−3.82682	9.69489	−89.28764	2.01737
87.0	−4.68688	10.92682	−5.71429	10.45105	−92.10149	2.52938
88.0	−7.11960	12.39766	−8.00671	11.70871	−95.27575	3.50546
89.0	−10.36981	14.75152	−11.03731	13.77489	−99.11568	5.22270
90.0	−15.22940	18.76141	−15.52791	17.35109	−104.27945	8.31744
91.0	−23.86848	26.58217	−23.45351	24.39590	−112.56471	14.56969
92.0	−45.28371	47.19756	−42.99776	43.07913	−131.41576	31.40885
93.0	−223.33072	224.45180	−205.00035	204.22816	−279.83417	177.82324
94.0	82.74416	−82.41949	73.31239	−74.94219	−27.78000	−76.23948
95.0	35.03911	−35.52502	29.85597	32.35879	−68.38082	−37.66392
96.0	21.83119	−23.15286	17.77299	−21.17605	−80.44943	−27.65004
97.0	15.38219	−17.57679	11.83412	−16.17759	−86.93335	−23.26424
98.0	11.38824	−14.50671	8.12380	−13.46268	−91.39506	−20.96001
99.0	8.54181	−12.65147	5.45141	−11.85836	−94.91897	−19.67202
100.0	6.30502	−11.49353	3.32605	−10.89563	−97.95336	−18.97555
105.0	−2.27858	−10.93197	−5.15805	−11.05845	−111.41627	−20.25816
110.0	−18.64727	−22.39438	−22.39386	−23.82640	−134.64042	−34.80506
115.0	43.02407	43.31527	44.42895	46.67750	−67.31047	40.48897
120.0	8.49007	13.26433	7.56941	13.90809	−111.72652	4.41311
125.0	−0.22153	11.25119	−1.27012	11.18360	−125.76901	0.56164
130.0	−10.79237	16.37749	−11.37344	15.41455	−140.49756	3.53136
135.0	−128.46206	129.90327	−119.17837	118.74818	−244.33740	97.12168
140.0	18.94186	−21.62416	15.09500	−19.97621	−127.09945	−30.34585
145.0	4.94643	−12.83385	1.87967	−12.38113	−144.65360	−24.28517
150.0	−3.44055	−13.18544	−6.50920	−13.42949	−158.11909	−26.43644
155.0	−18.24360	−23.04347	−22.08625	−24.48734	−179.58418	−39.16300
160.0	226.17654	225.45588	242.23778	243.56646	101.19138	249.63129
165.0	17.12223	20.58114	16.79373	21.94577	−147.00062	9.54529
170.0	4.89601	13.59887	4.02079	13.97170	−165.41749	0.12844
175.0	−3.22315	13.87458	−4.05763	13.57752	−178.52721	−1.33904
180.0	−15.87177	21.51773	−16.04976	20.16173	−194.96824	3.83429
185.0	−149.88442	151.41828	−139.07656	138.74487	−312.98166	111.62082
190.0	26.97543	−29.52415	22.49925	−27.24361	−170.44227	−41.11304
195.0	8.74693	−16.11474	5.41194	−15.36596	−191.53383	−31.07833
200.0	0.05082	−14.32644	−3.12213	−14.30047	−204.98265	−31.12933

BIBLIOGRAPHY

[1] BACH K. et al.: Pneus in Nature and Technics. Mitteilungen des Instituts für Leichte Flächentragwerke, IL 9, Stuttgart 1977.

[2] BERKA L. and MINSTER J.: Deformation properties of fabric membranes loaded by biaxial tension (in Czech). Stavebnícky časopis **26** (1978), 204—218.

[3] BUBNER E. et al.: Lufthallenhandbuch. Mitteilungen des Instituts für Leichte Flächentragwerke, IL 15, Stuttgart 1977.

[4] BULSON P. S.: The behaviour of some experimental inflated structures. In [29].

[5] BULSON P. S.: Structural behaviour of a 30 ft inflated bridge. In [29].

[6] BULSON P. S.: Design principles of pneumatic structures. Journal of the Institute of Structural Engineers **51** (6), June 1973.

[7] BUDINSKÝ B. and KEPR B.: Fundamentals of Differential Geometry with Technical Applications (in Czech). SNTL, Prague 1970.

[8] COLLATZ L.: Funktionalanalysis und numerische Mathematik. Springer-Verlag, Berlin—Göttingen—Heidelberg 1964.

[9] COURBON J.: Vibrations radialles d'une enveloppe sphérique. Annales des Ponts et Chaussées **141** (2) (1971), 59—79.

[10] CHUDNOVSKIJ V. G.: Methods of Calculation of Vibrations and Stability of Frame Systems (in Russian). Izd. AN SSSR, Kiev 1952.

[11] FIŘT V.: Static and Dynamics of Air-Supported Membrane Structures (in Czech). Academia, Prague 1976.

[12] FIŘT V.: Air-supported structures with automatic adjustment of internal overpressure. In [28].

[13] FIŘT V.: Device for the damping of dynamic effects of wind on masts. In [56].

[14] FIŘT V.: Stability and Vibrations of Structures with Non-Rigid Joints (in Czech). Academia, Prague 1974.

[15] FIŘT V.: Two-membrane suspended roofs. Int. Conference on Tension Roof Structures, Proc. IABSE, London 1974.

[16] FIŘT V.: Determining the shape of air-supported structures and their inlet openings. In [29].

[17] FIŘT V.: Czechoslovak patents No. 136 153 (1970), No. 136 564 (1970), No. 136 566 (1970), No. 136 567 (1970), No. 136 568 (1970), No. 150 309 (1973), No. 155 310 (1973), No. 153 671 (1973), No. 151 952 (1973), No. 155 372 (1973), No. 160 953 (1974).

[18] FIŘT V.: Formfinding of membrane structures, Int. Symposium "Plastics in Material and Structural Engineering", Proc. ICP, Prague 1981.

[19] FLÜGGE W.: Stresses in Shells. Springer-Verlag, Berlin—Heidelberg—New York 1966.

[20] GOLDENVEIZER A. L.: Theory of Elastic Thin Shells (in Russian). Gosizdat, Moscow 1953. Translated into English and published by Pergamon Press, Ltd., 1961.

[21] GRADSHTEJN I. S. and RYZHIK I. M.: Tables of Integrals of Sums, Series and Products (in Russian). Fizmatgiz, Moscow 1962.

[22] GREEN A. E. and ADKINS J. E.: Large elastic deformations and non-linear continuum mechanics. Oxford 1960.

[23] HAUG E.: Finite element analysis of pneumatic structures. In [28].

[24] HEARLE J. W. S., GROSBERG P. and BACKER S.: Structural Mechanics of Fibers, Yarns and Fabrics. Wiley-Interscience, New York—London—Sydney—Toronto 1969.

[25] HENRYCH J.: The Dynamics of Explosion and Its Use. Elsevier Scientific Publishing Company, Amsterdam—Oxford—New York 1979.

[26] HOSTINSKÝ B.: Differential Geometry of Curves and Surfaces (in Czech). Přírodovědecké nakladatelství, Prague 1950.

[27] International Conference on Tension Roof Structures. Proc. IABSE, London, April 1974.

[28] International Symposium on Pneumatic Structures. Proc. IASS, Delft 1972.

[29] International Symposium "Wide Span Surface Structures". Proc., University of Stuttgart, April 1976.

[30] ISONO Y.: Abstract of "Pneumatic structure design standard in Japan". In [28].

[31] KAMKE E.: Differentialgleichungen, Lösungsmethoden und Lösungen, Leipzig 1959.

[32] KAWAGUCHI M.: The shallowest possible pneumatic forms. Bulletin of the Int. Association for Shell and Special Structures, Vol. XVIII, No. 1 (1977), 3—11.

[33] KYDONIEFS A. D.: Finite axisymmetric deformations of an initially cylindrical membrane reinforced with inextensible cords. The Quarterly Journal of Mechanics and Applied Mathematics, Vol. XXIII, Part 4 (1970), 481—488.

[34] LEKHNITSKI S. G.: Anisotropic Plates (in Russian). Gosizdat, Moscow 1957. English translations published by Gordon and Breech Co. New York 1973.

[35] LEONARD R. W. et al.: Structural Considerations of Inflatable Reentry Vehicles. NASA TN D-457, Washington, D. C. 1960.

[36] LEONARD R. W. et al.: Analysis of Inflated Reentry and Space Structures. Proc. of Recovery of Space Vehicles Symposium, Los Angeles 1960. (Sponsored and published by the Institute of the Aeronautical Sciences).

[37] MCCOMB H. G., JR.: A Linear Theory for Inflatable Plates of Arbitrary Shape. NASA TN D-930, Washington 1961.

[38] MCCOMB H. G., JR. and LEONARD R. W.: Slanted Drop Cords in Inflatable Airmat Beams. Journal of the Aerospace Sciences, Vol. **29**, No. 4, April 1962.

[39] MINSTER J.: Strength and rheological characteristics of the fabric 824-550-743 (in Czech). Research report of the Inst. of Theor. and Appl. Mechanics, Czech. Acad. of Sci., Prague 1973.

[40] MINSTER J. and BERKA L., Strength of technical fabrics under biaxial state of stress (in Czech). Stavebnícky časopis **25** (1977), 37—48.

[41] MORTON W. E. and HEARLE J. W. S.: Physical Properties of Textile Fibres. Heinemann, London 1975.

[42] NASH W. A. and HO F. H.: Nonlinear free tranverse vibrations of inflatable shallow shells. In [50].

[43] NASH W. A. and HO F. H.: Mechanics of inflatable sandwich type shallow shells. Proc. of the

Symposium on the Theory of Shells to Honor. L. H. Donnell, University of Houston Press 1967.
[44] NIEMANN H. J.: Wind tunnel experiments on aeroelastic models of air-supported structures: results and conclusions. In [28].
[45] NOVOSHILOV V. V.: Thin Shell Theory. P. Noordhoff Ltd., Groningen, The Netherlands 1964.
[46] ODEN J. T.: Finite deformation of elastic plates, shells and membranes by the finite element method. In [56].
[47] OTTO F. and TROSTEL R.: Zugbeanspruchte Konstruktionen. Ullstein Fachverlag, Berlin 1966.
[48] Pacific Symposium on Tension Structures and Space Frames. Proc. IASS, Tokyo and Kyoto, October 1971.
[49] PRIS M. R.: Action du vent sur les bâtiments et constructions. Annales de l'ITBTP, No. 170, Février 1962.
[50] Proceedings of the 1st Int. Coloquium on Pneumatic Structures, IASS. University of Stuttgart, May 1967.
[51] RÜHLE H.: Räumliche Dachtragwerke Konstruktion and Anführung, Band 2. VEB Verlag für Bauwesen, Berlin 1970.
[52] SEBRING R. E., SHOPPER M. M. and FREESTON W. D.: Biaxial Tensil Tests of Coated Fabrics. Test-Reports, Natick Laboratoires, Natick 1969.
[53] SKELTON J.: Comparation and selection of materials for air supported structures. Journal of Coated Fibrous Materials, Vol. **1** (April 1972), 208—221.
[54] STEPANOV V. V.: Cours of Differential Equations (in Russian). Gosizdat, Moscow—Leningrad 1950.
[55] STROUD W. J.: Experimental and Theoretical Deflections and Natural Frequencies of an Inflatable Fabric Plate. NASA TN D-931, Washington, D. C. 1961.
[56] The First Canadian IASS Conference "Shell Structures and Climatic Influences", Proc. IASS. The University of Calgary, July 1972.
[57] TIMOSHENKO S. and WOINOWSKY—KRIEGER S.: Theory of plates and shells. McGraw-Hill, New York—Toronto—London 1959.
[58] TRYGGVASON B. V. and SYUMOVN.: Similarity requirements for inflatable structures. Proc. of the Third U. S. National Conference on Wind Engineering Research, University of Florida 1978.
[59] TUTT F. J. H.: The effect of concentrated loads on load bearing inflatable structures. In [28].
[60] TUTT F. J. H.: A survey of inflation requirements and methods for inflatable structures. In [28].
[61] TUTT F. J. H.: Inflatable jacking systems. In [28].
[62] VISHWANATH T. and GLOCKNER P. G.: Effects of temperature on inflatable shells. In [56].
[63] VLASOV V. Z.: Selected Works (in Russian), Vol. I. Izd. AN SSSR, Moscow 1962.
[64] VOLMIR A. S.: Stability of Elastic Systems (in Russian). Gosizdat, Moscow 1963.
[65] WANG C. C. and CROSS J. J.: Universal solutions for isotropic elastic membranes. Archive for Rational Mechanics and Analysis **65** (1977), 73—86.
[66] YANG W. H. and FENG W. E.: On axisymmetrical deformations of nonlinear membranes. Journal of Applied Mechanics, Trans. ASME, Vol. **37,** No. 4 (1970), 1002—1011.

INDEX